Theory and Design for Mechanical Measurements

Sixth Edition

Richard S. Figliola
Clemson University

Donald E. Beasley
Clemson University

VP & EXECUTIVE PUBLISHER	*Don Fowley*
EXECUTIVE EDITOR	*Linda Ratts*
SENIOR MARKETING MANAGER	*Christopher Ruel*
ASSOCIATE PRODUCTION MANAGER	*Joyce Poh*
SENIOR DESIGNER	Wendy Lai

This book was set in 10/12 TimesLTStd-Roman by Thomson Digital.

Founded in 1807, John Wiley & Sons, Inc. has been a valued source of knowledge and understanding for more than 200 years, helping people around the world meet their needs and fulfill their aspirations. Our company is built on a foundation of principles that include responsibility to the communities we serve and where we live and work. In 2008, we launched a Corporate Citizenship Initiative, a global effort to address the environmental, social, economic, and ethical challenges we face in our business. Among the issues we are addressing are carbon impact, paper specifications and procurement, ethical conduct within our business and among our vendors, and community and charitable support. For more information, please visit our website: www.wiley.com/go/citizenship.

Evaluation copies are provided to qualified academics and professionals for review purposes only, for use in their courses during the next academic year. These copies are licensed and may not be sold or transferred to a third party. Upon completion of the review period, please return the evaluation copy to Wiley. Return instructions and a free of charge return shipping label are available at www.wiley.com/go/returnlabel. Outside of the United States, please contact your local representative.

ISBN: 9781118881279 (cloth)

The inside back cover will contain printing identification and country of origin if omitted from this page. In addition, if the ISBN on the back cover differs from the ISBN on this page, the one on the back cover is correct.

Printed and bound in Singapore by Markono Print Media Pte Ltd
10 9 8 7 6 5 4 3 2 1

Preface

The sixth edition of *Theory and Design for Mechanical Measurements* continues to provide a well-founded, fundamental background in the theory of engineering measurements. Integrated throughout are the necessary elements to conduct engineering measurements through the design of measurement systems and measurement test plans, with an emphasis on the role of statistics and uncertainty analyses in that process. The measurements field is very broad, but through careful selection of topical coverage we establish the physical principles and practical techniques for quantifying a number of engineering variables that have many engineering applications. In this edition we continue to emphasize the conceptual design framework for selecting and specifying equipment and test procedures and for interpreting test results, which we feel are necessary and common bases for the practice of test engineering. Throughout the text, coverage of topics, applications, and devices has been updated and some out-of-date material eliminated in favor of more relevant and more current information. New in this edition, we have added several examples that illustrate either case studies or interesting vignettes related to the application of measurements in current practice.

The text is appropriate for undergraduate- and graduate-level study in engineering but is also suitably advanced and oriented to serve as a reference source for professional practitioners. The pedagogical approach invites independent study or use in related fields requiring an understanding of instrumentation and measurements.

The organization of the text develops from our view that certain aspects of measurements can be generalized, such as test plan design, signal analysis and reconstruction, and measurement system response. Topics such as statistics and uncertainty analysis require a basic development of principles but are then best illustrated by integrating these topics throughout the text material. Other aspects are better treated in the context of the measurement of a specific physical quantity, such as strain or temperature.

PEDAGOGICAL TOOLS TO AID LEARNING

In this textbook:

- Each chapter begins by defining a set of **learning outcomes**.
- The text develops an **intuitive understanding** of measurement concepts with its focus on test system modeling, test plan design, and uncertainty analysis.
- Each chapter includes carefully constructed **example problems** that illustrate new material and problems that build on prior material. Where appropriate case studies and vignettes illustrate current applications of measurement theory or design.
- Example problems make use of a **KNOWN, FIND, SOLVE** approach as an organizational aid towards solution. This methodology for problem solutions helps new users to link words and concepts with symbols and equations. Many problems contain **COMMENTS** that expand on the solution, provide a proper context for application of the principle, or offer design application insight.

iii

- **End-of-chapter practice problems** are included for each chapter to exercise new concepts.
  - ○ Practice problems range from those focused on concept development to building of advanced skills to open-ended design applications. Some problems require independent research for specific applications or measurement techniques.
  - ○ Within each chapter, we have added new practice problems and have also "refreshed" many problems from previous editions.
  - ○ We provide online access to a solutions manual for instructors who have adopted the book and registered with Wiley. Materials are posted on the textbook website at www.wiley.com/college/figliola.
  - ○ Answers to selected problems are also posted on the textbook website.
- Use of the software in problem solving allows in-depth exploration of key concepts that would be prohibitively time consuming otherwise. The text includes online access to **interactive software** of focused examples based on software using National Instruments Labview® or Matlab®. The Labview programs are available as executables so that they can be run directly without a Labview license. The software is available on both the Wiley student and instructor websites.

NEW TO THE SIXTH EDITION

With this sixth edition, we have new or expanded material on a number of topics:

- We have updated the end-of-chapter practice problems by removing some older problems and adding new problems. The number of end-of-chapter problems is roughly the same as in past editions, but we have replaced more than 20% of the problems in each chapter.
- We have added a number of case studies, vignettes, and anecdotes on applications of engineering measurements that reflect the topical coverage of the chapter.
- In Chapter 1, we have substantially increased coverage devoted to significant digits and rounding operations, and we devote more space to plotting formats. We find that these continue to be issues for engineering students; the refresher may be useful.
- In Chapter 4, we have expanded our treatment of hypothesis testing as it relates to differences in measured data sets.
- Treatment of uncertainty analysis in Chapter 5 has been updated using the most recent language and approaches. Where we deviate from the methodology of the standards, we do so for pedagogical reasons. We have also added more discussion on alternative terminology used for errors and uncertainties in the engineering mainstream and on how these are related to our use of systematic and random effects.
- In Chapter 7, we have updated the section on data acquisition hardware and on communication protocols between devices to reflect technology advancements.
- We have added a section introducing infrared imaging with a case study on full-field imaging in Chapter 8.
- We have added material regarding microphones, acoustical measurements, and acoustical signal weighting in Chapter 9.
- We have added an introduction to fiber optic–based strain measurement, specifically fiber Bragg grating strain gauges, in Chapter 11.

- Some materials, such as the appendix on technical writing, have been moved from the textbook to the book companion website, www.wiley.com/college/figliola.

SUGGESTED COURSE COVERAGE

To aid in course preparation, Chapters 1–5 provide an introduction to measurement theory with statistics and uncertainty analysis, Chapters 6 and 7 provide a broad treatment of analog and digital sampling methods, and Chapters 8–12 focus on instrumentation.

Many users report to us that they use different course structures—so many that it makes a preferred order of topical presentation difficult to anticipate. To accommodate this, we have written the text in a manner that allows any instructor to customize the order of material presentation. Although the material of Chapters 4 and 5 is integrated throughout the text and should be taught in sequence, the other chapters can stand on their own. The text is flexible and can be used in a variety of course structures at both the undergraduate and graduate levels.

For a complete measurements course, we recommend the study of Chapters 1–7 with use of the remaining chapters as appropriate. For a lab course sequence, we recommend using chapters as they best illustrate the course exercises while building complete coverage over the several lab courses normally within a curriculum. The manner of the text allows it to be a resource for a lab-only course with minimal lecture. Over the years we have used it in several forums, as well as in professional development courses, and we simply rearrange material and emphasis to suit our audience and objective.

We express our sincerest appreciation to the students, teachers, and engineers who have used our earlier editions. We really appreciate the input from the reviewers for this edition: Oleksandr Diloyan, Temple University, Rees Fullmer, Utah State University, Rorik Peterson, University of Alaska, Fairbanks, Stephen W. Sofie, Montana State University, Kenneth W. Van Treuren, Baylor University, and Mark Nagurka, Marquette University. We are indebted to the many who have written us with their constructive comments and encouragement.

Richard S. Figliola
Donald E. Beasley
Clemson, South Carolina

Contents

Chapter 1

Basic Concepts of Measurement Methods

1.1 INTRODUCTION

We take measurements every day. We routinely read the temperature of an outdoor thermometer to choose appropriate clothing for the day. We add exactly 5.3 gallons (about 20 liters) of fuel to our car fuel tank. We use a tire gauge to set the correct car tire pressures. We monitor our body weight. And we put little thought into the selection of instruments for these measurements. After all, the instruments and techniques are routine or provided, the direct use of the information is clear to us, and the measured values are assumed to be good enough. But as the importance and complexity increases, the selection of equipment and techniques and the quality of the results can demand considerable attention. Just contemplate the various types of measurements and tests needed to certify that an engine meets its stated design specifications.

The objective in any measurement is to assign numbers to variables so as to answer a question. The information acquired is based on the output of some measurement device or system. We might use that information to ensure a manufacturing process is executing correctly, to diagnose a defective part, to provide values needed for a calculation or a decision, or to adjust a process variable. There are important issues to be addressed to ensure that the output of the measurement device is a reliable indication of the true value of the measured variable. In addition, we must address some important questions:

1. How can a measurement or test plan be devised so that the measurement provides the unambiguous information we seek?

2. How can a measurement system be used so that the engineer can easily interpret the measured data and be confident in their meaning?

There are procedures that address these measurement questions.

At the onset, we want to stress that the subject of this text is real-life–oriented. Specifying a measurement system and measurement procedures represents an open-ended design problem. That means there may be several approaches to meeting a measurements challenge, and some will be better than others. This text emphasizes accepted procedures for analyzing a measurement challenge to aid selection of equipment, methodology, and data analysis.

Upon completion of this chapter, the reader will be able to

- Identify the major components of a general measurement system and state the function of each
- Develop an experimental test plan
- Distinguish between random and systematic errors
- Become familiar with the hierarchy of units standards, and with the existence and use of test standards and codes
- Understand the international system of units and other unit systems often found in practice
- Understand and work with significant digits

1.2 GENERAL MEASUREMENT SYSTEM

Measurement[1] is the act of assigning a specific value to a physical variable. That physical variable is the *measured variable*. A measurement system is a tool used to quantify the measured variable. Thus a measurement system is used to extend the abilities of the human senses, which, although they can detect and recognize different degrees of roughness, length, sound, color, and smell, are limited and relative: They are not very adept at assigning specific values to sensed variables.

A system is composed of components that work together to accomplish a specific objective. We begin by describing the components that make up a measurement system, using specific examples. Then we will generalize to a model of the generic measurement system.

Sensor and Transducer

A *sensor* is a physical element that employs some natural phenomenon to sense the variable being measured. To illustrate this concept, suppose we want to measure the profile of a surface at the nanometer scale. We discover that a very small cantilever beam placed near the surface is deflected by atomic forces. Let's assume for now that they are repulsive forces. As this cantilever is translated over the surface, the cantilever will deflect, responding to the varying height of the surface. This concept is illustrated in Figure 1.1; the device is called an atomic force microscope. The cantilever beam is a sensor. In this case, the cantilever deflects under the action of a force in responding to changes in the height of the surface.

A *transducer* converts the sensed information into a detectable signal. This signal might be mechanical, electrical, optical, or it may take any other form that can be meaningfully quantified. Continuing with our example, we will need a means to change the sensor motion into something that we can quantify. Suppose that the upper surface of the cantilever is reflective, and we shine a laser onto the upper surface, as shown in Figure 1.2. The movement of the cantilever will deflect the laser. Employing a number of light sensors, also shown in Figure 1.2, the changing deflection of the light

Cantilever and tip

Sample surface

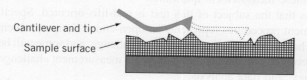

Figure 1.1 Sensor stage of an atomic-force microscope.

---

[1] A glossary of the italicized terms is located in the back of the text for your reference.

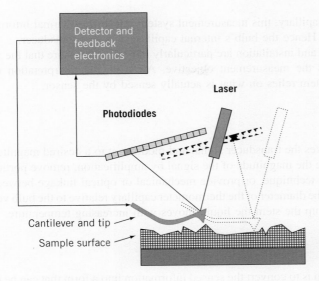

Figure 1.2   Atomic-force microscope with sensor and transducer stages.

can be measured as a time-varying current signal with the magnitude corresponding to the height of the surface. Together the laser and the light sensors (photodiodes) form the transducer component of the measurement system.

   A familiar example of a complete measurement system is the bulb thermometer. The liquid contained within the bulb of the common bulb thermometer shown in Figure 1.3 exchanges energy with its surroundings until the two are in thermal equilibrium. At that point they are at the same temperature. This energy exchange is the input signal to this measurement system. The phenomenon of thermal expansion of the liquid results in its movement up and down the stem, forming an output signal from which we determine temperature. The liquid in the bulb acts as the sensor. By forcing the

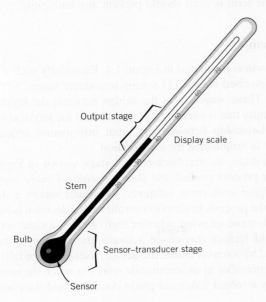

Figure 1.3   Components of bulb thermometer equivalent to sensor, transducer, and output stages.

expanding liquid into a narrow capillary, this measurement system transforms thermal information into a mechanical displacement. Hence the bulb's internal capillary acts as a transducer.

Sensor selection, placement, and installation are particularly important to ensure that the sensor output signal accurately reflects the measurement objective. After all, the interpretation of the information indicated by the system relies on what is actually sensed by the sensor.

Signal Conditioning Stage

The *signal-conditioning stage* takes the transducer signal and modifies it to a desired magnitude or form. It might be used to increase the magnitude of the signal by amplification, remove portions of the signal through some filtering technique, or provide mechanical or optical linkage between the transducer and the output stage. The diameter of the thermometer capillary relative to the bulb volume (Figure 1.3) determines how far up the stem the liquid moves with increasing temperature.

Output Stage

The goal of a measurement system is to convert the sensed information into a form that can be easily quantified. The *output stage* indicates or records the value measured. This might be by a simple readout display, a marked scale, or even a recording device such as computer memory from which it can later be accessed and analyzed. In Figure 1.3, the readout scale of the bulb thermometer serves as its output stage.

As an endnote, the term "transducer" is often used in reference to a packaged measuring device that contains the sensor and transducer elements described above, and even some signal conditioning elements, in one unit. This terminology differs from the term "transducer" when describing the function served by an individual stage of a measurement system. Both uses are correct, and we use both: one to refer to how a sensed signal is changed into another form and the other to refer to a packaged device. The context in which the term is used should prevent any ambiguity.

General Template for a Measurement System

A general template for a measurement system is illustrated in Figure 1.4. Essentially such a system consists of part or all of the previously described stages: (1) sensor–transducer stages, (2) signal-conditioning stage, and (3) output stage. These stages form the bridge between the input to the measurement system and the output, a quantity that is used to infer the value of the physical variable measured. We discuss later how the relationship between the input information acquired by the sensor and the indicated output signal is established by a calibration.

Some systems may use an additional stage, the feedback-control stage shown in Figure 1.4. Typical to measurement systems used for process control, the *feedback-control stage* contains a controller that compares the measured signal with some reference value and makes a decision regarding actions required in the control of the process. In simple controllers, this decision is based on whether the magnitude of the signal exceeds some set point, whether high or low—a value set by the system operator. For example, a household furnace thermostat activates the furnace as the local temperature at the thermostat, as determined by a sensor within the device, rises above or falls below the thermostat set point. The cruise speed controller in an automobile works in much the same way. A programmable logic controller (PLC) is a robust industrial-grade computer and data acquiring

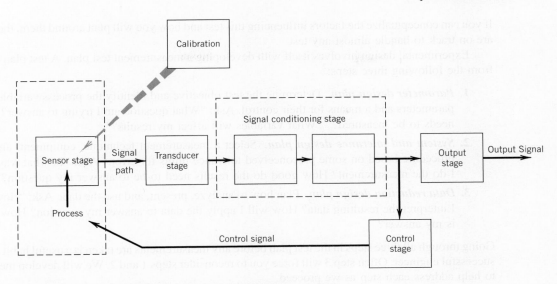

Figure 1.4 Components of a general measurement system.

device used to measure many variables simultaneously and to take appropriate corrective action per programmed instructions. We discuss some features of control systems in detail in Chapter 12.

1.3   EXPERIMENTAL TEST PLAN

An experimental test is designed to make measurements so as to answer a question. The test should be designed and executed to answer that question and that question alone. Consider an example.

Suppose you want to design a test to answer the question "What fuel use can my family expect from my new car?" In a test plan, you identify the variables that you will measure, but you also need to look closely at other variables that will influence the result. Two important measured variables here will be distance and fuel volume consumption. The accuracies of the odometer and station fuel pump affect these two measurements. Achieving a consistent final volume level when you fill your tank affects the estimate of the fuel volume added used. In fact, this effect, which can be classified as a zero error, could be significant. From this example, we can surmise that a consistent measurement technique must be part of the test plan.

What other variables might influence your results? The driving route, whether highway, city, rural, or a mix, affects the results and is an independent variable. Different drivers drive differently, so the driver becomes an independent variable. So how we interpret the measured data is affected by variables in addition to the primary ones measured. Imagine how your test would change if the objective were to establish the fleet average fuel use expected from a car model used by a rental company.

In any test plan, you need to consider just how accurate an answer you need. Is an accuracy within 2 liters per 100 kilometers (or, in the United States, 1 mile per gallon) good enough? If you cannot achieve such accuracy, then the test may require a different strategy. Last, as a concomitant check, is there a way to check whether your test results are reasonable, a sanity check to avoid subtle mistakes? Interestingly, this one example contains all the same elements of any sophisticated test.

If you can conceptualize the factors influencing this test and how you will plan around them, then you are on track to handle almost any test.

Experimental design involves itself with developing a measurement test plan. A test plan draws from the following three steps:[2]

1. ***Parameter design plan.*** Determine the test objective and identify the process variables and parameters and a means for their control. Ask: "What question am I trying to answer? What needs to be measured?" "What variables will affect my results?"

2. ***System and tolerance design plan.*** Select a measurement technique, equipment, and test procedure based on some preconceived tolerance limits for error.[3] Ask: "In what ways can I do the measurement? How good do the results need to be to answer my question?"

3. ***Data reduction design plan.*** Plan how to analyze, present, and use the data. Ask: "How will I interpret the resulting data? How will I apply the data to answer my question? How good is my answer?"

Going through all three steps in the test plan before any measurements are taken is a useful habit of the successful engineer. Often step 3 will force you to reconsider steps 1 and 2. We will develop methods to help address each step as we proceed.

Variables

A *variable* is a physical quantity whose value influences the test result. Variables can be continuous or discrete by nature. A *continuous variable* can assume any value within its range. A *discrete variable* can assume only certain values.

Functionally, variables can be classed as being dependent, independent, or extraneous. A variable whose value is affected by changes in the value of one or more other variables is known as a *dependent variable*. Variables that do not influence each other are *independent variables*. Dependent variables are functions of independent variables. The test result is a dependent variable whose value depends on the values of the independent variables. For example, the value of strain measured in a loaded cantilever beam (dependent variable) results from the value of the loading applied (independent variable).

The values of independent variables are either purposely changed or held fixed throughout a test to study the effect on the dependent variable. A *controlled variable* is a variable whose value is held fixed. A test *parameter* is a controlled variable or grouping of variables whose fixed value sets the behavior of the process being measured. The dimensions and material properties of the beam in the loaded cantilever beam are parameters. The value of the parameter is changed when you want to change the process behavior.

In engineering, groupings of variables, such as moment of inertia or Reynolds number, are also referred to as parameters when these relate to certain system behaviors. For example, the damping ratio of an automotive suspension system affects how the displacement velocity will change with differing input conditions; this parameter affects the variables that measure vehicle behavior, such as handling and ride quality. In our treatment of statistics, we use the term parameter to refer to

---

[2] These three strategies are similar to the bases for certain design methods used in engineering system design (1).

[3] The tolerance design plan strategy used in this text draws on uncertainty analysis, an extension of sensitivity analysis. Sensitivity methods are common in design optimization.

quantities that describe the behavior of the population of a measured variable, such as the true mean value or variance.

Variables that are not purposely manipulated or controlled during measurement but that do affect the test result are called *extraneous variables*. Dependent variables are affected by extraneous variables. If not treated properly within a test plan, extraneous variables can impose a false trend or impose variation onto the behavior of the measured variable. This influence can confuse the clear relation between cause and effect.

There are other uses of the term "control." Experimental control describes how well we can maintain the prescribed value of an independent variable within some level of confidence during a measurement. For example, if we set the loading applied in a bending beam test to 103 kN, does it stay exactly fixed during the entire test, or does it vary by some amount on repeated tries? Different fields use variations on this term. In statistics, a control group is one held apart from the treatment under study. But the nuance of holding something fixed is retained.

Example 1.1

Consider an introductory thermodynamics class experiment used to demonstrate phase change phenomena. A beaker of water is heated and its temperature measured over time to establish the temperature at which boiling occurs. The results, shown in Figure 1.5, are for three separate tests conducted on different days by different student groups using the same equipment and method. Why might the data from three seemingly identical tests show different results?

KNOWN   Temperature-time measurements for three tests conducted on three different days

FIND   Boiling point temperature

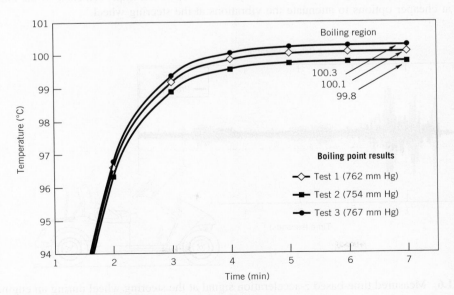

Figure 1.5   Results of a boiling point test for water.

SOLUTION   Each group of students anticipated a value of exactly 100.0 °C. Individually, each test result is close, but when compared, there is a clear offset between any two of the three test results. Suppose we determine that the measurement system accuracy and natural chance account for only 0.1 °C of the disagreement between tests—then something else is happening. A plausible contributing factor is an effect due to an extraneous variable.

Fortunately, an assistant recorded the local barometric pressure during each test. The boiling point (saturation) temperature is a function of pressure. Each test shows a different outcome, in part, because the local barometric pressure was not controlled (i.e., it was not held fixed between the tests).

COMMENT   One way to control the effects of pressure here might be to conduct the tests inside a barometric pressure chamber. Direct control of all variables is not often possible in a test. So another way to handle the extraneous variable applies a special strategy: Consider each test as a single block of data taken at the existing barometric pressure. Then consolidate the three blocks of test data. In that way, the measured barometric pressure becomes treated as if it were an independent variable, with its influence integrated into the entire data set. That is, we can actually take advantage of the differences in pressure so as to study its effect on saturation temperature. This is a valuable control treatment called randomization, a procedure we discuss later in this chapter. *Identify and control important variables, or be prepared to solve a puzzle!*

Example 1.2: Case Study

The golf cart industry within the United States records nearly $1 billion in sales annually. The manufacturer of a particular model of golf cart sought a solution to reduce undesirable vibrations in the cart steering column (Figure 1.6). Engineers identified the source of the vibrations as the gas-powered engine and consistent with operating the engine at certain speeds (revolutions/minute or rpm). The company management ruled out expensive suspension improvements, so the engineers looked at cheaper options to attenuate the vibrations at the steering wheel.

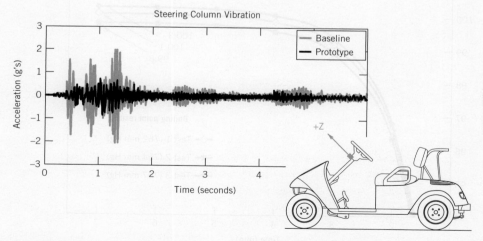

Figure 1.6   Measured time-based z-acceleration signal at the steering wheel during an engine run-up test: baseline and the attenuated prototype (proposed solution) signals.

Using test standard ANSI S2.70-2006 (17) as a guide for placing sensors and interpreting results, accelerometers (discussed in Chapter 12) were placed on the steering column, engine, suspension, and frame. The measurement chain followed Figure 1.4. The signal from each accelerometer sensor-transducer was passed through a signal conditioning charge amplifier (Chapter 6) and on to a laptop-based data acquisition system (DAS; Chapter 7). The amplifier converted the transducer signal to a 0–5 V signal that matched the input range of the DAS. The DAS served as a voltmeter, output display, and time-based recorder.

Measurements showed that the offensive vibration at the steering wheel was at the natural frequency of the steering column (~20 Hz). This vibration was excited at forcing function frequencies between 17–37 Hz, corresponding to normal operating speeds of the engine (1,000–2,200 rpm). From a lumped parameter model (Chapter 3) of the steering assembly, they determined that a slightly heavier steering column would reduce its natural frequency to below the engine idle speed and thereby attenuate the amplitude of the vibrations (Chapter 3) excited at higher engine speeds. The analysis was verified and the fix was refined based on the tests. The measured vibration signals during engine runup, both before (baseline) and with the proposed fix (prototype), are compared in Figure 1.6. Peak vibration amplitudes were shown to reduce to comfortable levels with the proposed and validated solution. The inexpensive fix was implemented by the manufacturer.

COMMENT   This example illustrates how measurements applied to a mechanical test provided information to diagnose a problem and then formulate and validate a proposed solution.

Noise and Interference

Just how extraneous variables affect measured data can be described in terms of noise and interference. *Noise* is a random variation in the value of the measured signal. Noise increases data scatter. Building vibrations, variations in ambient conditions, and random thermal noise of electrons passing through a conductor are examples of common extraneous sources for the random variations found in a measured signal.

Interference imposes undesirable deterministic trends on the measured signal. A common interference in electrical instruments comes from an AC power source and is seen as a sinusoidal wave superimposed onto the measured signal. Hum and acoustic feedback in public address systems are ready examples of interference effects superimposed onto a desirable signal. Sometimes the interference is obvious. But if the period of the interference is longer than the period over which the measurement is made, the false trend imposed may go unnoticed. A goal should be either to control the source of interference or to break up its trend.

Consider the effects of noise and interference on the signal, $y(t) = 2 + \sin 2\pi t$. As shown in Figure 1.7, noise adds to the scatter of the signal. Through statistical techniques and other means, we can sift through the fuzziness to extract desirable signal information. Interference imposes a false trend onto the signal. The test plan should be devised to break up such trends so that they become random variations in the data set. Although this increases the scatter in the measured values of a data set, noise may mask, but does not change, the deterministic aspects of a signal. It is far more important to eliminate false trends in the data set.

In Example 1.1, by combining the three tests at three random but measured values of barometric pressure into a single dataset, we incorporated the uncontrolled pressure effects into the results for boiling temperature while breaking up any offset in results found in any one test. This approach imparts some control over an otherwise uncontrolled variable.

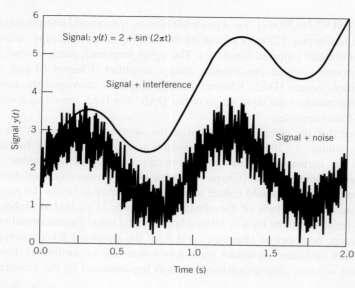

Figure 1.7 Effects of noise and interference superimposed on the signal $y(t) = 2 + \sin 2\pi t$.

Randomization

An engineering test purposely changes the values of one or more independent variables to determine the effect on a dependent variable. *Randomization* refers to test strategies that apply the changes in the independent variables in some random order. The intent is to neutralize effects that may not be accounted for in the test plan.

In general, consider the situation in which the dependent variable, y, is a function of several independent variables, $x_a, x_b, \ldots$. To find the dependence of y on the independent variables, they are varied in a prescribed manner. However, the measurement of y can also be influenced by extraneous variables, z_j, where $j = 1, 2, \ldots$, such that $y = f(x_a, x_b, \ldots; z_j)$. Although the influence of the z_j variables on these tests cannot be eliminated, the possibility of their introducing a false trend on y can be minimized by using randomization strategies. One approach is to add randomness to the order in which an independent variable is manipulated. This works well for continuous extraneous variables.

Example 1.3

In the pressure calibration system shown in Figure 1.8, a pressure transducer is exposed to a known pressure, p. Increasing pressure deforms the elastic diaphragm sensor of the device, which is sensed as a mechanical displacement. A transducer element converts the displacement into a voltage that is measured by the voltmeter. The test approach is to control the applied pressure through the measured displacement of a piston that is used to compress a gas contained within the piston–cylinder chamber. The gas closely obeys the ideal gas law. Hence, piston displacement, x, which sets the chamber volume, $\forall = (x \times \text{area})$, is easily related to chamber pressure. Accordingly, the known chamber pressure can be associated with the transducer output signal. Identify the independent and dependent variables in the calibration and possible extraneous variables.

KNOWN Pressure calibration system of Figure 1.8.

FIND Independent, dependent, and extraneous variables.

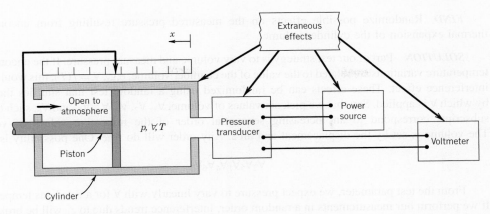

Figure 1.8 Pressure calibration system.

SOLUTION   The sensor is exposed to the chamber gas pressure. The pressure transducer output signal will vary with this chamber pressure. So the dependent variable is the chamber gas pressure, which is also the pressure applied to the transducer. An independent variable in the calibration is the piston displacement, which sets the volume. A parameter for this problem is formed from the ideal gas law: $p\forall/T = $ constant, where T is the gas temperature and $\forall = (x \times \text{area})$. This parameter shows that the gas pressure also depends on gas temperature, so gas temperature is an independent variable. Volume is to be varied through variation of piston displacement, but T can be controlled provided that a mechanism is incorporated into the scheme to maintain a constant chamber gas. However, T and $\forall$ are not in themselves independent. The chamber cross-sectional area is a function of temperature and $p = f(\forall, T)$, where $\forall = f_1(x, T)$. If chamber temperature is held constant, the applied variations in $\forall$ will be the only variable with an effect on pressure, as desired $p = f(\forall, T)$.

If gas temperature is not controlled, or if environmental effects allow uncontrolled variations in chamber and transducer temperature, then temperature behaves as an extraneous variable, z_1. Hence

$$p = f(\forall, z_1), \text{ where } \forall = f_1(x, T)$$

COMMENT   To a lesser extent, line voltage variations and electrical interference, z_2, affect the excitation voltage from the power supply to the transducer or affect the performance of the voltmeter. This list is not exhaustive.

Example 1.4

Develop a test matrix that could minimize interference owing to any uncontrolled temperature effects discussed in Example 1.3.

KNOWN   Control variable $\forall$ is changed through displacement x. Dependent variable p is measured.

FIND Randomize possible effects on the measured pressure resulting from uncontrolled thermal expansion of the cylinder volume.

SOLUTION Part of our test strategy is to vary volume and measure pressure. If the uncontrolled temperature variations are related to the value of the chamber volume, e.g., $\forall = f(T)$, this would be an interference effect. These effects can be randomized using a random test that shuffles the order by which $\forall$ is applied. Say that we pick six values of volume, $\forall_1$, $\forall_2$, $\forall_3$, $\forall_4$, $\forall_5$, and $\forall_6$, such that the subscripts correspond to an increasing sequential order of the respective values of volume. The volume is set by the displacement and area. Any order will do fine. One possibility is

$$\forall_2 \forall_5 \forall_1 \forall_4 \forall_6 \forall_3$$

From the test parameter, we expect pressure to vary linearly with $\forall$ for a fixed gas temperature. If we perform our measurements in a random order, interference trends due to z_1 will be broken up. This approach should serve also to break up any interference trends resulting from environmental effects that might influence the transducer, excitation voltage, and voltmeter.

The use of different test operators, different instruments, and different test operating conditions are examples of discrete extraneous variables that can affect the outcome of a measurement. Randomizing a test matrix to minimize discrete influences can be done efficiently through the use of experimental design methods using random blocks. A block consists of a data set comprised of the measured (dependent) variable in which the extraneous variable is fixed while the independent variable is varied. The extraneous variable will vary between blocks. This enables some amount of local control over the discrete extraneous variable through randomization.

In the fuel-usage discussion earlier, we might consider several blocks, each comprising a different driver (an extraneous variable) driving similar routes, and averaging the fuel consumption results for our car. In Example 1.1, we imposed the strategy of using several tests (blocks) under different values of barometric pressure to break up the interference effect found in a single test. Many strategies for randomized blocks exist, as do advanced statistical methods for data analysis (2–5).

Example 1.5

The manufacture of a particular composite material requires mixing a percentage by weight of binder with resin to produce a gel. The gel is used to impregnate a fiber to produce the composite material in a manual process called the lay-up. The strength, σ, of the finished material depends on the percent binder in the gel. However, the strength may also be lay-up operator–dependent. Formulate a test matrix by which the strength to percent binder–gel ratio under production conditions can be established.

KNOWN $\sigma = f$ (binder; operator)

ASSUMPTION Strength is affected only by binder and operator.

FIND Test matrix to randomize effects of operator.

SOLUTION The dependent variable, σ, is to be tested against the independent variable, percent binder–gel ratio. The operator is an extraneous variable in actual production. As a simple test, we could test the relationship between three binder–gel ratios, $A, B,$ and C and measure strength. We could also choose three typical operators (z_1, z_2, and z_3) to produce N separate composite test samples for each of the three binder–gel ratios. This gives the three-block test pattern:

Block				
1	z_1:	A	B	C
2	z_2:	A	B	C
3	z_3:	A	B	C

In the analysis of the test, all these data can be combined. The results of each block will include each operator's influence as a variation. We can assume that the order used within each block is unimportant. But if only the data from one operator are considered, the results will allow a trend consistent with the lay-up technique of that operator. The test matrix above will randomize the influence of any one operator on the strength test results by introducing the influence of several operators.

Example 1.6

Suppose that after lay-up, the composite material of Example 1.5 is allowed to cure at a controlled but elevated temperature. We wish to develop a relationship between the binder–gel ratio and the cure temperature and strength. Develop a suitable test matrix.

KNOWN $\sigma = f$ (binder, temperature, operator)

ASSUMPTION   Strength is affected only by binder, temperature, and operator.

FIND   Test matrix to randomize effect of operator.

SOLUTION   We develop a simple matrix to test for the dependence of composite strength on the independent variables of binder–gel ratio and cure temperature. We could proceed as in Example 1.5 and set up three randomized blocks for ratio and three for temperature for a total of 18 separate tests. Suppose instead we choose three temperatures $T_1, T_2,$ and T_3, along with three binder–gel ratios $A, B,$ and C and three operators $z_1, z_2,$ and z_3 and set up a 3×3 test matrix representing a single randomized block. If we organize the block such that no operator runs the same test combination more than once, we randomize the influence of any one operator on a particular binder–gel ratio, temperature test.

	z_1	z_2	z_3
A	T_1	T_2	T_3
B	T_2	T_3	T_1
C	T_3	T_1	T_2

COMMENT   The suggested test matrix not only randomizes the extraneous variable, but also has reduced the number of tests by half over the direct use of three blocks for ratio and for temperature. However, either approach is fine. The above matrix is referred to as a Latin square (2–5).

Replication and Repetition

In general, the estimated value of a measured variable improves with the number of measurements. Repeated measurements of the same variable made during any single test run or on a single batch are called *repetitions*. Repetition helps quantify the variation in a measured variable under a set of test conditions. For example, a bearing manufacturer would obtain a better estimate of the mean diameter and the variation in the diameter within a new batch containing thousands of bearings by measuring many bearings rather than just a few.

An independent duplication of a set of measurements or test is referred to as a *replication*. Replication allows for quantifying the variation in a measured variable as it occurs between duplicate tests conducted under similar conditions. If the bearing manufacturer was interested in how closely the bearing mean diameter was controlled in day-in and day-out operations with a particular machine or test operator, duplicate tests run on different days would be needed.

If the bearing manufacturer were interested in how closely bearing mean diameter was controlled when using different machines or different machine operators, duplicate tests using these different configurations holds the answer. Here, replication provides a means to randomize the interference effects of the different bearing machines or operators.

Example 1.7

Consider a room furnace thermostat. Set to some temperature, we can make repeated measurements (repetition) of room temperature and come to a conclusion about the average value and the fluctuations in room temperature achieved at that particular thermostat setting. Repetition permits an assessment of the set condition and how well we can maintain it.

Now suppose we change the set temperature to another arbitrary value but some time later return it to the original setting and duplicate the measurements. The two sets of test data are replications of each other. We might find that the average temperature in the second test differs from the first. The different averages suggest something about our ability to set and control the temperature in the room. Replication provides information to assess how well we can duplicate a set of conditions.

Concomitant Methods

Is my test working? What value of result should I expect from my measurements? To help answer these, a good strategy is to incorporate *concomitant methods* in a measurement plan. The goal is to obtain two or more estimates for the result, each estimate based on a different method, which can be compared as a check for agreement. This may affect the experimental design in that additional variables may need to be measured. Or the different method could be an analysis that estimates the expected value for the measurement. The two need not be of same accuracy as one method serves to check the reasonableness of the other. For example, suppose we want to establish the volume of a cylindrical rod of known material. We could measure the diameter and length of the rod to compute this. Alternatively, we could measure the weight of the rod and compute volume based on the specific weight of the material. The second method complements the first and provides an important check on the adequacy of the first estimate.

1.4   CALIBRATION

A *calibration* applies a known input value to a measurement system for the purpose of observing the system output value. It establishes the relationship between the input and output values. The known value used for the calibration is called the *standard*.

Static Calibration

The most common type of calibration is known as a static calibration. In this procedure, a value is applied (input) to the system under calibration and the system output is recorded. The value of the applied standard is known to some acceptable level. The term "static" means that the input value does not vary with time or space or that average values are used.

By applying a range of known input values and by observing the system output values, a direct calibration curve can be developed for the measurement system. On such a curve the applied input, x, is plotted on the abscissa against the measured output, y, on the ordinate, as indicated in Figure 1.9. In a calibration the input value is an independent variable, while the measured output value is the dependent variable of the calibration.

The static calibration curve describes the static input–output relationship for a measurement system and forms the logic by which the indicated output can be interpreted during an actual measurement. For example, the calibration curve is the basis for fixing the output display scale on a measurement system, such as that of Figure 1.3. Alternatively, a calibration curve can be used as part of developing a functional relationship, an equation known as a correlation, between input and output. A correlation will have the form $y = f(x)$ and is determined by applying physical reasoning and curve fitting techniques to the calibration curve. The correlation can then be used in later measurements to ascertain an unknown input value based on the measured output value, the value indicated by the measurement system.

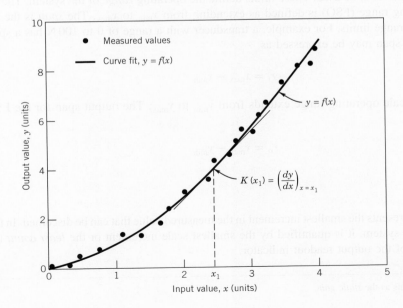

Figure 1.9   Representative static calibration curve.

Dynamic Calibration

When the variables of interest are time-dependent (or space-dependent) and such varying information is sought, we need dynamic information. In a broad sense, dynamic variables are time (or space) dependent in both their magnitude and amplitude and frequency content. A *dynamic calibration* determines the relationship between an input of known dynamic content and the measurement system output. For example, does the output signal track the input value exactly in time and space, is there lag, or is the output value input frequency dependent? Usually, such calibrations involve applying either a sinusoidal signal of known amplitude and frequency or a step change as the input signal. The dynamic response of measurement systems is explored fully in Chapter 3.

Static Sensitivity

The slope of a static calibration curve provides the static sensitivity[4] of the measurement system. As depicted graphically in the calibration curve of Figure 1.9, the static sensitivity, K, at any particular static input value, say x_1, is evaluated by

$$K = K(x_1) = \left(\frac{dy}{dx}\right)_{x=x_1}$$ (1.1)

where K can be a function of the applied input value x. The static sensitivity is a measure relating the change in the indicated output associated with a given change in a static input.

Range and Span

A calibration applies known inputs ranging from the minimum to the maximum values for which the measurement system is to be used. These limits define the operating *range* of the system. The input full scale operating range (FSO) is defined as extending from x_{min} to x_{max}. The *span* is the value difference of the range limits. For example, a transducer with a range of 0 to 100 N has a span of 100 N. The input span may be expressed as

$$r_i = x_{max} - x_{min}$$ (1.2)

The output full scale operating range extends from y_{min} to y_{max}. The output span for the FSO is expressed as

$$r_o = y_{max} - y_{min}$$ (1.3)

Resolution

The *resolution* represents the smallest increment in the measured value that can be discerned. In terms of a measurement system, it is quantified by the smallest scale increment or the *least count* (least significant digit) of the output readout indicator.

---

[4] Some texts refer to this as the *static gain*.

Accuracy and Error

The exact value of a variable is called the *true value*. The value of a variable as indicated by a measurement system is called the *measured value*. The *accuracy* of a measurement refers to the closeness of agreement between the measured value and the true value. Unfortunately, the true value is rarely known exactly in engineering practice, and various influences, called errors, have an effect on both of these values. This means that the concept of the accuracy of a measurement is a qualitative one.

Instead, an appropriate approach to stating this closeness of agreement is to identify the measurement errors and to quantify them by the value of their associated uncertainties, where an uncertainty is the estimated range of the value of an error. We define an *error*, *e*, as the difference between the measured value and the true value—that is,

$$e = \text{Measured value} - \text{True value} \tag{1.4}$$

The exact value for error is usually not known. So Equation 1.4 serves only as a reference definition. Errors exist, and they have a magnitude, as given by Equation 1.4. Sometimes we can correct a reading to account for an estimated amount of error, even if that correction is approximate. We'll discuss this concept next and then develop it extensively in Chapter 5.

Often an estimate for the value of error is based on a reference value used during the instrument's calibration, which becomes a surrogate for the true value. A relative error based on this reference value is estimated by

$$A = \frac{|e|}{\text{Reference value}} \times 100 \tag{1.5}$$

Some vendors may refer to this term as the "relative accuracy."

Random and Systematic Errors and Uncertainty

Errors cause a measured value to differ from its true value. *Random error* causes a random variation in measured values found during repeated measurements of the variable. *Systematic error* causes an offset between the mean value of the data set and its true value. Both random and systematic errors affect a system's accuracy.

The concept of accuracy and the effects of systematic and random errors in instruments and measurement systems can be illustrated by dart throws. Consider the dart boards in Figure 1.10; the goal is to throw the darts into the bullseye. For this analogy, the bullseye can represent the true value, and each throw can represent a measured value. The error in each throw can be calculated as the distance between the dart and the bullseye. In Figure 1.10a, the thrower displays good repeatability (i.e., small random error) in that each throw repeatedly hits nearly the same spot on the board, but the thrower is not accurate in that the dart misses the bullseye each time. The average value of the error gives an estimate of the systematic error in the throws. The random error is some average of the variation between each throw, which is near zero in Figure 1.10a. We see that a small amount of random error is not a complete measure of the accuracy of this thrower. This thrower has an offset to the left of the target, a systematic error. If the effect of this systematic error could be reduced, then this thrower's accuracy would improve. In Figure 1.10b,

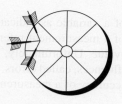

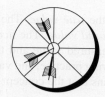

(a) High repeatability gives low random error but no direct indication of accuracy.

(b) High accuracy means low random and systematic errors.

(c) Systematic and random errors lead to poor accuracy.

Figure 1.10 Throws of a dart: illustration of random and systematic errors and accuracy.

the thrower displays a high accuracy, hitting within the bull's-eye on each throw. Both scatter and offset are near zero. High accuracy must be associated with small values of random and systematic errors as shown. In Figure 1.10c, the thrower does not show good accuracy, with errant throws scattered around the board. Each throw contains a different amount of error. While an estimate of the systematic error is the average of the errors in the throws, the estimate of the random error is related to the varying amount of error in the throws, a value that can be estimated using statistical methods. The estimates in the random and systematic errors of the thrower can be computed using the statistical methods that are discussed in Chapter 4 or the methods of comparison discussed in Chapter 5.

Suppose we used a measurement system to measure a variable whose value was kept constant and known almost exactly, as in a calibration. For example, 10 independent measurements are made with the results, as shown in Figure 1.11. The variations in the measurements, the observed scatter in the data, would be related to the random error associated with the measurement of the variable. This scatter is mainly owing to (1) the measurement system and the measurement method (2) any

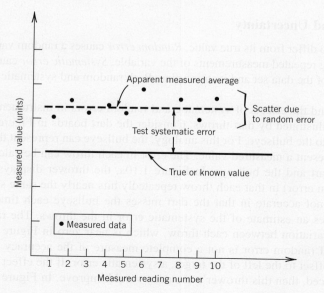

Figure 1.11 Effects of random and systematic errors on calibration readings.

natural variation in the variable measured, and (3) any uncontrolled variations in the variable. However, the offset between the apparent average of the readings and the true value would provide a measure of the systematic error to be expected from this measurement system.

Uncertainty

The *uncertainty* is the numerical estimate of the possible range of the error in the stated value of a variable. In any measurement, the error is not known exactly since the true value is rarely known exactly. But based on available information, the operator might feel confident that the error is within certain bounds, a plus or minus range of the indicated reading. This is the assigned uncertainty. Uncertainty is brought about by each of the errors that are present in a measurement—its system calibration, the data set statistics, and the measurement technique. Individual errors are properties of the instruments, the test method, the analysis, and the measurement system. Uncertainty is a property of the test result. In Figure 1.11, we see that we might assign an estimate to the random error, that is, the random uncertainty, based on the data scatter. The systematic uncertainty might be an estimate of the offset based on a comparison against a value found by a concomitant method. An uncertainty based on the estimates of all known errors would then be assigned to the stated result. A method of estimating the overall uncertainty in the test result is treated in detail in Chapter 5.

The uncertainty values assigned to an instrument or measurement system specification are usually the result of several interacting random and systematic errors inherent to the measurement system, the calibration procedure, and the standard used to provide the known value. An example of some known calibration errors affecting a typical pressure transducer is given in Table 1.1. The value assigned to each stated error is its uncertainty.

Sequential Test

A *sequential test* varies the input value sequentially over the desired input range. This may be accomplished by increasing the input value (upscale direction) or by decreasing the input value (downscale direction) over the full input range. The output value is the measured value.

Table 1.1   Manufacturer's Specifications: Typical Pressure Transducer

Operation	
Input range	0–1000 cm H$_2$O
Excitation	±15 V DC
Output range	0–5 V
Temperature range	0–50 °C
Performance	
Linearity error	±0.5% FSO
Hysteresis error	Less than ±0.15% FSO
Sensitivity error	±0.25% of reading
Thermal sensitivity error	±0.02%/°C of reading
Thermal zero drift	±0.02%/°C FSO

FSO, full-scale operating range.

Hysteresis

Hysteresis error refers to differences in the measured value between an upscale sequential test and a downscale sequential test. The sequential test is an effective diagnostic technique for identifying and quantifying hysteresis error in a measurement system. The effect of hysteresis in a sequential test calibration curve is illustrated in Figure 1.12a. The hysteresis error of the system is estimated by its uncertainty, $u_h = (y)_{\text{upscale}} - (y)_{\text{downscale}}$. Hysteresis is usually specified for a measurement system in terms of an uncertainty based on the maximum hysteresis error as a percentage of the full-scale output, r_o,

$$\%u_{h_{\max}} = \frac{u_{h_{\max}}}{r_o} \times 100 \tag{1.6}$$

such as the value indicated in Table 1.1. Hysteresis occurs when the output of a measurement system is dependent on the previous value indicated by the system. Such dependencies can be

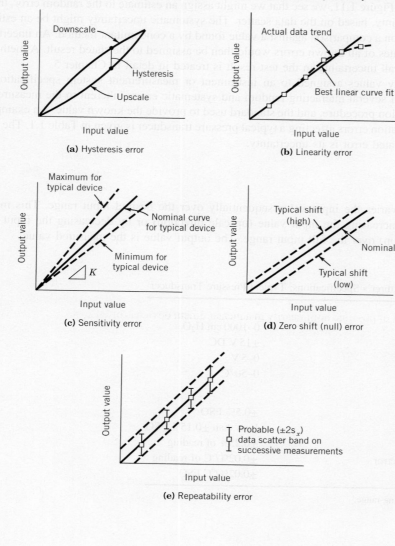

(a) Hysteresis error

(b) Linearity error

(c) Sensitivity error

(d) Zero shift (null) error

(e) Repeatability error

Figure 1.12 Examples of some common elements of instrument error.
(a) Hysteresis error.
(b) Linearity error.
(c) Sensitivity error.
(d) Zero shift (null) error.
(e) Repeatability error.

brought about through system limitations such as friction or viscous damping in moving parts or residual charge in electrical components. Some hysteresis is normal for any system and affects the repeatability of the system.

Random Test

A *random test* applies the input value in a random order over the intended calibration range. The random application of input tends to reduce the effects of interference. It breaks up hysteresis effects and observation errors. It ensures that each application of input value is independent of the previous. Thus it reduces calibration systematic error, converting it to random error. Generally, such a random variation in input value will more closely simulate the actual measurement situation.

A random test provides an important diagnostic for the delineation of several measurement system performance characteristics based on a set of random calibration test data. In particular, linearity error, sensitivity error, zero error, and instrument repeatability error, as illustrated in Figure 1.12b–e, can be quantified from a static random test calibration.

Linearity Error

Many instruments are designed to achieve a linear relationship between the applied static input and indicated output values. Such a linear static calibration curve would have the general form

$$y_L(x) = a_0 + a_1 x \qquad (1.7)$$

where the curve fit $y_L(x)$ provides a predicted output value based on a linear relation between x and y. As a measure of how well a linear relation is actually achieved, measurement device specifications usually provide a statement as to the expected linearity of the static calibration curve for the device. The relationship between $y_L(x)$ and measured value $y(x)$ is a measure of the nonlinear behavior of a system:

$$u_L(x) = y(x) - y_L(x) \qquad (1.8)$$

where the uncertainty $u_L(x)$ is a measure of the *linearity error* that arises in describing the actual system behavior by Equation 1.7. Such behavior is illustrated in Figure 1.12b, in which a linear curve has been fit through a calibration data set. For a measurement system that is essentially linear in behavior, the extent of possible nonlinearity in a measurement device is often specified in terms of the maximum expected linearity error as a percentage of full-scale output, r_o,

$$\%u_{L_{max}} = \frac{u_{L_{max}}}{r_o} \times 100 \qquad (1.9)$$

This is how the linearity error for the pressure transducer in Table 1.1 was estimated. Statistical methods of quantifying data scatter about a curve fit are discussed in Chapter 4.

Sensitivity and Zero Errors

The scatter in the data measured during a calibration affects the precision in predicting the slope of the calibration curve. As shown for the linear calibration curve in Figure 1.12c, in which the zero intercept is fixed, the scatter in the data about the curve fit indicates random errors. The *sensitivity error* as described by its uncertainty, u_K, is a statistical measure of the random error in the estimate of the slope of the calibration curve (we discuss the statistical estimate further in Chapter 4). In Table 1.1, the sensitivity error reflects calibration results at a constant reference ambient temperature, whereas the thermal sensitivity error was found by calibration at different temperatures.

A drift in the zero intercept introduces a vertical shift of the calibration curve, as shown in Figure 1.12d. This shift is known as the *zero error* with uncertainty, u_z. Zero error can usually be reduced by periodically nulling the output from the measurement system under a zero input condition. However, some random variation in the zero intercept is common, particularly with electronic and digital equipment subjected to external noise and temperature variations (e.g., thermal zero drift in Table 1.1). Zero error can be a major source of uncertainty in many measurements.

Instrument Repeatability

The ability of a measurement system to indicate the same value on repeated but independent application of the same input provides a measure of the instrument *repeatability*. Specific claims of repeatability are based on multiple calibration tests (replication) performed within a given lab on the particular unit. Repeatability, as shown in Figure 1.12e, is an uncertainty based on a statistical measure (developed in Chapter 4) called the standard deviation, s_x, a measure of the variation in the output for a given input. The value claimed is usually in terms of the maximum expected error as a percentage of full-scale output:

$$\%u_{R_{\max}} = \frac{2s_x}{r_o} \times 100 \tag{1.10}$$

The instrument repeatability reflects only the variations found under controlled calibration conditions.

Reproducibility

The term "reproducibility," when reported in instrument specifications, refers to the closeness of agreement in results obtained from duplicate tests carried out under similar conditions of measurement. As with repeatability, the uncertainty is based on statistical measures. Manufacturer claims of instrument reproducibility must be based on multiple tests (replication) performed in different labs on a single unit or model of instrument.

Instrument Precision

The term "instrument precision," when reported in instrument specifications, refers to a random uncertainty based on the results of separate repeatability tests. Manufacturer claims of instrument precision must be based on multiple tests (replication) performed on different units of the same manufacture, either performed in the same lab (same-lab precision) or, preferably, performed in different labs (between-lab precision).

Overall Instrument Error and Instrument Uncertainty

An estimate of the *overall instrument error* is made by combining the uncertainty estimates of all identified instrument errors into a term called the *instrument uncertainty*. The estimate is computed from the square root of the sum of the squares of all known uncertainty values. For M known errors, the overall instrument uncertainty, u_c, is calculated as

$$u_c = \left[u_1^2 + u_2^2 + \cdots + u_M^2 \right]^{1/2} \tag{1.11}$$

For example, for an instrument having known hysteresis, linearity, and sensitivity errors, the instrument uncertainty is estimated by

$$u_c = \left[u_h^2 + u_L^2 + u_K^2 \right]^{1/2} \tag{1.12}$$

An assumption in Eq. 1.11 is that each error contributes equally to u_c. If not, then sensitivity factors are used as described in Chapter 5.

Verification and Validation

V*erification* refers to executing the intended model correctly. Within a test or measurement system, this means verifying that the measurement system, along with each of its components, is functioning correctly. One approach is to place the measurement system into a state in which its output can be anticipated, a systems-level calibration. Unacceptable differences suggest a problem somewhere within the system. This is a debugging tool. A close analogy is checking a computer program for programming errors.

Validation refers to ensuring that the experimental model used is itself correct. To do this, we compare a set of test results to results obtained using another reliable method for the same conditions. For example, we can compare the results from a wind tunnel test of a vehicle to the results obtained from a road test. Gross differences indicate shortcomings in one or both models. A close analogy is determining whether the algorithm programmed within a computer model actually simulates the intended physical process. Verification and validation should always be a part of a test plan to ensure meaningful results.

1.5 STANDARDS

During calibration, the indicated value from the measurement system is compared directly with a reference or known value, which is applied as the input signal. This reference value forms the basis of the comparison and is known as the *standard*. Primary standards, as we shall discuss next, are exact but are not available for routine calibrations. So instead the standard used could be a close copy with direct traceability to the primary standard, or it could be based on the output from a trusted piece of equipment or even from a well-accepted technique known to produce a reliable value. Of course, the accuracy of the calibration is limited to the accuracy of the standard used. The following sections explore how certain standards are the foundation of all measurements and how other standards are derived from them.

Primary Unit Standards

A *dimension* defines a physical variable that is used to describe some aspect of a physical system. A *unit* defines a quantitative measure of a dimension. For example, mass, length, and time describe base dimensions with which we associate the units of kilogram, meter, and second. A *primary standard* defines the exact value of a unit. It provides the means to describe the unit with a unique number that can be understood throughout the world. In 1960, the General Conference on Weights and Measures (CGPM), the international agency responsible for maintaining exact uniform standards of measurements, formally adopted the International System of Units (SI) as the international standard of units (6). The SI system has been adopted worldwide and contains seven base units. All other units are derived from these seven.

Other unit systems are commonly used in the consumer market and so deserve mention. Examples of these include the inch-pound (I-P) unit system found in the United States and the gravitational mks (meter-kilogram-second or metric) unit system common in much of the world. The units used in these systems are directly related, through conversion factors, to SI units. Although SI defines our unit system exactly, practical units sometimes better fit our customs. For example, we use temperature units of Celsius and Fahrenheit in daily life rather than Kelvin.

Primary unit standards are necessary because the value assigned to a unit is actually quite arbitrary. For example, over 4,500 years ago, the Egyptian cubit was used as a standard of length and was based on the length from outstretched fingertips to the elbow. It was later codified with a master of marble, a stick about 52 cm in length, on which scratches were etched to define subunits of length. This standard served well for centuries.

So whether today's standard unit of length, the meter, is the length of a king's forearm or the distance light travels in a fraction of a second really only depends on how we want to define it. To avoid contradiction, the units of primary standards are defined by international agreement. Once agreed upon, a primary standard forms the exact definition of the unit until it is changed by some later agreement. Important features sought in any standard should include global availability, continued reliability, and stability with minimal sensitivity to external environmental sources. Next we examine dimensions and the primary standards that form the definition of the units used to describe them (6,7).

Base Dimensions and Their Units

Mass

The kilogram is the base unit of mass. Originally, the kilogram was defined by the mass of 1 liter of water at room temperature. The modern definition defines the kilogram exactly as the mass of a particular platinum-iridium cylindrical bar that is maintained under controlled conditions at the International Bureau of Weights and Measures (BIPM) located in Sevres, France. This particular bar (consisting of 90% platinum and 10% iridium by mass) forms the primary standard for the kilogram. It remains today as the only basic unit still defined in terms of a material artifact. Official copies exist at BIPM and a directly calibrated set of national prototypes serve as national reference standards throughout the world (6–8).

In the United States, the I-P unit system (also referred to as the U.S. customary units) remains widely used. In the I-P system, *mass* is defined by the pound-mass, lb_m, which is derived directly from the definition of the kilogram:

$$1\,lb_m = 0.4535924\,kg \tag{1.13}$$

Equivalent standards for the kilogram and other standards units are maintained by national labs around the globe. In the United States, this role is assigned to the National Institute of Standards and Technology (NIST) in Gaithersburg, Maryland.

Example 1.8: The Changing Kilogram

The kilogram remains the only basic unit defined in terms of a unique physical object, so defined in 1889. A problem with using a physical object is that it can change over time. For example, atmospheric pollutants add mass and periodic cleanings can remove metal molecules, reducing mass. This standard object has been removed from protected storage several times in the past century to compare with exact official copies (8). By 1989, there were differences detected between all of the official masses. Fortunately for consumers, this difference is only about 50 micrograms, but this is a large discrepancy for a primary standard. Accordingly, BIPM is considering options for changing the standard to one using a method based on a physical constant, such as the Planck constant (8).

Time and Frequency

The second is the base unit of time. One second (s) is defined as the time elapsed during 9,192,631,770 periods of the radiation emitted between two excitation levels of the fundamental state of cesium-133. Despite this seemingly unusual definition, this primary standard can be reliably reproduced at suitably equipped laboratories throughout the world to an uncertainty of within 2 parts in 10 trillion.

The Bureau International de l'Heure (BIH) in Paris maintains the primary standard for clock time. Periodically, adjustments to clocks around the world are made relative to the BIH clock so as to keep time synchronous.

The standard for cyclical frequency is a derived unit based on the time standard (s). The standard unit is the hertz (1 Hz = 1 cycle/s). The cyclical frequency is related to the circular frequency (radians/s) by

$$1 \text{ Hz} = \frac{2\pi \text{ rad}}{1 \text{ s}} \tag{1.14}$$

Length

The meter is the base unit for length. New primary standards are established when our ability to determine the new standard becomes more accurate (i.e., lower uncertainty) than the existing standard. In 1982, a new primary standard was adopted by the CGPM to define the unit of a meter. One meter (m) is now defined exactly as the length traveled by light in 1/299,792,458 of a second, a number derived from the velocity of light in a vacuum (defined as 299,792,458 m/s).

The I-P system unit of the inch and the related unit of the foot are derived exactly from the meter.

$$1 \text{ ft} = 0.3048 \text{ m}$$
$$1 \text{ in.} = 0.0254 \text{ m} \tag{1.15}$$

Temperature

The kelvin, K, is the base unit of thermodynamic temperature. It is the fraction 1/273.16 of the thermodynamic temperature of the triple point of water.

A temperature scale was devised by William Thomson, Lord Kelvin (1824–1907) that is based on polynomial interpolation between the equilibrium phase change points of a number of common pure substances from the triple point of equilibrium hydrogen (13.81 K) to the freezing point of pure gold (1337.58 K). Above 1337.58 K, the scale is based on Planck's law of radiant emissions. The details of the standard scale have been modified over the years but are governed by the International Temperature Scale–1990 (9).

The I-P unit system uses the absolute scale of Rankine (°R). This and the common scales of Celsius (°C), used in the metric system, and Fahrenheit (°F) are related to the Kelvin scale by the following:

$$(°C) = (K) - 273.15$$

$$(°F) = (°R) - 459.67$$

$$(°F) = 1.8 \times (°C) + 32.0$$

(1.16)

Current

The ampere is the base unit for electrical current. One ampere (A) is defined as the constant current that, if maintained in two straight parallel conductors of infinite length and of negligible circular cross section and placed 1 m apart in vacuum, would produce a force equal to 2×10^{-7} newton per meter of length between these conductors. The newton is a derived unit of force.

Measure of Substance

The mole is the base unit defining the quantity of a substance. One mole (mol) is the amount of substance of a system that contains as many elementary entities as there are atoms in 0.012 kilogram of carbon-12.

Luminous Intensity

The candela is the base unit of the intensity of light. One candela (cd) is the luminous intensity, in a given direction, of a source that emits monochromatic radiation of frequency 5.40×10^{14} hertz and that has a radiant intensity in that direction of 1/683 watt per steradian. A watt is a derived unit of power.

Derived Units

Other dimensions and their associated units are defined in terms of and derived from the base dimensions and units (7).

Force

From Newton's law, force equals mass times acceleration. Force is defined by the unit of the newton (N), which is derived from the base units for mass, length, and time:

$$1 \text{ N} = 1\frac{\text{kg-m}}{s^2} \tag{1.17}$$

One Newton is defined as the force applied to a 1 kg mass to accelerate it at 1 m/s^2.

When working within and between unit systems, it can be helpful to view force in terms of

$$\text{Force} = \frac{\text{Mass} \times \text{Acceleration}}{g_c}$$

where g_c is a proportionality constant used to maintain consistency between the units of force and mass. So for the SI system, the value of g_c must be 1.0 kg-m/s^2-N. The term g_c does not need to be explicitly stated, but the relationship between force and mass units holds.

However, in I-P units, the units of force and mass are related through the definition: One pound-mass (lb_m) exerts a force of 1 pound (lb) in a standard earth gravitational field. With this definition,

$$1 \text{ lb} = \frac{(1 \text{ lb}_m)(32.1740 \text{ ft/s}^2)}{g_c} \tag{1.18}$$

and g_c must take on the value of $32.1740 \text{ lb}_m\text{-ft/s}^2$-lb.

Similarly, in the gravitational mks (metric) system, which uses the kilogram-force (kg_f),

$$1 \text{ kg}_f = \frac{(1 \text{ kg})(9.80665 \text{ m/s}^2)}{g_c} \tag{1.19}$$

Here the value for g_c takes on a value of exactly 9.80665 kg-m/s^2-kg_f.

Many engineers have some difficulty distinguishing between units of mass and force in the non-SI unit systems. Actually, whenever force and mass appear in the same expression, just relate them using g_c through Newton's law:

$$g_c = \frac{mg}{F} = 1\frac{\text{kg-m/s}^2}{\text{N}} = 32.1740\frac{\text{lb}_m\text{-ft/s}^2}{\text{lb}} = 9.80665\frac{\text{kg-m/s}^2}{\text{kg}_f} \tag{1.20}$$

Shown in Figure 1.13 is a common example of multiple unit systems being cited on a consumer product marketed for use in different parts of the world: a bicycle tire with recommended inflation pressures shown in I-P, metric, and SI ($1 \text{ kPa} = 1000 \text{ N/m}^2$) units.

Other Derived Dimensions and Units

Energy is defined as force times length and uses the unit of the joule (J), which is derived from base units as

$$1 \text{ J} = 1\frac{\text{kg-m}^2}{s^2} = 1 \text{ N} - \text{m} \tag{1.21}$$

Figure 1.13   Bicycle tire pressure expressed in three different unit systems to meet different consumer preferences within a global market. (Photo courtesy of Richard Figliola.)

Power is defined as energy per unit time in terms of the unit of the watt (W), which is derived from base units as

$$1\text{ W} = 1\frac{\text{kg-m}^2}{\text{s}^3} = 1\frac{\text{J}}{\text{s}} \tag{1.22}$$

Stress and pressure are defined as force per unit area in terms of the pascal (Pa), which is derived from base units as

$$1\text{ Pa} = 1\frac{\text{kg}}{\text{m} - \text{s}^2} = 1\text{ N/m}^2 \tag{1.23}$$

Electrical Dimensions

The units for the dimensions of electrical potential, resistance, charge, and capacitance are based on the definitions of the absolute volt (V), ohm (Ω), coulomb (C), and farad (F), respectively. Derived from the ampere, 1 ohm is defined as 0.9995 times the resistance to current flow of a column of mercury that is 1.063 m in length and that has a mass of 0.0144521 kg at 273.15 K. The volt is derived from the units for power and current, 1 V = 1 N-m/s-A = 1 W/A. The ohm is derived from the units for electrical potential and current, $1\ \Omega = 1\text{ kg-m}^2/\text{s}^3\text{-A}^2 = 1$ V/A. The coulomb is derived from the units for current and time, 1 C = 1 A-s. One volt is the difference of potential between two points of an electrical conductor when a current of 1 ampere flowing between those points dissipates a power of 1 watt. The farad (F) is the standard unit for capacitance derived from the units for charge and electric potential, 1 F = 1 C/V.

On a practical level, working standards for resistance and capacitance take the form of certified standard resistors and capacitors or resistance boxes and are used as standards for comparison in the calibration of resistance measuring devices. The practical potential standard makes use of a standard cell consisting of a saturated solution of cadmium sulfate. The potential difference of two conductors connected across such a solution is set at 1.0183 V and at 293 K. The standard cell maintains constant voltage over very long periods of time, provided that it is not subjected to a current drain exceeding 100 µA for more than a few minutes. The standard cell is typically used as a standard for comparison for voltage measurement devices.

Table 1.2 Dimensions and Units[a]

Unit	Dimension	
	SI	I-P
Primary		
Length	meter (m)	inch (in.)
Mass	kilogram (kg)	pound-mass (lb$_\text{m}$)
Time	second (s)	second (s)
Temperature	kelvin (K)	rankine (°R)
Current	ampere (A)	ampere (A)
Substance	mole (mol)	mole (mol)
Light intensity	candela (cd)	candela (cd)
Derived		
Force	newton (N)	pound-force (lb)
Voltage	volt (V)	volt (V)
Resistance	ohm (Ω)	ohm (Ω)
Capacitance	farad (F)	farad (F)
Inductance	henry (H)	henry (H)
Stress, pressure	pascal (Pa)	pound-force/inch^2 (psi)
Energy	joule (J)	foot pound-force (ft-lb)
Power	watt (W)	foot pound-force/second (ft-lb/s)

[a]SI dimensions and units are the international standards. I-P units are presented for convenience.

A chart for converting between units is included inside the text cover. Table 1.2 lists the standard base and derived units used in SI and the corresponding units used in the I-P and gravitational metric systems.

Example 1.9: Units Conversion Issues

In 1999, communications were lost with the Mars Climate Orbiter during maneuvers intended to place the spacecraft into permanent Mars orbit. The fault was tracked to units conversion incompatibility between the ground-based software and the spacecraft-based software used in tandem to control the spacecraft (10). At the root of the problem was that the systems and software were each developed by different vendors. In one set of instructions, the thruster pulsation program was written to apply vehicle thrust in units of newtons, whereas the other set was written to apply thrust in units of pounds. Thruster execution placed the spacecraft in a much lower orbit than intended, and (presumably) it burned up in the atmosphere. This example illustrates the value of rigorous systems-level verification.

In 1983, Air Canada Flight 143 ran out of fuel during a flight requiring an emergency landing in Gimli, Manitoba. Although Canada had recently undergone a metrification program and this Boeing 767 was a metric aircraft, many aircraft and operations were still based on Imperial units (pounds and gallons). A sequence of conversion errors by ground fuel crews resulted in 22,300 lbs of fuel being loaded rather than the 22,300 kg specified. A compounding factor was a fuel computer glitch that also rendered the onboard fuel gauges inoperative. Fortunately, the pilots were able to glide the aircraft from 41,000 ft to a complicated but safe landing, thus earning the aircraft the nickname "The Gimli Glider" (11).

Hierarchy of Standards

The known value or reference value applied during calibration becomes the standard on which the calibration is based. So how do we pick this standard, and how good does it need to be? Obviously, actual primary standards are unavailable for normal calibration use. But they serve as a reference for exactness. So for practical reasons, there exists a hierarchy of secondary standards used to duplicate the primary standards. Just below the primary standard in terms of absolute accuracy are the national reference standards maintained by designated standards laboratories throughout the world. These provide a reasonable duplication of the primary standard while allowing for worldwide access to an extremely accurate standard. Next to these, we develop transfer standards. These are used to calibrate individual laboratory standards that might be used at various calibration facilities within a country. Laboratory standards serve to calibrate working standards. Working standards are used to calibrate everyday devices used in manufacturing and research facilities. In the United States, NIST maintains primary, national reference, and transfer standards and recommends standard procedures for the calibration of measurement systems.

Each subsequent level of the hierarchy is derived by calibration against the standard at the previous higher level. Table 1.3 lists an example of such a lineage for standards from a primary or reference standard maintained at a national standards lab down to a working standard used in a typical laboratory or production facility to calibrate everyday working instruments. If the facility does not maintain a local (laboratory or working) standard, then the instruments must be sent off and calibrated elsewhere. In such a case, a standards traceability certificate will be issued for the instrument that details the lineage of standards used toward its calibration and the uncertainty in the calibration. For measurements that do not require traceable accuracy, a trusted value can be used to calibrate another device.

Moving down through the standards hierarchy, the degree of exactness by which a standard approximates the primary standard deteriorates. That is, increasing amounts of error are introduced into the standard as we move down from one level of hierarchy to the next. As a common example, a company might maintain its own working standard used to calibrate the measurement devices found in the individual laboratories throughout the company. Periodic calibration of that working standard might be against the company's well-maintained local standard. The local standard would be periodically sent off to be calibrated against the NIST (or appropriate national standards lab) transfer

Table 1.3 Hierarchy of Standards[a]

Primary standard	Maintained as absolute unit standard
Reference standard	Duplicate of a primary standard at a national lab
Transfer standard	Used to calibrate local standards
Local standard	Used to calibrate working standards
Working standard	Used to calibrate local instruments
Calibration standard	Any reference value used in a calibration

[a]There may be additional intermediate standards between each hierarchy level.

Table 1.4 Example of a Temperature Standard Traceability

Level	Standard Method	Uncertainty [K][a]
Primary	Fixed thermodynamic points	0
Transfer	Platinum resistance thermometer	±0.005
Working	Platinum resistance thermometer	±0.05
Local	Thermocouple	±0.5

[a]Typical combined instrument systematic and random uncertainties.

standard (and traceability certificate issued). NIST will periodically calibrate its own transfer standard against its reference or primary standard. This idea is illustrated for a temperature standard traceability hierarchy in Table 1.4. The uncertainty in the approximation of the known value increases as one moves down the hierarchy. It follows, then, that the accuracy of the calibration will be limited by the accuracy of the standard used. But if typical working standards contain errors, how is accuracy ever determined? At best, this closeness of agreement is quantified by the estimates of the known uncertainties in the calibration.

Test Standards and Codes

The term "standard" is also applied in other ways in engineering. A *test standard* is a document, established by consensus, that provides rules, guidelines, methods or characteristics for activities or their results. They define appropriate use of certain terminology as used in commerce. The goal of a test standard is to provide consistency between different facilities or manufacturers in the conduct and reporting of a certain type of measurement on some test specimen. Similarly, a *test code* is a standard that refers to procedures for the manufacture, installation, calibration, performance specification, and safe operation of equipment.

Diverse examples of test standards and codes are illustrated in readily available documents (12–17) from professional societies, such as the American Society of Mechanical Engineers (ASME), the American Society of Testing and Materials (ASTM), the American National Standards Institute (ANSI), and the International Organization for Standardization (ISO). For example, ASME Power Test Code 19.5 provides detailed designs and operation procedures for flow meters, and ANSI S2.70-2006 details sensor placement and data reduction methods to evaluate and report human exposure to vibrations transmitted to the hand from operating machinery.

Engineering standards and codes are consensus documents agreed on by knowledgeable parties interested in establishing a common basis for comparing equipment performance between manufacturers or for consistent and safe operation. These are not binding legal documents unless specifically adopted and implemented as such by a government agency. Still, they present a convincing argument for best practice.

1.6 PRESENTING DATA

Data presentation conveys significant information about the relationship between variables. Software is readily available to assist in providing high-quality plots, or plots can be generated manually using graph paper. Several forms of plotting formats are discussed next.

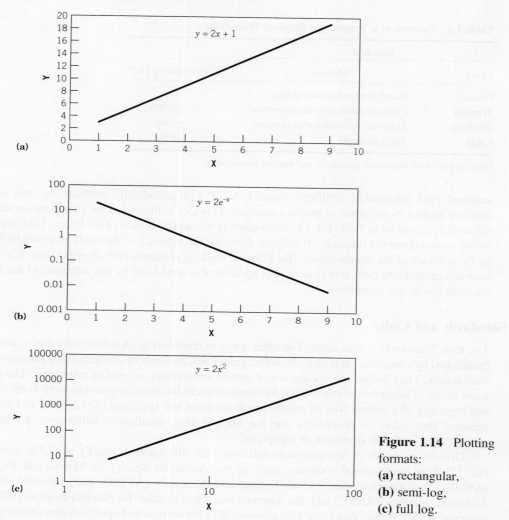

Figure 1.14 Plotting formats:
(a) rectangular,
(b) semi-log,
(c) full log.

Rectangular Coordinate Format

In rectangular grid format, such as Figure 1.14a, both the ordinate and the abscissa have uniformly sized divisions providing a linear scale. This is a common format used for constructing plots and establishing the form of the relationship between the independent and dependent variable.

Semilog Coordinate Format

In a semilog format, one coordinate has a linear scale and one coordinate has a logarithmic scale. Plotting values on a logarithmic scale performs a logarithmic operation on those values—for example, plotting $y = f(x)$ on a logarithmic x-axis is the same as plotting $y = \log f(x)$ on rectangular axes. Logarithmic scales are advantageous when one of the variables spans more than one order of magnitude. A linear curve results when the data follow a trend of the form $y = ae^x$, as in Figure 1.14b, or $y = a10^x$. These two logarithmic forms are related as $\ln y = 2.3 \log y$.

Full-Log Coordinate Format

The full-log or log-log format (Figure 1.14c) has logarithmic scales for both axes and is equivalent to plotting $\log y$ vs. $\log x$ on rectangular axes. Such a format is preferred when both variables contain data values that span more than one order of magnitude. With data that follow a trend of the form $y = ax^n$ as in Figure 1.14c, a linear curve will result.

Significant Digits

Just how many digits should you assign to a number when you report it? This number should reflect both your level of confidence and the level of relevance for the information it represents. In this section, we discuss how to identify significant digits and how to round to an appropriate number of significant digits, offering suggestions for assigning significant digits in calculated values. Despite the suggestions here, the number of significant digits reported should also reflect practical sense.

The term *digit* refers to a numerical figure between 0 and 9. The position or *place value* of a digit within a number provides its order of magnitude. A *significant digit* refers to any digit that is used together with its place value to represent a numerical value to a desired approximation. In writing numbers, the leftmost nonzero digit is the *most significant digit*. The rightmost digit is the *least significant digit*. Leading zeros refer to the leftmost zeros leading any nonzero digits. Trailing zeros are the rightmost zeros following nonzero digits.

All nonzero digits present in a number are significant. Zeros require special attention within the following guidelines: (1) All leading zeros, whether positioned before or after a decimal point, are not significant, as these serve only to set order of magnitude and not value; (2) zeros positioned between nonzero digits are significant; (3) all trailing zeros when situated to either side of a decimal point are significant; and (4) within an exact count, the zeros are significant. These guidelines are consistent with *ASTM E29—Standard Practice for Using Significant Digits* (16).

Writing a number using scientific notation is a convenient way to distinguish between meaningful digits and nonsignificant zero digits used only to hold place value. For example, 0.0042 can be written as 4.2×10^{-3}. From the guidelines above, this number has two significant digits, which are represented by the two nonzero digits. In this example, the leading zeros serve only to hold place values for the order of magnitude, which is explicit (i.e., 10^{-3}) when written in scientific notation.

When a number does not have a decimal point cited, trailing zeros can be problematic. If the zeros are intended to be significant, use a decimal point. Otherwise, they will be assumed not significant. To avoid ambiguity, use scientific notation. For example, writing 150,000 Pa as 1.5×10^4 Pa clearly indicates that two significant digits are intended.

Rounding

Rounding is a process in which the number of digits is reduced and the remaining least significant digit(s) adjusted appropriately. In reducing the number of digits, (1) if the digits to be discarded begin with a digit less than 5, the digit preceding the 5 is not changed; (2) if the digits to be discarded begin with a 5 and at least one of the following digits is greater than 0, the digit preceding the 5 is increased by 1;

(3) if the digits to be discarded begin with a 5 and all of the following digits are 0, the digit preceding the 5 is unchanged if it is an even digit but increased by 1 if it is an odd digit. When a result must meet a specification limit for conformance, do not round the number to meet the specification.

Numerical Operations

Here are suggestions for adjusting the significant digits of the result after common numerical operations:

1. In addition or subtraction, the number of digits following the decimal in the reported result should not be greater than the least number of digits found following a decimal in any of the data points used.

2. In other operations, the significant digits in the result should not be greater than the least number of significant digits in any of the data points or the operand used.

3. The number of significant digits in an exact count is not considered when establishing the number of significant digits to be reported.

4. Round your final result, but do not round intermediate calculations.

Computations within software, hand calculators, and spreadsheets carry large numbers of digits, usually many more than can be justified. However, rounding intermediate calculations introduces new truncation errors, so best practice is to round only the final result.

Example 1.10: Rounding Decisions

Rounding is a decision process and practical sense should prevail. As an example, when the most significant digit in the result of a numerical operation is increased in order of magnitude relative to any of the data points used, it may make sense to retain extra significant digit(s). Consider $5 \times 3 = 15$. Would you keep the 15 or round to 20 to maintain the one significant digit suggested in the guidelines? Consider $5.1 \times 2.1 = 10.71$. The rounded result might be better expressed as 10.7 rather than 11. One rationale is that the maximum rounding error in 5.1 is $0.05/5.1 = 0.94\%$, in 2.1 is $0.05/2.1 = 2.4\%$, in 11 is $0.5/11 = 4.5\%$, and in 10.7 is $0.05/10.7 = 0.46\%$. Rounding to 11, discards information present in the data, whereas rounding to 10.7 adds information. In rounding, the engineer must choose between removing known information and adding unwarranted information while keeping the quality of the known information and the intended use of the result in perspective (7). In short, the correct approach is at the discretion of the engineer.

Applications

How many significant digits should you carry when recording a measurement? For direct measurements using an analog readout device (e.g., dial indicator), the number of significant digits recorded should include all the digits known exactly, plus up to one more digit from interpolation when possible. This holds the same for data taken from charts. For direct measurements from a digital readout, record all known digits. With tables, interpolate as needed but retain the same number of digits as used in the table. When recording data by computer-based data acquisition methods, carry as many digits as possible and then round a final result to an appropriate number of digits.

Example 1.11

Round the following numbers to three significant digits and then write them in scientific notation.

49.0749   becomes   49.1 or 4.91×10^1

0.0031351   becomes   0.00314 or 3.14×10^{-3}

0.0031250   becomes   0.00312 or 3.12×10^{-3}

Example 1.12

A handheld appliance consumes 1.41 kW of power. So two identical units (exact count $N = 2$) consume 1.41 kW $\times$ 2 = 2.82 kW of power. The exact count does not affect the significant digits in the result.

Example 1.13

Many conversion factors can be treated as exact counts. Reference 7 provides a list. For example, in converting 1 hour into seconds, the result is exactly 3,600 s. Similarly, 1 ft equals exactly 0.3048 m. Otherwise, apply the rounding suggestions using discretion.

1.7 SUMMARY

During a measurement the input signal is not known but is inferred from the value of the output signal from the measurement system. We discussed the process of calibration as the means to relate the measured input value to the measurement system output value. We discussed the role of standards in that process. An important step in the design of a measurement system is the inclusion of a means for a reproducible calibration that closely simulates the type of signal to be input during actual measurements. A test is the process of "asking a question." The idea of a test plan was developed to allow measurements that answer that question. However, a measured output signal can be affected by many variables that will introduce variation and trends and confuse that answer. Careful test planning is required to reduce such effects. A number of test plan strategies were developed, including randomization. The popular term "accuracy" was explained in terms of the more useful concepts of random error, systematic error, random uncertainty, and systematic uncertainty and their effects on a measured value. We explored the idea of test standards and engineering codes, legal documents that influence practically every manufactured product around us.

REFERENCES

1. Peace, G. S., *Taguchi Methods*, Addison-Wesley, Reading, MA, 1993.
2. Lipsen, C., and N. J. Sheth, *Statistical Design and Analysis of Engineering Experimentation*, McGraw-Hill, New York, 1973.
3. Peterson, R. G., *Design and Analysis of Experiments*, Marcel-Dekker, New York, 1985.

4. Mead, R., *The Design of Experiments: Statistical Principles for Practical Application*, Cambridge Press, New York, 1988.

5. Montgomery, D., *Design and Analysis of Experiments*, 7th ed., Wiley, New York, 2009.

6. Bureau International des Poids et Mesures, *The International System of Units (SI)*, 8th ed., Organisation Intergouvernementale de la Convention du Mètre, Paris, 2006.

7. Taylor, B., and A. Thompson, Guide for the Use of the International System of Units, NIST Special Publication 811, 2008. http://physics.nist.gov/cuu/pdf/sp811.pdf

8. Davis, R., "The SI unit of mass," *Metrologia* **40** 299–305, 2003.

9. Committee Report, International Temperature Scale—1990 *Metrologia* **27** (1): 3–10, 1990.

10. Stephenson, A. (Chairman), "Mars Climate Orbiter Mishap Investigation Board Phase I Report," NASA, Washington, DC, November 1999.

11. Nelson, Wade, "The Gimli Glider," Soaring Magazine, October 1997, 19–22.

12. *ASME Power Test Codes*, American Society of Mechanical Engineers, New York.

13. *ASHRAE Handbook of Fundamentals*, American Society of Heating, Refrigeration and Air Conditioning Engineers, New York, 2009.

14. *ANSI Standard*, American National Standards Institute, New York, 2009.

15. *Annual Book of ASTM Standards*, American Society for Testing and Materials, Philadelphia, 2010.

16. *Standard Practice for Using Significant Digits, ASTM E29-13*, American Society for Testing and Materials, Philadelphia, 2013.

17. *Guide for the Measurement and Evaluation of Human Exposure to Vibration Transmitted to the Hand, ANSI S2.70-2006*, American National Standards Institute, New York, 2006.

NOMENCLATURE

e absolute error

p pressure ($ml^{-1}t^{-2}$)

r_i input span

r_o output span

s_x standard deviation of x

u_c overall instrument uncertainty

u_h hysteresis uncertainty; uncertainty assigned to hysteresis error

u_K sensitivity uncertainty; uncertainty assigned to sensitivity error

u_L linearity uncertainty; uncertainty assigned to linearity error

u_R repeatability uncertainty; uncertainty assigned to repeatability error

u_z zero uncertainty; uncertainty assigned to zero error

x independent variable; input value; measured variable

y dependent variable; output value

y_L linear polynomial

A relative error; relative accuracy

K static sensitivity

T temperature (°)

$\forall$ volume (l^3)

PROBLEMS

1.1   Select three different types of measurement systems with which you have experience, and identify which attributes of the system comprise the measurement system stages of Figure 1.4. Use sketches as needed.

1.2   For each of the following systems, identify the components that comprise each measurement system stage per Figure 1.4:

a. microphone/amplifier/speaker system

b. room heating thermostat

c. handheld micrometer

d. tire pressure (pencil-style) gauge

1.3 Consider Example 1.1. Discuss the effect of the extraneous variable barometric pressure in terms of noise and interference relative to any one test and relative to several tests. Explain how interference effects can be broken up into noise using randomization.

1.4 Cite three examples each of a continuous variable and a discrete variable.

1.5 Suppose you found a dial thermometer in a stockroom. Could you state how accurate it is? Discuss methods by which you might estimate random and systematic error in the thermometer?

1.6 Discuss how the resolution of the display scale of an instrument could affect its uncertainty. Suppose the scale was somehow offset by one least count of resolution: How would this affect its uncertainty? Explain in terms of random and systematic error.

1.7 A bulb thermometer hangs outside a house window. Comment on extraneous variables that might affect the difference between the actual outside temperature and the indicated temperature on the thermometer.

1.8 A synchronous electric motor test stand permits either the variation of input voltage or the output shaft load with the subsequent measurement of motor efficiency, winding temperature, and input current. Comment on the independent, dependent, and extraneous variables for a motor test.

1.9 The transducer specified in Table 1.1 is chosen to measure a nominal pressure of 500 cm H_2O. The ambient temperature is expected to vary between 18 °C and 25 °C during tests. Estimate the possible range (magnitude) of each listed elemental error affecting the measured pressure.

1.10 A force measurement system (weight scale) has the following specifications:

Range	0 to 1000 N
Linearity error	0.10% FSO
Hysteresis error	0.10% FSO
Sensitivity error	0.15% FSO
Zero drift	0.20% FSO

Estimate the overall instrument uncertainty for this system based on available information. Use the maximum possible output value over the FSO in your computations.

1.11 State the purpose of using randomization methods during a test. Develop an example to illustrate your point.

1.12 Provide an example of repetition and replication in a test plan from your own experience.

1.13 Develop a test plan that might be used to estimate the average temperature that could be maintained in a heated room as a function of the heater thermostat setting.

1.14 Develop a test plan that might be used to evaluate the fuel efficiency of a production model automobile. Explain your reasoning.

1.15 A race engine shop has just completed two engines of the same design. How might you, the team engineer, determine which engine would perform better for an upcoming race (a) based on a test stand data (engine dynamometer) and (b) based on race track data? Describe some measurements that you feel might be useful, and explain how you might use that information. Discuss possible differences between the two tests and how these might influence the interpretation of the results.

1.16 A thermodynamics model assumes that a particular gas behaves as an ideal gas: Pressure is directly related to temperature and density. How might you determine that the assumed model is correct (validation)?

1.17 Regarding the Mars Climate Orbiter spacecraft example presented, discuss how verification tests before launch could have identified the software units problem that led to the catastrophic spacecraft failure. Explain the purpose of verification testing?

1.18 A large batch of carefully made machine shafts can be manufactured on 1 of 4 lathes by 1 of 12 quality machinists. Set up a test matrix to estimate the tolerances that can be held within a production batch. Explain your reasoning.

1.19 Suggest an approach or approaches to estimate the linearity error and the hysteresis error of a measurement system.

1.20 Suggest a test matrix to evaluate the wear performance of four different brands of aftermarket passenger car tires of the same size, load, and speed ratings on a fleet of eight cars of the same make. If the cars were not of the same make, what would change?

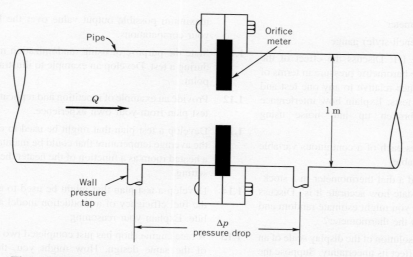

Figure 1.15 Orifice flow meter setup used for Problem 1.21.

1.21 The relation between the flow rate, Q, through a pipeline of area A and the pressure drop Δp across an orifice-type flow meter inserted in that line (Figure 1.15) is given by $Q = CA\sqrt{2\Delta p/\rho}$ where ρ is density and C is a coefficient. For a pipe diameter of 1 m and a flow range of 20 °C water between 2 and 10 m^3/min and $C = 0.75$, plot the expected form of the calibration curve for flow rate versus pressure drop over the flow range. Is the static sensitivity a constant? Incidentally, the device and test method is described by both ANSI/ASME Test Standard PTC 19.5 and ISO 5167.

1.22 The sale of motor fuel is an essential business in the global economy. Federal (U.S.) law requires the quantity of fuel delivered at a retail establishment to be accurate to within 0.5%. (i) Determine the maximum allowable error in the delivery of 25 gallons (or use 95 L). (b) Provide an estimate of the potential costs to a consumer at the maximum allowable error over 150,000 miles (240,000 km) based on an average fuel mileage of 30.2 mpg (7.8 L/100 km). (c) As a knowledgeable consumer, cite one or two ways for you to identify an inaccurate pump.

1.23 Using either the ASME 19.5 or ISO 5167 test standard, explain how to use a venturi flowmeter. What are the independent variable(s) and dependent variable(s) in engineering practice? Explain.

1.24 A simple thermocouple circuit is formed using two wires of different alloy: One end of the wires is twisted together to form the measuring junction, and the other ends are connected to a digital voltmeter, forming the reference junction. A voltage is set up by the difference in temperature between the two junctions. For a given pair of alloy material and reference junction temperature, the temperature of the measuring junction is inferred from the measured voltage difference. What are the dependent and independent variables during practical use? What about during a calibration?

1.25 A linear variable displacement transducer (LVDT) senses displacement and indicates a voltage output that is linear to the input. Figure 1.16 shows an LVDT setup used for static calibration. It uses a micrometer to apply the known displacement and a voltmeter for the output. A well-defined voltage powers the transducer. What are the independent and dependent variables in this calibration? Would these change in practical use?

1.26 For the LVDT calibration of the previous problem, what would be involved in determining the repeatability of the instrument? The reproducibility? What effects are different in the two tests? Explain.

1.27 A manufacturer wants to quantify the expected average fuel mileage of a product line of automobiles. It decides that either it can put one or more

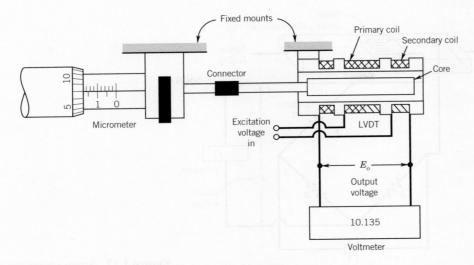

Figure 1.16 LVDT setup used for Problem 1.25.

cars on a chassis dynamometer and run the wheels at desired speeds and loads to assess this, or it can use drivers and drive the cars over some selected course instead. (a) Discuss the merits of either approach, considering the control of variables and identifying extraneous variables. (b) Can you recognize that somewhat different tests might provide answers to different questions? For example, discuss the difference in meanings possible from the results of operating one car on the dynamometer and comparing it to one driver on a course. Cite other examples. (c) Do these two test methods serve as examples of concomitant methods?

1.28 The coefficient of restitution of a volleyball is found by dropping the ball from a known height H and measuring the height of the bounce, h. The coefficient of restitution, C_R, is then calculated as $C_R = \sqrt{h/H}$. Develop a test plan for measuring C_R that includes the range of conditions expected in college-level volleyball play.

1.29 As described in a preceding problem, the coefficient of restitution of a volleyball, $C_R = \sqrt{h/H}$, is determined for a range of impact velocities. The impact velocity is $v_i = \sqrt{2\,gH}$ and is controlled by the dropping the ball from a known height H. Let the velocity immediately after impact be $v_f = \sqrt{2\,gh}$ where h is the measured return height.

List the independent and dependent variables, parameters, and measured variables.

1.30 Light gates may be used to measure the speed of projectiles, such as arrows shot from a bow. English longbows made of yew in the 1400s achieved launch speeds of 60 m/s. Determine the relationship between the distance between light gates and the accuracy required for sensing the times when the light gate senses the presence of the arrow.

1.31 You estimate your car's fuel use by recording regularly the fuel volume used over a known distance. Your brother, who drives the same model car, disagrees with your claimed results based on his own experience. Suggest reasons to justify the differences? How might you test to provide justification for each of these reasons?

1.32 When discussing concomitant methods, we used the example of estimating the volume of a rod. Identify another concomitant method that you might use to verify whether your first test approach to estimating rod volume is working.

1.33 When a strain gauge is stretched under uniaxial tension, its resistance varies with the imposed strain. A resistance bridge circuit is used to convert the resistance change into a voltage. Suppose a known tensile load were applied to a test specimen using the system shown in Figure 1.17. What are the

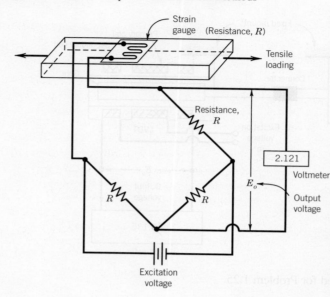

Strain gauge (Resistance, R)

Tensile loading

Resistance, R

2.121

Voltmeter

E_o

Output voltage

Excitation voltage

Figure 1.17 Strain gauge setup used for Problem 1.33.

independent and dependent variables in this calibration? How do these change during actual testing?

1.34 For the strain gauge calibration of the previous problem, what would be involved in determining the repeatability of the instrument? The reproducibility? What effects are different in the tests? Explain.

1.35 The acceleration of a cart down a plane inclined at an angle α to horizontal can be determined by measuring the change in speed of the cart at two points, separated by a distance s, along the inclined plane. Suppose two photocells are fixed at the two points along the plane. Each photocell measures the time for the cart, which has a length L, to pass it. Identify the important variables in this test. List any assumptions that you feel are intrinsic to such a test. Suggest a concomitant approach. How would you interpret the data to answer the question?

1.36 In general, what is meant by the term "standard"? Discuss your understanding of the hierarchy of standards used in calibration beginning with the primary standard. How do these differ from test standards and codes?

1.37 A common scenario: An engineer has two pencil-style pressure gauges in her garage for setting tire pressures. She notices that the two gauges disagree by about 14 kPa (2 psi) on the same tire. How does she choose the most accurate gauge to set car tire

pressures? Discuss possible strategies she might use to arrive at her best option.

1.38 Explain the potential differences in the following evaluations of an instrument's accuracy. Figure 1.11 will be useful, and you may refer to ASTM E177 if needed.

a. The closeness of agreement between the true value and the average of a large set of measurements.

b. The closeness of agreement between the true value and an individual measurement.

1.39 Research the following test standards and codes. Write a short summary to describe the intent, and give an overview of each code:

a. ASTM F 558 (Air Performance of Vacuum Cleaners)

b. ANSI Z21.86 (Gas Fired Space Heating Appliances)

c. ISO 10770-1 (Test Methods for Hydraulic Control Valves)

d. ANSI/ASME PTC19.1 (Test Uncertainty)

e. ISO 7401 (Road Vehicles: Lateral Response Test Methods)

f. ISO 5167 Parts 1–4

1.40 A hotel chain based in the United States contracts with a European vacuum cleaner manufacturer to supply a large number of upright cleaner units. After delivery, the hotel raises questions about manufacturer vacuum cleaner performance claims pointing out that the units should have been tested to meet ASTM 558, an American standard. The manufacturer notes the advertised performance is based on IEC 60312, a European standard, and the two test codes will yield similar, if not exact, results. Investigate both test codes and address similarities and differences. Is there is a legitimate claim here?

1.41 Test code ASTM 558-13 allows for the comparison of the maximum potential air power available between vacuum cleaners when tested under the prescribed conditions. The test requires using at least three units for each make/model tested with the units obtained at random. Explain what might be a reasonable way to meet this requirement. Explain a possible reason for this requirement.

1.42 Suggest a reasonable number of significant digits for reporting the following common values, and give some indication of your reasoning:

a. Your body weight for a passport

b. A car's fuel usage (use liters per 100 km)

c. The weight of a standard ("Good Delivery") bar of pure (at least 99.5%) gold

d. Distance (meters) traveled by a body in 1 second if moving at 1 m/s

1.43 Using spreadsheet software (such as Microsoft Excel), create a list of 1,000 numbers each randomly selected between 100 and 999. Divide each number by 10 to obtain a trailing digit; this will be Column 1. Each number in Column 1 will have three significant digits. In Column 2, have the spreadsheet round each number using the default *ROUND* function to two significant digits. In Column 3, have the spreadsheet round each number in Column 1 per ASTM E29 (e.g., *(=IF(MOD(A1,1)= 0.5,MROUND(A1,2),ROUND(A1,0))*. Compute the sum of each column and compare. Discuss the meaning of the operations and the results, and explain any differences due to rounding errors.

1.44 How many significant digits are present in the following numbers? Convert each to scientific notation.

(i) 10.020

(ii) 0.00034

(iii) 2500.

(iv) 042.02

(v) 0.1×10^{-3}

(vi) 999 kg/m^3

(vii) population 152,000 grams

(viii) 0.000001 grams

1.45 Round the following numbers to 3 significant digits.

(i) 15.963

(ii) 1232 kPa

(iii) 0.00315

(iv) 21.750

(v) 21.650

(vi) 0.03451

(vii) 1.82512314

(viii) 1.2350×10^{-4}

1.46 Express the result, rounding to an appropriate number of significant digits.

(i) 27.76 m + 4.907 m + 111.2 m =

(ii) 91.15 kg + 12.113 kg =

(iii) 101.2 J × 12.1 J =

(iv) 23. m + 42.15 m =

(v) 10. J − 5.1 J =

(vi) 10.0 kg − 5.1 kg =

(vii) 1.1 N + 5.47 N + 0.9178 N =

(viii) 7.81/81. =

1.47 Express the result by rounding to an appropriate number of significant digits.

(i) $\sin(nx) = $; $n = 0.010$ m^{-1}, $x = 5.73$ m

(ii) $e^{0.31} = $

(iii) $\ln(0.31) = $

(iv) $xy/z = $; $x = 423.$ J, $y = 33.42$ J, $z = 11.32$

(v) $(0.21^2 + 0.321^2 + 0.121^2)/3 = $

(vi) $107^2 + 6542 = $

(vii) $22.1^{1/2} = $

(viii) $(22.3 + 16.634) \times 59 = $

1.48 A car's speed is determined by the time it takes to travel between two marks. The marks are measured to be 21.0 m apart. A stopwatch measures the time of travel as 2.2 s. Using an appropriate number of significant digits, report the car's speed.

1.49 How much error could you tolerate in (1) book shelf length when two shelves are mounted one above the other, (2) differences between a car's tire pressures, (3) car tire pressure (any tire), (4) car speedometer, (5) home oven thermostat? Explain.

1.50 Apply the guidelines to determine the number of significant digits for these numbers that contain zero digits. Write the number in scientific notation using two significant digits.

(i) 0.042

(ii) 42.0250

(iii) 420.

(iv) 427,000

Table 1.5 Calibration Data

X [cm]	Y [V]	X [cm]	Y [V]
0.5	0.4	10.0	15.8
1.0	1.0	20.0	36.4
2.0	2.3	50.0	110.1
5.0	6.9	100.0	253.2

1.51 Using a tape measure having 1 mm graduations, the engineer measures an object and determines its length to be just longer than 52 mm but shorter than 53 mm. Using an appropriate number of significant digits, what might be an appropriate statement regarding this measured length?

1.52 Show how the following functions can be transformed into a linear curve of the form $Y = a_1 X + a_o$ where a_1 and a_0 are constants. Let $m, b,$ and c be constants.

    a. $y = bx^m$

    b. $y = be^{mx}$

    c. $y = b + c\sqrt[m]{x}$

1.53 For the calibration data of Table 1.5, plot the results using rectangular and log-log scales. Discuss the apparent advantages of either presentation.

1.54 For the calibration data of Table 1.5, determine the static sensitivity of the system at (a) $X = 5$, (b) $X = 10$, and (c) $X = 20$. For which input values is the system more sensitive? Explain what this might mean in terms of a measurement and in terms of measurement errors.

1.55 Consider the voltmeter calibration data in Table 1.6. Plot the data using a suitable scale. Specify the percent maximum hysteresis based on full-scale range. Input X is based on a standard known to be accurate to better than 0.05 mV.

1.56 Each of the following equations can be represented as a straight line on an x-y plot by choosing the appropriate axis scales. Plot each, and explain how the plot operation yields the straight line. Variable y has units of volts. Variable x has units of meters (use a range of $0.01 < x < 10.0$).

    a. $y = 5x^{-0.25}$

    b. $y = (0.4)(10^{-2x})$

Table 1.6 Voltmeter Calibration Data

Increasing Input [mV]		Decreasing Input [mV]	
X	Y	X	Y
0.0	0.1	5.0	5.0
1.0	1.1	4.0	4.2
2.0	2.1	3.0	3.2
3.0	3.0	2.0	2.2
4.0	4.1	1.0	1.2
5.0	5.0	0.0	0.2

    c. $y = 5e^{-0.5x}$

    d. $y = 2/x$

1.57 Plot $y = 10e^{-0.5x}$ volts on in semilog format (use three cycles). Determine the slope of the equation at $x = 0$, $x = 2$, and $x = 20$.

1.58 The following data have the form $y = ax^b$. Plot the data in an appropriate format to estimate the coefficients a and b. Estimate the static sensitivity K at each value of X. How is K affected by X?

Y [cm]	X [cm]
0.14	0.5
2.51	2.0
15.30	5.0
63.71	10.0

1.59 For the calibration data given, plot the calibration curve using suitable axes. Estimate the static sensitivity of the system at each X. Then plot K against X. Comment on the behavior of the static sensitivity with static input magnitude for this system.

Y [cm]	X [kPa]
4.76	0.05
4.52	0.1
3.03	0.5
1.84	1.0

Chapter 2

Static and Dynamic Characteristics of Signals

2.1 INTRODUCTION

A measurement system takes an *input* quantity and transforms it into an *output* quantity that can be observed or recorded, such as the movement of a pointer on a dial or the magnitude of a digital display. This chapter discusses the characteristics of both the input signals to a measurement system and the resulting output signals. The shape and form of a signal are often referred to as its *waveform*. The waveform contains information about the *magnitude*, which indicate the size of the input quantity, and the *frequency*, which indicates the rate at which the signal changes in time. An understanding of waveforms is required for the selection of measurement systems and the interpretation of measured signals.

2.2 INPUT/OUTPUT SIGNAL CONCEPTS

Two important tasks that engineers face in the measurement of physical variables are (1) selecting a measurement system and (2) interpreting the output from a measurement system. A simple example of selecting a measurement system might be the selection of a tire gauge for measuring the air pressure in a bicycle tire or in a car tire, as shown in Figure 2.1. The gauge for the car tire would be required to indicate pressures up to 275 kPa (40 lb/in.^{2}), but the bicycle tire gauge would be required to indicate higher pressures, perhaps up to 700 kPa (100 lb/in.^{2}). This idea of the *range* of an instrument, its lower to upper measurement limits, is fundamental to all measurement systems and demonstrates that some basic understanding of the nature of the input signal, in this case the *magnitude*, is necessary in evaluating or selecting a measurement system for a particular application.

A much more difficult task is the evaluation of the output of a measurement system when the time or spatial behavior of the input is not known. The pressure in a tire does not change while we are trying to measure it, but what if we wanted to measure pressure in a cylinder in an automobile engine? Would the tire gauge or another gauge based on its operating principle work? We know that the pressure in the cylinder varies with time. If our task was to select a measurement system to determine this time-varying pressure, information about the pressure variations in the cylinder would be necessary. From thermodynamics and the speed range of the engine, it may be possible to estimate the magnitude of pressures to be expected and the rate with which they change. From that we can

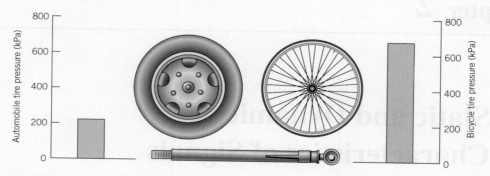

Figure 2.1 Measurement system selection based on input signal range.

select an appropriate measurement system. But to do this we need to develop a way to express this idea of the magnitude and rate of change of a measured variable.

Upon completion of this chapter, the reader will be able to

- Define the range of an instrument
- Classify signals as analog, discrete time, or digital
- Compute the mean and rms values for time-varying signals
- Characterize signals in the frequency domain

Generalized Behavior

Many measurement systems exhibit similar responses under a variety of conditions, which suggests that the performance and capabilities of measuring systems may be described in a generalized way. To examine further the generalized behavior of measurement systems, we first examine the possible forms of the input and output signals. We will associate the term "signal" with the "transmission of information." A *signal* is the physical information about a measured variable being transmitted between a process and the measurement system, between the stages of a measurement system, or as the output from a measurement system.

Classification of Waveforms

Signals may be classified as analog, discrete time, or digital. *Analog* describes a signal that is continuous in time. Because physical variables tend to be continuous, an analog signal provides a ready representation of their time-dependent behavior. In addition, the magnitude of the signal is continuous and thus can have any value within the operating range. An analog signal is shown in Figure 2.2a; a similar continuous signal would result from a recording of the pointer rotation with time for the output display shown in Figure 2.2b. Contrast this continuous signal with the signal shown in Figure 2.3a. This format represents a *discrete time signal*, for which information about the magnitude of the signal is available only at discrete points in time. A discrete time signal usually results from the sampling of a continuous variable at repeated finite time intervals. Because information in the signal shown in Figure 2.3a is available only at discrete times, some assumption must be made about the behavior of the measured variable during the times when it is not available.

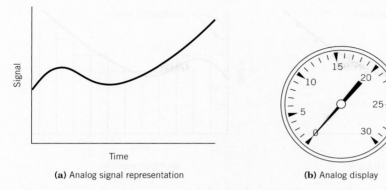

(a) Analog signal representation   **(b)** Analog display

Figure 2.2   Analog signal concepts.

One approach is to assume the signal is constant between samples—a sample and hold method. The resulting waveform is shown in Figure 2.3b. Clearly, as the time between samples is reduced, the difference between the discrete variable and the continuous signal it represents decreases.

Digital signals are particularly useful when data acquisition and processing are performed using a computer, a device that handles information in digital form. A digital signal has two important characteristics. First, a digital signal exists at discrete values in time, like a discrete time signal. Second, the magnitude of a digital signal is discrete, determined by a process known as quantization at each discrete point in time. *Quantization* assigns a single number to represent a range of magnitudes of a continuous signal. As an example of quantization, consider a digital clock that displays time in hours and minutes. For the entire duration of 1 minute a single numerical value is displayed until it is updated at the next discrete time step. As such, the continuous physical variable of time is quantized in its conversion to a digital display.

Figure 2.4a shows digital and analog forms of the same signal where the magnitude of the digital signal can have only certain discrete values. Thus, a digital signal provides a quantized magnitude at discrete times. The waveform that would result from assuming that the signal is constant between sampled points in time is shown in Figure 2.4b.

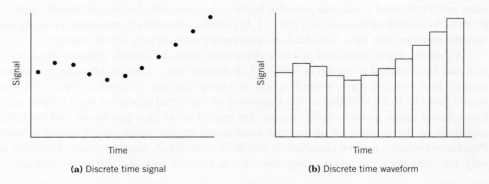

(a) Discrete time signal   **(b)** Discrete time waveform

Figure 2.3   Discrete time signal concepts.

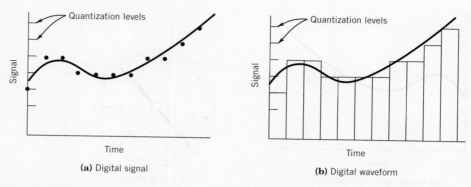

(a) Digital signal   **(b)** Digital waveform

Figure 2.4   Digital signal representation and waveform.

Sampling of an analog signal to produce a digital signal can be accomplished by using an *analog-to-digital (A/D) converter*[1], a solid-state device that converts an analog voltage signal to a binary number system representation. The limited resolution of the binary number that corresponds to a range of voltages creates the quantization levels and establishes the range.

For example, digital music recording technology relies on the conversion of a continuously available signal, such as the voltage signal from a microphone, into a digital form such as an MP3 file. However, because headphones and the human ear are analog devices, the digital information is converted back into a continuous voltage signal for playback. Broadcast of digital signals makes possible high-definition television service in our homes and satellite radio in our vehicles.

Signal Waveforms

In addition to classifying signals as analog, discrete time, or digital, some description of the waveform associated with a signal is useful. Signals, represented here by a function $y(t)$, may be characterized as either static or dynamic. A *static signal* does not vary with time. The diameter of a shaft is an example. Many physical variables change slowly enough in time, compared to the process with which they interact, that for all practical purposes, these signals may be considered static in time. For example, consider measuring temperature by using an outdoor thermometer; since the outdoor temperature does not change significantly in a matter of minutes, this input signal might be considered static when compared to our time period of interest. A mathematical representation of a static signal is given by a constant, as indicated in Table 2.1. In contrast, often we are interested in how the measured variable changes with time. This leads us to consider time-varying signals further.

A *dynamic signal* is defined as a time-dependent signal. In general, dynamic signal waveforms, $y(t)$, may be classified as shown in Table 2.1. A *deterministic signal* varies in time in a predictable manner, such as a sine wave, a step function, or a ramp function, as shown in Figure 2.5. A signal is *steady periodic* if the variation of the magnitude of the signal repeats at regular intervals in time. Examples of steady periodic behaviors are the motion of an ideal pendulum, and the temperature variations in the cylinder of an internal combustion engine under steady operating conditions. Periodic waveforms may be classified as simple or complex. A *simple periodic waveform* contains only one frequency, ω, and an *amplitude*, C. A *complex periodic waveform* contains multiple

---

[1] The processes and devices employed in A/D conversion are detailed in Chapter 7.

Table 2.1 Classification of Waveforms

I	Static	$y(t) = A_0$
II	Dynamic	
	Periodic waveforms	
	Simple periodic waveform	$y(t) = A_0 + C \sin(\omega t + \phi)$
	Complex periodic waveform	$y(t) = A_0 + \sum_{n=1}^{\infty} C_n \sin(n\omega t + \phi_n)$
	Aperiodic waveforms	
	Step[a]	$y(t) = A_0 U(t)$ $= A_0 \text{ for } t > 0$
	Ramp	$y(t) = A_0 t \text{ for } 0 < t < t_f$
	Pulse[b]	$y(t) = A_0 U(t) - A_0 U(t - t_1)$
III	Nondeterministic waveform[c]	$y(t) \approx A_0 + \sum_{n=1}^{\infty} C_n \sin(n\omega t + \phi_n)$

[a]$U(t)$ represents the unit step function, which is zero for $t < 0$ and 1 for $t \geq 0$.
[b]t_1 represents the pulse width.
[c]Nondeterministic waveforms can only be approximated in closed form.

frequencies and amplitudes, and is represented as a superposition of multiple simple periodic waveforms. *Aperiodic* is the term used to describe deterministic signals that do not repeat at regular intervals, such as a step function.

Also described in Figure 2.5 is a *nondeterministic signal* that has no discernible pattern of repetition. A nondeterministic signal cannot be prescribed before it occurs, although certain characteristics of the signal may be known in advance. As an example, consider the transmission of data files from one computer to another. Signal characteristics such as the rate of data transmission and the possible range of signal magnitude are known for any signal in this system. However, it would not be possible to predict future signal characteristics based on existing information in such a signal. Such a signal is properly characterized as nondeterministic. Nondeterministic signals are generally described by their statistical characteristics or by a model signal that represents the statistics of the actual signal.

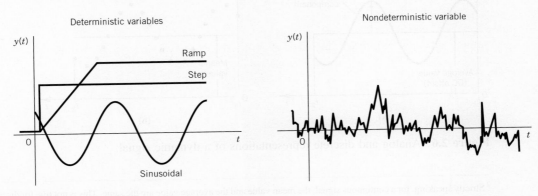

Figure 2.5 Examples of dynamic signals.

2.3 SIGNAL ANALYSIS

In this section, we consider concepts related to the characterization of signals. A measurement system produces a signal that may be analog, discrete time, or digital. An analog signal is continuous with time and has a magnitude that is analogous to the magnitude of the physical variable being measured. Consider the analog signal shown in Figure 2.6a, which is continuous over the recorded time period from t_1 to t_2. The average or mean value[2] of this signal is found by

$$\bar{y} \equiv \frac{\int_{t_1}^{t_2} y(t)dt}{\int_{t_1}^{t_2} dt} \tag{2.1}$$

The mean value, as defined in Equation 2.1, provides a measure of the static portion of a signal over the time period from t_1 to t_2. It is sometimes called the *DC component* or DC offset of the signal.

The mean value does not provide any indication of the amount of variation in the dynamic portion of the signal. The characterization of the dynamic portion, or *AC* component, of the signal may be illustrated by considering the average power dissipated in an electrical resistor through which a fluctuating current flows. The power dissipated in a resistor due to the flow of a current is

$$P = I^2 R$$

where

P = power dissipated = time rate of energy dissipated
I = current
R = resistance

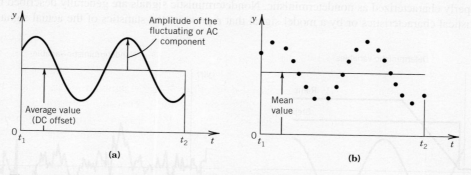

(a)   **(b)**

Figure 2.6 Analog and discrete representations of a dynamic signal.

---

[2] Strictly speaking, for a continuous signal, the mean value and the average value are the same. This is not true for discrete time signals.

If the current varies in time, the total electrical energy dissipated in the resistor over the time t_1 to t_2 would be

$$\int_{t_2}^{t_2} P\,dt = \int_{t_1}^{t_2} [I(t)]^2 R\,dt \tag{2.2}$$

The current $I(t)$ would, in general, include both a DC component and a changing AC component.

Signal Root-Mean-Square Value

Consider finding the magnitude of a constant effective current, I_e, that would produce the same total energy dissipation in the resistor as the time-varying current, $I(t)$, over the time period t_1 to t_2. Assuming that the resistance, R, is constant, this current would be determined by equating $(I_e)^2 R(t_2 - t_1)$ with Equation 2.2 to yield

$$I_e = \sqrt{\frac{1}{t_2 - t_1} \int_{t_1}^{t_2} [I(t)]^2 \, dt} \tag{2.3}$$

This value of the current is called the root-mean-square (rms) value of the current. Based on this reasoning, the rms value of any continuous analog variable $y(t)$ over the time, $t_2 - t_1$, is expressed as

$$y_{\text{rms}} = \sqrt{\frac{1}{t_2 - t_1} \int_{t_1}^{t_2} [y(t)]^2 \, dt} \tag{2.4}$$

Discrete-Time or Digital Signals

A time-dependent analog signal, $y(t)$, can be represented by a discrete set of N numbers over the time period from t_1 to t_2 through the conversion

$$y(t) \rightarrow \{y(r\delta t)\} \quad r = 0, 1, \dots, (N-1)$$

which uses the sampling convolution

$$\{y(r\delta t)\} = y(t)\delta(t - r\delta t) = \{y_i\} \quad i = 0, 1, 2, \dots, (N-1)$$

Here, $\delta(t - r\delta t)$ is the delayed unit impulse function, δt is the sample time increment between each number, and $N\delta t = t_2 - t_1$ gives the total sample period over which the measurement of $y(t)$ takes place. The effect of discrete sampling on the original analog signal is demonstrated in Figure 2.6b in which the analog signal has been replaced by $\{y(r\delta t)\}$, which represents N values of a discrete time signal representing $y(t)$.

For either a discrete time signal or a digital signal, the mean value can be estimated by the discrete equivalent of Equation 2.1 as

$$\bar{y} = \frac{1}{N} \sum_{t=0}^{N-1} y_i \tag{2.5}$$

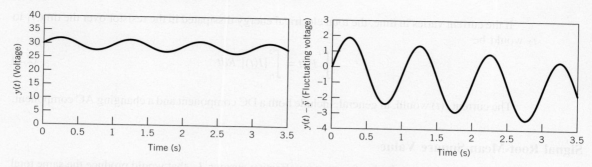

Figure 2.7 Effect of subtracting DC offset for a dynamic signal.

where each y_i is a discrete number in the data set of $\{y(r\delta t)\}$. The mean approximates the static component of the signal over the time interval t_1 to t_2. The rms value can be estimated by the discrete equivalent of Equation 2.4 as

$$y_{\text{rms}} = \sqrt{\frac{1}{N} \sum_{i=0}^{N-1} y_i^2} \tag{2.6}$$

The rms value takes on additional physical significance when either the signal contains no DC component or the DC component has been subtracted from the signal. The rms value of a signal having a zero mean is a statistical measure of the magnitude of the fluctuations in the signal.

Direct Current Offset

When the AC component of the signal is of primary interest, the DC component can be removed. This procedure often allows a change of scale in the display or plotting of a signal such that fluctuations that were a small percentage of the DC signal can be more clearly observed without the superposition of the large DC component. The enhancement of fluctuations through the subtraction of the average value is illustrated in Figure 2.7.

Figure 2.7a contains the entire complex waveform consisting of both static (DC) and dynamic (AC) parts. When the DC component is removed, the characteristics of the fluctuating component of the signal are readily observed, as shown in Figure 2.7b. The statistics of the fluctuating component of the signal may contain important information for a particular application.

Example 2.1

Suppose the current passing through a resistor can be described by

$$I(t) = 10 \sin t$$

where I represents the time-dependent current in amperes, and t represents time in seconds. Establish the mean and rms values of current over a time from 0 to t_f, with $t_f = \pi$ and then with $t_f = 2\pi$. How do the results relate to the power dissipated in the resistor?

KNOWN $I(t) = 10 \sin t$ A

FIND $\bar{I}$ and I_{rms} with $t_f = \pi$ and 2π sec

SOLUTION   The average value for a time from 0 to t_f is found from Equation 2.1 as

$$\bar{I} = \frac{\int_0^{t_f} I(t)dt}{\int_0^{t_f} dt} = \frac{\int_0^{t_f} 10 \sin t dt}{t_f}$$

Evaluation of this integral yields

$$\bar{I} = \frac{1}{t_f}[-10 \cos t]_0^{t_f}$$

With $t_f = \pi$, the average value, $\bar{I}$ is $20/\pi$. For $t_f = 2\pi$, the evaluation of the integral yields an average value of zero.

The rms value for the time period 0 to t_f is given by the application of Equation 2.4, which yields

$$I_{rms} = \sqrt{\frac{1}{t_f}\int_0^{t_f} I(t)^2 dt} = \sqrt{\frac{1}{t_f}\int_0^{t_f} (10 \sin t)^2 dt}$$

This integral is evaluated as

$$I_{rms} = \sqrt{\frac{100}{t_f}\left(-\frac{1}{2}\cos t \sin t + \frac{t}{2}\right)\Big|_0^{t_f}}$$

For $t_f = \pi$ the rms value is $\sqrt{50}$. Evaluation of the integral for the rms value with $t_f = 2\pi$ also yields $\sqrt{50}$.

COMMENT   Although the average value over the period 2π is zero, the power dissipated in the resistor must be the same over both the positive and negative half-cycles of the sine function. Thus, the rms value of current is the same for the time period of π and 2π and is indicative of the rate at which energy is dissipated.

2.4   SIGNAL AMPLITUDE AND FREQUENCY

A key factor in measurement system behavior is the nature of the input signal to the system. A means is needed to classify waveforms for both the input signal and the resulting output signal relative to their magnitude and frequency. It would be very helpful if the behavior of measurement systems could be defined in terms of their response to a limited number and type of input signals. This is, in fact, exactly the case. A very complex signal, even one that is nondeterministic in nature, can be approximated as an infinite series of sine and cosine functions, as suggested in Table 2.1. The method of expressing such a complex signal as a series of sines and cosines is called *Fourier analysis*.

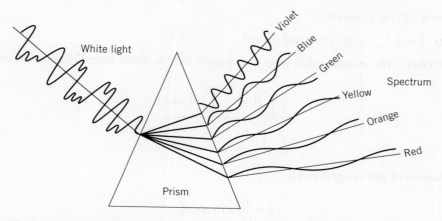

Figure 2.8 Separation of white light into its color spectrum. Color corresponds to a particular frequency or wavelength; light intensity corresponds to varying amplitudes.

Nature provides some experiences that support our contention that complex signals can be represented by the addition of a number of simpler periodic functions. For example, combining a number of different pure tones can generate rich musical sound. And an excellent physical analogy for Fourier analysis is provided by the separation of white light through a prism. Figure 2.8 illustrates the transformation of a complex waveform, represented by white light, into its simpler components, represented by the colors in the spectrum. In this example, the colors in the spectrum are represented as simple periodic functions that combine to form white light. Fourier analysis is roughly the mathematical equivalent of a prism and yields a representation of a complex signal in terms of simple periodic functions.

The representation of complex and nondeterministic waveforms by simple periodic functions allows measurement system response to be reasonably well defined by examining the output resulting from a few specific input waveforms, one of which is a simple periodic. As represented in Table 2.1, a simple periodic waveform has a single, well-defined amplitude and a single frequency. Before a generalized response of measurement systems can be determined, an understanding of the method of representing complex signals in terms of simpler functions is necessary.

Periodic Signals

The fundamental concepts of frequency and amplitude can be understood through the observation and analysis of periodic motions. Although sines and cosines are by definition geometric quantities related to the lengths of the sides of a right triangle, for our purposes sines and cosines are best thought of as mathematical functions that describe specific physical behaviors of systems. These behaviors are described by differential equations that have sines and cosines as their solutions. As an example, consider a mechanical vibration of a mass attached to a linear spring, as shown in Figure 2.9. For a linear spring, the spring force F and displacement y are related by $F = ky$, where k is the constant of proportionality, called the spring constant. Application of

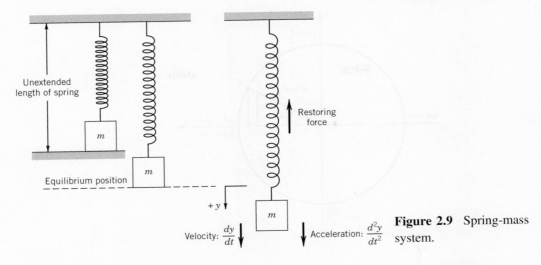

Unextended length of spring

Equilibrium position

$+y$

Velocity: $\dfrac{dy}{dt}$

Restoring force

Acceleration: $\dfrac{d^2y}{dt^2}$

Figure 2.9 Spring-mass system.

Newton's second law to this system yields a governing equation for the displacement y as a function of time t as

$$m\frac{d^2y}{dt^2} + ky = 0 \tag{2.7}$$

This linear, second-order differential equation with constant coefficients describes the motion of the idealized spring mass system when there is no external force applied. The general form of the solution to this equation is

$$y = A\cos\omega t + B\sin\omega t \tag{2.8}$$

where $\omega = \sqrt{k/m}$. Physically we know that if the mass is displaced from the equilibrium point and released, it will oscillate about the equilibrium point. The time required for the mass to finish one complete cycle of the motion is called the *period* and is generally represented by the symbol T.

Frequency is related to the period and is defined as the number of complete cycles of the motion per unit time. This *frequency*, f, is measured in cycles per second (Hz; 1 cycle/s = 1 Hz). The term ω is also a frequency, but instead of having units of cycles per second it has units of radians per second. This frequency, ω, is called the *circular frequency* because it relates directly to cycles on the unit circle, as illustrated in Figure 2.10. The relationship among ω, f, and the period, T, is

$$T = \frac{2\pi}{\omega} = \frac{1}{f} \tag{2.9}$$

In Equation 2.8, the sine and cosine terms can be combined if a phase angle is introduced such that

$$y = C\cos(\omega t - \phi) \tag{2.10a}$$

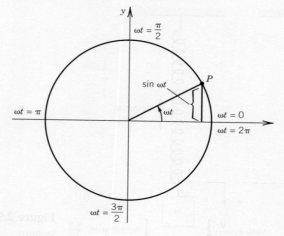

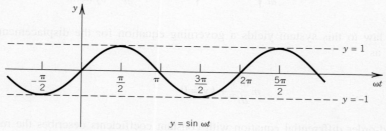

Figure 2.10 Relationship between cycles on the unit circle and circular frequency.

or

$$y = C \sin(\omega t + \phi^*) \tag{2.10b}$$

The values of C, ϕ, and ϕ^* are found from the following trigonometric identities:

$$A \cos \omega t + B \sin \omega t = \sqrt{A^2 + B^2} \cos(\omega t - \phi)$$
$$A \cos \omega t + B \sin \omega t = \sqrt{A^2 + B^2} \sin(\omega t + \phi^*) \tag{2.11}$$
$$\phi = \tan^{-1} \frac{B}{A} \qquad \phi^* = \tan^{-1} \frac{A}{B} \qquad \phi^* = \frac{\pi}{2} - \phi$$

The size of the maximum and minimum displacements from the equilibrium position, or the value C, is the *amplitude* of the oscillation. The concepts of amplitude and frequency are essential for the description of time-dependent signals.

Frequency Analysis

Many signals that result from the measurement of dynamic variables are nondeterministic in nature and have a continuously varying rate of change. These signals, having *complex waveforms*, present difficulties in the selection of a measurement system and in the interpretation of an output signal. However, it is possible to separate a complex signal, or any signal for that matter, into a number of sine and cosine functions. *In other words, any complex signal can be thought of as consisting of sines and cosines of differing periods and amplitudes, which are added together in an infinite trigonometric series.* This representation of a signal as a series of sines and cosines is called a *Fourier series*. When a signal is broken down into a series of periodic functions, the importance of each frequency can be easily determined. This information about frequency content allows proper choice of a measurement system and precise interpretation of output signals.

In theory, Fourier analysis allows essentially all mathematical functions of practical interest to be represented by an infinite series of sines and cosines.[3]

The following definitions relate to Fourier analysis:

1. A function $y(t)$ is *a periodic function* if there is some positive number T such that

$$y(t + T) = y(t)$$

The *period* of $y(t)$ is T. If both $y_1(t)$ and $y_2(t)$ have period T, then

$$ay_1(t) + by_2(t)$$

also has a period of T (a and b are constants).

2. A *trigonometric series* is given by

$$A_0 + A_1 \cos t + B_1 \sin t + A_2 \cos 2t + B_2 \sin 2t + \ldots$$

where A_n and B_n are the coefficients of the series.

Example 2.2

As a physical (instead of mathematical) example of frequency content of a signal, consider stringed musical instruments, such as guitars and violins. When a string is caused to vibrate by strumming, plucking, or bowing, the sound is a result of the motion of the string and the resonance of the instrument. (The concept of resonance is explored in Chapter 3.) The musical pitch for such an instrument is the lowest frequency of the string vibrations. Our ability to recognize differences in musical instruments is primarily a result of the higher frequencies present in the sound, which are usually integer multiples of the fundamental frequency. These higher frequencies are called *harmonics*.

The motion of a vibrating string is best thought of as composed of several basic motions that together create a musical tone. Figure 2.11 illustrates the vibration modes associated with a string plucked at its center. The string vibrates with a fundamental frequency and odd-numbered harmonics,

---

[3] A periodic function may be represented as a Fourier series if the function is piecewise continuous over the limits of integration and the function has a left- and right-hand derivative at each point in the interval. The sum of the resulting series is equal to the function at each point in the interval except points where the function is discontinuous.

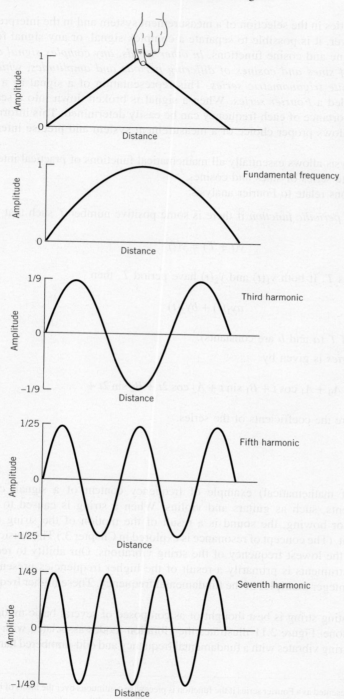

Figure 2.11 Modes of vibration for a string plucked at its center.

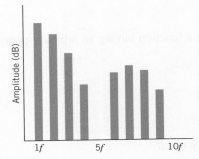

Figure 2.12   Amplitude-frequency plot (spectrum) for a string plucked one-fifth of the distance from a fixed end.

each having a specific phase relationship with the fundamental. The relative strength of each harmonic is graphically illustrated through its amplitude in Figure 2.11. For a string plucked one-fifth of the distance from a fixed end, Figure 2.12 shows the resulting frequencies. Notice that the fifth and tenth harmonics are missing from the resulting sound.

Musical sound from a vibrating string is analogous to a measurement system input or output signal, which may contain few or many frequency components. Fourier analysis is a method used to express such a time-based signal in terms of its amplitude, signal power or phase versus its frequency content. A resulting plot of this type of information is referred to broadly to as a *frequency spectrum*, but when the ordinate value is amplitude it is specifically called an *amplitude spectrum*, such as the plot in Figure 2.12. So while a time-based signal superposes this information, a frequency spectrum delineates the information as a function of frequency. Fourier analysis provides an insightful and practical means for decomposing complex signals into a combination of simple waveforms. Next we explore the frequency and amplitude analysis of complex signals.

The companion software's Labview®[4] program *Sound from Microphone* uses your computer's microphone and sound board to sample ambient sounds and to decompose them into their amplitude-frequency components (try humming a tune).

Fourier Series and Coefficients

A periodic function $y(t)$ with a period $T = 2\pi$ is to be represented by a trigonometric series, such that for any t,

$$y(t) = A_0 + \sum_{n=1}^{\infty} (A_n \cos nt + B_n \sin nt) \qquad (2.12)$$

With $y(t)$ known, the coefficients A_n and B_n are to be determined.

The trigonometric series corresponding to $y(t)$ is called the *Fourier series for $y(t)$*, and the coefficients A_n and B_n are called the *Fourier coefficients* of $y(t)$. When $n = 1$, the corresponding terms in the Fourier series are called *fundamental* and have the lowest frequency in the series, termed the *fundamental frequency*. Frequencies corresponding to $n = 2, 3, 4, \ldots$ are known as *harmonics*, with, for example, $n = 2$ representing the second harmonic.

---

[4] LabView® is a registered trademark of National Instruments.

Fourier Coefficients

The coefficients of a trigonometric series representing a function having an arbitrary period T are given by

$$A_0 = \frac{1}{T} \int_{-T/2}^{T/2} y(t)dt$$

$$A_n = \frac{2}{T} \int_{-T/2}^{T/2} y(t)\cos n\omega t\, dt \qquad (2.13)$$

$$B_n = \frac{2}{T} \int_{-T/2}^{T/2} y(t)\sin n\omega t\, dt$$

where $n = 1, 2, 3, \ldots$, and $T = 2\pi/\omega$ is the period of $y(t)$. The trigonometric series that results from these coefficients is a Fourier series and may be written as

$$y(t) = A_0 + \sum_{n=1}^{\infty} (A_n \cos n\omega t + B_n \sin n\omega t) \qquad (2.14)$$

A series of sines and cosines may be written as a series of either sines or cosines through the introduction of a phase angle, so that the Fourier series in Equation 2.14,

$$y(t) = A_0 + \sum_{n=1}^{\infty} (A_n \cos n\omega t + B_n \sin n\omega t)$$

may be written as

$$y(t) = A_0 + \sum_{n=1}^{\infty} C_n \cos(n\omega t - \phi_n) \qquad (2.15)$$

or

$$y(t) = A_0 + \sum_{n=1}^{\infty} C_n \sin(n\omega t + \phi_n^*) \qquad (2.16)$$

where

$$C_n = \sqrt{A_n^2 + B_n^2}$$

$$\tan \phi_n = \frac{B_n}{A_n} \quad \text{and} \quad \tan \phi_n^* = \frac{A_n}{B_n} \qquad (2.17)$$

Special Case: Functions with $T = 2\pi$

A classical mathematical result occurs for a function $y(t)$ with a period 2π. In this case, the coefficients of the trigonometric series representing $y(t)$ are given by the Euler formulas:

$$A_0 = \frac{1}{2\pi} \int_{-\pi}^{\pi} y(t)$$

$$A_n = \frac{1}{\pi} \int_{-\pi}^{\pi} y(t)\cos ntdt \qquad (2.18)$$

$$B_n = \frac{1}{\pi} \int_{-\pi}^{\pi} y(t)\sin ntdt$$

Even and Odd Functions

A function $g(t)$ is even if it is symmetric about the y-axis, which may be stated, for all t, as

$$g(-t) = g(t)$$

A function $h(t)$ is odd if, for all t,

$$h(-t) = -h(t)$$

For example, $\cos nt$ is even, whereas $\sin nt$ is odd. A particular function or waveform may be even, odd, or neither even nor odd.

Fourier Cosine Series

If $y(t)$ is even, its Fourier series will contain only cosine terms:

$$y(t) = \sum_{n=0}^{\infty} A_n \cos\frac{2\pi nt}{T} = \sum_{n=0}^{\infty} A_n \cos n\omega t \qquad (2.19)$$

Fourier Sine Series

If $y(t)$ is odd, its Fourier series will contain only sine terms:

$$y(t) = \sum_{n=1}^{\infty} B_n \sin\frac{2\pi nt}{T} = \sum_{n=1}^{\infty} B_n \sin n\omega t \qquad (2.20)$$

Note: Functions that are neither even nor odd result in Fourier series that contain both sine and cosine terms.

Example 2.3

Determine the Fourier series that represents the function shown in Figure 2.13.

KNOWN $T = 10$ (i.e., -5 to $+5$)

$$A_0 = 0$$

FIND The Fourier coefficients $A_1, A_2, \ldots$ and $B_1, B_2, \ldots$

SOLUTION Because the function shown in Figure 2.13 is odd, the Fourier series will contain only sine terms (see Eq. 2.20):

$$y(t) = \sum_{n=1}^{\infty} B_n \sin \frac{2\pi nt}{T}$$

where

$$B_n = \frac{2}{10} \left[\int_{-5}^{0} (-1)\sin\left(\frac{2\pi nt}{10}\right) dt + \int_{0}^{5} (1)\sin\left(\frac{2\pi nt}{10}\right) dt \right]$$

$$B_n = \frac{2}{10} \left\{ \left[\frac{10}{2n\pi} \cos\left(\frac{2\pi nt}{10}\right) \right]_{-5}^{0} + \left[\frac{-10}{2n\pi} \cos\left(\frac{2\pi nt}{10}\right) \right]_{0}^{5} \right\}$$

$$B_n = \frac{2}{10} \left\{ \frac{10}{2n\pi} [1 - \cos(-n\pi) - \cos(n\pi) + 1] \right\}$$

The reader can verify that all $A_n = 0$ for this function. For even values of n, B_n is identically 0, and for odd values of n,

$$B_n = \frac{4}{n\pi}$$

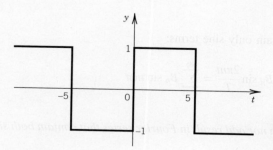

Figure 2.13 Function represented by a Fourier series in Example 2.3.

The resulting Fourier series is then

$$y(t) = \frac{4}{\pi}\sin\frac{2\pi}{10}t + \frac{4}{3\pi}\sin\frac{6\pi}{10}t + \frac{4}{5\pi}\sin\frac{10\pi}{10}t + \ldots$$

Note that the fundamental frequency is $\omega = \frac{2\pi}{10}$ rad/s, and the subsequent terms are the odd-numbered harmonics of ω.

COMMENT   Consider the function given by

$$y(t) = 1 \quad 0 < t < 5$$

We may represent $y(t)$ by a Fourier series if we extend the function beyond the specified range (0 to 5) either as an even periodic extension or as an odd periodic extension. (Because we are interested only in the Fourier series representing the function over the range $0 < t < 5$, we can impose any behavior outside that domain that helps to generate the Fourier coefficients!) Let's choose an odd periodic extension of $y(t)$; the resulting function remains identical to the function shown in Figure 2.13.

Example 2.4

Find the Fourier coefficients of the periodic function

$$y(t) = -5 \text{ when } -\pi < t < 0$$
$$y(t) = +5 \text{ when } 0 < t < \pi$$

and $y(t + 2\pi) = y(t)$. Plot the resulting first four partial sums for the Fourier series.

KNOWN   Function $y(t)$ over the range $-\pi$ to π

FIND   Coefficients A_n and B_n

SOLUTION   The function as stated is periodic, having a period of $T = 2\pi$ (i.e., $\omega = 1$ rad/s), and is identical in form to the odd function examined in Example 2.3. Since this function is also odd, the Fourier series contains only sine terms, and

$$B_n = \frac{1}{\pi}\int_{-\pi}^{\pi} y(t)\sin n\omega t \, dt = \frac{1}{\pi}\left[\int_{-\pi}^{0}(-5)\sin nt \, dt + \int_{0}^{\pi}(+5)\sin nt \, dt\right]$$

which yields upon integration

$$\frac{1}{\pi}\left\{\left[(5)\frac{\cos nt}{n}\right]_{-\pi}^{0} - \left[(5)\frac{\cos nt}{n}\right]_{0}^{\pi}\right\}$$

Thus,

$$B_n = \frac{10}{n\pi}(1 - \cos n\pi)$$

which is zero for even n, and $20/n\pi$ for odd values of n. The Fourier series can then be written[5]

$$y(t) = \frac{20}{\pi}\left(\sin t + \frac{1}{3}\sin 3t + \frac{1}{5}\sin 5t + \ldots\right)$$

Figure 2.14 shows the first four partial sums of this function as they compare with the function they represent. Note that at the points of discontinuity of the function, the Fourier series takes on the arithmetic mean of the two function values.

   COMMENT   As the number of terms in the partial sum increases, the Fourier series approximation of the square wave function becomes even better. For each additional term in the series, the number of "humps" in the approximate function in each half-cycle corresponds to the number of terms included in the partial sum.

   The Matlab[®6] program file *FourSeries* with the companion software illustrates the behavior of the partial sums for several waveforms. The LabView program *Waveform-Generation* reconstructs signals from several selected trigonometric series.

Example 2.5

As an example of interpreting the frequency content of a given signal, consider the output voltage from a rectifier. A rectifier functions to "flip" the negative half of an alternating current (AC) into the positive half plane, resulting in a signal that appears as shown in Figure 2.15. For the AC signal the voltage is given by

$$E(t) = 120 \sin 120\pi t$$

Where $E(t)$ is in volts and t is in seconds. The period of the signal is $1/60$ s, and the frequency is 60 Hz.

---

[5] If we assume that the sum of this series most accurately represents the function y at $t = \pi/2$, then

$$y\left(\frac{\pi}{2}\right) = 5 = \frac{20}{\pi}\left(1 - \frac{1}{3} + \frac{1}{5} - + \ldots\right)$$

or

$$\frac{\pi}{4} = 1 - \frac{1}{3} + \frac{1}{5} - \frac{1}{7} + - \ldots$$

This series approximation of π was first obtained by Gottfried Wilhelm Leibniz (1646–1716) in 1673 from geometrical reasoning.
[6] Matlab[®] is a registered trademark of Mathworks, Inc.

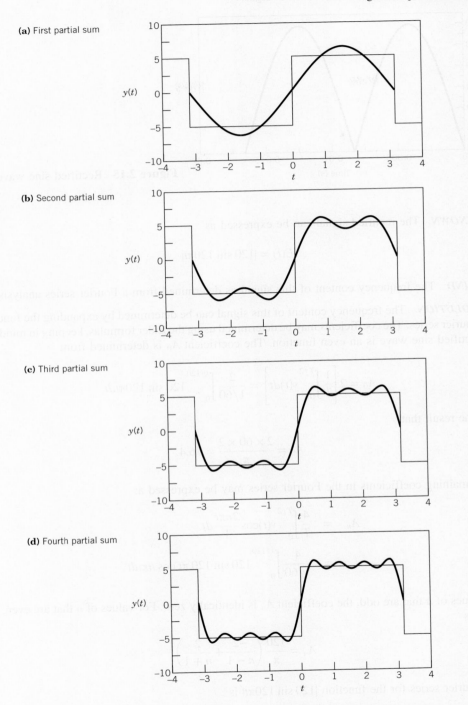

(a) First partial sum

(b) Second partial sum

(c) Third partial sum

(d) Fourth partial sum

Figure 2.14 First four partial sums of the Fourier series $(20/\pi)(\sin t + 1/3 \sin 3t + 1/5 \sin 5t + \ldots)$ in comparison with the exact waveform.

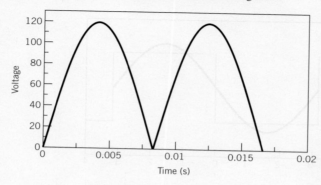

Figure 2.15 Rectified sine wave.

KNOWN The rectified signal can be expressed as

$$E(t) = |120 \sin 120\pi t|$$

FIND The frequency content of this signal as determined from a Fourier series analysis.

SOLUTION The frequency content of this signal can be determined by expanding the function in a Fourier series. The coefficients may be determined using the Euler formulas, keeping in mind that the rectified sine wave is an even function. The coefficient A_0 is determined from

$$A_0 = 2\left[\frac{1}{T}\int_0^{T/2} y(t)dt\right] = \frac{2}{1/60}\int_0^{1/120} 120 \sin 120\pi t \, dt$$

with the result that

$$A_0 = \frac{2 \times 60 \times 2}{\pi} = 76.4$$

The remaining coefficients in the Fourier series may be expressed as

$$\begin{aligned} A_n &= \frac{4}{T}\int_0^{T/2} y(t)\cos\frac{2n\pi t}{T}\,dt \\ &= \frac{4}{1/60}\int_0^{1/120} 120 \sin 120\,\pi t \cos n\pi t \, dt \end{aligned}$$

For values of n that are odd, the coefficient A_n is identically zero. For values of n that are even, the result is

$$A_n = \frac{120}{\pi}\left(\frac{-2}{n-1} + \frac{2}{n+1}\right)$$

The Fourier series for the function $|120 \sin 120\pi t|$ is

$$76.4 - 50.93 \cos 240\pi t - 10.10 \cos 480\pi t - 4.37 \cos 720\,\pi t \ldots$$

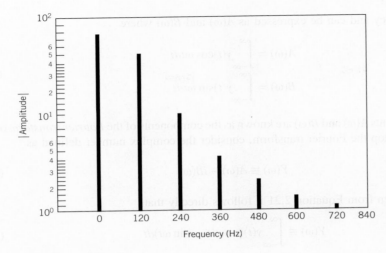

Figure 2.16 Frequency content of the function $y(t) = |120 \sin 20\pi t|$ displayed as a frequency (amplitude) spectrum.

Figure 2.16 shows the amplitude versus frequency content of the rectified signal, based on a Fourier series expansion.

COMMENT   The frequency content of a signal is determined by examining the amplitude of the various frequency components that are present in the signal. For a periodic mathematical function, expanding the function in a Fourier series and plotting the amplitudes of the contributing sine and cosine terms can illustrate these frequency contributions.

The LabView program *WaveformGeneration* follows Examples 2.4 and 2.5 and provides the amplitude spectra for a number of signals. Several Matlab programs are provided (e.g., *FourSeries*, *FunSpect*, *DataSpect*) to explore the concept of superposition of simple periodic signals to create a more complex signal. *DataSpect* allows you to import your own files (or use the provided files) to explore their frequency and amplitude content.

2.5   FOURIER TRANSFORM AND THE FREQUENCY SPECTRUM

The previous discussion of Fourier analysis demonstrates that an arbitrary, but known, function can be expressed as a series of sines and cosines known as a Fourier series. The coefficients of the Fourier series specify the amplitudes of the sines and cosines, each having a specific frequency. Unfortunately, in most practical measurement applications the input signal may not be known in functional form. Therefore, although the theory of Fourier analysis demonstrates that any mathematical function can be expressed as a Fourier series, the analysis presented so far has not provided a specific technique for analyzing measured signals. Such a technique for the decomposition of a measured dynamic signal in terms of amplitude and frequency is now described.

The fundamental mathematics required to develop a method for analyzing measured data requires taking the limit of a Fourier series as the period of the signal approaches infinity. In the limit as T approaches infinity, the Fourier series becomes an integral. The spacing between frequency components becomes infinitesimal. This means that the coefficients A_n and B_n become continuous functions of frequency and can be expressed as $A(\omega)$ and $B(\omega)$ where

$$A(\omega) = \int_{-\infty}^{\infty} y(t)\cos \omega t \, dt$$

$$B(\omega) = \int_{-\infty}^{\infty} y(t)\sin \omega t \, dt \qquad (2.21)$$

The Fourier coefficients $A(\omega)$ and $B(\omega)$ are known as the components of the *Fourier transform* of $y(t)$. To further develop the Fourier transform, consider the complex number defined as

$$Y(\omega) \equiv A(\omega) - iB(\omega) \qquad (2.22)$$

where $i = \sqrt{-1}$. Then from Equation 2.21 it follows directly that

$$Y(\omega) \equiv \int_{-\infty}^{\infty} y(t)(\cos \omega t - i\sin \omega t)dt \qquad (2.23)$$

Introducing the identity

$$e^{-i\theta} = \cos \theta - i \sin \theta$$

leads to

$$Y(\omega) \equiv \int_{-\infty}^{\infty} y(t)e^{-i\omega t}dt \qquad (2.24)$$

Alternately, recalling from Equation 2.9 that the cyclical frequency f, in hertz, is related to the circular frequency and its period by

$$f = \frac{\omega}{2\pi} = \frac{1}{T}$$

Equation 2.24 is rewritten as

$$Y(f) \equiv \int_{-\infty}^{\infty} y(t)e^{-i2\pi ft}dt \qquad (2.25)$$

Equation 2.24 or 2.25 provides the two-sided *Fourier transform* of $y(t)$. If $y(t)$ is known, then its Fourier transform will provide the amplitude-frequency properties of the signal, $y(t)$, which otherwise are not readily apparent in its time-based form. We can think of the Fourier transform as a decomposition of $y(t)$ into amplitude versus frequency information. This property is analogous to the optical properties displayed by the prism in Figure 2.8.

If $Y(f)$ is known or measured, we can recover the signal $y(t)$ from

$$y(t) = \int_{-\infty}^{\infty} Y(f)e^{i2\pi ft}df \qquad (2.26)$$

Equation 2.26 describes the *inverse Fourier transform* of $Y(f)$. It suggests that given the amplitude-frequency properties of a signal, we can reconstruct the original signal $y(t)$. The Fourier transform is a complex number having a magnitude and a phase,

$$Y(f) = |Y(f)|e^{i\phi(f)} = A(f) - iB(f) \tag{2.27}$$

The magnitude of $Y(f)$, also called the modulus, is given by

$$|Y(f)| = \sqrt{\mathrm{Re}[Y(f)]^2 + \mathrm{Im}[Y(f)]^2} \tag{2.28}$$

and the phase by

$$\phi(f) = \tan^{-1}\frac{\mathrm{Im}[Y(f)]}{\mathrm{Re}[Y(f)]} \tag{2.29}$$

As noted earlier, the Fourier coefficients are related to cosine and sine terms. Each of these terms assigns amplitude and phase information to its unique frequency. The general form of this information is called a frequency spectrum. When amplitude is expressed in terms of frequency, the frequency spectrum is referred to as the amplitude spectrum. This is given by

$$C(f) = \sqrt{A(f)^2 + B(f)^2} \tag{2.30}$$

When phase angle is expressed in terms of frequency, the frequency spectrum is referred to as the phase spectrum. This is given by

$$\phi(f) = \tan^{-1}\frac{B(f)}{A(f)} \tag{2.31}$$

Thus the introduction of the Fourier transform provides a method to decompose a measured signal $y(t)$ into its amplitude, phase and frequency components. Later we will see how important this method is, particularly when digital sampling is used to measure and interpret an analog signal.

Another variation of the frequency spectrum is the power spectrum, in which the ordinate is $C(f)^2/2$. Further details concerning the properties of the Fourier transform and spectrum functions can be found in Bracewell (1) and Champeney (2). An excellent historical account and discussion of the wide-ranging applications are found in Bracewell (3).

Discrete Fourier Transform

As a practical matter, it is likely that if $y(t)$ is measured and recorded, then it will be stored in the form of a digital signal. A computer-based data-acquisition system is the most common method for recording data. These data are acquired over a finite period of time rather than the mathematically convenient infinite period of time. A discrete data set containing N values representing a time interval from 0 to t_f will accurately represent the signal provided that the measuring period has been properly chosen and is sufficiently long. We deal with the details for such period selection in a discussion on sampling concepts in Chapter 7. The preceding analysis is now extended to accommodate a discrete series.

Consider the time-dependent portion of the signal $y(t)$, which is measured N times at equally spaced time intervals δt. In this case, the continuous signal $y(t)$ is replaced by the discrete time signal given by $y(r\delta t)$ for $r = 0, 1, \ldots, (N-1)$. In effect, the relationship between $y(t)$ and $\{y(r\delta t)\}$ is described by a set of impulses of an amplitude determined by the value of $y(t)$ at each time step $r\delta t$. This transformation from a continuous to discrete time signal is described by

$$\{y(r\delta t)\} = y(t)\delta(t - r\delta t) \ r = 0, 1, 2, \ldots, N - 1 \qquad (2.32)$$

where $\delta(t - r\delta t)$ is the delayed unit impulse function and $\{y(r\delta t)\}$ refers to the discrete data set given by $y(r\delta t)$ for $r = 0, 1, 2, \ldots, N - 1$.

An approximation to the Fourier transform integral of Equation 2.25 for use on a discrete data set is the discrete Fourier transform (DFT). The DFT is given by

$$Y(f_k) = \frac{2}{N}\sum_{r=0}^{N-1} y(r\delta t)e^{-i2\pi rk/N} \text{ with } Y(f_0) = \frac{1}{N}\sum_{r=0}^{N-1} y(r\delta t)e^{-i2\pi r0/N} = \frac{1}{N}\sum_{r=0}^{N-1} y(r\delta t)$$

$$f_k = k\delta f \qquad k = 0, 1, 2, \ldots, \left(\frac{N}{2} - 1\right) \qquad (2.33)$$

$$\delta f = 1/N\delta t$$

Here, δf is the frequency resolution of the DFT with each value of $Y(f_k)$ spaced at frequency increments of δf. In developing Equation 2.33 from Equation 2.25, t was replaced by $r\delta t$ and f replaced by $k/N\delta t$. The factor $2/N$ scales the transform when it is obtained from a data set of finite length (use $1/N$ for $k = 0$ only, which returns the DC or mean value at $0\,\text{Hz}$).

The DFT as expressed by Equation 2.33 yields $N/2$ discrete values of the Fourier transform of $\{y(r\delta t)\}$. This is the so-called one-sided or half-transform, because it assumes that the data set is one-sided, extending from 0 to t_f, and it returns only positive valued frequencies.

Equation 2.33 performs the numerical integration required by the Fourier integral. Equations 2.30, 2.31, and 2.33 demonstrate that the application of the DFT on the discrete series of data, $y(r\delta t)$, permits the decomposition of the discrete data in terms of frequency and amplitude content. Hence, by using this method, a measured discrete signal of unknown functional form can be reconstructed as a Fourier series through Fourier transform techniques.

Software for computing the Fourier transform of a discrete signal is included in the companion software. The time required to compute directly the DFT algorithm described in this section increases at a rate that is proportional to N^2. This makes it inefficient for use with data sets of large N. A fast algorithm for computing the DFT, known as the *fast Fourier transform* (FFT), was developed by Cooley and Tukey (4). This method is widely available and is the basis for most Fourier analysis software packages. The FFT algorithm is discussed in most advanced texts on signal analysis (5,6). The accuracy of discrete Fourier analysis depends on the frequency content[7] of $y(t)$ and on the Fourier transform frequency resolution. An extensive discussion of these interrelated parameters is given in Chapter 7.

Example 2.6

Convert the continuous signal described by $y(t) = 10 \sin 2\pi t$ V into a discrete set of eight numbers using a time increment of 0.125 s.

---

[7] The value of $1/\delta t$ must be greater than twice the highest frequency contained in $y(t)$.

Table 2.2   Discrete Data Set for $y(t) = 10 \sin 2\pi t$

r	$y(r\delta t)$	r	$y(r\delta t)$
0	0.000	4	0.000
1	7.071	5	−7.071
2	10.000	6	−10.000
3	7.071	7	−7.071

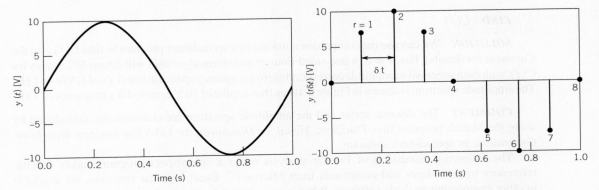

Figure 2.17   Representation of a simple periodic function as a discrete signal.

KNOWN   The signal has the form $y(t) = C_1 \sin 2\pi f_1 t$ where

$$f_1 = \omega_1/2\pi = 1 \text{ Hz}$$
$$C(f_1 = 1 \text{ Hz}) = 10 \text{ V}$$
$$\phi(f_1) = 0$$
$$\delta t = 0.125 \text{ s}$$
$$N = 8$$

FIND $\{y(r\delta t)\}$

SOLUTION   Measuring $y(t)$ every 0.125 s over 1 s produces the discrete data set $\{y(r\delta t)\}$ given in Table 2.2. Note that the measurement produces four complete periods of the signal and that the signal duration is given by $N\delta t = 1$ s. The signal and its discrete representation as a series of impulses in a time domain are plotted in Figure 2.17. See Example 2.7 for more on how to create this discrete series by using common software.

Example 2.7

Estimate the amplitude spectrum of the discrete data set in Example 2.6.

KNOWN   Discrete data set $\{y(r\delta t)\}$ of Table 2.2

$$\delta t = 0.125 \text{ s}$$
$$N = 8$$

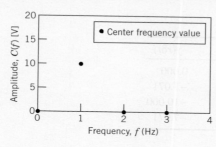

Figure 2.18 Amplitude as a function of frequency for a discrete representation of $10 \sin 2\pi t$ resulting from a discrete Fourier transform algorithm.

FIND $C(f)$

SOLUTION   We can use the companion software or a spreadsheet program to find $C(f)$ (see the *Comment* for details). For $N = 8$, a one-sided Fourier transform algorithm will return $N/2$ values for $C(f)$, with each successive amplitude corresponding to a frequency spaced at intervals of $1/N\delta t = 1$ Hz. The amplitude spectrum is shown in Figure 2.18 and has a spike of 10 V centered at a frequency of 1 Hz.

COMMENT   The discrete series and the amplitude spectrum are conveniently reproduced by using the Matlab program files *FunSpect, Signal,* or *DataSpect,* the LabView program *Waveform-Generation,* or spreadsheet software.

The following discussion of Fourier analysis using a spreadsheet program makes specific references to procedures and commands from Microsoft® Excel;[8] similar functions are available in other engineering analysis software. When a spreadsheet is used as shown, the $N = 8$ data point sequence $\{y(r\delta t)\}$ is created as in column 3. Under Data/Data Analysis, select Fourier Analysis. At the prompt, define the N cells containing $\{y(r\delta t)\}$ and define a cell destination (column 4). The analysis executes the DFT of Equation 2.33 by using the FFT algorithm, and it returns N complex numbers, the Fourier coefficients $Y(f)$ of Equation 2.27, to column 4. The analysis returns the two-sided transform. However, we are interested in only the first $N/2$ Fourier coefficients of column 4 (the second $N/2$ coefficients just mirror the first $N/2$). To find the coefficient magnitude as given in Equation 2.28, use the Excel function $\text{IMABS}\left(= \sqrt{A^2 + B^2}\right)$ on each of the first $N/2$ coefficients, and then scale each coefficient magnitude by dividing by $N/2$ (for $k = 0$, divide by N). The $N/2$ scaled coefficients (column 5) now represent the discrete amplitudes corresponding to $N/2$ discrete frequencies (column 6) extending from $f = 0$ to $(N/2 - 1)/N\delta t$ Hz, with each frequency separated by $1/N\delta t$.

			Column		
1	2	3	4	5	6
r	$t(s)$	$y(r\delta t)$	$Y(f) = A - Bi$	$C(f)$	f(Hz)
0	0	0	0	0	0
1	0.125	7.07	$-40i$	10	1
2	0.25	10	0	0	2
3	0.375	7.07	0	0	3
4	0.5	0	0		
5	0.625	-7.07	0		
6	0.75	-10	0		
7	0.875	-7.07	$40i$		

[8] Microsoft® and Excel are either registered trademarks or trademarks of Microsoft Corporation in the United States and/or other countries.

These same operations in Matlab code are as follows:

$t = 1/8 : 1/8 : 1$	defines time from 0.125 s to 1 s in increments of 0.125 s
$y = 10*sin(2*pi*t)$	creates the discrete time series with $N = 8$
ycoef = fft(y)	performs the Fourier analysis; returns N coefficients
c = coef/4	divides by $N/2$ to scale and determine the magnitudes

When signal frequency content is not known before conversion to a discrete signal, it may be necessary to experiment with the parameters of frequency resolution, N and δt, to obtain an unambiguous representation of the signal. Techniques for this are discussed in Chapter 7.

Analysis of Signals in Frequency Space

Fourier analysis is a tool for extracting details about the frequencies that are present in a signal. Frequency analysis is routinely used in vibration analysis and isolation, determining the condition of bearings, and a wide variety of acoustic applications. An example from acoustics follows.

Example 2.8

Examine the spectra of representative brass and woodwind instruments to illustrate why the characteristic sound of these instruments is so easily distinguished.

KNOWN Figures 2.19 and 2.20 provide a representative frequency spectrum for a clarinet, a woodwind instrument having a reed, and for brass instruments.

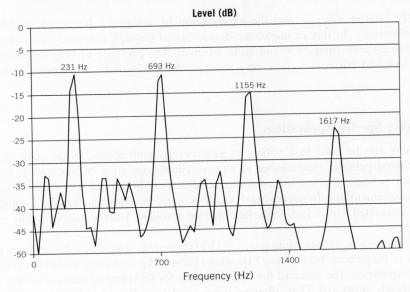

Figure 2.19 Frequency spectrum for a clarinet playing the note one whole tone below middle C. (Adapted from Campbell, D. M., Nonlinear Dynamics of Musical Reed and Brass Wind Instruments, *Contemporary Physics*, 40(6) November 1999, pages 415–431.)

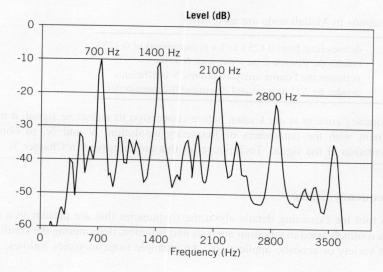

Figure 2.20 Frequency spectrum for the first three seconds of Aaron Copland's "Fanfare for the Common Man," played by the brass section of the orchestra.

DISCUSSION Figure 2.19 displays the frequency spectrum for a clarinet playing a B flat. The base tone, f_o, is at 231 Hz, with the harmonics having frequencies of $3f_o$, $5f_o$, etc. Contrast this with the frequency spectrum, shown in Figure 2.20, of the first few seconds of Aaron Copland's "Fanfare for the Common Man," which is played by an ensemble of brass instruments, such as French horns, and which has a fundamental frequency of approximately 700 Hz. This spectrum contains harmonics that are $2f_o$, $3f_o$, $4f_o$, and so forth.

COMMENT Fourier analysis and the examination of a signal's frequency content have myriads of applications. In this example, we demonstrated through frequency analysis a clear reason why clarinets and trumpets sound quite different. The presence of only odd harmonics, or both even and odd harmonics, is a very recognizable difference, easily perceived by the human ear.

Example 2.9: Frequency Spectra Applications

Frequency spectra can be used in a variety of engineering applications. In the past, engineers employed scheduled periodic maintenance to ensure reliable operation of machinery, including components such as bearings. Now the operating condition or "health" of a bearing is monitored through the measurement and frequency analysis of acoustic or vibration signals. Characteristic frequencies are associated with defects in the bearing race or rollers. Thus bearings are only replaced when necessary.

Figure 2.21 shows a power spectrum (i.e., $C^2(f)/2$) measured on a race car traveling on an oval track. The range of frequencies from 30 to 37 Hz shows increased power compared to the trend in the data for other frequencies. The standard tire circumference for this class of race car is 88.6 inches. Therefore the speeds at 30 and 37 revolutions per second are 151 and 186 mph. This anomaly corresponds to the tire rotation frequency and indicates that this tire has a defect, most likely a flat spot in the tire created by skidding during braking.

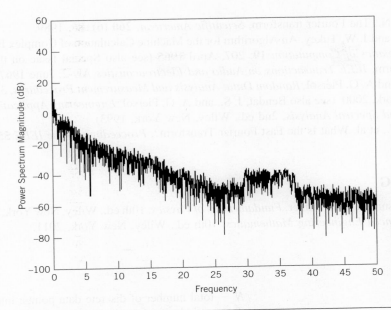

Figure 2.21 Power spectrum of suspension displacement for a race car on an oval track.

2.6 SUMMARY

This chapter has provided a fundamental basis for the description of signals. The capabilities of a measurement system can be properly specified when the nature of the input signal is known. The descriptions of general classes of input signals will be seen in Chapter 3 to allow universal descriptions of measurement system dynamic behavior.

Any signal can be represented by a static magnitude and a series of varying frequencies and amplitudes. This being the case, measurement system selection and design must consider the frequency content of the input signals the system is intended to measure. Fourier analysis was introduced to allow a precise definition of the frequencies and the phase relationships among various frequency components within a particular signal. In Chapter 7, it will be shown as a tool for the accurate interpretation of discrete signals.

Signal characteristics form an important basis for the selection of measurement systems and the interpretation of measurement system output. In Chapter 3 these ideas are combined with the concept of a generalized set of measurement system behaviors. The combination of generalized measurement system behavior and generalized descriptions of input waveforms provides for an understanding of a wide range of instruments and measurement systems.

REFERENCES

1. Bracewell, R. N., *The Fourier Transform and Its Applications*, 3rd ed., rev., McGraw Hill, New York, 1999.
2. Champeney, D. C., *Fourier Transforms and Their Physical Applications*, Academic, London, 1973.

3. Bracewell, R. N., The Fourier transform, *Scientific American*, **260** (6): 86, 1989.

4. Cooley, J. W., and J. W. Tukey, An Algorithm for the Machine Calculation of Complex Fourier Series, *Mathematics of Computation* **19**: 207, April 1965 (see also Special Issue on the fast Fourier transform, *IEEE Transactions on Audio and Electroacoustics* AU-2, June 1967).

5. Bendat, J. S., and A. G. Piersol, *Random Data: Analysis and Measurement Procedures*, 3rd ed., Wiley, New York, 2000 (see also Bendat, J. S., and A. G. Piersol, *Engineering Applications of Correlation and Spectral Analysis*, 2nd ed., Wiley, New York, 1993).

6. Cochran, W. T., et al. What Is the Fast Fourier Transform?, *Proceedings of the IEEE* **55** (10): 1664, 1967.

SUGGESTED READING

Halliday, D., R. Resnick, and J. Walker. *Fundamentals of Physics*, 10th ed., Wiley, New York, 2013.

Kreyszig, E., *Advanced Engineering Mathematics*, 10th ed., Wiley, New York, 2011.

NOMENCLATURE

f	frequency, in Hz (t^{-1})		N	total number of discrete data points; integer
k	spring constant (mt^{-2})		T	period (t)
m	mass (m)		U	unit step function
t	time (t)		Y	Fourier transform of y
y	dependent variable		α	angle (rad)
y_m	discrete data points		β	angle (rad)
A	amplitude		δf	frequency resolution (t^{-1})
B	amplitude		δt	sample time increment (t)
C	amplitude		ϕ	phase angle (rad)
F	force (mlt^{-2})		ω	circular frequency in rad/s (t^{-1})

PROBLEMS

2.1 Define the term "signal" as it relates to measurement systems. Provide two examples of static and dynamic input signals to particular measurement systems.

2.3 List the important characteristics of input and output digital signals and define each.

2.3 Research and describe the importance of multiplexing and data compression in the transmission of voice data as in phone conversations.

2.4 Research and describe the importance of data compression in the storage and transmission of image files.

2.5 Determine the average and rms values for the function

$$y(t) = 25 + 10 \sin 6\pi t$$

over the time periods (a) 0 to 0.1 s, (b) 0.4 to 0.5 s, (c) 0 to 1/3 s, and (d) 0 to 20 s. Comment on the nature and meaning of the results in terms of analysis of dynamic signals.

2.6 The following values are obtained by sampling two time-varying signals once every 0.4 s:

t	$y_1(t)$	$y_2(t)$	t	$y_1(t)$	$y_2(t)$
0	0	0			
0.4	11.76	15.29	2.4	−11.76	−15.29
0.8	19.02	24.73	2.8	−19.02	−24.73
1.2	19.02	24.73	3.2	−19.02	−24.73
1.6	11.76	15.29	3.6	−11.76	−15.29
2.0	0	0	4.0	0	0

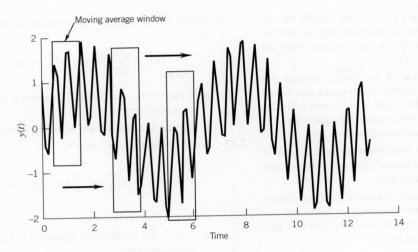

Figure 2.22 Moving average and windowing.

Determine the mean and the rms values for this discrete data. Discuss the significance of the rms value in distinguishing these signals.

2.7 A *moving average* is an averaging technique that can be applied to an analog, discrete time, or digital signal. A moving average is based on the concept of windowing, as illustrated in Figure 2.22. The portion of the signal that lies inside the window is averaged and the average values plotted as a function of time as the window moves across the signal. A 10-point moving average of the signal in Figure 2.22 is plotted in Figure 2.23.

a. Discuss the effects of employing a moving average on the signal depicted in Figure 2.22.

b. Develop a computer-based algorithm for computing a moving average, and determine the effect of the width of the averaging window on the signal described by

$$y(t) = \sin 5t + \cos 11t$$

This signal should be represented as a discrete time signal by computing the value of the function at equally spaced time intervals. An appropriate time interval for this signal would be 0.05 s. Examine the signal with averaging windows of 4 and 30 points.

2.8 The data file in the companion software *noisy.txt* provides a discrete time-varying signal that contains

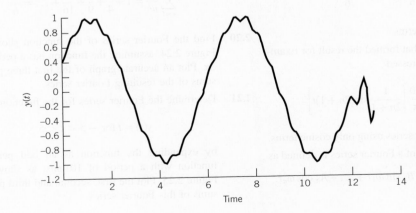

Figure 2.23 Effect of moving average on signal illustrated in Figure 2.22.

random noise. Apply a 2-, 3-, and 4-point moving average (see Problem 2.7) to these data, and plot the results. How does the moving average affect the noise in the data? Why?

2.9 Determine the value of the spring constant that would result in a spring-mass system that would execute one complete cycle of oscillation every 2 s, for a mass of 1 kg. What natural frequency does this system exhibit in radians/second?

2.10 A spring with $k = 5000$ N/cm supports a mass of 1 kg. Determine the natural frequency of this system in radians/second and hertz.

2.11 For the following sine and cosine functions, determine the period, the frequency in hertz, and the circular frequency in radians/second. (Note: t represents time in seconds).

a. $\sin 10\pi t/5$

b. $8 \cos 8t$

c. $\sin 5n\pi t$ for $n = 1$ to ∞

2.12 Express the function

$$y(t) = 5 \sin 4t + 3 \cos 4t$$

in terms of (a) a cosine term only and (b) a sine term only.

2.13 Express the function

$$y(t) = 4 \sin 2\pi t + 15 \cos 2\pi t$$

in terms of (a) a cosine term only and (b) a sine term only.

2.14 Express the Fourier series given by

$$y(t) = \sum_{n=1}^{\infty} \frac{2\pi n}{6} \sin n\pi t + \frac{4\pi n}{6} \cos n\pi t$$

using only cosine terms.

2.15 The Fourier series that formed the result for Example 2.4 may be expressed

$$y(t) = \sum_{n=1}^{\infty} \frac{20}{\pi} \left[\frac{1}{2n + 1} \sin(2n + 1)t \right]$$

Express this Fourier series using only cosine terms.

2.16 The nth partial sum of a Fourier series is defined as

$$A_0 + A_1 \cos \omega_1 t + B_1 \sin \omega_1 t + \ldots + A_n \cos \omega_n t + B_n \sin \omega_n t$$

For the third partial sum of the Fourier series given by

$$y(t) = \sum_{n=1}^{\infty} \frac{3n}{2} \sin nt + \frac{5n}{3} \cos nt$$

a. What is the fundamental frequency and the associated period?

b. Express this partial sum as cosine terms only.

2.17 For the Fourier series given by

$$y(t) = 4 + \sum_{n=1}^{\infty} \frac{2\pi n}{10} \cos \frac{n\pi}{4} t + \frac{120n\pi}{30} \sin \frac{n\pi}{4} t$$

where t is time in seconds:

a. What is the fundamental frequency in hertz and radians/second?

b. What is the period T associated with the fundamental frequency?

c. Express this Fourier series as an infinite series containing sine terms only.

2.18 Determine the Fourier series for the function $y(t) = \sin t$.

2.19 Show that $y(t) = t^2 (-\pi < t < \pi)$, $y(t + 2\pi) = y(t)$ has the Fourier series

$$y(t) = \frac{\pi^2}{3} - 4 \left(\cos t - \frac{1}{4} \cos 2t + \frac{1}{9} \cos 3t - + \ldots \right)$$

By setting $t = \pi$ in this series, show that a series approximation for π, first discovered by Euler, results as

$$\sum_{n=1}^{\infty} \frac{1}{n^2} = 1 + \frac{1}{4} + \frac{1}{9} + \frac{1}{16} + \ldots = \frac{\pi^2}{6}$$

2.20 Find the Fourier series of the function shown in Figure 2.24, assuming the function has a period of 2π. Plot an accurate graph of the first three partial sums of the resulting Fourier series.

2.21 Determine the Fourier series for the function

$$y(t) = t \text{ for } -5 < t < 5$$

by expanding the function as an odd periodic function with a period of 10 units, as shown in Figure 2.25. Plot the first, second, and third partial sums of this Fourier series.

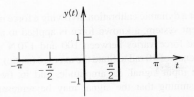

Figure 2.24 Function to be expanded in a Fourier series in Problem 2.20.

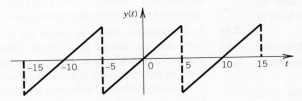

Figure 2.25 Sketch for Problem 2.21.

a. $\sin 10t$ V

b. $5 + 2\cos 2t$ m

c. $5t$ s

d. 2 V

2.22 Determine the Fourier series that represents the function $y(t)$ where

$$y(t) = t \text{ for } 0 < t < 1$$

and

$$y(t) = 2 - t \text{ for } 0 < t < 1$$

Clearly explain your choice for extending the function to make it periodic.

2.23 Consider the triangle wave shown in Figure 2.26 as a periodic function with period 2π. Determine the Fourier series that represents this function. Plot the first three partial sums over the range from $-\pi$ to π.

2.24 Classify the following signals as static or dynamic, and identify any that may be classified as periodic:

2.25 A particle executes linear harmonic motion around the point $x = 0$. At time zero the particle is at the point $x = 0$ and has a velocity of 5 cm/s. The frequency of the motion is 1 Hz. Determine (a) the period, (b) the amplitude of the motion, (c) the displacement as a function of time, and (d) the maximum speed.

2.26 Define the following characteristics of signals: (a) frequency content, (b) amplitude, (c) magnitude, (d) period.

2.27 Construct an amplitude spectrum plot for the Fourier series in Problem 2.16 for $y(t) = t$. Discuss

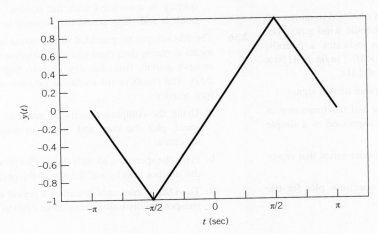

Figure 2.26 Plot of triangle wave for Problem 2.2423.

the significance of this spectrum for measurement or interpretation of this signal. Hint: The plot can be done by inspection or using software such as *DataSpect*.

2.28   Construct an amplitude spectrum plot for the Fourier series in Problem 2.17 for $y(t) = t^2$. Discuss the significance of this spectrum for selecting a measurement system. Hint: The plot can be done by inspection or using software such as *DataSpect* or *WaveformGeneration.vi*.

2.29   Sketch representative waveforms of the following signals, and represent them as mathematical functions (if possible):

a. The output signal from the thermostat on a refrigerator.

b. The electrical signal to a spark plug in a car engine.

c. The input to a cruise control from an automobile.

d. A pure musical tone (e.g., 440 Hz is the note A).

e. The note produced by a guitar string.

f. AM and FM radio signals.

2.30   Represent the function

$$e(t) = 5 \sin 31.4t + 2 \sin 44t$$

as a discrete set of $N = 128$ numbers separated by a time increment of $(1/N)$. Use an appropriate algorithm to construct an amplitude spectrum from this data set. (Hint: A spreadsheet program or the program file *DataSpect* will handle this task.)

2.31   Repeat Problem 2.30 using a data set of 256 numbers at $\delta t = (1/N)$ and $\delta t = (1/2N)$ seconds. Compare and discuss the results.

2.32   A particular strain sensor is mounted to an aircraft wing that is subjected to periodic wind gusts. The strain measurement system indicates a periodic strain that ranges from 3250×10^{-6} in./in. to 4150×10^{-6} in./in. at a frequency of 1 Hz.

a. Determine the average value of this signal.

b. Determine the amplitude and the frequency of this output signal when expressed as a simple periodic function.

c. Determine a one-term Fourier series that represents this signal.

d. Construct an amplitude spectrum plot for the output signal.

2.33   For a dynamic calibration involving a force measurement system, a known force is applied to a sensor. The force varies between 100 and 170 N at a frequency of 10 rad/s. State the average (static) value of the input signal, its amplitude, and its frequency. Assuming that the signal may be represented by a simple periodic waveform, express the signal as a one-term Fourier series and create an amplitude spectrum from an equivalent discrete time series.

2.34   A displacement sensor is placed on a dynamic calibration rig known as a shaker. This device produces a known periodic displacement that serves as the input to the sensor. If the known displacement is set to vary between 2 and 5 mm at a frequency of 100 Hz, express the input signal as a one-term Fourier series. Plot the signal in the time domain, and construct an amplitude spectrum plot.

2.35   Consider the upward flow of water and air in a tube having a circular cross section, as shown in Figure 2.27. If the water and air flow rates are within a certain range, there are slugs of liquid and large gas bubbles flowing upward together. This type of flow is called "slug flow." The data file *gas_liquid_data.txt* provided in the online supplements contains measurements of pressure made at the wall of a tube in which air and water were flowing. The data were acquired at a sample frequency of 300 Hz. The average flow velocity of the air and water is 1 m/s.

a. Construct an amplitude spectrum from the data, and determine the dominant frequency.

b. Using the frequency information from part (a), determine the length L shown in the drawing in Figure 2.27. Assume that the dominant frequency is associated with the passage of the bubbles and slugs across the pressure sensor.

2.36   The file *sunspot.txt* provided in the online supplements contains data from monthly observations of sunspot activity from January 1746 to September 2005. The numbers are a relative measure of sunspot activity.

a. Using the companion software program Dataspect, plot the data and create an amplitude spectrum.

b. From the spectrum, identify any cycles present in the sunspot activity and determine the period(s).

c. The Dalton minimum describes a period of low sunspot activity lasting from about 1790 to 1830;

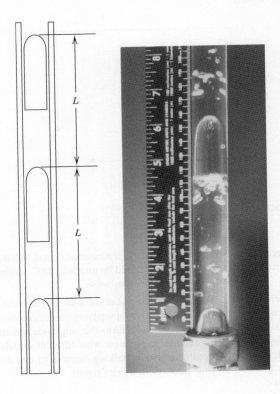

Figure 2.27 Upward gas–liquid flow: slug flow regime. (Photo courtesy of Donald Beasley.)

this is apparent in your plot of the data. The Dalton minimum occurred near the end of the Little Ice Age and coincided with a period of lower-than-average global temperatures. Research the "year without a summer" and its relationship to the Dalton minimum.

2.37 Classify the following signals as completely as possible:

a. Clock face having hands

b. Morse code

c. Musical score, as input to a musician

d. Flashing neon sign

e. Telephone conversation

f. Fax transmission

2.38 Describe the signal defined by the amplitude and phase spectra of Figure 2.28 in terms of its Fourier series. What is the frequency resolution of these plots? What was the sample time increment used?

2.39 For the even-functioned triangle wave signal defined by

$$y(t) = (4C/T)t + C \quad -T/2 \le t \le 0$$
$$y(t) = (-4C/T)t + C \quad 0 \le t \le T/2$$

where C is an amplitude and T is the period,

a. Show that this signal can be represented by the Fourier series

$$y(t) = \sum_{n=1}^{\infty} \frac{4C(1 - \cos n\pi)}{(\pi n)^2} \cos \frac{2\pi n t}{T}$$

b. Expand the first three nonzero terms of the series. Identify the terms containing the fundamental frequency and its harmonics.

c. Plot each of the first three terms on a separate graph. Then plot the sum of the three. For this, set $C = 1$ V and $T = 1$ s. Note: The program *FourCoef* or *Waveform Generation* is useful in this problem.

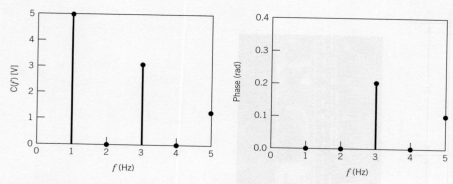

Figure 2.28   Amplitude and phase spectra for Problem 2.39.

d. Sketch the amplitude spectrum for the first three terms of this series, first separately for each term and then combined.

2.40 Figure 2.14 illustrates how the inclusion of higher-frequency terms in a Fourier series refine the accuracy of the series representation of the original function. However, measured signals often display an amplitude spectrum for which amplitudes decrease with frequency. Using Figure 2.14 as a resource, discuss the effect of high-frequency, low-amplitude noise on a measured signal. What signal characteristics would be unaffected? Consider both periodic and random noise.

2.41 The program *Sound.vi* samples the ambient room sounds using your laptop computer microphone and sound board and returns the amplitude spectrum of the sounds. Experiment with different sounds (e.g., tapping, whistling, talking, humming), and describe your findings in a brief report.

Chapter 3

Measurement System Behavior

3.1 INTRODUCTION

This chapter introduces the concept of simulating measurement system behavior through mathematical modeling. Modeling provides details on how a measurement system's design affects its response to an input signal. Each measurement system responds differently to different types of input signals and to the dynamic content within these signals. If the response is insufficient, the output signal will not be the intended copy of the input signal. Thus a particular system may not be suitable for measuring certain signals—or at least portions of some signals. To explore this concept, this chapter discusses system response behavior to certain input signals.

Throughout this chapter, we use the term "measurement system" in a generic sense. It can be used to refer either to the response of the measurement system as a whole or to the response of any individual component or instrument that makes up that system. Either is important, and both are interpreted in similar ways. Each individual stage of the measurement system has its own response to a given input. The overall system response is affected by the response of each stage of the complete system.

Upon completion of this chapter, the reader will be able to

- Relate generalized measurement system models to dynamic response
- Describe and analyze models of zero-, first-, and second-order measurement systems and predict their general behavior
- Calculate static sensitivity, magnitude ratio, and phase shift for a range of systems and input waveforms
- State the importance of phase linearity in reducing signal distortion
- Analyze the response of a measurement system to a complex input waveform
- Determine the response of coupled measurement systems

3.2 GENERAL MODEL FOR A MEASUREMENT SYSTEM

As pointed out in Chapter 2, all input and output signals can be broadly classified as being static, dynamic, or some combination of the two. For a static signal, only the signal magnitude is needed to reconstruct the input signal based on the indicated output signal. Consider measuring the length of a board using a ruler. Once the ruler is positioned, the indication (output) of the magnitude of length is immediately displayed, because the board length (input) does not change over the time required to make the measurement. The board length represents a static input signal that is interpreted through the static output magnitude indicated by the ruler. But consider measuring the vibration displacement of a motor. Vibration signals vary in amplitude and time and at one or more frequencies. These provide a dynamic input signal to the measuring instrument. But a device intended for measuring static information is not very useful in determining dynamic information, so we need different instruments that can follow the input time signal faithfully.

Dynamic Measurements

For dynamic signals, signal amplitudes and frequencies, or waveform information are needed to reconstruct the input signal. Because dynamic signals vary with time, the measurement system must be able to respond fast enough to keep up with the input signal. The ability of any measurement system to follow dynamic signals is a characteristic of the design of the measuring system components. Consider the time response of a common bulb thermometer for measuring body temperature. The thermometer, initially at approximately room temperature, is placed under the tongue. But even after several seconds, the thermometer does not indicate the expected value of body temperature, and its display continually changes. What has happened? Surely your body temperature is not changing. Experience shows that within a few minutes, the correct body temperature will be indicated, so we wait. Experience also tells us that if we need accurate information faster, we would need a different type of temperature sensor. In this example, body temperature itself is constant (static) during the measurement, but the input signal to the thermometer is suddenly changed from room temperature to body temperature—that is, mathematically, the applied waveform is a step change between these two temperatures. This is a dynamic event as the thermometer (the measurement system) sees it! The thermometer must gain energy from its new environment to reach thermal equilibrium, and this takes a finite amount of time.

Now consider the task of assessing the ride quality of an automobile suspension system. A simplified view for one wheel of this system is shown in Figure 3.1. As a tire moves along the road, the road surface provides the time-dependent input signal, $F(t)$, to the suspension at the tire contact

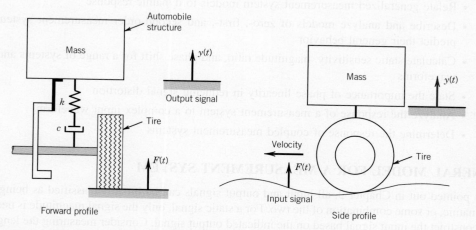

Figure 3.1 Lumped parameter model of an automobile suspension showing input and output signals.

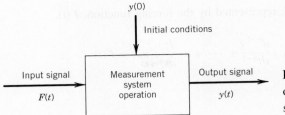

Figure 3.2 Measurement system operation on an input signal, $F(t)$, provides the output signal, $y(t)$.

point. The motion sensed by the passengers, $y(t)$, is a basis for the ride quality and can be described by a waveform that depends on the input from the road and the behavior of the suspension. An engineer must anticipate the form of the input signals so as to design the suspension to attain a desirable output signal.

Measurement systems play a key role in documenting ride quality. But just as the road and car interact to provide ride quality, the input signal and the measurement system interact in creating the output signal. In many situations, the goal of the measurement is to deduce the input signal based on the output signal. Either way, we see that it is important to understand how a measurement system responds to different forms of input signals.

The general behavior of measurement systems to a few common input signal types provides the input–output signal relationships necessary to correctly interpret measured signals. We will show that only a few measurement system parameters (specifications) are needed to predict the system response.

With the previous discussion in mind, consider that the primary task of a measurement system is to sense an input signal and to translate that information into a readily understandable and quantifiable output form. We can reason that a measurement system performs some mathematical operation on a sensed input. In fact, a general measurement system can be represented by a differential equation that describes the operation that a measurement system performs on the input signal. This concept is illustrated in Figure 3.2. For an input signal, $F(t)$, the system performs some operation that yields the output signal, $y(t)$. Then we use $y(t)$ to infer $F(t)$. We will propose a general mathematical model for a measurement system. Then, by representing a typical input signal as some function that acts as an input to the model, we can study just how the measurement system would behave by solving the model equation. In essence, we perform the analytical equivalent of a system calibration. This information can then be used to determine those input signals for which a particular measurement system is best suited.

Measurement System Model

In this section, we apply lumped parameter modeling to measurement systems. In lumped parameter modeling, the spatially distributed physical attributes of a system are modeled as discrete elements. The automotive suspension model discussed earlier is a lumped parameter model. As a simpler example, the mass, stiffness, and damping of a coil spring are properties spatially distributed along its length, but these can be replaced by the discrete elements of a mass, spring, and damper. An advantage is that the governing equations of the models reduce from partial (time and space dependent) to ordinary differential equations (time dependent).

Consider the following general model of a measurement system, which consists of an nth-order linear ordinary differential equation in terms of a general output signal, represented by variable $y(t)$, and subject to a general input signal, represented by the forcing function, $F(t)$:

$$a_n \frac{d^n y}{dt^n} + a_{n-1} \frac{d^{n-1} y}{dt^{n-1}} + \cdots + a_1 \frac{dy}{dt} + a_0 y = F(t) \tag{3.1}$$

where

$$F(t) = b_m \frac{d^m x}{dt^m} + b_{m-1} \frac{d^{m-1} x}{dt^{m-1}} + \cdots + b_1 \frac{dx}{dt} + b_0 x \quad m \le n$$

The coefficients $a_0, a_1, a_2, \ldots, a_n$ and $b_0, b_1, \ldots, b_m$ represent physical system parameters whose properties and values depend on the measurement system itself. Real measurement systems can be modeled this way by considering their governing system equations. These equations are generated by application of pertinent fundamental physical laws of nature to the measurement system. Our discussion is limited to measurement system concepts, but a general treatment of systems can be found in text books dedicated to that topic (1–3).

Example 3.1

As an illustration, consider the seismic accelerometer depicted in Figure 3.3a. Various configurations of this instrument are used in seismic and vibration engineering to determine the motion of large bodies to which the accelerometer is attached. Basically, as the small accelerometer mass reacts to motion, it places the piezoelectric crystal into compression or tension, causing a surface charge to develop on the crystal. The charge is proportional to the motion. As the large body moves, the mass of

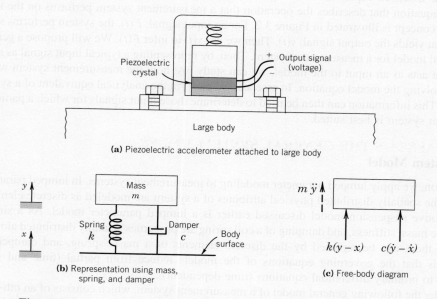

Figure 3.3 Lumped parameter model of accelerometer (Ex. 3.1).

the accelerometer will move with an inertial response. The stiffness of the spring, k, provides a restoring force to move the accelerometer mass back to equilibrium while internal frictional damping, c, opposes any displacement away from equilibrium. A model of this measurement device in terms of ideal lumped elements of stiffness, mass, and damping is given in Figure 3.3b and the corresponding free-body diagram in Figure 3.3c. Let y denote the position of the small mass within the accelerometer and x denote the displacement of the body. Solving Newton's second law for the free body yields the second-order linear, ordinary differential equation

$$m\frac{d^2y}{dt^2} + c\frac{dy}{dt} + ky = c\frac{dx}{dt} + kx$$

Because the displacement y is the pertinent output from the accelerometer due to displacement x, the equation has been written such that all output terms—that is, all the y terms, are on the left side. All other terms are to be considered as input signals and are placed on the right side. Comparing this to the general form for a second-order equation ($n = 2; m = 1$) from Equation 3.1,

$$a_2\frac{d^2y}{dt^2} + a_1\frac{dy}{dt} + a_0y = b_1\frac{dx}{dt} + b_0x$$

we can see that $a_2 = m, a_1 = b_1 = c, a_0 = b_0 = k$ and that the forces developed due to the velocity and displacement of the body become the input signals to the accelerometer. If we could anticipate the waveform of x, for example, $x(t) = x_0 \sin \omega t$, we could solve for $y(t)$, which gives the measurement system time response.

Fortunately, many measurement systems can be modeled by zero-, first-, or second-order linear, ordinary differential equations. More complex systems can usually be simplified to these lower orders. Our intention here is only to attempt to understand how systems behave and how such response is closely related to the design features of a measurement system. The exact input–output relationship is found from calibration. But modeling guides us in choosing specific instruments and measuring methods by predicting system response to signals and in determining the type, range, and specifics of calibration. Next, we examine several special cases of Equation 3.1 that model the most important concepts of measurement system behavior.

3.3   SPECIAL CASES OF THE GENERAL SYSTEM MODEL

Zero-Order Systems

The simplest model of a measurement system and one used with static signals is the zero-order system model. This is represented by the zero-order differential equation:

$$a_0y = F(t)$$

Dividing through by a_0 gives

$$y(t) = KF(t) \tag{3.2}$$

where $K = 1/a_0$. K is called the static sensitivity or steady gain of the system. This system property was introduced in Chapter 1 as the relation between the change in output associated with a change in static input. In a zero-order model, the system output is considered to respond to the input signal instantaneously. If an input signal of magnitude $F(t) = A$ were applied, the instrument would indicate KA, as modeled by Equation 3.2. The scale of the measuring device would be calibrated to indicate A directly.

For real systems, the zero-order system concept is used to model the non–time-dependent measurement system response to static inputs. In fact, the zero-order concept appropriately models any system during a static calibration. When dynamic input signals are involved, a zero-order model is valid only at static equilibrium. Real measurement systems possess inertial or storage capabilities that require higher-order differential equations to correctly model their time-dependent response to dynamic input signals.

Determination of K

The static sensitivity is found from the static calibration of the measurement system. It is the slope of the calibration curve, $K = dy/dx|_x$, evaluated at a pertinent value of x.

Example 3.2

A pencil-type pressure gauge commonly used to measure tire pressure can be modeled at static equilibrium by considering the force balance on the gauge sensor, a piston that slides up and down a cylinder. An exploded view of an actual gauge is shown in Figure 3.4c. In Figure 3.4a, we model the

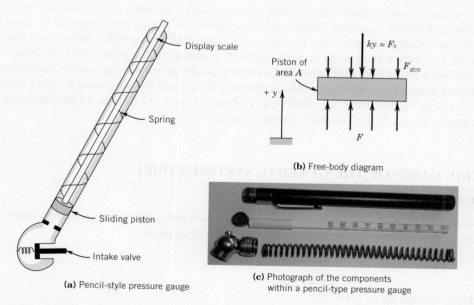

(a) Pencil-style pressure gauge

(b) Free-body diagram

(c) Photograph of the components within a pencil-type pressure gauge

Figure 3.4 Lumped parameter model of pressure gauge (Ex. 3.2). (Photo courtesy of Richard S. Figliola.)

piston motion[1] as being restrained by an internal spring of stiffness, k, so that at static equilibrium the absolute pressure force, F, bearing on the piston equals the force exerted on the piston by the spring, F_s, plus the atmospheric pressure force, F_{atm}. In this manner, the transduction of pressure into mechanical displacement occurs. By keeping the piston of very small mass, we can neglect inertia. Considering the piston free-body at static equilibrium in Figure 3.4b, the static force balance, $\Sigma F = 0$, gives

$$ky = F - F_{atm}$$

where y is measured relative to some static reference position marked as zero on the output display. Pressure is simply the force acting inward over the piston surface area, A. Dividing through by area provides the zero-order response equation between output displacement and input pressure and gives

$$y = (A/k)(p - p_{atm})$$

The term $(p - p_{atm})$ represents the pressure relative to atmospheric pressure. It is the pressure indicated by this gauge. Direct comparison with Equation 3.2 shows that the input pressure magnitude equates to the piston displacement through the static sensitivity, $K = A/k$. Drawing from the concept of Figure 3.2, the system operates on pressure so as to bring about the relative displacement of the piston, the magnitude of which is used to indicate the pressure. The equivalent of spring stiffness and piston area affect the magnitude of this displacement—factors considered in its design. The exact static input–output relationship is found through calibration of the gauge. Because elements such as piston inertia and frictional dissipation were not considered, this model would not be appropriate for studying the dynamic response of the gauge.

First-Order Systems

Measurement systems that contain storage elements do not respond instantaneously to changes in input. The bulb thermometer discussed in Section 3.2 is a good example. The bulb exchanges energy with its environment until the two are at the same temperature, storing or releasing energy during the exchange. The temperature of the bulb sensor changes with time until this equilibrium is reached, which accounts physically for its lag in response. The rate at which temperature changes with time can be modeled with a first-order derivative and the thermometer behavior modeled as a first-order equation. In general, systems with a storage or dissipative capability but negligible inertial forces may be modeled using a first-order differential equation of the form

$$a_1\dot{y} + a_0 y = F(t) \tag{3.3}$$

with $\dot{y} = dy/dt$. Dividing through by a_0 gives

$$\tau\dot{y} + y = KF(t) \tag{3.4}$$

---

[1] In a common gauge, there may or may not be the mechanical spring shown.

where $\tau = a_1/a_0$. The parameter τ is called the *time constant* of the system. Regardless of the physical dimensions of a_0 and a_1, their ratio will always have the dimensions of time. The time constant provides a measure of the speed of system response and thus is an important specification in measuring dynamic input signals. To explore this concept more fully, consider the response of the general first-order system to the following two forms of an input signal: the step function and the simple periodic function.

Step Function Input

The step function, $AU(t)$, is defined as

$$AU(t) = 0 \quad t \le 0^-$$
$$AU(t) = A \quad t \ge 0^+$$

where A is the amplitude of the step function and $U(t)$ is defined as the unit step function as depicted in Figure 3.5. Physically, this function describes a sudden change in the input signal from a constant value of one magnitude to a constant value of some other magnitude, such as a sudden change in loading, displacement, or any physical variable. When we apply a step function input to a measurement system, we obtain information about how quickly a system will respond to a sudden change in input signal. To illustrate this, let us apply a step function as an input to the general first-order system. Setting $F(t) = AU(t)$ in Equation 3.4 gives

$$\tau\dot{y} + y = KAU(t) = KF(t)$$

with an arbitrary initial condition denoted by, $y(0) = y_0$. Solving for $t \ge 0^+$ yields

$$\underbrace{y(t)}_{\text{Time response}} = \underbrace{KA}_{\text{Steady response}} + \underbrace{(y_0 - KA)e^{-t/\tau}}_{\text{Transient response}} \tag{3.5}$$

The solution of the differential equation, $y(t)$, is the time response (or simply the response) of the system. Equation 3.5 describes the behavior of the system to a step change in input value. The value $y(t)$ is in fact the output value indicated by the display stage of the system. It should represent

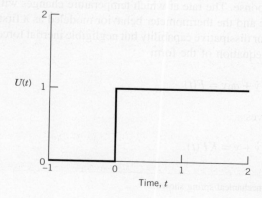

Time, t

Figure 3.5 The unit step function, $U(t)$.

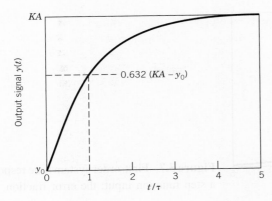

Figure 3.6 First-order system time response to a step function input: the time response, $y(t)$.

the time variation of the output display of the measurement system if an actual step change were applied to the system. We have simply used mathematics to simulate this response.

Equation 3.5 consists of two parts. The first term is known as the steady response because, as $t \to \infty$, the response of $y(t)$ approaches this steady value. The steady response is that portion of the output signal that remains after the transient response has decayed to zero. The second term on the right side of Equation 3.5 is known as the transient response of $y(t)$ because it is both time-dependent and, as $t \to \infty$, the magnitude of this term eventually reduces to zero.

For illustrative purposes, let $y_0 < A$ so that the time response becomes as shown in Figure 3.6. Over time, the indicated output value rises from its initial value, the value at the instant the change in input is applied, to an eventual constant value, $y_\infty = KA$, the value at steady response. As an example, compare this general time response to the recognized behavior of the bulb thermometer when measuring body temperature as discussed earlier. We see a qualitative similarity. In fact, in using a bulb thermometer to measure body temperature, this is a real step function input to the thermometer, itself a first-order measuring system.

Suppose we rewrite the response Equation 3.5 in the form

$$\Gamma(t) = \frac{y(t) - y_\infty}{y_0 - y_\infty} = e^{-t/\tau} \tag{3.6}$$

The term $\Gamma(t)$ is called the *error fraction* of the output signal. Equation 3.6 is plotted in Figure 3.7, where the time axis is nondimensionalized by the time constant. We see that the error fraction decreases from a value of 1 and approaches a value of 0 with increasing t/τ. At the instant the step change in input is introduced, $\Gamma = 1.0$, so that the indicated output from the measurement system is 100% in error. This means that the system has responded 0% to the input change—that is, it has not changed. From Figure 3.7, it is apparent that as time moves forward, the system does respond, and with decreasing error in its indicated value. Let the percent response of the system to a step change be given as $(1 - \Gamma) \times 100$. Then by $t = \tau$, where $\Gamma = 0.368$, the system will have responded to 63.2% of the step change. Furthermore, when $t = 2.3\tau$, the system will have responded ($\Gamma = 0.10$) to 90% of the step change; by $t = 5\tau$, we find the response to be 99.3%. Values for the percent response and the corresponding error as functions of t/τ are summarized in Table 3.1. The time required for a system to respond to a value that is 90% of the step input, $y_\infty - y_0$, is an instrument specification and is called the *rise time* of the system.

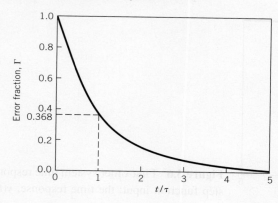

Figure 3.7 First-order system time response to a step function input: the error fraction, Γ.

Table 3.1 First-Order System Response and Error Fraction

t/τ	% Response	Γ	% Error
0	0.0	1.0	100.0
1	63.2	0.368	36.8
2	86.5	0.135	13.5
2.3	90.0	0.100	10.0
3	95.0	0.050	5.0
5	99.3	0.007	0.7
∞	100.0	0.0	0.0

Based on this behavior, we can see that the time constant is in fact a measure of how quickly a first-order measurement system will respond to a change in input value. A smaller time constant indicates a shorter time between the instant that an input is applied and when the system reaches an essentially steady output. We define the *time constant* as the time required for a first-order system to achieve 63.2% of the step change magnitude, $y_\infty - y_0$. The time constant is a system property.

Determination of τ From the development above, the time constant can be experimentally determined by recording the system's response to a step function input of a known magnitude. In practice, it is best to record that response from $t = 0$ until steady response is achieved. The data can then be plotted as error fraction versus time on a semilog plot, shown in Figure 3.8. This type of plot is equivalent to the transformation

$$\ln \Gamma = 2.3 \log \Gamma = -(1/\tau)t \tag{3.7}$$

which is of the linear form $Y = mX + B$ (where $Y = \ln \Gamma$, $m = -(1/\tau)$, $X = t$, and $B = 0$ here). A linear curve fit through the data will provide a good estimate of the slope, m, of the resulting plot. From Equation 3.7, we see that $m = -1/\tau$, which yields the estimate for τ.

This method offers advantages over attempting to compute τ directly from the exact time required to achieve 63.2% of the step-change magnitude. First, real systems may deviate somewhat

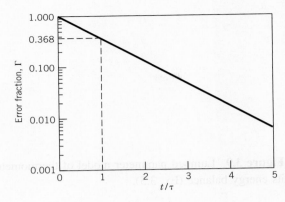

Figure 3.8 The error fraction plotted on semilog coordinates.

from perfect first-order behavior. On a semilog plot (Figure 3.8), such deviations are readily apparent as trends away from a straight line. Modest deviations do not pose a problem. But strong deviations indicate that the system is not behaving as expected, thus requiring a closer examination of the system operation, the step function experiment used, or the assumed form of the system model. Second, acquiring data during the step function experiment is prone to some random error in each data point. The use of a data curve fit to determine τ uses all the data over time so as to minimize the influence of an error in any one data point. Third, the method eliminates the need to determine the $\Gamma = 1.0$ and 0.368 points exactly, which are difficult to establish in practice and so are prone to systematic error.

Example 3.3

Suppose a bulb thermometer originally indicating $20\,°C$ is suddenly exposed to a fluid temperature of $37\,°C$. Develop a model that simulates the thermometer output response.

   KNOWN
$$T(0) = 20°C$$
$$T_\infty = 37°C$$
$$F(t) = [T_\infty - T(0)]U(t)$$

   ASSUMPTIONS   To keep things simple, assume the following: no installation effects (neglect conduction and radiation effects); sensor mass is mass of liquid in bulb only; uniform temperature within bulb (lumped mass); thermometer scale is calibrated to indicate temperature

   FIND $T(t)$

   SOLUTION   Consider the energy balance developed in Figure 3.9. According to the first law of thermodynamics, the rate at which energy is exchanged between the sensor and its environment through convection, $\dot{Q}$, must be balanced by the storage of energy within the thermometer, dE/dt. This conservation of energy is written as

$$\frac{dE}{dt} = \dot{Q}$$

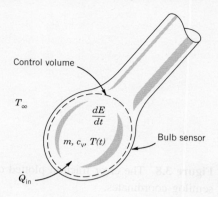

Figure 3.9 Lumped parameter model of thermometer and its energy balance (Ex. 3.3).

Energy storage in the bulb is manifested by a change in bulb temperature so that for a constant mass bulb, $dE(t)/dt = mc_v dT(t)/dt$. Energy exchange by convection between the bulb at $T(t)$ and an environment at T_∞ has the form $\dot{Q} = hA_s\Delta T$. The first law can be written as

$$mc_v \frac{dT(t)}{dt} = hA_s[T_\infty - T(t)]$$

This equation can be written in the form

$$mc_v \frac{dT(t)}{dt} + hA_s[T(t) - T(0)] = hA_s F(t) = hA_s[T_\infty - T(0)]U(t)$$

with initial condition $T(0)$ and
where

  m = mass of liquid within thermometer

  c_v = specific heat of liquid within thermometer

  h = convection heat transfer coefficient between bulb and environment

  A_s = thermometer surface area

The term hA_s controls the rate at which energy can be transferred between a fluid and a body; it is analogous to electrical conductance. By comparison with Equation 3.3, $a_0 = hA_s$, $a_1 = mc_v$, and $b_0 = hA_s$. Rewriting for $t \geq 0^+$ and simplifying yields

$$\frac{mc_v}{hA_s}\frac{dT(t)}{dt} + T(t) = T_\infty$$

From Equation 3.4, this determines that the time constant and static sensitivity are

$$\tau = \frac{mc_v}{hA_s} \quad K = \frac{hA_s}{hA_s} = 1$$

Direct comparison with Equation 3.5 yields this thermometer response:

$$T(t) = T_\infty + [T(0) - T_\infty]e^{-t/\tau}$$
$$= 37 - 17e^{-t/\tau}[°C]$$

COMMENT Two interactive examples of the thermometer problem (from Exs. 3.3–3.5) are available using the online supplements. In the Matlab program *Firstord*, the user can choose different forcing functions and adjust system parameters to study the system response. In the LabView program *Thermal_response*, the user can apply interactively a step change in temperature and vary the time constant to view the first-order system response of a thermal sensor.

Clearly the time constant, τ, of the thermometer can be reduced by decreasing its mass-to-area ratio or by increasing h (for example, increasing the fluid velocity around the sensor). Without modeling, such information could be ascertained only by trial and error, a time-consuming and costly method with no assurance of success. Also, it is significant that we found that the response of the temperature measurement system in this case depends on the environmental conditions of the measurement that control h, because the magnitude of h affects the magnitude of τ. If h is not controlled during response tests (i.e., if it is an extraneous variable), ambiguous results will occur. For example, the curve of Figure 3.8 could become nonlinear, or replications will not yield the same values for τ.

Review of this example too should make it apparent that the results of a well-executed step response calibration may not be directly indicative of an instrument's performance during a measurement if the measurement conditions differ from those existing during the step calibration. A carefully thought out test plan can minimize such differences.

Example 3.4

For the thermometer in Example 3.3 subjected to a step change in input, calculate the 90% rise time in terms of t/τ.

KNOWN Same as Example 3.3

ASSUMPTIONS Same as Example 3.3

FIND 90% response time in terms of t/τ

SOLUTION The percent response of the system is given by $(1 - \Gamma) \times 100$ with the error fraction, Γ, defined by Equation 3.6. From Equation 3.5, we note that at $t = \tau$, the thermometer will indicate $T(t) = 30.75 \,°C$, which represents only 63.2% of the step change from 20 °C to 37 °C. The 90% rise time represents the time required for Γ to drop to a value of 0.10. Then

$$\Gamma = 0.10 = e^{-t/\tau}$$

or $t/\tau = 2.3$.

COMMENT In general, a time equivalent to 2.3τ is required to achieve 90% of the applied step input for a first-order system.

Example 3.5

A particular thermometer is subjected to a step change, such as in Example 3.3, in an experimental exercise to determine its time constant. The temperature data are recorded with time and presented in Figure 3.10. Determine the time constant for this thermometer. In the

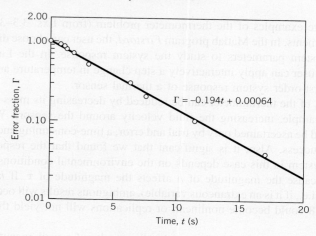

Figure 3.10 Temperature–time history of Example 3.5.

experiment, the heat transfer coefficient, h, is estimated to be 6 W/m²-°C from engineering handbook correlations.

KNOWN Data of Figure 3.10

$$h = 6 \text{ W/m}^2\text{-}°C$$

ASSUMPTIONS First-order behavior using the model of Example 3.3; constant properties

FIND τ

SOLUTION According to Equation 3.7, the time constant should be the negative reciprocal of the slope of a line drawn through the data of Figure 3.10. Aside from the first few data points, the data appear to follow a linear trend, indicating a nearly first-order behavior and validating our model assumption. The data is fit to the first-order equation[2]

$$2.3 \log \Gamma = (-0.194)t + 0.00064$$

With $m = -0.194 = -1/\tau$, the time constant is calculated to be $\tau = 5.15$ seconds.

COMMENT If the experimental data were to deviate significantly from first-order behavior, this would be a clue either that our assumptions do not fit the real-problem physics or that the test conduct has control or execution problems.

Simple Periodic Function Input

Periodic signals are commonly encountered in engineering processes. Examples include vibrating structures, vehicle suspension dynamics, biological circulations, and reciprocating pump flows. When periodic inputs are applied to a first-order system, the input signal frequency has an important

---

[2] The least-squares approach to curve fitting is discussed in detail in Chapter 4.

influence on measuring system time response and affects the output signal. This behavior can be studied effectively by applying a simple periodic waveform to the system. Consider the first-order measuring system to which an input of the form of a simple periodic function, $F(t) = A\, sin\, \omega t$, is applied for $t \geq 0^+$:

$$\tau \dot{y} + y = KA \sin \omega t$$

with initial conditions $y(0) = y_0$. Note that ω in [rad/s] $= 2\pi f$ with f in [Hz]. The general solution to this differential equation yields the measurement system's output signal—that is, the time response to the applied input, $y(t)$:

$$y(t) = Ce^{-t/\tau} + \frac{KA}{\sqrt{1 + (\omega\tau)^2}} \sin\left(\omega t - \tan^{-1}\omega\tau\right) \tag{3.8}$$

where the value for C depends on the initial conditions.

So what has happened? The output signal, $y(t)$, of Equation 3.8 consists of a transient and a steady periodic response. The first term on the right side is the transient response. As t increases, this term decays to zero and no longer influences the output signal. Transient response is important only during the initial period following the application of the new input. We already have information about the system transient response from the step function study, so we focus our attention on the second term, the steady response. The steady response is time-dependent but periodically repeats itself. This term persists for as long as the periodic input is maintained. From Equation 3.8, we see that the frequency of the steady response term remains the same as the input signal frequency, but note that the amplitude of the steady response depends on the value of the applied frequency, ω. Also, the phase angle of the periodic function has changed.

Equation 3.8 can be rewritten in a general form:

$$y(t) = Ce^{-t/\tau} + B(\omega)\sin[\omega t + \Phi]$$

$$B(\omega) = \frac{KA}{\sqrt{1 + (\omega\tau)^2}} \tag{3.9}$$

$$\Phi(\omega) = -\tan^{-1}(\omega\tau)$$

where $B(\omega)$ represents the amplitude of the steady response and the angle $\Phi(\omega)$ represents the *phase shift*. A relative illustration between the input signal and the system output response is given in Figure 3.11 for an arbitrary frequency and system time constant. From Equation 3.9, both B and Φ are frequency-dependent. Hence the exact form of the output response depends on the value of the frequency of the input signal. The steady response of any system to which a periodic input of frequency, ω, is applied is known as the *frequency response* of the system. The frequency affects the magnitude of amplitude B and also can bring about a time delay. This time delay, β_1, is seen in the phase shift, $\Phi(\omega)$, of the steady response. For a phase shift given in radians, the time delay in units of time is

$$\beta_1 = \frac{\Phi}{\omega}$$

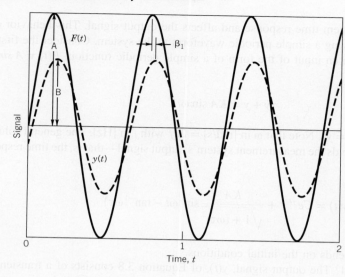

Figure 3.11 Relationship between a sinusoidal input and output: amplitude, frequency, and time delay.

that is, we can write

$$\sin(\omega t + \Phi) = \sin\left[\omega\left(t + \frac{\Phi}{\omega}\right)\right] = \sin\left[\omega\left(t + \beta_1\right)\right]$$

The value for β_1 will be negative, indicating that the time shift is a delay between the output and input signals. Since Equation 3.9 applies to all first-order measuring systems, the magnitude and phase shift by which the output signal differs from the input signal are predictable.

We define a *magnitude ratio*, $M(\omega)$, as the ratio of the output signal amplitude to the input signal amplitude, $M(\omega) = B/KA$. For a first-order system subjected to a simple periodic input, the magnitude ratio is

$$M(\omega) = \frac{B}{KA} = \frac{1}{\sqrt{1 + (\omega\tau)^2}} \tag{3.10}$$

The magnitude ratio for a first-order system is plotted in Figure 3.12, and the corresponding phase shift is plotted in Figure 3.13. The effects of both system time constant and input signal frequency on frequency response are apparent in both figures. This behavior can be interpreted in the following manner. For those values of $\omega\tau$ for which the system responds with values of $M(\omega)$ near unity, the measurement system transfers all or nearly all of the input signal amplitude to the output and with very little time delay; that is, B will be nearly equal to KA in magnitude, and $\Phi(\omega)$ will be near zero degrees. At large values of $\omega\tau$ the measurement system filters out any frequency information of the input signal by responding with very small amplitudes, which is seen by the small $M(\omega)$, and by large time delays, as evidenced by increasingly nonzero β_1.

Any combination of ω and τ produces the same results. If we wanted to measure signals with high-frequency content, then we would need a system having a small τ. On the other hand, systems of large τ may be adequate to measure signals of low-frequency content. Often the trade-offs compete available technology against cost.

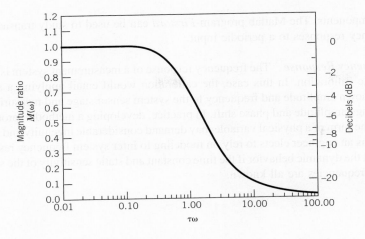

Figure 3.12 First-order system frequency response: magnitude ratio.

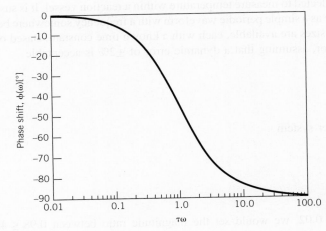

Figure 3.13 First-order system frequency response: phase shift.

The *dynamic error*, $\delta(\omega)$, of a system is defined as $\delta(\omega) = M(\omega) - 1$. It is a measure of the inability of a system to adequately reconstruct the amplitude of the input signal for a particular input frequency. We normally want measurement systems to have a magnitude ratio at or near unity over the anticipated frequency band of the input signal to minimize $\delta(\omega)$. Perfect reproduction of the input signal is not possible, so some dynamic error is inevitable. We need some way to state this. For a first-order system, we define a *frequency bandwidth* as the frequency band over which $M(\omega) \geq 0.707$; in terms of the decibel (plotted in Figure 3.12) defined as

$$\text{dB} = 20 \log M(\omega) \tag{3.11}$$

this is the band of frequencies within which $M(\omega)$ remains above -3 dB, that is, $-3\,\text{dB} \leq M(\omega) \leq 0\,\text{dB}$.

   The functions $M(\omega)$ and $\Phi(\omega)$ represent the frequency response of a measurement system to periodic inputs. These equations and universal curves provide guidance in selecting measurement systems and system components. The Matlab program *Firstord* can be used to study transient and steady time and frequency responses to a periodic input.

Determination of Frequency Response The frequency response of a measurement system is found physically by a dynamic calibration. In this case, the calibration would entail applying a simple periodic waveform of known amplitude and frequency to the system sensor stage and measuring the corresponding output stage amplitude and phase shift. In practice, developing a method to produce a periodic input signal in the form of a physical variable may demand considerable ingenuity and effort. Hence, in many situations an engineer elects to rely on modeling to infer system frequency response behavior. We can predict the dynamic behavior if the time constant and static sensitivity of the system and the range of input frequencies are all known.

Example 3.6

A temperature sensor is to be selected to measure temperature within a reaction vessel. It is suspected that the temperature will behave as a simple periodic waveform with a frequency somewhere between 1 and 5 Hz. Sensors of several sizes are available, each with a known time constant. Based on time constant, select a suitable sensor, assuming that a dynamic error of $\pm 2\%$ is acceptable.

KNOWN $1 \leq f \leq 5\,\text{Hz}$

$$|\delta(\omega)| \leq 0.02$$

ASSUMPTIONS First-order system

$$F(t) = A \sin \omega t$$

FIND Time constant, τ

SOLUTION With $|\delta(\omega)| \leq 0.02$, we would set the magnitude ratio between $0.98 \leq M \leq 1$. Then,

$$0.98 \leq M(\omega) = \frac{1}{\sqrt{1 + (\omega\tau)^2}} \leq 1$$

From Figure 3.12, this constraint is maintained over the range $0 \leq \omega\tau \leq 0.2$. We can also see in this figure that for a system of fixed time constant, the smallest value of $M(\omega)$ will occur at the largest frequency. So with $\omega = 2\pi f = 2\pi(5)\text{rad/s}$ and solving for $M(\omega) = 0.98$ yields, $\tau \leq 6.4\,\text{ms}$. Accordingly, a sensor having a time constant of 6.4 ms or less will work.

Second-Order Systems

Systems possessing inertia contain a second-derivative term in their model equation (e.g., see Ex. 3.1). A system modeled by a second-order differential equation is called a second-order system.

Examples of some second-order instruments include accelerometers and pressure transducers (including microphones and loudspeakers).

In general, a second-order measurement system subjected to an arbitrary input, $F(t)$, can be described by an equation of the form

$$a_2 \ddot{y} + a_1 \dot{y} + a_0 y = F(t) \tag{3.12}$$

where a_0, a_1, and a_2 are physical parameters used to describe the system and $\ddot{y} = d^2 y/dt^2$. This equation can be rewritten as

$$\frac{1}{\omega_n^2} \ddot{y} + \frac{2\zeta}{\omega_n} \dot{y} + y = KF(t) \tag{3.13}$$

where

$$\omega_n = \sqrt{\frac{a_0}{a_2}} = \text{natural frequency of the system}$$

$$\zeta = \frac{a_1}{2\sqrt{a_0 a_2}} = \text{damping ratio of the system}$$

Consider the homogeneous solution to Equation 3.13. Its form depends on the roots of the characteristic equation of Equation 3.13:

$$\frac{1}{\omega_n^2} \lambda^2 + \frac{2\zeta}{\omega_n} \lambda + 1 = 0$$

This quadratic equation has two roots,

$$\lambda_{1,2} = -\zeta \omega_n \pm \omega_n \sqrt{\zeta^2 - 1}$$

Depending on the value for ζ three forms of homogeneous solution are possible:

$0 \leq \zeta < 1$ (underdamped system solution)

$$y_h(t) = Ce^{-\zeta \omega_n t} \sin\left(\omega_n \sqrt{1 - \zeta^2} t + \Theta\right) \tag{3.14a}$$

$\zeta = 1$ (critically damped system solution)

$$y_h(t) = C_1 e^{\lambda_1 t} + C_2 t e^{\lambda_2 t} \tag{3.14b}$$

$\zeta > 1$ (overdamped system solution)

$$y_h(t) = C_1 e^{\lambda_1 t} + C_2 e^{\lambda_2 t} \tag{3.14c}$$

The homogeneous solution determines the transient response of a system. The damping ratio, ζ, is a measure of system damping, a property of a system that enables it to dissipate energy internally. For systems with $0 \leq \zeta \leq 1$, the transient response will be oscillatory, whereas for $\zeta \geq 1$, the transient response will not oscillate. The critically damped solution, $\zeta = 1$, denotes the demarcation between oscillatory and nonoscillatory behavior in the transient response.

Step Function Input

Again, the step function input is applied to determine the general behavior and speed at which the system will respond to a change in input. The response of a second-order measurement system to a step function input is found from the solution of Equation 3.13, with $F(t) = AU(t)$, to be

$$y(t) = KA - KAe^{-\zeta \omega_n t}\left[\frac{\zeta}{\sqrt{1-\zeta^2}}\sin\left(\omega_n t\sqrt{1-\zeta^2}\right) + \cos\left(\omega_n t\sqrt{1-\zeta^2}\right)\right] \quad 0 \leq \zeta < 1 \quad (3.15a)$$

$$y(t) = KA - KA(1 + \omega_n t)e^{-\omega_n t} \qquad \zeta = 1 \quad (3.15b)$$

$$y(t) = KA - KA\left[\frac{\zeta + \sqrt{\zeta^2-1}}{2\sqrt{\zeta^2-1}}e^{(-\zeta+\sqrt{\zeta^2-1})\omega_n t} - \frac{\zeta - \sqrt{\zeta^2-1}}{2\sqrt{\zeta^2-1}}e^{(-\zeta-\sqrt{\zeta^2-1})\omega_n t}\right] \qquad \zeta > 1 \quad (3.15c)$$

where we set the initial conditions, $y(0) = \dot{y}(0) = 0$ for convenience.

Equations 3.15a–c are plotted in Figure 3.14 for several values of ζ. The interesting feature is the transient response. For underdamped systems, the transient response is oscillatory about the steady value and occurs with a period

$$T_d = \frac{2\pi}{\omega_d} = \frac{1}{f_d} \tag{3.16}$$

$$\omega_d = \omega_n\sqrt{1-\zeta^2} \tag{3.17}$$

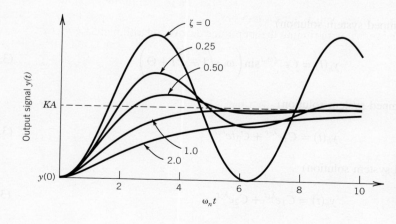

Figure 3.14 Second-order system time response to a step function input.

where ω_d is called the *ringing frequency*. In instruments, this oscillatory behavior is called "ringing." The ringing phenomenon and the associated ringing frequency are properties of the measurement system and are independent of the input signal. It is the free oscillation frequency of a system displaced from its equilibrium.

The duration of the transient response is controlled by the $\zeta\omega_n$ term. In fact, its influence is similar to that of a time constant in a first-order system, such that we could define a second-order time constant as $\tau = 1/\zeta\omega_n$. The system settles to KA more quickly when it is designed with a larger $\zeta\omega_n$ (i.e., smaller τ). Nevertheless, for all systems with $\zeta > 0$, the response eventually indicates the steady value of $y_\infty = KA$ as $t \to \infty$.

Recall that the *rise time* is defined as the time required to achieve a value within 90% of the step input. For a second-order system, the rise time is the time required to first achieve 90% of $(KA - y_0)$. Rise time is reduced by decreasing the damping ratio, as seen in Figure 3.14. However, the severe ringing associated with very lightly damped systems actually delays the time to achieve a steady value compared to systems of higher damping. This is demonstrated by comparing the response at $\zeta = 0.25$ with the response at $\zeta = 1$ in Figure 3.14. With this in mind, the time required for a measurement system's oscillations to settle to within $\pm 10\%$ of the steady value, KA, is defined as its *settling time*. The settling time is an approximate measure of the time to achieve a steady response. A damping ratio of about 0.7 appears to offer a good compromise between ringing and settling time. If an error fraction (Γ) of a few percent is acceptable, then a system with $\zeta = 0.7$ will settle to steady response in about half the time of a system having $\zeta = 1$. For this reason, most measurement systems intended to measure sudden changes in input signal are typically designed such that parameters a_0, a_1, and a_2 provide a damping ratio of between 0.6 and 0.8.

Determination of Ringing Frequency and Rise and Settling Times   The experimental determination of the ringing frequency associated with underdamped systems is performed by applying a step input to the second-order measurement system and measuring the response. This type of calibration also yields information concerning the time to steady response of the system, which includes rise and settling times. Example 3.8 describes such a test. Typically, measurement systems suitable for dynamic signal measurements have specifications that include 90% rise time and settling time. The Labview program *Adjust Second Order System* enables the user to vary system parameters (ζ, ω_n, K) and compute the time response to a step function.

Determination of Natural Frequency and Damping Ratio   From the underdamped system response to a step function test, values for the damping ratio and natural frequency can be extracted. From Figure 3.14, we see that with ringing the amplitude decays logarithmically with time towards a steady-state value. Let y_{max} represent the peak amplitude occurring with each cycle. Then for the first two successive peak amplitudes, let $y_1 = (y_{max})_1 - y_\infty$ and $y_2 = (y_{max})_2 - y_\infty$. The damping ratio is found from

$$\zeta = \frac{1}{\sqrt{1 + \left(2\pi/\ln\left(\frac{y_1}{y_2}\right)\right)^2}} \tag{3.18}$$

From the calculation to find the ringing frequency using Equation 3.16, the natural frequency is found using Equation 3.17. Alternately, we could perform the step function test with $y(0) = KA$ and $y_\infty = 0$ and still use the same approach.

Example 3.7

Determine the physical parameters that affect the natural frequency and damping ratio of the accelerometer of Example 3.1.

KNOWN   Accelerometer shown in Figure 3.3

ASSUMPTIONS   Second-order system as modeled in Example 3.1

FIND ω_n, ζ

SOLUTION   A comparison between the governing equation for the accelerometer in Example 3.1 and Equation 3.13 gives

$$\omega_n = \sqrt{\frac{k}{m}} \quad \zeta = \frac{c}{2\sqrt{km}}$$

The physical parameters of mass, spring stiffness, and frictional damping control the natural frequency and damping ratio of this measurement system.

Example 3.8

The curve shown in Figure 3.15 is the recorded voltage output signal of a diaphragm pressure transducer subjected to a step change in input. From a static calibration, the pressure–voltage relationship was found to be linear over the range 1 atmosphere (atm) to 4 atm with a static sensitivity of 1 V/atm. For the step test, the initial pressure was atmospheric pressure, p_a, and the final pressure was $2p_a$. Estimate the rise time, the settling time, and the ringing frequency of this measurement system.

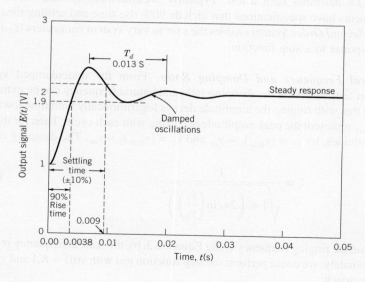

Figure 3.15   Pressure transducer time response to a step input for Example 3.8.

KNOWN $p(0) = 1$ atm

$p_\infty = 2$ atm

$K = 1$ V/atm

ASSUMPTIONS Second-order system behavior

FIND Rise and settling times; ω_d

SOLUTION The ringing behavior of the system noted on the trace supports the assumption that the transducer can be described as having a second-order behavior. From the given information,

$$E(0) = Kp(0) = 1 \text{ V}$$
$$E_\infty = Kp_\infty = 2 \text{ V}$$

so that the step change observed on the trace should appear as a magnitude of 1 V. The 90% rise time occurs when the output first achieves a value of 1.9 V. The 90% settling time will occurs when the output settles between $1.9 < E(t) < 2.1$ V. From Figure 3.15, the rise occurs in about 4 ms, and the settling time is about 9 ms. The period of the ringing behavior, T_d, is judged to be about 13 ms for an $\omega_d \approx 485$ rad/s.

Simple Periodic Function Input

The response of a second-order system to a simple periodic function input of the form $F(t) = A \sin \omega t$ is given by

$$y(t) = y_h + \frac{KA \sin[\omega t + \Phi(\omega)]}{\left\{ \left[1 - (\omega/\omega_n)^2 \right]^2 + \left[2\zeta\omega/\omega_n \right]^2 \right\}^{1/2}} \tag{3.19}$$

with frequency-dependent phase shift

$$\Phi(\omega) = \tan^{-1} \left(-\frac{2\zeta\omega/\omega_n}{1 - (\omega/\omega_n)^2} \right) \tag{3.20}$$

The exact form for y_h is found from Equations 3.14a–c and depends on the value of ζ. The steady response, the second term on the right side, has the general form

$$y_{steady}(t) = y(t \to \infty) = B(\omega)\sin[\omega t + \Phi(\omega)] \tag{3.21}$$

with amplitude $B(\omega)$. Comparing Equations 3.19 and 3.21 shows that the amplitude of the steady response of a second-order system subjected to a sinusoidal input is also dependent on ω. So the

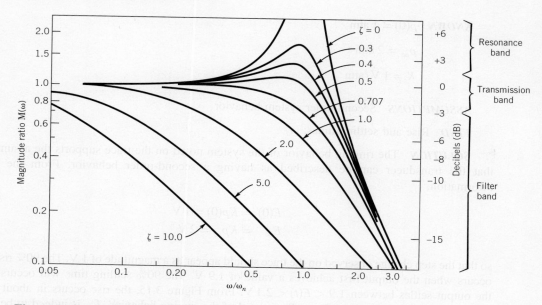

Figure 3.16 Second-order system frequency response: magnitude ratio.

amplitude of the output signal is frequency-dependent. In general, we can define the magnitude ratio, $M(\omega)$, for a second-order system as

$$M(\omega) = \frac{B(\omega)}{KA} = \frac{1}{\left\{ \left[1 - (\omega/\omega_n)^2\right]^2 + \left[2\zeta\omega/\omega_n\right]^2 \right\}^{1/2}} \tag{3.22}$$

The frequency response of a second order system is given by Equations 3.20 and 3.22 and this is plotted in Figures 3.16 and 3.17. The magnitude ratio-frequency dependence for a second-order system is plotted in Figure 3.16 for several values of damping ratio. A corresponding plot of the phase-shift dependency on input frequency and damping ratio is shown in Figure 3.17. For an ideal measurement system, $M(\omega)$ would equal unity and $\Phi(\omega)$ would equal zero for all values of measured frequency. Instead, $M(\omega)$ approaches zero and $\Phi(\omega)$ approaches $-\pi$ as ω/ω_n becomes large. Keep in mind that ω_n is a property of the measurement system, whereas ω is a property of the input signal.

System Characteristics

Several tendencies are apparent in Figures 3.16 and 3.17. For a system of zero damping, $\zeta = 0, M(\omega)$ will approach infinity and $\Phi(\omega)$ jumps to $-\pi$ in the vicinity of $\omega = \omega_n$. This behavior is characteristic of system resonance. Real systems possess some amount of damping, which modifies the abruptness and magnitude of resonance, but underdamped systems may still achieve a damped resonance. This region around $\omega = \omega_n$ on Figures 3.16 and 3.17 is called the *resonance band* of the system, referring to the range of frequencies over which the

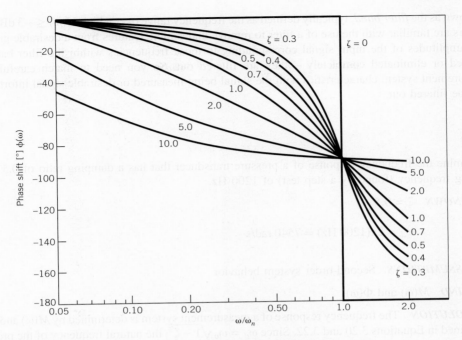

Figure 3.17   Second-order system frequency response: phase shift.

system shows $M(\omega) > 1$. Peak resonance in underdamped systems occurs at the *resonance frequency*,

$$\omega_R = \omega_n \sqrt{1 - 2\zeta^2} \tag{3.23}$$

The resonance frequency is a property of the measurement system. Resonance behavior is excited by a periodic input signal frequency. The resonance frequency differs from the ringing frequency of free oscillation. Resonance behavior results in considerable phase shift. For most applications, operating at frequencies within the resonance band is confusing and could even be damaging to some delicate sensors. Resonance behavior is very nonlinear and results in distortion of the signal. On the other hand, systems having $\zeta > 0.707$ do not resonate.

At small values of ω/ω_n, $M(\omega)$ remains near unity and $\Phi(\omega)$ near zero. This means that information concerning the input signal of frequency ω will be passed through to the output signal with little alteration in the amplitude or phase shift. This region on the frequency response curves is called the *transmission band*. The actual extent of the frequency range for near unity gain depends on the system damping ratio. The transmission band of a system is either specified by its frequency bandwidth, typically defined for a second-order system as -3 dB $\leq M(\omega) \leq 3$ dB, or otherwise specified explicitly. One needs to operate within the transmission band of a measurement system to measure the dynamic content of the input signal accurately.

At large values of ω/ω_n, $M(\omega)$ approaches zero. In this region, the measurement system attenuates the amplitude information in the input signal. A large phase shift occurs. This region is known as the *filter band*, typically defined as the frequency range over which $M(\omega) \leq -3$ dB. Most readers are familiar with the use of a filter to remove undesirable features from a desirable product. The amplitudes of the input signal corresponding to those frequencies within the filter band are reduced or eliminated completely – they are filtered out. So you need to match carefully the measurement system characteristics with the signal being measured or desirable signal information may be filtered out.

Example 3.9

Determine the frequency response of a pressure transducer that has a damping ratio of 0.5 and a ringing frequency (found by a step test) of 1200 Hz.

    KNOWN $\zeta = 0.5$

$$\omega_d = 2\pi(1200 \text{ Hz}) = 7540 \text{ rad/s}$$

    ASSUMPTIONS   Second-order system behavior

    FIND $M(\omega)$ and $\Phi(\omega)$

    SOLUTION   The frequency response of a measurement system is determined by $M(\omega)$ and $\Phi(\omega)$ as defined in Equations 3.20 and 3.22. Since $\omega_d = \omega_n \sqrt{1 - \zeta^2}$, the natural frequency of the pressure transducer is found to be $\omega_n = 8706$ rad/s. The frequency response at selected frequencies is computed from Equations 3.20 and 3.22:

ω (rad/s)	$M(\omega)$	$\Phi(\omega)$ [°]
500	1.00	−3.3
2,600	1.04	−18.2
3,500	1.07	−25.6
6,155	1.15	−54.7
7,540	1.11	−73.9
8,706	1.00	−90.0
50,000	0.05	−170.2

    COMMENT   The resonance behavior in the transducer response peaks at $\omega_R = 6155$ rad/s. As a rule, resonance effects can be minimized by operating at input frequencies of less than ~30% of the system's natural frequency. The time and frequency responses of second-order systems to periodic forcing functions can be studied in more detail using the companion software Matlab program *SecondOrd*.

Example 3.10

An accelerometer is to be selected to measure a time-dependent motion. In particular, input signal frequencies below 100 Hz are of prime interest. Select a set of acceptable parameter specifications for the instrument assuming a dynamic error of ±5%.

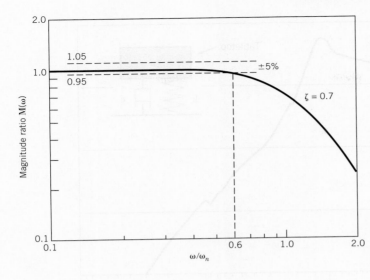

Figure 3.18 Magnitude ratio for second-order system with $\zeta = 0.7$ for Example 3.10.

KNOWN $f \leq 100\,\text{Hz}$ (i.e., $\omega \leq 628\,\text{rad/s}$)

ASSUMPTIONS   Second-order system
Dynamic error of $\pm 5\%$ acceptable

FIND   Select ω_n and ζ

SOLUTION   To meet a $\pm 5\%$ dynamic error constraint, we want $0.95 \leq M(\omega) \leq 1.05$ over the frequency range $0 \leq \omega \leq 628\,\text{rad/s}$. This solution is open ended in that a number of instruments with different ω_n will do the task. So as one solution, let us set $\zeta = 0.7$ and then solve for the required ω_n using Equation 3.22:

$$0.95 \leq M(\omega) = \frac{1}{\left\{ \left[1 - (\omega/\omega_n)^2 \right]^2 + \left[2\zeta\omega/\omega_n \right]^2 \right\}^{1/2}} \leq 1.05$$

With $\omega = 628\,\text{rad/s}$, these equations give $\omega_n \geq 1047\,\text{rad/s}$. We could plot Equation 3.22 with $\zeta = 0.7$, as shown in Figure 3.18. In Figure 3.18, we find that $0.95 \leq M(\omega) \leq 1.05$ over the frequency range $0 \leq \omega/\omega_n \leq 0.6$. Again, this makes $\omega_n \geq 1047\,\text{rad/s}$ acceptable. So as one solution, an instrument having $\zeta = 0.7$ and $\omega_n \geq 1047\,\text{rad/s}$ meets the problem constraints.

Example 3.11:   Measurements Isolation Table

A common approach to vibration control in state-of-the-art metrology and for performing any measurement sensitive to building vibrations is to mount test equipment on an isolation table. These are practical solutions when using sensitive microscopes, interferometers, delicate instruments and balance scales. An isolation table is simply a table surface that is isolated from floor vibrations through a pneumatic suspension (mass-spring-damper) system mounted within each of the table

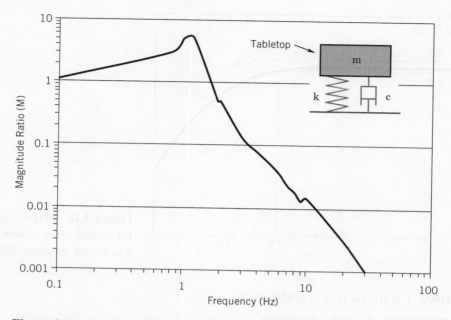

Figure 3.19 Frequency response of an isolation table to vertical ground induced vibrations. Inset depicts a single leg of the isolation system.

legs. The goal is to design a suspension system that has a very low resonance frequency so as to effectively operate in the filter band of system response for the anticipated excitation frequencies. The frequency response of a commercially available table is shown in Figure 3.19. Its resonance frequency is about 1 Hz. As such, excitation floor vibrations above 2 Hz will be highly attenuated at the table surface.

COMMENT As discussed in Example 1.2, engineers added more mass, m, to the steering column of the golf cart. The steering column natural frequency is described by $\omega_n = \sqrt{k/m}$. Adding mass lowered its natural frequency. In this way, the engineers were able to move the steering column response to the engine's speed-based excitation frequencies well into its filter band, thus attenuating the vibrations felt by the driver.

3.4 TRANSFER FUNCTIONS

Consider the schematic representation in Figure 3.20. The measurement system operates on the input signal, $F(t)$, by some function, $G(s)$, so as to indicate the output signal, $y(t)$. This operation can be explored by taking the Laplace transform of both sides of Equation 3.4, which describes the general first-order measurement system. One obtains

$$Y(s) = \frac{1}{\tau s + 1} KF(s) + \frac{y_0}{\tau s + 1}$$

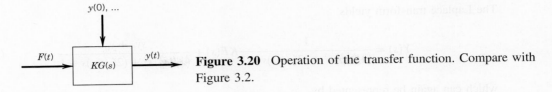

Figure 3.20 Operation of the transfer function. Compare with Figure 3.2.

where $y_0 = y(0)$. This can be rewritten as

$$Y(s) = G(s)KF(s) + G(s)Q(s) \qquad (3.24)$$

where $G(s)$ is the transfer function of the first-order system given by

$$G(s) = \frac{1}{\tau s + 1} \qquad (3.25)$$

and $Q(s)$ is the system initial state function. Because it includes $KF(s)$, the first term on the right side of Equation 3.24 contains the information that describes the steady response of the measurement system to the input signal. The second term describes its transient response. Included in both terms, the transfer function $G(s)$ plays a role in the complete time response of the measurement system. As indicated in Figure 3.20, the *transfer function* defines the mathematical operation that the measurement system performs on the input signal $F(t)$ to yield the time response (output signal) of the system. A table of common Laplace transforms with their application to the solution of ordinary differential equations can be found in Appendix B.

The system frequency response, which has been shown to be given by $M(\omega)$ and $\Phi(\omega)$, can be found by finding the value of $G(s)$ at $s = i\omega$. This yields the complex number

$$G(s = i\omega) = G(i\omega) = \frac{1}{\tau i\omega + 1} = M(\omega)e^{i\Phi(\omega)} \qquad (3.26)$$

where $G(i\omega)$ is a vector on the real–imaginary plane having a magnitude, $M(\omega)$, and inclined at an angle, $\Phi(\omega)$, relative to the real axis as indicated in Figure 3.21. For the first-order system, the magnitude of $G(i\omega)$ is simply that given by $M(\omega)$ from Equation 3.10 and the phase shift angle by $\Phi(\omega)$ from Equation 3.9.

For a second-order or higher system, the approach is the same. The governing equation for a second-order system is defined by Equation 3.13, with initial conditions $y(0) = y_0$ and $\dot{y}(0) = \dot{y}_0$.

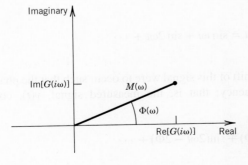

Figure 3.21 Complex plane approach to describing frequency response.

The Laplace transform yields

$$Y(s) = \frac{1}{(1/\omega_n^2)s^2 + (2\zeta/\omega_n)s + 1} KF(s) + \frac{s\dot{y}_0 + y_0}{(1/\omega_n^2)s^2 + (2\zeta/\omega_n)s + 1} \tag{3.27}$$

which can again be represented by

$$Y(s) = G(s)KF(s) + G(s)Q(s) \tag{3.28}$$

By inspection, the transfer function is given by

$$G(s) = \frac{1}{(1/\omega_n^2)s^2 + (2\zeta/\omega_n)s + 1} \tag{3.29}$$

Solving for $G(s)$ at $s = i\omega$, we obtain for a second-order system

$$G(s = i\omega) = \frac{1}{(i\omega)^2/\omega_n^2 + 2\zeta i\omega/\omega_n + 1} = M(\omega)e^{i\Phi(\omega)} \tag{3.30}$$

which gives exactly the same magnitude ratio and phase shift relations as given by Equations 3.20 and 3.22.

3.5   PHASE LINEARITY

We can see from Figures 3.16 and 3.17 that systems having a damping ratio near 0.7 possess the broadest frequency range over which $M(\omega)$ will remain at or near unity and that over this same frequency range the phase shift essentially varies in a linear manner with frequency. Although it is not possible to design a measurement system without accepting some amount of phase shift, it is desirable to design a system such that the phase shift varies linearly with frequency. This is because a nonlinear phase shift is accompanied by a significant distortion in the waveform of the output signal. *Distortion* refers to a notable change in the shape of the waveform from the original, as opposed to simply an amplitude alteration or relative time delay. To minimize distortion, many measurement systems are designed with $0.6 \leq \zeta \leq 0.8$.

Signal distortion can be illustrated by considering a particular complex waveform represented by a general function, $u(t)$:

$$u(t) = \sum_{n=1}^{\infty} \sin n\omega t = \sin \omega t + \sin 2\omega t + \cdots \tag{3.31}$$

Suppose during a measurement a phase shift of this signal were to occur such that the phase shift remained linearly proportional to the frequency; that is, the measured signal, $v(t)$, could be represented by

$$v(t) = \sin(\omega t - \Phi) + \sin(2\omega t - 2\Phi) + \cdots \tag{3.32}$$

Or, by setting

$$\theta = (\omega t - \Phi) \tag{3.33}$$

we write

$$v(t) = \sin \theta + \sin 2\theta + \cdots \tag{3.34}$$

We see that $v(t)$ in Equation 3.22 is equivalent to the original signal, $u(t)$. If the phase shift were not linearly related to the frequency, this would not be so. This is demonstrated in Example 3.12.

Example 3.12

Consider the effect of the variations in phase shift with frequency on a measured signal by examination of the signal defined by the function

$$u(t) = \sin t + \sin 5t$$

Suppose this signal is measured in such a way that a phase shift that is linearly proportional to the frequency occurs in the form

$$v(t) = \sin(t - 0.35) + \sin[5t - 5(0.35)]$$

Both $u(t)$ and $v(t)$ are plotted in Figure 3.22. We can see that the two waveforms are identical except that $v(t)$ lags $u(t)$ by some time increment.

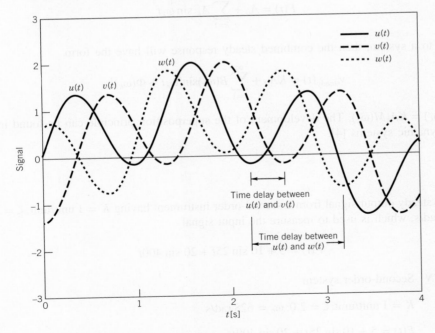

Figure 3.22 Waveforms for Example 3.12.

Now suppose this signal is measured in such a way that the relation between phase shift and frequency was nonlinear, such as in the signal output form

$$w(t) = \sin(t - 0.35) + \sin(5t - 5)$$

The $w(t)$ signal is also plotted in Figure 3.22. It behaves differently from $u(t)$, and this difference is the signal distortion. In comparing $u(t)$, $v(t)$, and $w(t)$, it is apparent that a nonlinear relation between phase shift and frequency brings about signal distortion. This is an undesirable result.

3.6   MULTIPLE-FUNCTION INPUTS

So far we have discussed measurement system response to a signal containing only a single frequency. What about measurement system response to multiple input frequencies? Or to an input that consists of both a static and a dynamic part, such as a periodic strain signal from an oscillating beam? When using models that are linear, such as ordinary differential equations subjected to inputs that are linear in terms of the dependent variable, the principle of superposition of linear systems applies in the solution of these equations. *The principle of superposition* states that a linear combination of input signals applied to a linear measurement system produces an output signal that is simply the linear addition of the separate output signals that would result if each input term had been applied separately. Because the form of the transient response is not affected by the input function, we can focus on the steady response. In general, we can write that if the forcing function of a form

$$F(t) = A_0 + \sum_{k=1}^{\infty} A_k \sin \omega_k t \tag{3.35}$$

is applied to a system, then the combined steady response will have the form

$$y_{\text{steady}}(t) = KA_0 + \sum_{k=1}^{\infty} B(\omega_k)\sin[\omega_k t + \Phi(\omega_k)] \tag{3.36}$$

where $B(\omega_k) = KA_kM(\omega_k)$. The development of the superposition principle can be found in basic texts on dynamic systems [4].

Example 3.13

Predict the steady output signal from a second-order instrument having $K = 1$ unit/unit, $\zeta = 2$, and $\omega_n = 628$ rad/s, which is used to measure the input signal

$$F(t) = 5 + 10 \sin 25t + 20 \sin 400t$$

KNOWN   Second-order system

$K = 1$ unit/unit; $\zeta = 2.0$; $\omega_n = 628$ rad/s

$F(t) = 5 + 10 \sin 25t + 20 \sin 400t$

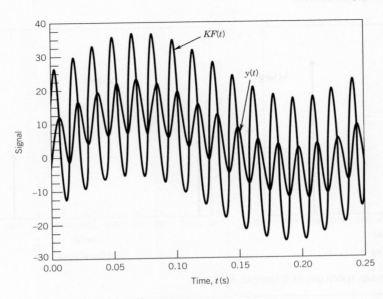

Figure 3.23   Input and output signals for Example 3.13.

ASSUMPTIONS   Linear system (superposition holds)

FIND $y(t)$

SOLUTION   Because $F(t)$ has a form consisting of a linear addition of multiple input functions, the steady response signal will have the form of Equation 3.33 of $y_{steady}(t) = KF(t)$ or

$$y(t) = 5K + 10\,KM(25\text{ rad/s})\sin[25t + \Phi(25\text{ rad/s})] + 20\,KM(400\text{ rad/s})\sin[400t + \Phi(400\text{ rad/s})]$$

Using Equations 3.20 and 3.22, or, alternatively, using Figures 3.16 and 3.17, with $\omega_n = 628$ rad/s and $\zeta = 2.0$, we calculate

$$M(25\text{ rad/s}) = 0.99 \qquad \Phi(25\text{ rad/s}) = -9.1°$$
$$M(400\text{ rad/s}) = 0.39 \qquad \Phi(400\text{ rad/s}) = -77°$$

so that the steady output signal will have the form

$$y(t) = 5 + 9.9\sin(25t - 9.1°) + 7.8\sin(400t - 77°)$$

The output signal is plotted against the input signal in Figure 3.23. The amplitude spectra for both the input signal and the output signal are shown in Figure 3.24. Spectra can also be generated from signals by using the accompanying Matlab programs *FunSpect* and, for your custom data files, *DataSpect*. The magnitude ratio-frequency response can be found by plotting the ratio of the output to input magnitudes.

COMMENT   The static magnitude and the amplitude of the 25 rad/s component of the input signal are passed along to the output signal closely. However, the amplitude of the 400-rad/s component is severely reduced (down 61%). This is a filtering effect due to the measurement system frequency response.

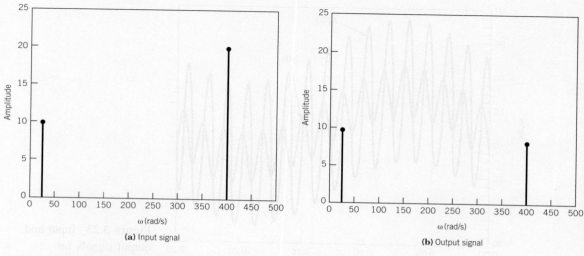

Figure 3.24 Amplitude spectrum of Example 3.13.

3.7   COUPLED SYSTEMS

As instruments in each stage of a measurement system are connected (transducer, signal conditioner, output device, etc.), the output from one stage becomes the input to the next stage to which it is connected, and so forth. The overall measurement system will have a coupled output response to the original input signal that is a combination of each individual response to the input. However, the system concepts of zero-, first-, and second-order systems studied previously can still be used for a case-by-case study of the coupled measurement system.

This concept is easily illustrated by considering a first-order sensor – transducer that may be connected to a second-order recording device. Suppose the input to the sensor is a simple periodic waveform, $F_1(t) = A \sin \omega t$. The transducer responds with an output signal of the form of Equation 3.8:

$$y_t(t) = Ce^{-t/\tau} + \frac{K_t A}{\sqrt{1 + (\omega\tau)^2}} \sin(\omega t + \Phi_t)$$

$$\Phi_t = -\tan^{-1}\omega\tau$$

(3.37)

where the subscript t refers to the transducer. However, the sensor-transducer output signal now becomes the input signal, $F_2(t) = y_t$, to the second-order recording device. The output from the second-order device, $y_s(t)$, will be a second-order response appropriate for input $F_2(t)$,

$$y_s(t) = y_h(t) + \frac{K_t K_s A \sin\left[\omega t + \Phi_t + \Phi_s\right]}{[1 + (\omega\tau^2)]^{1/2} \left\{\left[1 - (\omega/\omega_n)^2\right]^2 + [2\zeta\omega/\omega_n]^2\right\}^{1/2}}$$

$$\Phi_s = -\tan^{-1}\frac{2\zeta\omega/\omega_n}{1 - (\omega/\omega_n)^2}$$

(3.38)

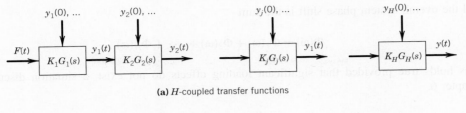

(a) *H*-coupled transfer functions

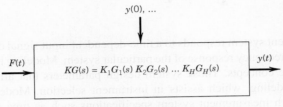

(b) Equivalent system transfer function

Figure 3.25 Coupled systems: describing the system transfer function.

where subscript s refers to the signal recording device and $y_h(t)$ is the transient response. The output signal displayed on the recorder, $y_s(t)$, is the measurement system response to the original input signal to the transducer, $F_1(t) = A \sin \omega t$. The steady amplitude of the recording device output signal is the product of the static sensitivities and magnitude ratios of the first- and second-order systems. The phase shift is the sum of the phase shifts of the two systems.

Based on the concept behind Equation 3.28, we can make a general observation. Consider the schematic representation in Figure 3.25, which depicts a measurement system consisting of H interconnected devices, $j = 1, 2, \ldots, H$, each device described by a linear system model. The overall transfer function of the combined system, $G(s)$, is the product of the transfer functions of each of the individual devices, $G_j(s)$, such that

$$KG(s) = K_1 G_1(s) K_2 G_2(s) \ldots K_H G_H(s) \tag{3.39}$$

At $s = i\omega$, Equation 3.37 becomes

$$KG(i\omega) = (K_1 K_2 \ldots K_H) \times [M_1(\omega) M_2(\omega) \ldots M_H(\omega)] e^{i[\Phi_1(\omega) + \Phi_2(\omega) + \cdots + \Phi_H(\omega)]} \tag{3.40}$$

According to Equation 3.40, given an input signal to device 1, the system steady output signal at device H will be described by the system frequency response $G(i\omega) = M(\omega) e^{i\Phi(\omega)}$, with an overall system static sensitivity described by

$$K = K_1 K_2 \ldots K_H \tag{3.41}$$

The overall system magnitude ratio is the product

$$M(\omega) = M_1(\omega) M_2(\omega) \ldots M_H(\omega) \tag{3.42}$$

and the overall system phase shift is the sum

$$\Phi(\omega) = \Phi_1(\omega) + \Phi_2(\omega) + \cdots + \Phi_H(\omega) \tag{3.43}$$

This holds true provided that significant loading effects do not exist, a situation discussed in Chapter 6.

3.8 SUMMARY

Just how a measurement system responds to a time-dependent input signal depends on the properties of the signal and the frequency response of the particular system. Modeling has enabled us to develop and to illustrate these concepts. Those system design parameters that affect system response are exposed through modeling, which assists in instrument selection. Modeling has also suggested the methods by which measurement system specifications such as time constant, response time, frequency response, damping ratio, and resonance frequency can be determined both analytically and experimentally. Interpretation of these system properties and their effect on system performance was determined.

The response rate of a system to a change in input is estimated by use of the step function input. The system parameters of time constant, for first-order systems, and natural frequency and damping ratio, for second-order systems, are used as indicators of system response rate. The frequency response of a system, through the magnitude ratio and phase shift, and are found by applying an input signal of a periodic waveform. Figures 3.12, 3.13, 3.16, and 3.17 are universal frequency response curves for first- and second-order systems. These curves can be found in most engineering and mathematical handbooks and can be applied to any first- or second-order system, as appropriate.

REFERENCES

1. Close, C. M., D. K. Frederick, J. Newell, *Modeling and Analysis of Dynamic Systems*, 3rd ed., Wiley, Boston, 2001.
2. Doebelin, E. O., *Measurement Systems: Applications and Design*, McGraw-Hill, New York, 2003.
3. Ogata, K., *System Dynamics*, 4th ed., Prentice-Hall, Englewood Cliffs, NJ, 2003.
4. Palm, W. J., III, *Modeling, Analysis and Control of Dynamic Systems*, 2nd ed., Wiley, New York, 2000.
5. Burgess, J.C., A quick estimation of damping from free damped oscillograms, Wright Air Development Center, WADC TR 59–676, March 1961, pp. 457–460.

NOMENCLATURE

$a_0, a_1, \ldots, a_n$	physical coefficients	f	cyclical frequency ($f = \omega/2\pi$) (Hz)
$b_0, b_1, \ldots, b_m$	physical coefficients	k	spring constant or stiffness (mt^{-2})
c	damping coefficient	m	mass (m)
		$p(t)$	pressure $(ml^{-1}t^{-2})$

t	time (t)	$T(t)$	temperature $(°)$
$x(t)$	independent variable	T_d	ringing period (t)
$y(t)$	dependent variable	$U(t)$	unit step function
y^n	nth time derivative of $y(t)$	β_1	time lag (t)
$y^n(0)$	initial condition of y^n	$\delta(\omega)$	dynamic error
A	input signal amplitude	τ	time constant (t)
$B(\omega)$	output signal amplitude	$\Phi(\omega)$	phase shift
C	constant	ω	signal (circular) frequency (t^{-1})
$E(t)$	voltage (V) or energy	ω_n	natural frequency magnitude (t^{-1})
F	force $(m\,l\,t^{-2})$	ω_d	ringing frequency (t^{-1})
$F(t)$	forcing function	ω_R	resonance frequency (t^{-1})
$G(s)$	transfer function	ζ	damping ratio
K	static sensitivity	Γ	error fraction
$M(\omega)$	magnitude ratio, B/KA		

Subscripts

0	initial value
∞	final or steady value
h	homogeneous solution

PROBLEMS

Note: Although not required, the companion software can be used for solving many of these problems. We encourage the reader to explore the software provided.

3.1 An electronic scale has a static sensitivity of 2 V/kg. A voltmeter is used to display the measurement. For an input range of 1 to 10 kg, what range of voltmeter is required. What is the significance of increasing the static sensitivity?

3.2 Determine the 75%, 90%, and 95% response time for each of the systems given (assume zero initial conditions):

   a. $0.4\dot{T} + T = 4U(t)$

   b. $\ddot{y} + 2\dot{y} + 4y = U(t)$

   c. $2\ddot{P} + 8\dot{P} + 8P = 2U(t)$

   d. $5\dot{y} + 5y = U(t)$

3.3 A special sensor is designed to sense the percent vapor present in a liquid–vapor mixture. If during a static calibration the sensor indicates 80 units when in contact with 100% liquid, 0 units with 100% vapor, and 40 units with a 50:50% mixture, determine the static sensitivity of the sensor.

3.4 A measurement system can be modeled by the equation

$$0.5\dot{y} + y = F(t).$$

Initially, the output signal is steady at 75 volts. The input signal is then suddenly increased to 100 volts.

   a. Determine the response equation.

   b. On the same graph, plot both the input signal and the system time response from $t = 0$ s through steady response.

3.5 A thermal sensor having a time constant of 1 s is used to record a step change of 0 °C to 49 °C in process temperature. Show the sensor time response, $T(t)$. Determine the 90% rise time.

3.6 A student establishes the time constant of a temperature sensor by first holding it immersed in hot water and then suddenly removing it and holding it immersed in cold water. Several other students perform the same test with similar sensors. Overall, their results are inconsistent, with estimated time constants differing by as much as a factor of 1.2. Offer suggestions about why this might happen. Hint: Try this yourself and think about control of test conditions.

3.7   Which would you expect to be better suited to measure a time-dependent temperature, a thermal sensor (e.g., a thermocouple) having a small diameter spherical bead or one having a large diameter spherical bead? Why?

3.8   A thermistor is a thermal sensor made of sintered semiconductor material that shows a large change in resistance for a small temperature change. Suppose one thermistor has a calibration curve given by $R(T) = 0.5e^{-10/T} \Omega$ where T is absolute temperature. What is the static sensitivity $[\Omega/°]$ at (i) 283 K, (ii) 350 K?

3.9   An RC low-pass filter responds as a first-order instrument. The time constant is given by the product RC. The time constant for this filter is 2 ms. Determine the time required to reach 99.3% of its steady value.

3.10   During a step function calibration, a first-order instrument is exposed to a step change of 100 units. If after 1.2 s the instrument indicates 80 units, estimate the instrument time constant. Estimate the error in the indicated value after 1.5 s. $y(0) = 0$ units; $K = 1$ unit/unit.

3.11   A lumped parameter model of the human systemic circulation includes the systemic vascular resistance and systemic vascular compliance. Compliance is the inverse of mechanical stiffness and is estimated by $C = \Delta \forall / \Delta p$, that is, the compliance is related to the pressure change required to accommodate a volume change. The typical stroke volume of each heart beat might eject 80 mL of blood into the artery during which the end-diastolic pressure of 80 mm Hg rises to a peak systolic pressure of 120 mm Hg. Similarly, the resistance to the 6 L/min of blood flow requires an average pressure over the pulse cycle of about 100 mm Hg to drive it. Estimate lumped parameter values for the systemic vascular compliance and resistance. Incidentally, these same relations are used in describing the physical properties of pressure measuring systems.

3.12   A signal expected to be of the form $F(t) = 10 \sin 15.7t$ is to be measured with a first-order instrument having a time constant of 50 ms. Write the expected indicated steady response output signal. Is this instrument a good choice for this measurement? What is the expected time lag between input and output signal? Plot the output amplitude spectrum. Use $y(0) = 0$ and $K = 1$ V/V.

3.13   A first-order instrument with a time constant of 2 s is to be used to measure a periodic input signal. If a dynamic error of ±2% can be tolerated, determine the maximum frequency of periodic input that can be measured. What is the associated time lag (in seconds) at this frequency?

3.14   A process temperature is known to fluctuate with a frequency of 0.1 Hz. The engineer needs to monitor this process using a thermocouple and wishes to have no more than 2% amplitude reduction. What time constant is needed?

3.15   Steady state response refers to the response of the measurement system after a harmonic input has been applied for a period of time and the transients have decayed to zero. Determine the length of time that a first-order system needs to achieve a steady response (to within 1% or better).

3.16   Determine the frequency response $[M(\omega)$ and $\phi(\omega)]$ for an instrument having a time constant of 10 ms. Estimate the instrument's usable frequency range to keep its dynamic error within 10%.

3.17   A temperature measuring device with a time constant of 0.15 s outputs a voltage that is linearly proportional to temperature. The device is used to measure an input signal of the form $T(t) = 115 + 12 \sin 2t$ °C. Plot the input signal and the predicted output signal with time, assuming first-order behavior and a static sensitivity of 5 mV/°C. Determine the dynamic error and time lag in the steady response. $T(0) = 115$ °C.

3.18   A first-order sensor is to be installed into a reactor vessel to monitor temperature. If a sudden rise in temperature greater than 100 °C should occur, shutdown of the reactor will need to begin within 5 s after reaching 100 °C. Determine the maximum allowable time constant for the sensor.

3.19   During a dynamic calibration, a system indicates $y(t) = 3 \sin 200t$ in mV when a signal of $F(t) = 4 \sin 200t$ mV is applied. With $K = 1$ mV/mV, what is the dynamic error (%) of the system at 200 rad/s?

3.20   The output from a temperature system indicates a steady, time-varying signal having an amplitude that varies between 30 °C and 40 °C with a single frequency of 10 Hz. Express the output signal as a waveform equation, $y(t)$. If the dynamic error is to be less than 1%, what must be the system's time constant?

3.21   A signal of frequency $1/\pi$ Hz is passed through a first-order low-pass RC filter having a time constant of $RC = 5$ s. What is the expected phase lag?

3.22   Determine the frequency bandwidth in hertz, the low cut-off frequency, and the high cut-off frequency of a first-order thermal sensor having a time constant of 0.10 s when subjected to the sinusoidal temperature variation, $T(t) = 20 \sin \omega t$ °C.

3.23   A measuring system has a natural frequency of 0.5 rad/s, a damping ratio of 0.5, and a static sensitivity of 0.5 m/V. Estimate its 90% rise time and settling time if $F(t) = 2 U(t)$ and the initial condition is zero. Plot the response $y(t)$ and indicate its transient and steady responses.

3.24   Plot the frequency response $M(\omega)$ and $\Phi(\omega)$ versus ω/ω_n for an instrument having a damping ratio of 0.6. Determine the frequency range over which the dynamic error remains within 5%. Repeat for a damping ratio of 0.9 and 2.0.

3.25   The 98% settling time is the time required for a second-order system to respond to within 2% of the steady response. An approximate formula for calculating this time is $t_{98} = 4/\zeta\omega_n$. Estimate t_{98} for $\zeta = 0.707$ and $\omega_n = 1$ rad/s and verify the adequacy of the approximation.

3.26   A cantilever beam instrumented with strain gauges is used as a force scale. A step-test on the beam provides a measured damped oscillatory signal with time. If the signal behavior is of second-order, show how a data-reduction design plan could use the information in this signal to determine the natural frequency and damping ratio of the cantilever beam. (Hint: Consider the shape of the decay of the peak values in the oscillation.)

3.27   A step test of a transducer shows a signal having a damped oscillation decaying to a steady value. If the period of oscillation is 0.577 ms, what is the transducer ringing frequency?

3.28   An input signal oscillates sinusoidally between 12 and 24 V with a frequency of 120 Hz. It is measured with an instrument having a damping ratio of 0.7, ringing frequency of 1000 Hz, and static sensitivity of 1 V/V. Determine and plot the amplitude spectrum of the output signal at steady response.

3.29   An application demands that a sinusoidal pressure variation of 250 Hz be measured with no more than 2% dynamic error. In selecting a suitable pressure transducer from a vendor catalog, you note that a desirable line of transducers has a fixed natural frequency of 600 Hz but that you have a choice of transducer damping ratios of between 0.5 and 1.5 in increments of 0.05. Select a suitable transducer.

3.30   A computer housing is to be isolated from room vibrations by placing it on an isolation pad. Using lumped parameters, the isolation pad can be considered as a board of mass m and a foam mat of stiffness k and damping coefficient c, where stiffness and damping act in parallel. For expected vibrations in the frequency range of between 2 and 40 Hz, select reasonable values for $m, k,$ and c such that the room vibrations are attenuated by at least 50%. Assume that the only degree of freedom is in the vertical direction.

3.31   A single-loop RCL electrical circuit can be modeled as a second-order system in terms of current. Show that the differential equation for such a circuit subjected to a forcing function potential $E(t)$ is given by

$$L\frac{d^2I}{dt^2} + R\frac{dI}{dt} + \frac{I}{C} = \dot{E}(t)$$

Determine the natural frequency and damping ratio for this system. For a forcing potential, $E(t) = 1 + 0.5 \sin 2000t$ V, determine the system steady response when $L = 2 H, C = 1\mu F,$ and $R = 10,000 \Omega$. Plot the steady output signal and input signal versus time. $I(0) = \dot{I}(0) = 0$.

3.32   A transducer that behaves as a second-order instrument has a damping ratio of 0.7 and a natural frequency of 1000 Hz. It is to be used to measure a signal containing frequencies as large as 750 Hz. If a dynamic error of $\pm 10\%$ can be tolerated, show whether or not this transducer is a good choice.

3.33   A strain-gauge measurement system is mounted on an airplane wing to measure wing oscillation and strain during wind gusts. The strain system has a 90% rise time of 100 ms, a ringing frequency of 1200 Hz, and a damping ratio of 0.8. Estimate the dynamic error in measuring a 1-Hz oscillation. Also, estimate any time lag. Explain in words the meaning of this information.

3.34   A measuring system has a known natural frequency of 2000 Hz and damping ratio of 0.8. Would it be satisfactory to use it to measure a signal expected to have frequency content at up to 500 Hz? Demonstrate your answer by finding the expected dynamic error and phase shift at 500 Hz. What is the system frequency bandwidth?

3.35   Select one set of appropriate values for damping ratio and natural frequency for a second-order instrument used to measure frequencies up to 100 rad/s with no more than $\pm 10\%$ dynamic error. A catalog offers models with damping ratios of 0.4, 1, and 2 and natural frequencies of 200 and 500 rad/s. Explain your reasoning.

3.36   A first-order measurement system with time constant of 25 ms and $K = 1$ V/N measures a signal of the form $F(t) = \sin 2\pi t + 0.8 \sin 6\pi t$ N. Determine the steady response (steady output signal) from the system. Plot the output signal amplitude spectrum. Discuss the transfer of information from the input to the output. Can the input signal be resolved based on the output?

3.37   Demonstrate for the second-order system ($\omega_n = 100$ rad/s, $\zeta = 0.4$) subjected to step function input, $U(t)$, that the damping ratio and natural frequency can be found from the logarithmic amplitude decay. Show that this is possible whether $F(t) = KAU(t)$ with $y(0) = 0$ or $F(t) = -KAU(t)$ with $y(0) = KA$. Use $K = 1$ mV/mV and $A = 600$ mV. Why would this technique be useful in practice?

3.38   A force transducer system is needed to measure a sinusoidal signal. Derived from the logarithmic amplitude decay test, the system has a damping ratio of 0.6 and natural frequency of 1500 Hz. Estimate the frequency bandwidth of this system and plot $M(f)$. What is the upper cut-off frequency?

3.39   An accelerometer has a damping ratio of 0.4 and a natural frequency of 18, 000 Hz. It is used to sense the relative displacement of a beam to which it is attached. If an impact to the beam imparts a vibration at 4500 Hz, estimate the dynamic error and phase shift in the accelerometer output. Estimate its resonance frequency.

3.40   Derive the equations for the magnitude ratio and phase shift of the seismic accelerometer of Example 3.1. This response differs from the classic second-order response described in this chapter. For what type of measurement would you suppose this instrument would be best suited?

3.41   A measuring system has a frequency response defined by $M(\omega)$. The system has a damping ratio of 0.6 and natural frequency of 1385 Hz. Show how $M(\omega)$ varies with frequency. At which frequency does $M(\omega)$ reach a maximum?

3.42   A pressure transducer is attached to a stiff-walled catheter, an open-ended long, small diameter tube. The catheter is filled with saline from a small balloon attached at its tip. The initial system pressure is 50 mm Hg. At $t = 0$ s, the balloon is popped, forcing a step function change in pressure from 50 to 0 mm Hg. The time-based signal is recorded and the ringing period determined to be 0.03 s. Find the natural frequency and damping ratio attributed to the pressure-catheter system; $K = 1$ mV/mm Hg; $y_1(0) = 50.0$; $y_2(0.03) = 2.108$.

3.43   A catheter is inserted into the vena cava of an animal to measure the venous blood pressure variation with breathing. The other end of the catheter is attached to a pressure transducer and conditioning system to record the measured pressure. The catheter has a known natural frequency and damping ratio of 4 rad/s and 0.2, respectively, while these values for the pressure transducer are 2000 rad/s and 0.7. The venous pressure is expected to vary as $p(t) = 10 + 1.5 \sin 2t$ in mm Hg. Find the time-dependent magnitude attenuation and time delay of the output pressure signal. Assume $K_{system} = 1$ mm Hg/mm Hg and linear behavior of the system.

3.44   The output stage of a first-order transducer is to be connected to a second-order display stage device. The transducer has a known time constant of 1.4 ms and static sensitivity of 2 V/°C while the display device has values of sensitivity, damping ratio, and natural frequency of 1 V/V, 0.9, and 5000 Hz, respectively. Determine the steady response of this measurement system to an input signal of the form $T(t) = 10 + 50 \sin 628t$ °C.

3.45   The displacement of a solid body is to be monitored by a transducer (second-order system) with signal output displayed on a recorder (second-order system). The displacement is expected to vary sinusoidally between 2 and 5 mm at a rate of

85 Hz. Select appropriate design specifications for the measurement system for no more than 5% dynamic error (i.e., specify an acceptable range for natural frequency and damping ratio for each device).

3.46 The input signal

$$F(t) = 2 + \sin 15.7t + \sin 160t \ \text{N}$$

is measured by a force measurement system. The system has a known $\omega_n = 100$ rad/s, $\zeta = 0.4$, and $K = 1$ V/N. Write the expected form of the steady output signal in volts. Plot the resulting amplitude spectrum.

3.47 A signal suspected to be of the nominal form

$$y(t) = 5 \sin 1000t \ \text{mV}$$

is measured by a first-order instrument having a time constant of 100 ms and $K = 1$ V/V. It is then to be passed through a second-order amplifier having a $K = 100$ V/V, a natural frequency of 15,000 Hz, and a damping ratio of 0.8. What is the expected form of the output signal, $y(t)$? Estimate the dynamic error and phase lag in the output. Is this system a good choice here? If not, do you have any suggestions?

3.48 A typical modern DC audio amplifier has a frequency bandwidth of 0 to 20,000 Hz ± 1 dB. Research this and explain the meaning of this specification and its relation to music reproduction.

3.49 The displacement of a rail vehicle chassis as it rolls down a track is measured using a transducer ($K = 10$ mV/mm, $\omega_n = 10\,000$ rad/s, $\zeta = 0.6$) and a recorder ($K = 1$ mm/mV, $\omega_n = 700$ rad/s, $\zeta = 0.7$). The resulting amplitude spectrum of the output signal consists of a spike of 90 mm at 2 Hz and a spike of 50 mm at 40 Hz. Are the measurement system specifications suitable for the displacement signal? (If not, suggest changes.) Estimate the actual displacement of the chassis. State any assumptions.

3.50 The amplitude spectrum of the time-varying displacement signal from a vibrating U-shaped tube used in a Coriolis mass flow meter is expected to have a spike at each of 85, 147, 220, and 452 Hz. Each spike is related to an important vibrational mode of the tube. Displacement transducers available for this application have a range of natural

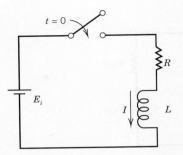

Figure 3.26 Figure for Problem 3.53.

frequencies from 500 to 1000 Hz, with a fixed damping ratio of about 0.5. The transducer output is to be monitored on a spectrum analyzer that has a frequency bandwidth extending from 0.1 Hz to 250 kHz. Within the stated range of availability, select a suitable transducer for this measurement from the given range.

3.51 A sensor mounted to a cantilever beam indicates beam motion with time. When the beam is deflected and released (step test), the sensor signal indicates that the beam oscillates as an underdamped second-order system with a ringing frequency of 10 rad/s. The maximum displacement amplitudes are measured at three different times corresponding to the 1st, 16th, and 32nd cycles and found to be 17, 9, and 5 mV, respectively. Estimate the damping ratio and natural frequency of the beam based on this measured signal, $K = 1$ mm/mV.

3.52 Burgess (5) reports that the damping ratio can be approximated from the system response to an impulse test by counting the number of cycles, n, required for the ringing amplitudes to fall to within 1% of steady state by $\zeta = 4.6/2\pi n$. The estimate is further improved if n is a noninteger. Investigate the quality of this estimate for a second-order system subjected instead to a step function input. Discuss your findings.

3.53 The starting transient of a DC motor can be modeled as an RL circuit, with the resistor and inductor in series (Figure 3.26). Let $R = 4\,\Omega$, $L = 0.1$ H, and $E_i = 50$ V with $t(0) = 0$. Find the current draw with time for $t > 0^+$.

3.54 A camera flash light is driven by the energy stored in a capacitor. Suppose the flash operates off a 6-V

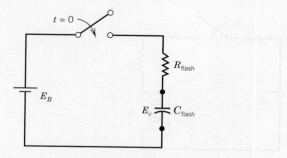

Figure 3.27 Figure for Problem 3.54.

battery. Determine the time required for the capacitor stored voltage to reach 90% of its maximum energy $\left(=\frac{1}{2}CE_B^2\right)$. Model this as an RC circuit (Fig. 3.27). For the flash: $C = 1000\ \mu F$, $R = 1\ k\Omega$, and $E_c(0) = 0$.

3.55 Run program *Temperature Response.vi*. The program allows the user to vary time constant τ and signal amplitude. To impose a step change, move the input value up/down. Discuss the effect of time constant on the time required to achieve a steady signal subsequent to a new step change.

Chapter 4

Probability and Statistics

4.1 INTRODUCTION

Suppose we had a large box containing a population of thousands of similar-sized round bearings. To get an idea of the size of the bearings, we might measure two dozen from these thousands. The resulting sample of diameter values forms a data set, which we then use to imply something about the entire population of the bearings in the box, such as the average diameter size and size variation. But how close are these values taken from our small data set to estimating the actual average size and variation of all the bearings in the box? If we selected a different two dozen bearings, should we expect the size values to be exactly the same? These are questions that surround most engineering measurements. We make some measurements that we then use to infer things about the variable measured. What is the mean value of the variable based on a finite number of measurements and how well does this value represent the entire population? Do the variations found ensure that we can meet tolerances for the population as a whole? How good are these results? These questions have answers in probability and statistics.

For a given set of measurements, we want to be able to quantify (1) a single representative value that best characterizes the average of the measured data set, (2) a representative value that provides a measure of the variation in the measured data set, and (3) how well the average of the data set represents the true average value of the entire population of the measured variable. The value of (1) can vary with repeated data sets, so the difference between it and the true average value of the population is a type of random error. Additionally, (3) requires establishing an interval within which the true average value of the population is expected to lie. This interval quantifies the probable range of this random error and is called a random uncertainty.

This chapter presents an introduction to the concepts of probability and statistics at a level sufficient to provide information for a large class of engineering judgments. Such material allows for the reduction of raw data into results. We use terminology common to engineering, and it is not meant to replace a sound study of inferential statistics. Upon completion of this chapter, the reader will be able to

- Quantify the statistical characteristics of a data set as it relates to the population of the measured variable
- Explain and use probability density functions to describe the behavior of variables
- Quantify a confidence interval about the measured mean value at a given probability and to apply hypothesis tests about the behavior of engineering measurements

- Perform regression analysis on a data set and quantify the confidence intervals for the parameters of the resulting curve fit or response surface
- Execute a Monte Carlo simulation that predicts the behavior expected in a result due to variations in the variables involved in computing that result

4.2 STATISTICAL MEASUREMENT THEORY

Sampling refers to obtaining a set of data through repeated measurements of a variable under fixed operating conditions. This variable is known as the *measured variable* or, in statistical terms, the *measurand*. We take the sampling from the population of all possible values of the measured variable. In engineering practice, truly fixed operating conditions are often difficult to maintain, so we use the term to mean that conditions are maintained as well as possible and deviations from these conditions will show up as variations in the data set.

This chapter considers the effects of random errors and how to quantify them. Recall that random errors are manifested through data scatter and by the statistical limitations of a finite sampling to predict the behavior of a population. In this chapter, we will assume that any systematic errors in the measurement are negligible.[1] Recall from Chapter 1 that this is the case when the average error in a data set is zero. We begin by considering the following measurement problem: Estimate the true value x' based on the information derived from the repeated measurement of variable x. In the absence of systematic error, the true value of x is the mean value of all possible measured values of x. This is the value we want to estimate from the measurement. A sampling of the variable x under controlled, fixed operating conditions renders a sample, a finite number of data points. We use these limited data to infer x' based on the calculated sample mean value, $\bar{x}$. We can imagine that if the number of data points, N, is very small, then the estimation of x' from the data set could be heavily influenced by the value of any one data point. If the data set were larger, then the influence of any one data point would be offset by the larger influence of the other data. As $N \to \infty$ or toward the total number in the population, all the possible variations in x become included in the data set. From a practical point of view, finite-sized data sets are common, in which case the measured data can provide only an estimate of the true value.

From a statistical analysis of the data set and an analysis of sources of error that influence these data, we can estimate x' as

$$x' = \bar{x} \pm u_{\bar{x}} \quad (P\%) \tag{4.1}$$

where $\bar{x}$ represents the most probable estimate of x' based on the available data and $\pm u_{\bar{x}}$ represents the *uncertainty interval* in that estimate at some probability level, $P\%$. *Uncertainties are numbers that quantify the possible range of the effects of the errors.* The uncertainty interval is the range about $\bar{x}$ within which we expect x' to lie. It combines the uncertainty estimates of the random error and of systematic error in the measurement of x.[2] This chapter discusses ways to estimate the uncertainty arising from the effects of random error, called the random uncertainty. Chapter 5 discusses systematic errors and how to combine the uncertainties of both random and systematic errors into the uncertainty interval of Equation 4.1.

---

[1] Systematic error does not vary with repeated measurements and thus does not affect the statistics of a measurement. Systematic errors are discussed in Chapter 5.

[2] Texts that ignore systematic (bias) error refer to this uncertainty interval as a "confidence interval."

Probability Density Functions

Random scatter of the measured data occurs regardless of the care taken to obtain the set from independent measurements under identical conditions. Thus the measured variable behaves as a random variable. A *random variable* is one whose value is subject to the effects of random chance. A continuous random variable is one that is continuous in time or space, such as the value of a motor's speed. A discrete random variable is one that is composed of discrete values, such as the values of the diameters for each of the individual bearings mentioned in Section 4.1. During repeated measurements of a variable, each data point may tend to assume one preferred value or lie within some interval about this value more often than not, when all the data are compared. This tendency toward one central value about which all the other values are scattered is known as a *central tendency* of a random variable.[3] Probability deals with the concept that certain values of a variable may be measured more frequently relative to other values.

The central value and those values scattered about it can be determined from the probability density of the measured variable. The frequency with which the measured variable assumes a particular value or interval of values is described by its probability density. Consider a sample of x shown in Table 4.1, which consists of N individual measurements, x_i, where $i = 2, \ldots, N$, each measurement taken at random but under identical test operating conditions. The measured values of this variable are plotted as data points along a single axis as shown in Figure 4.1.

In Figure 4.1, we can see a portion of the axis on which the data points tend to clump; this region contains the central value. Such behavior is typical of most engineering variables. We expect that the true mean value of the population of variable x is contained somewhere in this clump.

Table 4.1   Sample of Random Variable x

i	x_i	i	x_i
1	0.98	11	1.02
2	1.07	12	1.26
3	0.86	13	1.08
4	1.16	14	1.02
5	0.96	15	0.94
6	0.68	16	1.11
7	1.34	17	0.99
8	1.04	18	0.78
9	1.21	19	1.06
10	0.86	20	0.96

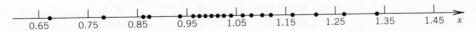

Figure 4.1   Concept of density in reference to a measured variable (from Ex. 4.1).

---

[3] Not all random variables display a central tendency; the value in a fair roll of a die, for example, would show values from 1 through 6, each with an equal frequency of 1/6. But the fair roll of two dice will show a central tendency to the combined value of 7. Try it!

This description for variable x can be extended. Suppose we plot the data of Table 4.1 in a different way. The abscissa will be divided between the maximum and minimum measured values of x into K small intervals. Let the number of times, n_j, that a measured value assumes a value within an interval defined by $x - \delta x \leq x \leq x + \delta x$ be plotted on the ordinate. The resulting plot of n_j versus x is called a *histogram* of the variable (1). The histogram is just another way of viewing both the tendency and the probability density of a variable. The ordinate can be nondimensionalized as $f_j = n_j/N$, converting the histogram into a *frequency distribution*. For small N, the interval number K should be conveniently chosen with a good rule that $n_j \geq 5$ for at least one interval. One suggestion for setting the minimum number of intervals K uses a correlation developed from Bendat and Piersol (2) as

$$K = 1.87(N - 1)^{0.40} + 1 \qquad (4.2)$$

As N becomes very large, a value of $K \approx N^{1/2}$ works reasonably well (1, 2). The concept of the histogram is illustrated in Example 4.1.

Example 4.1

Construct a histogram and frequency distribution for the data in Table 4.1.

KNOWN Data of Table 4.1

$N = 20$

ASSUMPTIONS Fixed operating conditions

FIND Histogram and frequency distribution

SOLUTION To develop the histogram, compute a reasonable number of intervals for this data set. For $N = 20$, a convenient estimate of K is found from Equation 4.2 to be

$$K = 1.87(N - 1)^{0.40} + 1 = 7$$

Next, determine the maximum and minimum values of the data set and divide this range into K intervals. For a minimum of 0.68 and a maximum of 1.34, a value of $\delta x = 0.05$ is chosen. The intervals are as follows:

j	Interval	n_j	$f_j = n_j/N$
1	$0.65 \leq x_i < 0.75$	1	0.05
2	$0.75 \leq x_i < 0.85$	1	0.05
3	$0.85 \leq x_i < 0.95$	3	0.15
4	$0.95 \leq x_i < 1.05$	7	0.35
5	$1.05 \leq x_i < 1.15$	4	0.20
6	$1.15 \leq x_i < 1.25$	2	0.10
7	$1.25 \leq x_i < 1.35$	2	0.10

The results are plotted in Figure 4.2. The plot displays a definite central tendency seen as the maximum frequency of occurrence falling within the interval 0.95 to 1.05.

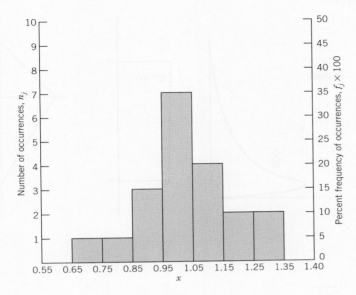

Figure 4.2 Histogram and frequency distribution for data in Table 4.1.

COMMENT   The total number of measurements, N, equals the sum of the number of occurrences,

$$N = \sum_{j=1}^{K} n_j$$

The area under the percent frequency distribution curve equals 100%; that is,

$$100 \times \sum_{j=1}^{K} f_j = 100\% \tag{4.3}$$

Probability-density.vi and *Running-histogram.vi* demonstrate the influence of population size and interval numbers on the histogram.

As $N \to \infty$, the probability density function, $p(x)$, of the population of variable x is developed. The *probability density function* is a statistical model of the population of a variable. It defines the probability that a variable might assume a particular value upon any individual measurement. It also describes the central tendency of the variable and its variation. This central tendency is the desired representative value that gives the best estimate of the true mean value.

The actual shape that the probability density function takes depends on the nature of the variable it represents and the circumstances surrounding the process in which the variable is involved. There are a number of standard distribution shapes that suggest how a variable could be distributed. Some common engineering examples are shown in Table 4.2. The specific values of the variable and the width of the distribution depend on the actual process, but the overall shape of the plot will most likely fit one of these. Often, experimentally determined

Table 4.2 Standard Statistical Distributions and Relations Common to Measurements

Distribution	Applications	Density Function	Shape
Normal	Most physical properties that are continuous or regular in time or space with variations due to random error	$p(x) = \dfrac{1}{\sigma(2\pi)^{1/2}} \exp\left[-\dfrac{1}{2}\dfrac{(x - x')^2}{\sigma^2}\right]$	
Log normal	Failure or durability projections; events whose outcomes tend to be skewed toward the extremity of the distribution	$p(x) = \dfrac{1}{x\sigma(2\pi)^{1/2}} \exp\left[-\dfrac{1}{2}\dfrac{((\ln x) - x')^2}{\sigma^2}\right]$	
Rectangular	Processes in which a likely outcome falls in the range between minimum value a and maximum value b occurring with equal probability	$p(x) = \dfrac{1}{b - a}$ where $a \leq x \leq b$, otherwise $p(x) = 0$	
Triangular	Process in which the likely outcome x falls between known lower bound a and upper bound b, with a peak value or mode c; used when population information is sparse	$p(x) = \dfrac{2(x - a)}{(b - a)(c - a)}$ for $a \leq x \leq c$ $= \dfrac{2(b - x)}{(b - a)(b - c)}$ for $c < x \leq b$	

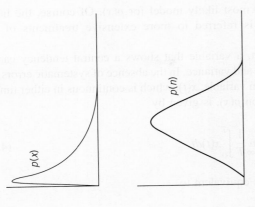

| Poisson | Events randomly occurring in time; $p(x)$ refers to probability of observing x events in time t; here λ refers to x' | $p(x) = \dfrac{e^{-\lambda}\lambda^x}{x!}$ |
| Binomial | Situations describing the number of occurrences, n, of a particular outcome during N independent tests where the probability of any outcome, P, is the same | $p(n) = \left[\dfrac{N!}{(N-n)!n!}\right]P^n(1-P)^{N-n}$ |

histograms are used to infer guidance on the most likely model for $p(x)$. Of course, the list in Table 4.2 is not complete, and the reader is referred to more extensive treatments of this subject (3–5).

Regardless of its probability density form, a variable that shows a central tendency can be described and quantified through its mean value and variance. In the absence of systematic errors, the *true mean value* or central tendency of a random variable, $x(t)$, which is continuous in either time or space, and having a probability density function $p(x)$, is given by

$$x' = \lim_{T \to \infty} \frac{1}{T} \int_0^T x(t)dt \tag{4.4a}$$

which for any continuous random variable x is equivalent to

$$x' = \int_{-\infty}^{\infty} xp(x)dx \tag{4.4b}$$

If the measured variable is described by discrete data, the mean value of measured variable, x_i, where $i = 1, 2 \ldots, N$, is given by

$$x' = \lim_{N \to \infty} \frac{1}{N} \sum_{i=1}^{N} x_i \tag{4.5}$$

Physically, the width of the density function reflects the data variation. For a continuous random variable, the *true variance* is given by

$$\sigma^2 = \lim_{T \to \infty} \frac{1}{T} \int_0^T [x(t) - x']^2 dt \tag{4.6a}$$

which is equivalent to

$$\sigma^2 = \int_{-\infty}^{\infty} (x - x')^2 p(x)dx \tag{4.6b}$$

or for discrete data, the variance is given by

$$\sigma^2 = \lim_{N \to \infty} \frac{1}{N} \sum_{i=1}^{N} (x_i - x')^2 \tag{4.7}$$

The *standard deviation*, σ, a commonly used statistical parameter, is defined as the square root of the variance—that is, $\sigma = \sqrt{\sigma^2}$.

The fundamental difficulty in using Equations 4.4 to 4.7 is that we assume a knowledge of the values of the entire population of the variable. But what if the data set represents only a finite portion of that population? Do these relations change? Real data set sizes may range from a single value to a large finite number. The next section discusses the interpretation of the behavior of the entire population of a variable with a connection between probability and statistics. Then we focus on trying to infer the behavior of a population based on the statistics of finite data sets.

4.3 DESCRIBING THE BEHAVIOR OF A POPULATION

This section discusses the relation between probability and statistics. To do this, we assume a particular distribution for $p(x)$ to characterize the behavior of the population of x. Table 4.2 lists a few of the many distributions we could use, but to develop our relations, we will use the *normal* (or *gaussian*) *distribution*.[4] Much of the existing theory of statistics was developed using this distribution. It is particularly suited to describing the behavior of the continuous random variables common to engineering measurements. The normal distribution predicts that the scatter seen in a measured data set will be distributed symmetrically about some central tendency. Its shape is the familiar bell curve, as seen in Table 4.2.

The probability density function for a random variable, x, having a normal distribution is defined by

$$p(x) = \frac{1}{\sigma\sqrt{2\pi}} \exp\left[-\frac{1}{2}\frac{(x - x')^2}{\sigma^2}\right] \tag{4.8}$$

where x' is defined as the true mean value of x and σ^2 as the true variance of x. Hence, $p(x)$ depends on the specific values for x' and σ^2. A maximum in $p(x)$ occurs at $x = x'$, the true mean value. This indicates that in the absence of systematic error the central tendency of a random variable having a normal distribution is toward its true mean value. The variance reflects the width or range of variation of $p(x)$.

Given $p(x)$, how can we predict the probability that any future measurement will fall within some stated interval of x values? The probability that x will assume a value within the interval $x' \pm \delta x$ is given by the area under $p(x)$, which is found by integrating over this interval. Thus, this probability is given by

$$P(x' - \delta x \le x \le x' + \delta x) = \int_{x' - \delta_x}^{x' + \delta_x} p(x)dx \tag{4.9}$$

Using Equation 4.9 is made easier with the following transformations. Begin by defining the terms $\beta = (x - x')/\sigma$ as the *standardized normal variate* for any value x, with

$$z_1 = (x_1 - x')/\sigma \tag{4.10}$$

as the z variable, which refers to a specific interval on $p(x)$ between x' and x_1. It follows that $dx = \sigma d\beta$, so that Equation 4.9 can be written as

$$P(-z_1 \le \beta \le z_1) = \frac{1}{\sqrt{2\pi}} \int_{-z_1}^{z_1} e^{-\beta^2/2} d\beta \tag{4.11}$$

---

[4] This distribution was independently suggested in the 18th century by Gauss, LaPlace, and DeMoivre. However, Gauss retains the eponymous honor.

Because for a normal distribution, $p(x)$ is symmetrical about x', we can write

$$\frac{1}{\sqrt{2\pi}}\int_{-z_1}^{z_1} e^{-\beta^2/2}d\beta = 2 \times \left[\frac{1}{\sqrt{2\pi}}\int_{0}^{z_1} e^{-\beta^2/2}d\beta\right] \tag{4.12}$$

The value in brackets in Equation 4.12 is the *normal error function*. The value $P(z_1) = \frac{1}{\sqrt{2\pi}}\int_{0}^{z_1} e^{-\beta^2/2}d\beta$ is tabulated in Table 4.3 for the interval defined by z_1 shown in Figure 4.3. As

Table 4.3 Probability Values for Normal Error Function: One-Sided Integral Solutions for $P(z_1) = \frac{1}{(2\pi)^{1/2}}\int_{0}^{z_1} e^{-\beta^2/2}d\beta$

$z_1 = \frac{x_1 - x'}{\sigma}$	0.00	0.01	0.02	0.03	0.04	0.05	0.06	0.07	0.08	0.09
0.0	0.0000	0.0040	0.0080	0.0120	0.0160	0.0199	0.0239	0.0279	0.0319	0.0359
0.1	0.0398	0.0438	0.0478	0.0517	0.0557	0.0596	0.0636	0.0675	0.0714	0.0753
0.2	0.0793	0.0832	0.0871	0.0910	0.0948	0.0987	0.1026	0.1064	0.1103	0.1141
0.3	0.1179	0.1217	0.1255	0.1293	0.1331	0.1368	0.1406	0.1443	0.1480	0.1517
0.4	0.1554	0.1591	0.1628	0.1664	0.1700	0.1736	0.1772	0.1809	0.1844	0.1879
0.5	0.1915	0.1950	0.1985	0.2019	0.2054	0.2088	0.2123	0.2157	0.2190	0.2224
0.6	0.2257	0.2291	0.2324	0.2357	0.2389	0.2422	0.2454	0.2486	0.2517	0.2549
0.7	0.2580	0.2611	0.2642	0.2673	0.2794	0.2734	0.2764	0.2794	0.2823	0.2852
0.8	0.2881	0.2910	0.2939	0.2967	0.2995	0.3023	0.3051	0.3078	0.3106	0.3133
0.9	0.3159	0.3186	0.3212	0.3238	0.3264	0.3289	0.3315	0.3340	0.3365	0.3389
1.0	0.3413	0.3438	0.3461	0.3485	0.3508	0.3531	0.3554	0.3577	0.3599	0.3621
1.1	0.3643	0.3665	0.3686	0.3708	0.3729	0.3749	0.3770	0.3790	0.3810	0.3830
1.2	0.3849	0.3869	0.3888	0.3907	0.3925	0.3944	0.3962	0.3980	0.3997	0.4015
1.3	0.4032	0.4049	0.4066	0.4082	0.4099	0.4115	0.4131	0.4147	0.4162	0.4177
1.4	0.4192	0.4207	0.4292	0.4236	0.4251	0.4265	0.4279	0.4292	0.4306	0.4319
1.5	0.4332	0.4345	0.4357	0.4370	0.4382	0.4394	0.4406	0.4418	0.4429	0.4441
1.6	0.4452	0.4463	0.4474	0.4484	0.4495	0.4505	0.4515	0.4525	0.4535	0.4545
1.7	0.4554	0.4564	0.4573	0.4582	0.4591	0.4599	0.4608	0.4616	0.4625	0.4633
1.8	0.4641	0.4649	0.4656	0.4664	0.4671	0.4678	0.4686	0.4693	0.4699	0.4706
1.9	0.4713	0.4719	0.4726	0.4732	0.4738	0.4744	0.4750	0.4758	0.4761	0.4767
2.0	0.4772	0.4778	0.4803	0.4788	0.4793	0.4799	0.4803	0.4808	0.4812	0.4817
2.1	0.4821	0.4826	0.4830	0.4834	0.4838	0.4842	0.4846	0.4850	0.4854	0.4857
2.2	0.4861	0.4864	0.4868	0.4871	0.4875	0.4878	0.4881	0.4884	0.4887	0.4890
2.3	0.4893	0.4896	0.4898	0.4901	0.4904	0.4906	0.4909	0.4911	0.4913	0.4916
2.4	0.4918	0.4920	0.4922	0.4925	0.4927	0.4929	0.4931	0.4932	0.4934	0.4936
2.5	0.4938	0.4940	0.4941	0.4943	0.4945	0.4946	0.4948	0.4949	0.4951	0.4952
2.6	0.4953	0.4955	0.4956	0.4957	0.4959	0.4960	0.4961	0.4962	0.4963	0.4964
2.7	0.4965	0.4966	0.4967	0.4968	0.4969	0.4970	0.4971	0.4972	0.4973	0.4974
2.8	0.4974	0.4975	0.4976	0.4977	0.4977	0.4978	0.4979	0.4979	0.4980	0.4981
2.9	0.4981	0.4982	0.4982	0.4983	0.4984	0.4984	0.4985	0.4985	0.4986	0.4986
3.0	0.49865	0.4987	0.4987	0.4988	0.4988	0.4988	0.4989	0.4989	0.4989	0.4990

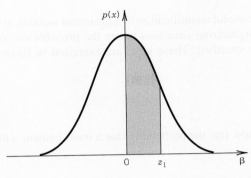

Figure 4.3 Integration terminology for the normal error function and Table 4.3.

written and tabulated, the integral is one-sided (that is, it is evaluated from 0 to z_1), so the normal error function provides half the probability expressed by Equation 4.11.

The parameters defined by Equations 4.4 to 4.7 are statements associated with the probability of a population. The area under the portion of the probability density function curve, $p(x)$, defined by the interval $x' - z_1\sigma \leq x \leq x' + z_1\sigma$, gives the probability that a measurement will assume a value within that interval. Direct integration of $p(x)$ for a normal distribution between the limits $x' \pm z_1\sigma$ with $z_1 = 1$ covers 68.26% of the area under $p(x)$. This equates to a 68.26% chance that a measurement of x will have a value within the interval $x' \pm 1\sigma$. As the interval defined by z_1 is increased, the probability of occurrence increases. For

$$z_1 = 1, \quad 68.26\% \text{ of the area under } p(x) \text{ lies within } \pm z_1\sigma \text{ of } x'.$$
$$z_1 = 2, \quad 95.45\% \text{ of the area under } p(x) \text{ lies within } \pm z_1\sigma \text{ of } x'.$$
$$z_1 = 3, \quad 99.73\% \text{ of the area under } p(x) \text{ lies within } \pm z_1\sigma \text{ of } x'.$$

This concept is illustrated in Figure 4.4.

It follows directly that the representative value that characterizes a measure of the variation in a measured data set is the standard deviation. The probability that some ith measured value of x will have a value between $x' \pm z_1\delta$ is $2 \times P(z_1) \times 100 = P\%$. This is written as

$$x_i = x' \pm z_1 \sigma \quad (P\%) \tag{4.13}$$

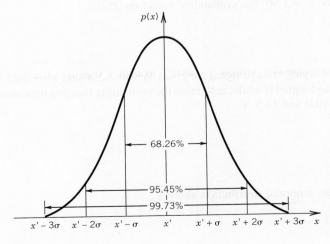

Figure 4.4 Relationship between the probability density function and its statistical parameters, x' and σ, for a normal (gaussian) distribution.

Thus, simple statistical analyses provide useful quantification of a measured variable in terms of probability. This in turn can be useful in engineering situations where the probable outcome of a measured variable needs to be predicted or specified. These ideas are exercised in Examples 4.2 and 4.3.

Example 4.2

Using the probability values in Table 4.3, show that the probability that a measurement will yield a value within $x' \pm \sigma$ is 0.6826 or 68.26%.

KNOWN Table 4.3

$$z_1 = 1$$

ASSUMPTIONS Data follow a normal distribution.

FIND $P(x' - \sigma \leq x \leq x' + \sigma)$

SOLUTION To estimate the probability that a single measurement will have a value within some interval, we need to solve the integral

$$\frac{1}{\sqrt{2\pi}} \int_0^{z_1=1} e^{-\beta^2/2} d\beta$$

over the interval defined by z_1. Table 4.3 lists the solutions for this integral. Using Table 4.3 for $z_1 = 1$, we find $P(z_1) = 0.3413$. However, because $z_1 = (x_1 - x')/\sigma$, the probability that any measurement of x will produce a value within the interval $0 \leq x \leq x' + \sigma$ is 34.13%. Since the normal distribution is symmetric about x', the probability that x will fall within the interval defined between $-z_1 \delta$ and $+z_1 \delta$ for $z_1 = 1$ is $(2)(0.3413) = 0.6826$ or 68.26%. Accordingly, if we made a single measurement of x, the probability that the value found would lie within the interval $x' - \sigma \leq x \leq x' + \sigma$ would be 68.26%.

COMMENT Similarly, for $z_1 = 1.96$, the probability would be 95.0%.

Example 4.3

The statistics of a well-defined varying voltage signal are given by $x' = 8.5$ V and $\sigma^2 = 2.25$ V^2. If a single measurement of the voltage signal is made, determine the probability that this measured value will have a value of between 10.0 and 11.5 V.

KNOWN $x' = 8.5$ V

$$\sigma^2 = 2.25 \text{ V}^2$$

ASSUMPTIONS Signal has a normal distribution about x'.

FIND $P(10.0 \leq x \leq 11.5)$

SOLUTION   To find the probability that x will fall into the interval $10.0 \leq x \leq 11.5$ requires finding the area under $p(x)$ bounded by this interval. The standard deviation of the variable is $\sigma = \sqrt{\sigma^2} = 1.5$ V, so our interval falls under the portion of the $p(x)$ curve bounded by $z_1 = (10.0 - 8.5)/1.5 = 1$ and $z_1 = (11.5 - 8.5)/1.5 = 2$. From Table 4.3, the probability that a value will fall between $8.5 \leq x \leq 10.0$ is $P(8.5 \leq x \leq 10.0) = P(z_1 = 1) = 0.3413$. For the interval defined by $8.5 \leq x \leq 11.5$, $P(8.5 \leq x \leq 11.5) = P(z_1 = 2) = 0.4772$. The area we need is just the overlap of these two intervals, so

$$P(10.0 \leq x \leq 11.5) = P(8.5 \leq x \leq 11.5) - P(8.5 \leq x \leq 10.0)$$
$$= 0.4772 - 0.3413 = 0.1359$$

So there is a 13.59% probability that the measurement will yield a value between 10.0 and 11.5 V.

COMMENT   In general, the probability that a measured value will lie within an interval defined by any two values of z_1, such as z_a and z_b, is found by integrating $p(x)$ between z_a and z_b. For a normal density function, this probability is identical to the operation, $P(z_b) - P(z_a)$.

4.4   STATISTICS OF FINITE-SIZED DATA SETS

We now try to predict the behavior of a population of variable x based on knowledge obtained from a finite-sized random sample of x. For example, if we recall the box of bearings discussed in Section 4.1, some two dozen bearings were measured, each having been randomly selected from a population numbering in the thousands. Within the constraints imposed by probability, and if we assume a probability density function for the population, it is possible to *estimate* the true mean and true variance the population of all the bearings from the statistics of the small sample. The science behind using a small sample to characterize the behavior of an entire population is called *inferential statistics*.

Suppose we examine the case where we obtain N measurements of x, each measurement represented by x_i, where $i = 1, 2, \ldots, N$, and N is a finite value. In cases where N does not represent the total population, the statistical values calculated from such finite data sets are only estimates of the true statistics of the population of x. We will refer to such statistical estimates as the *finite statistics*. An important point: Whereas infinite statistics describe the true behavior of the population of a variable, finite statistics describe only the behavior of the sampled data set, which we then use to infer the behavior of the population.

Finite-sized data sets provide the statistical estimates known as the *sample mean value* $(\bar{x})$, the *sample variance* (s_x^2), and its outcome, the *sample standard deviation* (s_x), defined by

$$\bar{x} = \frac{1}{N} \sum_{i=1}^{N} x_i \tag{4.14a}$$

$$s_x^2 = \frac{1}{N-1} \sum_{i=1}^{N} (x_i - \bar{x})^2 \tag{4.14b}$$

$$s_x = \sqrt{s_x^2} = \left(\frac{1}{N-1} \sum_{i=1}^{N} (x_i - \bar{x})^2 \right)^{1/2} \tag{4.14c}$$

where $(x_i - \bar{x})$ is called the *deviation* of x_i. The sample mean value provides a probable estimate of the true mean value, x'. The sample variance represents a probable measure of the variation found in a data set. The *degrees of freedom*, ν, in a statistical estimate equate to the number of data points minus the number of previously determined statistical parameters used in estimating that value. For example, the degrees of freedom in the sample variance is $\nu = N - 1$, as seen in denominator of Equations 4.14b and c. These equations are robust and are used regardless of the actual probability density function of the population.

The relations between probability and statistics previously developed can be extended to data sets of finite sample size with only some modification. When the parameters of mean value and variance are based on a sample rather than the entire population, the z variable does not give a reliable weighting of the true probability of the population. Instead we introduce the Student's t variable. For a normal distribution of x about some sample mean value, $\bar{x}$, we can state that statistically

$$x_i = \bar{x} \pm t_{\nu,P}s_x \quad (P\%) \tag{4.15}$$

where the variable $t_{\nu,P}$ provides a coverage factor used for finite data sets and which replaces the z variable. The Student's t variable is defined as

$$t = \frac{\bar{x} - x'}{s_x/\sqrt{N}} \tag{4.16}$$

The interval $\pm t_{\nu,P}s_x$ represents a precision interval, given at probability $P\%$, within which one should expect any measured value to fall.

The value for the t estimator provides a coverage factor, similar to the use of a z-variable, that is a function of the probability, P, and the degrees of freedom, $\nu = N - 1$. These t values can be obtained from Table 4.4, a tabulation from the *Student's t distribution* as developed by William S. Gosset[5] (1876–1937). Gossett recognized that the use of the z variable with s_x in place of σ did not yield accurate estimates of the precision interval, particularly at small degrees of freedom. Careful inspection of Table 4.4 shows that the t value inflates the size of the interval required to attain a percent probability, $P\%$. That is, it has the effect of increasing the magnitude of $t_{\nu,P}s_x$ relative to $z_1\sigma$ at a desired probability. However, as the value of N increases, t approaches those values given by the z variable just as the value of s_x must approach σ.

We point out that for very small sample sizes ($N \le 10$), the calculated statistics can be quite misleading. In that situation, other information regarding the measurement may be required, including additional measurements.

Standard Deviation of the Means

If we were to take a second batch of two dozen of the bearings discussed in Section 4.1 and measure each bearing's diameter, we would expect the statistics from this new sample to differ somewhat from the first sample. This is due to the variation in each bearing's diameter arising from manufacturing

---

[5] At the time, Gosset was employed as a brewer and statistician by a well-known Irish brewery. You might pause to reflect on his multifarious contributions.

Table 4.4   Student's t Distribution (Two-Sided)

ν	t_{50}	t_{90}	t_{95}	t_{99}
1	1.000	6.314	12.706	63.657
2	0.816	2.920	4.303	9.925
3	0.765	2.353	3.182	5.841
4	0.741	2.132	2.770	4.604
5	0.727	2.015	2.571	4.032
6	0.718	1.943	2.447	3.707
7	0.711	1.895	2.365	3.499
8	0.706	1.860	2.306	3.355
9	0.703	1.833	2.262	3.250
10	0.700	1.812	2.228	3.169
11	0.697	1.796	2.201	3.106
12	0.695	1.782	2.179	3.055
13	0.694	1.771	2.160	3.012
14	0.692	1.761	2.145	2.977
15	0.691	1.753	2.131	2.947
16	0.690	1.746	2.120	2.921
17	0.689	1.740	2.110	2.898
18	0.688	1.734	2.101	2.878
19	0.688	1.729	2.093	2.861
20	0.687	1.725	2.086	2.845
21	0.686	1.721	2.080	2.831
30	0.683	1.697	2.042	2.750
40	0.681	1.684	2.021	2.704
50	0.680	1.679	2.010	2.679
60	0.679	1.671	2.000	2.660
∞	0.674	1.645	1.960	2.576

tolerances (i.e., due to random chance). The difference between the calculated mean diameter of either batch and the mean diameter of the entire box is a random error due to using small data sets to infer the statistics of an entire population. So how can we quantify how good our estimate is of the true mean based on a sample mean?

Suppose we were to take N measurements from a large population of variable x under fixed operating conditions. If we were to duplicate this procedure M times, we would calculate slightly different values for the sample mean and sample variance for each data set. Why? The chance occurrence of events in any finite-sized sample affects the estimate of sample statistics; this is easily demonstrated. From the M replications of the N measurements, we can compute a set of mean values. We would find that the population of mean values would themselves each be normally distributed about some central value. In fact, regardless of the shape of $p(x)$ assumed, the distribution of mean values obtained from all possible random samples will follow a normal distribution, $p(\bar{x})$.[6] This

---

[6] This is a consequence of what is proved in the *central limit theorem* (3, 4).

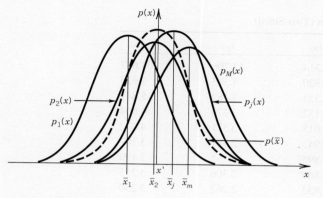

Figure 4.5 The normal distribution tendency of the sample means about a true value in the absence of systematic error.

process is visualized in Figure 4.5. The amount of variation possible in the sample means would depend on two values: the sample variance, s_x^2, and the sample size, N. The discrepancy tends to increase with variance and decrease with $N^{1/2}$.

The variation in the sample statistics gathered from each data set is characterized by a normal distribution of the sample mean values about the true mean. This variation is estimated from a single finite data set using the *standard deviation of the means*, $s_{\bar{x}}$:

$$s_{\bar{x}} = \frac{s_x}{\sqrt{N}} \tag{4.17}$$

An illustration of the relation between the standard deviation of a data set and the standard deviation of the means is given in Figure 4.6. The *standard deviation of the means provides a measure of how well a measured mean value from a small sample size represents the true mean of the entire population*. The range over which the possible values of the true mean value might lie at some probability level P based on the information from a sample data set is given as

$$\bar{x} \pm t_{\nu,P} s_{\bar{x}} \quad (P\%) \tag{4.18}$$

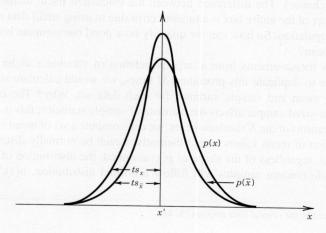

Figure 4.6 Relationships between s_x and the distribution of x and between $s_{\bar{x}}$ and the true value x'.

where $\pm t_{\nu,P}s_{\bar{x}}$ expresses a confidence interval about the mean value, with coverage factor t at the assigned probability, $P\%$, within which one should expect the true value of x to fall. This *confidence interval* is a quantified measure of the random error in the estimate of the true value of variable x. The value $s_{\bar{x}}$ represents the random standard uncertainty in the mean value, and the value $t_{\nu,P}s_{\bar{x}}$ represents the *random uncertainty* in the mean value at $P\%$ confidence due to variation in the measured data set. *In the absence of systematic error in a measurement*, the confidence interval assigns the true value within a likely range about the sample mean value (one due solely on the small sample size). The estimate of the true mean value based on the statistics of a finite data set is then stated as

$$x' = \bar{x} \pm t_{\nu,P}s_{\bar{x}} \quad (P\%) \tag{4.19}$$

Equation 4.19 is an important and powerful equation in engineering measurements.

Example 4.4

Consider the data of Table 4.1. (a) Compute the sample statistics for this data set. (b) Estimate the interval of values over which 95% of the measurements of x should be expected to lie. (c) Estimate the true mean value of x at 95% probability based on this finite data set.

KNOWN Table 4.1

$N = 20$

ASSUMPTIONS Data set follows a normal distribution.
No systematic errors.

FIND $\bar{x}, \bar{x} \pm t_{\nu,95}s_x$, and $\bar{x} \pm t_{\nu,95}s_{\bar{x}}$

SOLUTION The sample mean value is computed for the $N = 20$ values by the relation

$$\bar{x} = \frac{1}{20}\sum_{i=1}^{20} x_i = 1.02$$

This, in turn, is used to compute the sample standard deviation

$$s_x = \left[\frac{1}{19}\sum_{i=1}^{20}(x_i - \bar{x})^2\right]^{1/2} = 0.16$$

The degrees of freedom in the data set is $\nu = N - 1 = 19$. From Table 4.4 at 95% probability, $t_{19,95} = 2.093$. Then, the interval of values in which 95% of the measurements of x should lie is given by Equation 4.15:

$$x_i = 1.02 \pm (2.093 \times 0.16) = 1.02 \pm 0.33 \quad (95\%)$$

Accordingly, if a twenty-first data point were taken, there is a 95% probability that its value would lie between 0.69 and 1.35.

The true mean value is estimated by the sample mean value. However, the random uncertainty at 95% probability for this estimate is $t_{19,95}s_{\bar{x}}$, where

$$s_{\bar{x}} = \frac{s_x}{\sqrt{N}} = \frac{0.16}{\sqrt{20}} = 0.036$$

Then, in the absence of systematic errors, we write, from Equation 4.19,

$$x' = \bar{x} \pm t_{19,95}s_{\bar{x}} = 1.02 \pm 0.08 \quad (95\%)$$

So at 95% confidence, the true mean value lies between 0.94 and 1.10, assuming that the only error is due to random variations of the measured value. Program *Finite-population.vi* demonstrates the effect of sample size on the histogram and the statistics of the data set.

Repeated Tests and Pooled Data

Repeated tests provide separate data sets comprised of independent measurements of the same measured variable. These data can be combined to provide a better statistical estimate of a measured variable than are obtained from a single sampling. The assumption we make here is that the population measured remains the same between the tests.

Consider M repeated tests that provide M independent data sets of a measurement of variable x, each of N repeated measurements so as to yield the data set x_{ij}, where $i = 1, 2, \ldots, N$ and $j = 1, 2, \ldots, M$. The *pooled mean* of x is calculated by combining all of the measurements as

$$\langle \bar{x} \rangle = \frac{1}{MN} \sum_{j=1}^{M} \sum_{i=1}^{N} x_{ij} = \frac{1}{M} \sum_{j=1}^{M} \bar{x}_j \tag{4.20}$$

When the measurements are not maintained equal between tests, then a weighted average is used. In that case, let N_j be the number of measurements in the *jth* repeated test, then

$$\langle \bar{x} \rangle = \frac{\sum_{j=1}^{M} N_j \bar{x}_j}{\sum_{j=1}^{M} N_j} \tag{4.21}$$

In order to study the variability *between* the tests, a *pooled standard deviation* is calculated by (7)

$$\langle s_x \rangle = \sqrt{\frac{1}{(M-1)} \sum_{j=1}^{M} (\bar{x}_j - \langle \bar{x} \rangle)^2} \tag{4.22}$$

with degrees of freedom between the tests $\nu = M - 1$. A statistic used to test the significance of between-test variations is the mean-square value

$$\langle s_x \rangle_B^2 = \frac{1}{(M-1)} \sum_{j=1}^{M} N_j \left(\overline{x}_j - \langle \overline{x} \rangle \right)^2 \tag{4.23}$$

We use the subscript B to emphasize between-test variation. The *pooled standard deviation of the means* is a measure of the precision of the sample means. It is calculated by (7)

$$\langle s_{\overline{x}} \rangle = \frac{\langle s_x \rangle}{\sqrt{M}} \tag{4.24}$$

For example, in Example 1.1, we would combine the data from the three repeated tests used to determine boiling point temperature to obtain a pooled mean value. We would then use Equation 4.24 to obtain a measure of how well this pooled mean represents the true mean value of the measured population, which would be a measure of random uncertainty used within an uncertainty analysis.

As a measure of the variability *within* the repeated tests, a within-test pooled standard deviation is calculated by

$$\langle s_x \rangle_W = \sqrt{\frac{\nu_1 s_{x_1}^2 + \nu_2 s_{x_2}^2 + \cdots + \nu_M s_{x_M}^2}{\nu_1^2 + \nu_2^2 + \cdots + \nu_M^2}} \tag{4.25}$$

with degrees of freedom $\nu = \sum_{j=1}^{M} \nu_j = \sum_{j=1}^{M} (N_j - 1)$. Here the subscript W is used to emphasize within-test variation.

The ratio $F = \langle s_x \rangle_B^2 / \langle s_x \rangle_W^2$ is used as the test statistic for confirming the assumed equality of the population means under the null hypothesis test introduced in the next section. More details regarding how to analyze further the pooled data obtained from repeated tests are discussed in entry-level statistics texts (4, 6).

4.5 HYPOTHESIS TESTING

A hypothesis is a statement that something is believed true. A hypothesis test uses the tools of statistics to test it. The hypothesis might be that the difference between some observed (measured) behavior of a population and an assumed behavior is small enough to be considered as due to random variations (i.e., chance) alone. This is the null hypothesis, H_0. The alternative hypothesis, H_a, is that the difference results from something other than random variation, such as a systematic effect. So we might test whether measured differences in values between sample data sets arise from variations alone or whether something has changed between the data sets. The hypothesis test is used to decide whether the null hypothesis should be rejected in favor of the alternative hypothesis.

Suppose we want to test whether a measured sample mean is a reasonable proxy for the population mean comparing information of the observed scatter in the data set against the assumed distribution of the population. For this test, the null hypothesis is expressed as $H_0: x' = x_0$, where x_0 is

Table 4.5 Critical Values Used in the z-test

Level of significance, α	0.10	0.05	0.01
Critical values of z One-tailed tests	−1.28 or 1.28	−1.645 or 1.645	−2.33 or 2.33
Critical values of z Two-tailed tests	−1.645 and 1.645	−1.96 and 1.96	−2.58 and 2.58

a number. Now the alternative hypothesis will depend on the exact purpose of the hypothesis test. When we want to determine whether the sample mean, $\bar{x}$, differs from the population mean, the alternative hypothesis is expressed as $H_a: x' \neq x_0$. This is called the two-tailed test, because H_a could be true if either $x' < x_0$ or $x' > x_0$. If the sample data are inconsistent with the null hypothesis, we reject it in favor of the alternative hypothesis. Otherwise, we do not reject the null hypothesis. In other situations, we might be interested in knowing whether $x' < x_0$, in which case we set $H_a: x' < x_0$. If the interest is in knowing whether $x' > x_0$, we set $H_a: x' > x_0$. Each of these is called a single-tailed test. We assume in each of these tests that $p(\bar{x})$ has a normal distribution.

z-test The z-test is used when the population σ is known. This occurs in situations where a value for σ is specified, such as a value we might assign for quality control in manufacturing, or when the population distribution is already established. The test statistic is given by the z-variable. With $x' = x_0$, the observed z-variable is

$$z_0 = \frac{\bar{x} - x_0}{\sigma/\sqrt{N}} \tag{4.26}$$

This statistic is evaluated against critical values of z at a desired *level of significance*, α, where $P(z) \equiv 1 - \alpha$. We do not reject the null hypothesis if the values of the test statistic lie within a defined acceptance region. For the two-tailed test, the acceptance region is defined by the interval $P(-z_{\alpha/2} \leq z_0 \leq z_{\alpha/2})$ where $\pm z_{\alpha/2}$ are the critical values. Similarly, for the one-tailed tests, the acceptance region is defined as $P(-z_\alpha \leq z_0)$ or $P(z_0 \leq z_\alpha)$. We reject the null hypothesis if the value lies outside the acceptance region, a region we call the rejection region. The concept is shown in Figure 4.7 with common critical values listed in Table 4.5. When applying a hypothesis test to most engineering situations, testing at a two-tailed level of significance of 0.05 is adequate.

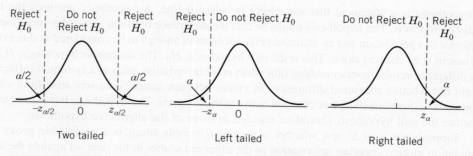

Figure 4.7 Critical values for a hypothesis test at level of significance, α.

t-test   The *t*-test is used when σ is not known. The test statistic is given by the *t*-variable. With $x' = x_0$, the observed *t*-variable is

$$t_0 = \frac{\bar{x} - x_0}{s_x/\sqrt{N}} \tag{4.27}$$

This statistic is evaluated against critical values for *t* at the desired level of significance, α. For the two-tailed test, the acceptance region is given by the interval $P(-t_{\alpha/2} \leq t_0 \leq t_{\alpha/2})$. Here $\pm t_{\alpha/2}$ are the critical values. Similarly for the one-tailed tests, the acceptance region is $P(-t_\alpha \leq t_0)$ or $P(t_0 \leq t_\alpha)$. The concept is also shown in Figure 4.7.

p-value   The *p-value* reports the probability that the difference between two tested values is due only to random chance. It is the observed level of significance. A very small *p*-value indicates that effects other than random chance are influencing the variations in a sampling. To calculate the *p*-value, assume that the null hypothesis is true and compute the observed level of significance corresponding to the test statistic. If this *p*-value is less than α, we reject H_0, otherwise we do not reject H_0.

Hypothesis test   (1) Establish the null hypothesis, such as $H_0: x' = x_0$, where x_0 is the population or a target value. Then, establish the appropriate alternative hypothesis, H_a. (2) Assign a level of significance, α, to determine the critical values; this sets the rejection region. (3) Calculate the observed value of the test statistic. (4) Compare the observed test statistic to the critical values. If the observed statistic is in the rejection region, reject the null hypothesis; otherwise, do not reject.

Example 4.5

From experience, a bearing manufacturer knows that it can produce roller bearings to the stated dimensions of 2.00 mm within a standard deviation of 0.061 mm. As part of its quality assurance program, engineering takes samples from boxes of bearings before shipment and measures them. On one occasion, a sample size of 25 bearings is obtained with the result $\bar{x} = 2.03$ mm. Do the measurements support the hypothesis that $x' = 2.00$ mm for the whole box at a 5% ($\alpha = 0.05$) level of significance (i.e., a 95% confidence level)?

KNOWN $\bar{x} = 2.03$ mm; $x_0 = 2.00$ mm; $\sigma = 0.061$ mm; $N = 25$

FIND   Apply hypothesis test at $\alpha = 0.05$

SOLUTION   Use the four steps for the hypothesis test: (1) We set the null hypothesis at $H_0: x' = x_0 = 2.00$ mm. The alternative hypothesis will be $H_a : x' \neq 2.00$ mm. (2) With $\alpha = 0.05$, the acceptance region is the area under the probability curve, $P(-z_{0.025} \leq z_0 \leq z_{0.025}) = 1 - \alpha = 0.95$. Assuming that the bearing population is normally distributed, we can write $P(-z_{0.025} \leq z \leq z_{0.025}) = 2P(0 \leq z \leq z_{0.025}) = 2 \times 0.475$. Then, using the one-sided Table 4.3, we find *z* by matching with the probability, $P(z = 1.960) = 0.475$. So the two-tailed critical value is $|\pm z_c| = 1.96$ (this is also noted in Table 4.5). (3) With $x_0 = 2.00$ mm, $\bar{x} = 2.03$ mm, $N = 25$, and $\sigma = 0.061$ mm; then $z_0 = (\bar{x} - x_0)/\sigma/\sqrt{N} = 2.459$. (4) We find that $z_0 > z_c$. So z_0 falls outside the acceptance region, and we reject H_0 at the 5% level of significance. The difference between the sample mean and the target population mean is larger than would likely occur by chance alone. Conclusion: This batch of bearings has a manufacturing defect!

COMMENT We know that random variations in a small sample can cause the sample mean to differ from the population mean. Here we find that the difference is larger than random variation alone can explain at $\alpha = 0.05$ or, with $P(z) \equiv 1 - \alpha$, at the 95% probability confidence level. There is likely a systematic error affecting the mean bearing diameter. We can also compute a p-value. The value of $z_0 = 2.459$ corresponds to a two-tailed, observed level of significance of $p = 0.0138$. That is, from Table 4.3, $2P(2.459) = 0.9862$, so $p = 1 - 0.9862 = 0.0138$. Because $p < \alpha$, we can reject H_0.

Example 4.6

Golf clubs and balls are tested for performance using a robotic swing device called the Iron Byron. A golf tour professional requests a 6-iron golf club from a manufacturer to be tuned to carry the ball a distance of 180 yards on average when struck with a club head speed of 85 mph. Accuracy is important, so the club should meet the requirement within 5 yards to either side of 180 yards. The company engineer tests the selected club using a tour-style golf ball. Six tests of the club provide a mean carry of 182.7 yards with a sample standard deviation of 2.7 yards. Should the engineer certify that the club meets the tour professional's specifications at the 5% level of significance (1 yard = 0.914 m)?

KNOWN $\bar{x} = 182.7$ yards; $x_0 = 180$ yards; $s_x = 2.7$ yards; $N = 6$

FIND Apply hypothesis test at $\alpha = 0.05$

SOLUTION (1) We test the null hypothesis, $H_0 : x' = x_0 = 180$ yards against the alternative hypothesis, $H_a : x' \neq 180$ yards. (2) For a two-sided t-test with $\alpha = 0.05$, $P = 1 - \alpha = 0.95$. So with $N = 6$, $\nu = N - 1 = 5$, the critical values (Table 4.4) will be $|\pm t_{5,95}| = |\pm t_c| = 2.571$. (3) The test statistic is $t_0 = (\bar{x} - x_0)/s_x/\sqrt{N} = (182.7 - 180.0)/2.7/\sqrt{6} = 2.449$. (4) The observed test statistic, t_0, lies within the acceptance region, the area under the probability curve defined between $\pm t_c$. We do not reject the null hypothesis at the 5% level of significance. The data do not provide sufficient evidence to conclude that the distance of carry of this golf club will differ from 180 yards when striking a ball at 85 mph. Conclusion: The club meets specifications.

COMMENT Using more extensive t-tables (e.g., 4), we can also find a two-tailed $t = 2.449$ and $\nu = 5$, which corresponds to $p = 0.0580$. As $p > \alpha$, do not reject H_0.

4.6 CHI-SQUARED DISTRIBUTION

If we plotted the sample standard deviation for many data sets, each having N data points, we would generate the probability density function, $p(\chi^2)$. The $p(\chi^2)$ follows the so-called *chi-squared* (χ^2) *distribution* depicted in Figure 4.8.

For the normal distribution, the χ^2 statistic (1, 3, 4) is

$$\chi^2 = \nu s_x^2/\sigma^2 \tag{4.28}$$

with degrees of freedom $\nu = N - 1$. We can use this to test how well s_x^2 predicts σ^2.

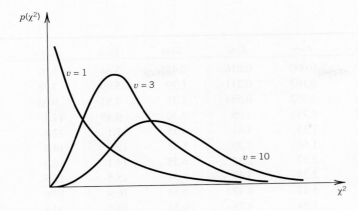

Figure 4.8 The χ^2 distribution with its dependency on degrees of freedom.

Precision Interval in a Sample Variance

A precision interval for the sample variance can be formulated by the probability statement

$$P\left(\chi^2_{1-\alpha/2} \leq \chi^2 \leq \chi^2_{\alpha/2}\right) = 1 - \alpha \tag{4.29}$$

with a probability of $P(\chi^2) = 1 - \alpha$. Combining Equations 4.28 and 4.29 gives

$$P\left(\nu s_x^2/\chi^2_{\alpha/2} \leq \sigma^2 \leq \nu s_x^2/\chi^2_{1-\alpha/2}\right) = 1 - \alpha \tag{4.30}$$

For example, the 95% precision interval by which s_x^2 estimates σ^2 is given by

$$\nu s_x^2/\chi^2_{0.025} \leq \sigma^2 \leq \nu s_x^2/\chi^2_{0.975} \quad (95\%) \tag{4.31}$$

Note that this interval is bounded by the 2.5% and 97.5% levels of significance (for 95% coverage).

The χ^2 distribution estimates the discrepancy expected due to random chance. Values for χ^2_α are tabulated in Table 4.6 as a function of the degrees of freedom. The $P(\chi^2)$ value equals the area under $p(\chi^2)$ as measured from the left, and the α value is the area as measured from the right, as noted in Figure 4.9. The total area under $p(\chi^2)$ is equal to unity.

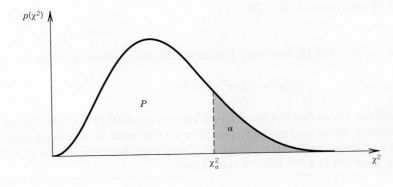

Figure 4.9 The χ^2 distribution as it relates to probability P and to the level of significance, $\alpha (= 1 - P)$.

Table 4.6   Values for χ_α^2

ν	$\chi_{0.99}^2$	$\chi_{0.975}^2$	$\chi_{0.95}^2$	$\chi_{0.90}^2$	$\chi_{0.50}^2$	$\chi_{0.05}^2$	$\chi_{0.025}^2$	$\chi_{0.01}^2$
1	0.000	0.000	0.000	0.016	0.455	3.84	5.02	6.63
2	0.020	0.051	0.103	0.211	1.39	5.99	7.38	9.21
3	0.115	0.216	0.352	0.584	2.37	7.81	9.35	11.3
4	0.297	0.484	0.711	1.06	3.36	9.49	11.1	13.3
5	0.554	0.831	1.15	1.61	4.35	11.1	12.8	15.1
6	0.872	1.24	1.64	2.20	5.35	12.6	14.4	16.8
7	1.24	1.69	2.17	2.83	6.35	14.1	16.0	18.5
8	1.65	2.18	2.73	3.49	7.34	15.5	17.5	20.1
9	2.09	2.70	3.33	4.17	8.34	16.9	19.0	21.7
10	2.56	3.25	3.94	4.78	9.34	18.3	20.5	23.2
11	3.05	3.82	4.57	5.58	10.3	19.7	21.9	24.7
12	3.57	4.40	5.23	6.30	11.3	21.0	23.3	26.2
13	4.11	5.01	5.89	7.04	12.3	22.4	24.7	27.7
14	4.66	5.63	6.57	7.79	13.3	23.7	26.1	29.1
15	5.23	6.26	7.26	8.55	14.3	25.0	27.5	30.6
16	5.81	6.91	7.96	9.31	15.3	26.3	28.8	32.0
17	6.41	7.56	8.67	10.1	16.3	27.6	30.2	33.4
18	7.01	8.23	9.39	10.9	17.3	28.9	31.5	34.8
19	7.63	8.91	10.1	11.7	18.3	30.1	32.9	36.2
20	8.26	9.59	10.9	12.4	19.3	31.4	34.2	37.6
30	15.0	16.8	18.5	20.6	29.3	43.8	47.0	50.9
60	37.5	40.5	43.2	46.5	59.3	79.1	83.3	88.4

Example 4.7

A manufacturer knows from experience that the variance in the diameter of the roller bearings used in its bearings is $3.15\ \mu m^2$. Rejecting bearings drives up the unit cost. However, the manufacturer rejects any batch of roller bearings if the sample variance of 20 pieces selected at random exceeds $5.00\ \mu m^2$. Assuming a normal distribution, what is the probability that any given batch might be rejected even though its true variance is actually within the tolerance limits?

KNOWN $\sigma^2 = 3.15\ \mu m^2$
$s_x^2 = 5.00\ \mu m^2$ based on $N = 20$

FIND χ_a^2

SOLUTION   For $\nu = N - 1 = 19$, and using Equation 4.28, the observed χ^2 value is

$$\chi_\alpha^2(\nu) = \nu s_x^2/\sigma^2 = 30.16$$

For this value of χ^2 with $\nu = 19$, we find that the observed value of $\alpha \approx 0.05$ (Table 4.6). Taking $\chi_{0.05}^2$ as a measure of discrepancy due to random chance, we interpret this result as there being only a 5% chance that a new batch within tolerance might be rejected. Conclusion: Rejecting a batch on the basis of $s_x^2 > 5.00\ \mu m^2$ is a good test at a 5% level of significance.

Goodness-of-Fit Test

Just how well does a set of measurements follow an assumed distribution function? In Example 4.4, we assumed that the data of Table 4.1 followed a normal distribution based only on the rough form of its histogram (Fig. 4.2). A more rigorous approach would apply the chi-squared test using the chi-squared distribution. The chi-squared test provides a measure of the discrepancy between the measured variation of a data set and the variation due to chance as predicted by the assumed density function.

To begin, construct a histogram of K intervals from a data set of N measurements. This establishes the number of measured values, n_j, that lie within the jth interval. Then calculate the degrees of freedom in the variance for the data set, $\nu = N - m$, where m is the number of restrictions imposed. From ν, estimate the predicted number of occurrences, n'_j, to be expected from the distribution function. For this test, the χ^2 value is calculated from the entire histogram by

$$\chi^2 = \frac{\sum_j \left(n_j - n'_j \right)^2}{n'_j} \quad j = 1, 2, \ldots, K \tag{4.32}$$

The goodness-of-fit test evaluates the null hypotheses that the data are described by the assumed distribution against the alternative that the data are not sampled from the assumed distribution. For the given degrees of freedom, the better a data set fits the assumed distribution function, the lower its observed χ^2 value (left side of Table 4.6), whereas the higher the observed χ^2 value (right side of Table 4.6), the more dubious the fit.

Example 4.8

Test the hypothesis that the variable x as given by the measured data of Table 4.1 is described by a normal distribution at the 5% level of significance ($\alpha = 0.05$).

> **KNOWN**   Table 4.1 and histogram of Figure 4.2
> From Example 4.4: $\bar{x} = 1.02$; $s_x = 0.16$; $N = 20$; $K = 7$

FIND   Apply the chi-squared test to the data set to test for normal distribution.

SOLUTION   Figure 4.2 (Ex. 4.1) provides a histogram for the data of Table 4.1 giving the values for n_j for $K = 7$ intervals. To evaluate the hypothesis, we must find the predicted number of occurrences, n'_j, for each of the seven intervals based on a normal distribution. To do this, substitute $x' = \bar{x}$ and $\sigma = s_x$ and compute the probabilities based on the corresponding z values.

For example, consider the second interval ($j = 2$). Using Table 4.3, the predicted probabilities are

$$P(0.75 \leq x_i \leq 0.85) = P(0.75 \leq x_i < x') - P(0.85 \leq x_i < x')$$

$$= P(z_a) - P(z_b)$$

$$= P(1.6875) - P(1.0625)$$

$$= 0.454 - 0.356 = 0.098$$

Table 4.7 χ^2 Test for Example 4.8

j	n_j	n_j'	$\left(n_j - n_j'\right)^2 / n_j'$
1	1	0.92	0.01
2	1	1.96	0.47
3	3	3.76	0.15
4	7	4.86	0.93
5	4	4.36	0.03
6	2	2.66	0.16
7	2	1.51	0.16
Totals	20	20	$\chi_\alpha^2 = 1.91$

So for a normal distribution, we should expect the measured value to lie within the second interval for 9.8% of the measurements. With

$$n_2' = N \times P(0.75 \leq x_i < 0.85) = 20 \times 0.098 = 1.96$$

which is to say that 1.96 occurrences are expected out of 20 measurements in this second interval. The actual measured data set shows $n_2 = 1$.

The results are summarized in Table 4.7 with χ^2 based on Equation 4.32. Because two calculated statistical values ($\overline{x}$ and s_x) are used in the computations, the degrees of freedom in χ^2 are restricted by 2. So, $\nu = K - 2 = 7 - 2 = 5$ and, from Table 4.5, for the observed $\chi_\alpha^2(\nu = 5) = 1.91$, so $\alpha \approx 0.86$ (note: $\alpha = 1 - P$ is found here by interpolation between columns). At the tested level of significance, the critical value is $\chi_{0.05}^2(5) = 11.1$. As $\chi^2 = 1.98 < 11.1$, we do not reject the null hypothesis. The difference between the measured distribution and a normal distribution is not statistically significant. Conclusion: The data set is described by a normal distribution at the 5% level of significance.

4.7   REGRESSION ANALYSIS

A measured variable is often a function of one or more independent variables. We can use regression analysis to establish a functional relationship between the dependent variable and the independent variable. A regression analysis assumes that the variation found in the dependent measured variable follows a normal distribution about each fixed value of the independent variable. This behavior is illustrated in Figure 4.10 by considering the dependent variable $y_{i,j}$ consisting of N measurements, $i = 1, 2, \ldots, N$, of y at each of n values of independent variable, $x_j, j = 1, 2, \ldots, n$. This type of behavior is common during calibrations and in many types of measurements in which the dependent variable y is measured under controlled values of x. Repeated measurements of y yield a normal distribution with variance $s_y^2(x_j)$, about some mean value, $\overline{y}(x_j)$.

Most spreadsheet and engineering software packages can perform a regression analysis on a data set. The following discussion presents the concepts of a particular type of regression analysis, its

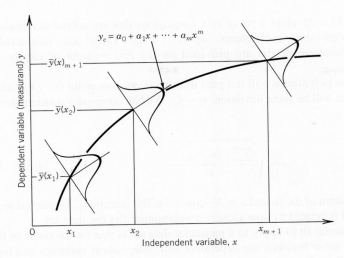

Figure 4.10 Distribution of measured value y about each fixed value of independent variable x. The curve y_c represents a possible functional relationship.

interpretation, and its limitations. More information on regression analysis, including multiple variable regression, can be found elsewhere (4, 6).

Least-Squares Regression Analysis

The regression analysis for a single variable of the form $y = f(x)$ provides an mth-order polynomial fit of the data in the form

$$y_c = a_0 + a_1 x + a_2 x^2 + \cdots + a_m x^m \tag{4.33}$$

where y_c refers to the value of y predicted by the polynomial equation for a given value of x. For n different values of the independent variable included in the analysis, the highest order, m, of the polynomial that can be determined is restricted to $m \leq n - 1$. The values of the m coefficients $a_0, a_1, \ldots, a_m$ are determined by the analysis. A common regression analysis for engineering applications is the method of least-squares. The *method of least-squares* attempts to minimize the sum of the squares of the deviations between the actual data and the polynomial fit of a stated order by adjusting the values of the polynomial coefficients.

Consider the situation in which there are N values of x and y, referred to as x_i, y_i, where $i = 1, 2, \ldots, N$. We seek an mth-order polynomial based on a set of N data points of the form (x, y) in which x and y are the independent and dependent variables, respectively. The task is to find the $m + 1$ coefficients, $a_0, a_1, \ldots, a_m$, of the polynomial of Equation 4.33. We define the deviation between any dependent variable y_i and the polynomial as $y_i - y_{c_i}$, where y_{c_i} is the value of the polynomial evaluated at the data point (x_i, y_i). The sum of the squares of this deviation for all values of $y_i, i = 1, 2, \ldots, N$, is

$$D = \sum_{i=1}^{N} \left(y_i - y_{c_i} \right)^2 \tag{4.34}$$

The goal of the method of least-squares is to reduce D to a minimum for a given order of polynomial. This is done by inserting Equation 4.33 into 4.34 and equating the total derivative of D to zero (i.e., $dD = 0$). This develops a set of $m + 1$ equations that are solved simultaneously to yield the unknown regression coefficients, $a_0, a_1, \ldots, a_m$. These are reinserted into Equation 4.33 for y_c. Intermediate details are provided on the text website or can be found in (1, 3, 4, 6).

In general, the regression polynomial will not pass through every data point (x_i, y_i) exactly. So with any data pair (x_i, y_i) there will be some deviation, $y_i - y_{c_i}$. We can compute a standard deviation based on these differences by

$$s_{yx} = \sqrt{\frac{\sum_{i=1}^{N} \left(y_i - y_{c_i}\right)^2}{\nu}} \tag{4.35}$$

where ν is the degrees of freedom of the fit and $\nu = N - (m + 1)$. The statistic s_{yx} is referred to as the *standard error of the fit* and is related to how closely a polynomial fits the data set.

The best order of polynomial fit to apply to a particular data set is that *lowest* order of fit that maintains a logical physical sense between the dependent and independent variables and reduces s_{yx} to an acceptable value. This first point is important. If the underlying physics of a problem implies that a certain order relationship should exist between dependent and independent variables, there is no sense in forcing a curve fit using higher order polynomials. Because the method of least-squares tries to minimize the sum of the squares of the deviations, it forces inflections in the curve fit that may not be real. Consequently, although higher-order curve fits generally reduce s_{yx}, they likely do not reflect the physics behind the data set very well. In any event, it is good practice to have at least two independent data points for each order of polynomial attempted—that is, at least two data points for a first-order curve fit, four for a second-order, and so on.

If we consider variability in both the independent and dependent variables, then the random uncertainty arising from random data scatter about the curve fit at any value of x is estimated by (1, 4)

$$t_{\nu, P} s_{yx} \left[\frac{1}{N} + \frac{(x - \bar{x})^2}{\sum_{i=1}^{N} (x_i - \bar{x})^2}\right]^{1/2} \quad (P\%) \tag{4.36}$$

which is known as the Scheffe band (10), and where

$$\bar{x} = \sum_{i=1}^{N} x_i / N$$

and x is the value used to estimate y_c in Equation 4.33. Hence the curve fit, y_c, with confidence interval is given as

$$y_c(x) \pm t_{\nu, P} s_{yx} \left[\frac{1}{N} + \frac{(x - \bar{x})^2}{\sum_{i=1}^{N} (x_i - \bar{x})^2}\right]^{1/2} \quad (P\%) \tag{4.37}$$

The effect of the second term in the brackets of either Equations 4.36 or 4.37 increases the confidence interval toward the outer limits of the polynomial. Often in engineering measurements, the independent variable is a well-controlled value and the principal source of variation in the curve fit is due to the random error in the dependent (measured) variable. A simplification to Equation 4.36 reduces the random uncertainty to

$$t_{\nu,P} \frac{s_{yx}}{\sqrt{N}} \quad (P\%) \tag{4.38}$$

The curve fit with its confidence interval is then approximated by

$$y_c \pm t_{\nu,P} \frac{s_{yx}}{\sqrt{N}} \quad (P\%) \tag{4.39}$$

The engineer should compare the values of Equations 4.36 and 4.38 to determine whether the approximation is acceptable. The simplification (Equation 4.38) should only be used when the curve fit uncertainty is not the dominant one in a complete uncertainty analysis.

Multiple regression analysis involving multiple variables of the form $y = f(x_{1_i}, x_{2_i}, \ldots)$ is also possible, leading to a multidimensional response surface. The details are not discussed here, but the concepts generated for the single-variable analysis are carried through for multiple-variable analysis. The interested reader is referred elsewhere (3, 4, 6).

The companion LabView program *Polynomial_Fit* performs a least-squares regression analysis. It allows the user to enter data points manually or to read data from a file. Other software packages can also do this processing.

Example 4.9

The following data are suspected to follow a linear relationship. Find an appropriate equation of the first-order form.

x (cm)	y (V)
1.0	1.2
2.0	1.9
3.0	3.2
4.0	4.1
5.0	5.3

KNOWN Independent variable x

Dependent measured variable y

$N = 5$

FIND $y_c = a_0 + a_1 x$

SOLUTION We seek a polynomial of the form $y_c = a_0 + a_1 x$, that minimizes the term $D = \sum_{i=1}^{N} (y_i - y_{c_i})^2$. With the total derivative expressed as $dD = \frac{\partial D}{\partial a_0} da_0 + \frac{\partial D}{\partial a_1} da_1$, then setting each partial derivative to zero gives

$$\frac{\partial D}{\partial a_0} = 0 = -2\left\{\sum_{i=1}^{N}\left[y_i - (a_0 + a_1 x_i)\right]\right\}$$

$$\frac{\partial D}{\partial a_1} = 0 = -2\left\{\sum_{i=1}^{N}\left[y_i - (a_0 + a_1 x_i)\right]\right\}$$

so that $dD = 0$. Solving simultaneously for the coefficients a_0 and a_1 yields

$$a_0 = \frac{\sum x_i \sum x_i y_i - \sum x_i^2 \sum y_i}{\left(\sum x_i\right)^2 - N \sum x_i^2}$$

$$a_1 = \frac{\sum x_i \sum y_i - N \sum x_i y_i}{\left(\sum x_i\right)^2 - N \sum x_i^2}$$

(4.40)

Substituting the data set into Equation 4.40 gives $a_0 = 0.02$ and $a_1 = 1.04$. Hence

$$y_c = 0.02 + 1.04x \quad \text{V}$$

Linear Polynomials

For linear polynomials a correlation coefficient r can be found by

$$r = \frac{N \sum_{i=1}^{N} x_i y_i - \sum_{i=1}^{N} x_i \sum_{i=1}^{N} y_i}{\sqrt{N \sum_{i=1}^{N} x_i^2 - \left(\sum_{i=1}^{N} x_i\right)^2} \sqrt{N \sum_{i=1}^{N} y_i^2 - \left(\sum_{i=1}^{N} y_i\right)^2}}$$

(4.41)

The correlation coefficient provides a measure of the association between x and y as predicted by the form of the curve fit equation. It is bounded by ± 1, which represents perfect correlation; the sign indicates that y increases or decreases with x. For $\pm 0.9 < r \leq \pm 1$, a linear regression can be considered a reliable relation between y and x. Alternatively, the r^2 value (called the coefficient of determination) is often reported, which is indicative of how well the variance in y is accounted for by the fit. The correlation coefficient only indicates the hypothesis that y and x are associated; correlation does not imply cause and effect. The r and r^2 values are not effective estimators of the random error in y_c; instead the s_{yx} value is used.

The precision estimate in the slope of the fit can be calculated by its standard error

$$s_{a_1} = s_{yx}\sqrt{\frac{1}{\sum_{i=1}^{N}(x_i - \bar{x})^2}}$$

(4.42)

The precision estimate of the zero intercept can be calculated by its standard error

$$s_{a_0} = s_{yx} \sqrt{\frac{\sum_{i=1}^{N} x_i^2}{N \sum_{i=1}^{N} (x_i - \bar{x})^2}} \tag{4.43}$$

An error in a_0 would offset a calibration curve from its y intercept. The derivation and further discussion on Equations 4.41 to 4.43 can be found elsewhere (1, 3, 4, 6).

Example 4.10

Compute the correlation coefficient and the standard error of the fit for the data in Example 4.9. Estimate the random uncertainty associated with the fit. State the correlation with its 95% confidence interval.

KNOWN $y_c = 0.02 + 1.04x$ V

ASSUMPTIONS   No systematic errors

FIND r and s_{yx}

SOLUTION   Direct substitution of the data set into Equation 4.41 yields the correlation coefficient of $r = 0.997$. An equivalent estimator is r^2. Here $r^2 = 0.99$, which indicates that 99% of the variance in y is accounted for by the fit. The correlation supports the linear relation between x and y.

The random uncertainty between the data and this fit is given by the standard error of the fit, s_{yx}. Using Equation 4.35,

$$s_{yx} = \sqrt{\frac{\sum_{i=1}^{N} (y_i - y_{c_i})^2}{\nu}} = 0.16$$

with degrees of freedom $\nu = N - (m + 1) = 3$. The t estimator, $t_{3,95} = 3.18$, establishes a random uncertainty about the fit of $t_{3,95} s_{yx} / \sqrt{N} = 0.23$. Applying Equation 4.39, the polynomial fit can be stated at 95% confidence as

$$y_c = 1.04x + 0.02 \pm 0.23 \text{ V} (95\%)$$

This curve is plotted in Figure 4.11 with its 95% confidence interval both using Equation 4.37 and the approximation given by Equation 4.39. Equation 4.37 shows a slightly broader uncertainty band with uncertainties of ± 0.39 at $x = 1$ and $x = 5$.

Example 4.11

A velocity probe provides a voltage output that is related to velocity, U, by the form $E = a + bU^m$. A calibration is performed, and the data ($N = 5$) are recorded below. Find an appropriate curve fit.

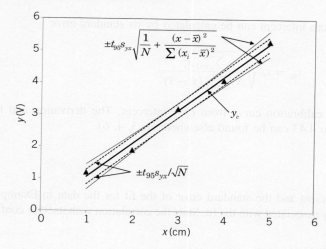

$$\pm t_{95}s_{yx}\sqrt{\frac{1}{N}+\frac{(x-\bar{x})^2}{\sum(x_i-\bar{x})^2}}$$

y_c

$\pm t_{95}s_{yx}/\sqrt{N}$

Figure 4.11 Results of the regression analysis of Example 4.10.

U (m/s)	E (V)
0.0	3.19
10.0	3.99
20.0	4.30
30.0	4.48
40.0	4.65

KNOWN $N = 5$

ASSUMPTIONS Data related by $E = a + bU^m$

FIND a and b

SOLUTION The suggested functional equation can be transformed into

$$\log(E - a) = \log b + m \log U$$

which has the linear form

$$Y = B + mX$$

Because at $U = 0$ m/s, $E = 3.19$ V, the value of a must be 3.19 V. The values for Y and X are computed below with the corresponding deviations from the resulting fit of 4 values:

U	$Y = \log(E-a)$	$X = \log U$	$E_i - E_{c_i}'$
0.0	—	—	0.0
10.0	−0.097	1.000	−0.01
20.0	0.045	1.301	0.02
30.0	0.111	1.477	0.0
40.0	0.164	1.602	−0.01

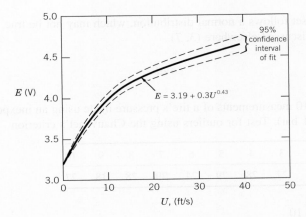

Figure 4.12   Curve fit for Example 4.11.

Substituting the values for Y and X into Equation 4.40 gives $a_0 = B = -0.525$ and $a_1 = m = 0.43$. The standard error of the fit is found using Equation 4.35 to be $s_{yx} = 0.007$. From Table 4.4, $t_{2,95} = 4.30$ so that $t_{2,95}s_{yx}/\sqrt{N} = 0.015$. This gives the curve fit with confidence interval as

$$Y = -0.525 + 0.43X \pm 0.015 \quad (95\%)$$

The polynomial is transformed back to the form $E = a + bU^m$:

$$E = 3.19 + 0.30U^{0.43} \text{ V}$$

The curve fit with its 95% confidence interval is shown on Figure 4.12.

4.8   DATA OUTLIER DETECTION

It is not uncommon to find a spurious data point that does not appear to fit the tendency of the data set. Data that lie outside the probability of normal variation incorrectly offset the sample mean value estimate, inflate the random uncertainty estimates, and influence a least-squares correlation. Statistical techniques can be used to detect such data points, which are known as *outliers*. Outliers may be the result of simple measurement glitches, such as a random spike in operating condition, or may reflect a more fundamental problem with test variable controls. Once detected, the decision to remove the data point from the data set must be made carefully. If outliers are removed from a data set, the statistics are recomputed using the remaining data.

One approach to outlier detection is Chauvenet's criterion, which identifies outliers having less than a $1/2N$ probability of occurrence. To apply this criterion, let the test statistic be $z_0 = |x_i - \bar{x}|/s_x$ where x_i is a suspected outlier in a data set of N values. In terms of the probability values of Table 4.3, the data point is a potential outlier if

$$(1 - 2 \times P(z_0)) < \frac{1}{2N} \tag{4.44}$$

For large data sets, another approach, the *three-sigma test*,[7] identifies those data points that lie outside the range of 99.73% probability of occurrence, $\bar{x} \pm t_{v,99.7}s_x$, as being potential outliers. Either

---

[7] So named for as $v \to \infty$, the value of t approaches 3 for 99.7% probability.

approach assumes that the sample set follows a normal distribution, which may not be true. Other methods for outlier detection are discussed elsewhere (3, 7).

Example 4.12

Consider the data given below for 10 measurements of a tire's pressure made using an inexpensive handheld gauge (note: 14.5 psi = 1 bar). Test for outliers using the Chauvenet's criterion.

i	1	2	3	4	5	6	7	8	9	10
x_i (psi)	28	31	27	28	29	24	29	28	18	27

KNOWN Data values for $N = 10$

ASSUMPTIONS Each measurement is obtained under fixed conditions

FIND Apply outlier detection tests

SOLUTION Based on the 10 data points, the sample mean and sample standard deviation are found to be $\bar{x} = 27$ psi with $s_x = 3.8$. But the tire pressure should not vary much between two readings beyond the precision capabilities of the measurement system and technique, so data point 9 ($x = 18$) is a potential outlier.

Apply Chauvenet's criterion to this data point. For $N = 10$, Equation 4.44 gives $(1 - 2 \times P(z_0)) < \frac{1}{2N} = \frac{1}{20} = 0.050$. So the criterion identifies the data point as an outlier if it lies outside $1 - 0.050 = 0.950$, which is the 95% probability spread of the data. For $x_i = 18$, $z_0 = 2.368$. Then $P(z_0) = P(2.368) = 0.4910$, so that $(1 - 2 \times 0.4910) = 0.018$. As $0.018 < 0.050$, this test identifies the data point $x = 18$ as a potential outlier.

4.9 NUMBER OF MEASUREMENTS REQUIRED

We can use the previous discussions to assist in the design and planning of a test program. For example, how many measurements, N, are required to reduce the estimated value for random error in the sample mean to an acceptable level? To answer this question, begin with Equation 4.19:

$$x' = \bar{x} \pm t_{v,P} s_{\bar{x}} \quad (P\%) \tag{4.19}$$

Let CI be the confidence interval in Equation 4.18; that is,

$$\mathrm{CI} = \pm t_{v,P} s_{\bar{x}} = \pm t_{v,P} \frac{s_x}{\sqrt{N}} \quad (P\%) \tag{4.45}$$

This interval is two-sided about the mean, defining a range from $-t_{v,P}\frac{s_x}{\sqrt{N}}$ to $+t_{v,P}\frac{s_x}{\sqrt{N}}$. We introduce the precision value d as

$$d = \frac{\mathrm{CI}}{2} = t_{v,P}\frac{s_x}{\sqrt{N}} \tag{4.46}$$

For example, if the confidence interval is ±1 units, an interval of width 2 units, then $d = 1$ unit. It follows that the required number of measurements is estimated by

$$N \approx \left(\frac{t_{v,P}s_x}{d}\right)^2 \quad (P\%) \tag{4.47}$$

Because the degrees of freedom in t depends on N, solving Equation 4.47 requires iteration. Equation 4.47 provides a first estimate for the number of measurements needed. How closely it reduces the range of random error to the constraint depends on how well the assumed value of s_x approximates the population σ.

A shortcoming of this method is the need to estimate s_x. In some cases, it can be based on experience of the population by using a known σ. Alternatively, the approach is to make a preliminary number of measurements, N_1, to obtain a sample estimate of s_1. Then use this s_1 to estimate the number of measurements required. The total number of measurements, N_T, will be estimated by

$$N_T \approx \left(\frac{t_{N-1,95}s_1}{d}\right)^2 \quad (95\%) \tag{4.48}$$

This establishes that $N_T - N_1$ additional measurements are required.

Example 4.13

Determine the number of measurements required to reduce the 95% confidence interval of the mean value of a variable to within ±1 unit if the variance of the population is estimated to be 64 units.

KNOWN   CI $= \pm 1$ units $= 2$ units $P = 95\%$
        $d = 1$ $\sigma^2 = 64$ units

FIND N required

SOLUTION Equation 4.47 has two unknowns in N and t

$$N \approx \left(\frac{t_{v,95}s_x}{d}\right)^2 \quad (95\%)$$

We iterate by using a trial-and-error approach until convergence. Suppose we begin by guessing that $N = 61$ so that $t_{v,95} = 2.00$. Then we solve Equation 4.47 with $t_{v,95} = 2.00$ and $s_x = 8$ units to find $N = 256$. We now use $N = 256$, so that $v = 255$ and $t_{v,95} = 1.96$. This gives the new estimate of $N = 245$. Repeat again with $N = 245$, so that $v = 244$, $t_{v,95} = 1.96$. We again get $N = 245$. The analysis is converged at 245 measurements. Check the results after 245 measurements to ensure that the variance level used was representative of the actual population.

COMMENT Since the confidence interval is reduced as $N^{1/2}$, the procedure of increasing N to decrease this interval becomes one of diminishing returns.

Example 4.14

From 21 preliminary measurements of a variable, the standard deviation of the data set is 160 units. We want to reduce the 95% confidence interval in the mean value to ±30 units. Estimate the total number of measurements required.

KNOWN $s_1 = 160$ units $N_1 = 21$
$d = CI/2 = 30$ $t_{20,95} = 2.093$

FIND N_T

SOLUTION   At 21 measurements, the confidence interval in the mean value is $\pm(2.093)(160)/\sqrt{21} = \pm 73$ units. We need to reduce this to ±30 units. The total number of measurements required is estimated by

$$N_T \approx \left(\frac{t_{N-1,95}s_1}{d}\right)^2 = \left(\frac{2.093 \times 160}{30}\right)^2 = 125 \quad (95\%)$$

Thus, as a first guess, a total of 125 measurements are estimated to be necessary. Take an additional 104 measurements, then reanalyze to be certain the constraint is met.

4.10   MONTE CARLO SIMULATIONS

When a result is computed from the values of one or more independent random variables, the variability in each of the independent variables directly affects the variability in the result. *Monte Carlo simulations* provide one way to incorporate such variability into predicting the behavior of the result. The simulation outcome is the predicted probability density function of the result, $p(R)$, and its associated statistics. This outcome makes it a very useful sampling method.

To illustrate a Monte Carlo simulation, consider result R that is a known function of two variables through the parametric relationship, $R = f(x, y)$, as in Figure 4.13. Each variable is defined by its own probability density function, $p(x)$ and $p(y)$. Each iteration of a Monte Carlo simulation

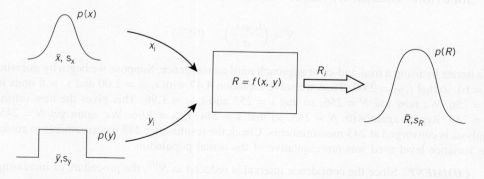

Figure 4.13   Elements of a Monte Carlo simulation of $R = f(x, y, \ldots)$

randomly draws one probable value for $x = x_i$ and for $y = y_i$ from their respective density functions to compute a value for $R = R_i$. This iteration process continues updating the data set for R until the predicted standard deviation for R converges to an asymptotic value (8). The quality of convergence must be traded against the cost of the simulation, but values within 1% to 5% suffice for many applications.

A Monte Carlo simulation is based on assumed distributions of variables, and so it is an approximation. Yet if the assumed distributions are reasonable, then the result of the approximation will be very good. The drawback is that the typical number of iterations required can be on the order of 10^4 to 10^6. The number of iterations can be reduced using improved sampling techniques (9).

Simulations can be run within spreadsheets, Matlab, or similar programs using built-in routines to facilitate sampling from a density function. For example, in spreadsheets, the RAND function samples from a rectangular distribution to generate a random number between 0 and 1, which can then be scaled to a population range. The NORMINV function samples from a normal distribution. In Matlab, these same operations use the RAND and NORMRAND functions, respectively.

Example 4.15

A small current is passed through a resistance circuit board to achieve a prescribed voltage for a certain application. From experience with one particular board design, a manufacturer knows that it can expect a mean resistance of 1000 Ω with a standard deviation of 100 Ω with a population best described by a normal distribution. A nominal 100 mA current is passed through the circuit, which can be set to within 5 mA, a process described by a rectangular distribution. Each circuit is tested for tolerance control. Model the expected population for this voltage test using a Monte Carlo simulation.

KNOWN $\overline{R} = 1000 \, \Omega$; $s_R = 100 \, \Omega$; normal distribution

$\overline{I} = 0.100 \, \text{A}$; $I_{max} = 0.105 \, \text{A}$; $I_{min} = 0.095 \, \text{A}$; rectangular distribution

SOLUTION   The parametric model is given by Ohm's law

$$E = f(I, \mathcal{R}) = I\mathcal{R}$$

where E is the result and $\mathcal{R}$ represents the resistance. The simulation starts by generating a random value for I and $\mathcal{R}$ based on a sampling of their respective density functions. A value for result E is then computed. This process repeats itself throughout the simulation.

For a rectangular function, sampling is done using a random number generator set between I_{max} and I_{min}. The standard deviation of a rectangular distribution is $(I_{max} - I_{min})/\sqrt{12}$. In a spreadsheet, each ith random sample of current is given by

$$I_i = I_{min} + RAND(\,)*(I_{max} - I_{min})$$

Here $(I_{max} - I_{min})$ appropriately scales the RAND value to the population. Similarly, the resistance is determined by sampling from a normal distribution. In a spreadsheet, this can be done with the

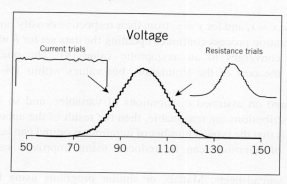

Figure 4.14 Predicted histogram for voltage (V) based on assumed distributions in current and resistance. Results of 100,000 Monte Carlo trials in Example 4.15.

NORMINV function. Each ith random sample for resistance is given by

$$\mathcal{R}_i = NORMINV(RAND(\,), \overline{\mathcal{R}}, S_{\mathcal{R}})$$

As new values for I and $\mathcal{R}$ are created, a new voltage is computed as

$$E_i = I_i \mathcal{R}_i$$

This builds the population for voltage. In Figure 4.14, we show the result from 100,000 iterations showing a normal distribution with the following statistics:

Variable, x	$\overline{x}$	s_x
$E\ [V]$	100.003	10.411
$I\ [A]$	0.100	0.0029
$\mathcal{R}\ [\Omega]$	999.890	99.940

So the voltage test population is normally distributed about a mean of 100.003 V with a standard deviation of 10.411 V.

4.11 SUMMARY

The behavior of a random variable is defined by its unique probability density function, which provides exact information about its mean value and variability. The purpose of measurements is to infer the parameters of this density function from the acquired limited data set. The normal scatter of data about some central mean value arises from several contributing factors, including the process variable's own variation under nominally fixed operating conditions and random errors in the measurement system and in the measurement procedure. Furthermore, a finite number of measurements of a variable can only go so far in estimating the behavior of an entire population of values. We discuss how estimates based on a limited data set introduce another random error into predicting the true value of the measurement, something addressed by the sample statistics.

Table 4.8   Summary Table for a Sample of N Data Points

Sample mean	$\bar{x} = \dfrac{1}{N} \displaystyle\sum_{i=1}^{N} x_i$
Sample standard deviation	$s_x = \sqrt{\dfrac{1}{N-1} \displaystyle\sum_{i=1}^{N} (x_i - \bar{x})^2}$
Standard deviation of the means^a	$s_{\bar{x}} = \dfrac{s_x}{\sqrt{N}}$
Precision interval for a single data point, x_i	$\pm t_{v,P} s_x \quad (P\%)$
Confidence interval^{b,c} for a mean value, $\bar{x}$	$\pm t_{v,P} s_{\bar{x}} \quad (P\%)$
Confidence interval^{b,d} for curve fit, $y = f(x)$	$\pm t_{v,P} s_{yx} \sqrt{\dfrac{1}{N} + \dfrac{(x-\bar{x})^2}{\sum_{i=1}^{N}(x_i - \bar{x})^2}} \quad (P\%)$

^{a}Measure of random standard uncertainty in x.
^{b}In the absence of systematic errors.
^{c}Measure of random uncertainty in $\bar{x}$.
^{d}Measure of random uncertainty in curve fit (see conditions of Eqs. 4.37–4.39).

Statistics is a powerful tool used to interpret and present data. In this chapter, we developed the most basic methods used to understand and quantify finite-sized data sets. Methods of estimating the true mean value based on a limited number of data points and the resulting random uncertainty in such estimates were presented along with treatment of data curve fitting. A summary table of these statistical estimators is given as Table 4.8. Monte Carlo methods were presented as an alternative tool to predict how variations in independent variables will affect the variation in a result. Throughout this chapter we have assumed negligible systematic error in the data set. In the next chapter, we expand our error considerations in our estimate of the true value of a measurement.

REFERENCES

1. Kendal, M. G., and A. Stuart, *Advanced Theory of Statistics*, Vol. 2, Griffin, London, 1961.
2. Bendat, J., and A. Piersol, *Random Data Analysis*, Wiley, New York, 1971.
3. Lipson, C., and N. J. Sheth, *Statistical Design and Analysis of Engineering Experiments*, McGraw-Hill, New York, 1973.
4. Miller, I., and J. E. Freund, *Probability and Statistics for Engineers*, 3rd ed., Prentice Hall, Englewood Cliffs, NJ, 1985.
5. Bury, K., *Statistical Distributions in Engineering*, Cambridge Press, Cambridge, UK, 1999.
6. Vardeman, S. B., *Statistics for Engineering Problem Solving*, PWS Publishing Company, Boston, 1994.
7. ASME/ANSI Power Test Codes, *Test Uncertainty PTC 19.1–2013*, American Society of Mechanical Engineers, New York, 2013.
8. Joint Committee for Guides in Metrology (JCGM), *Propagation of Distributions Using the Monte Carlo Method*, JCGM 101:2008, JCGM, Sevres Cedex, France, 2008.
9. Devroye, L., *Non-Uniform Random Variate Generation*, Springer, 1986.
10. Scheffe, H., *The Analysis of Variance*, Wiley Classics Series, Wiley Interscience, 1999.

NOMENCLATURE

$a_0, a_1, \ldots, a_m$	polynomial regression coefficients	$\langle s_{\bar{x}} \rangle$	pooled standard deviation of the means of x
f_j	frequency of occurrence of a value		
x	measured variable; measurand	$\langle s_x \rangle^2$	pooled sample variance of x
x_i	ith measured value in a data set	s_{yx}	standard error of the (curve) fit
x'	true mean value of the population of x		between y and x; standard random uncertainty in a curve fit
$\bar{x}$	sample mean value of x	t	Student's t variable
$\langle \bar{x} \rangle$	pooled sample mean of x	z	z variable
$p(x)$	probability density function of x	β	normalized standard variate
s_x	sample standard deviation x	σ	true standard deviation of the population of x
$s_{\bar{x}}$	sample standard deviation of the means of x; standard random uncertainty in $\bar{x}$	σ^2	true variance of the population of x
s_x^2	sample variance of x	χ^2	chi-squared value
$\langle s_x \rangle$	pooled sample standard deviation of x	ν	degrees of freedom

PROBLEMS

4.1 A large population of variable x is characterized by its known mean value of 6.1 units and a standard deviation of 1.0 units and a normal distribution. Determine the range of values containing 70% of the population of x.

4.2 From a very large data set, variable x is known to have a mean value of 192 units with a standard deviation of 10.0 units. Determine the range of values within which we would expect 90% of the population of x to fall. Assume that x is normally distributed.

4.3 At a fixed operating setting, the pressure in a line downstream of a reciprocating compressor is known to maintain a mean value of 12.0 bar with a standard deviation of 1.0 bar based on its history. What is the probability that the line pressure will exceed 13.0 bar during any measurement?

4.4 Consider the toss of four coins. There are 2^4 possible outcomes of a single toss. Develop a histogram of the number of heads (one side of the coin) that can appear on any toss. Does it look like a normal distribution? Should this be expected? What is the probability that three heads will appear on any toss?

4.5 As a game, slide a matchbook across a table, trying to make it stop at some predefined point on each attempt. Measure the distance from the starting point to the stopping point. Repeat this 10 times, 20 times, up through 50 times. Plot the frequency distribution from each set. Would you expect them to look like a normal distribution? What statistical outcomes would distinguish a better player? Would the distribution look different if the goal was to get the matchbox to stop as close as possible to the edge of the table?

4.6 The large population of a measured variable has a mean value of 23.0 and a standard deviation of 1.0. What percentage of the population falls between (a) 22 and 24? (b) 22.5 and 23.5? (c) 21 and 25? (d) 21 and 23?

4.7 A student conducts a lab exercise to calculate the acceleration of gravity, g, and gets a value of 9.5 m/s^2 with a standard deviation of 0.1. He knows that the answer at his altitude and earth coordinates should be 9.8 m/s^2. What is the probability of getting this result due to random chance (random error) alone? Neglecting outright blunders (he is a careful experimenter), give a possible cause for the result.

4.8 From a large database, the Highway Patrol estimates that the average speed on a particular busy stretch of route I-26 on a holiday weekend is 67 mph (miles per hour) with a standard deviation of 4 mph. The speed limit is 60 mph. The highway patrol will ticket anyone driving faster than 72 mph. What

Table 4.9 Measured Force Data for Exercise Problems

F (N) Set 1	F (N) Set 2	F (N) Set 3
51.9	51.9	51.1
51.0	48.7	50.1
50.3	51.1	51.4
49.6	51.7	50.5
51.0	49.9	49.7
50.0	48.8	51.6
48.9	52.5	51.0
50.5	51.7	49.5
50.9	51.3	52.4
52.4	52.6	49.5
51.3	49.4	51.6
50.7	50.3	49.4
52.0	50.3	50.8
49.4	50.2	50.8
49.7	50.9	50.2
50.5	52.1	50.1
50.7	49.3	52.3
49.4	50.7	48.9
49.9	50.5	50.4
49.2	49.7	51.5

percentage of drivers can they expect to exceed 72 mph? Note: 0.62 kph = 1 mph.

Problems 4.9 through 4.15 refer to the three measured data sets in Table 4.9. *Assume that the data have been recorded from three repeated tests (replications) from the same process under the same fixed operating condition.*

4.9 Develop the histograms of the three data sets represented under columns 1, 2, and 3. If these are taken from the same process, why might the histograms vary?

4.10 For the data in each column, determine the sample mean value, standard deviation, and standard deviation of the means. State the degrees of freedom in each.

4.11 Explain the concept of "central tendency" by comparing the range of the measured values, histograms, and the sample mean values from each of the three data sets.

4.12 From the data in column 1, estimate the range of values for which you would expect 95% of all possible measured values for this operating condition to fall. Repeat for columns 2 and 3. Discuss these outcomes in terms of what you might expect from finite statistics.

4.13 From the data in column 1, determine the best estimate of the mean value at a 95% probability level. How does this estimate differ from the estimates made in problem 4.12? Repeat for columns 2 and 3. Why do the estimates vary for each data set? Discuss these outcomes in terms of what you might expect from finite statistics if these are measuring the same measured variable during the same process.

4.14 Compute a pooled sample mean value for the process. State the range for the best estimate in mean force at 95% probability based on these data sets. Compute measures for the between-test and the within-test variations. What are the degrees of freedom for each measure. Compute the F value. Discuss the limitations of the inference of population statistics based on sample statistics.

4.15 Apply the χ^2 goodness-of-fit test to the data in column 1 and test the assumption of a normal distribution.

4.16 From 19 measurements of a pressure controller, the mean operating pressure is 4.97 kPa with a standard deviation of 0.0461 kPa. If more measurements are taken, estimate the expected range of pressure for 95% of the measurements.

4.17 From 19 measurements of a pressure controller, the mean operating pressure is 4.97 kPa with a standard deviation of 0.0461 kPa. Neglecting other sources or error, state the true mean pressure at 95% confidence.

4.18 When the parameters of a population are known, the likelihood of obtaining a certain mean value, $\bar{x}$, from a small sample can be predicted by the z-variable with $z = (\bar{x} - x')/\sigma/\sqrt{N}$. Consider a stamping process in which the applied load has a known true mean of 400 N with standard deviation of 25 N. If 25 measurements of applied load are taken at random, what is the probability that this sampling will have a mean value between 390 and 410?

4.19 A professor grades students on a normal curve. For any grade x, based on a course mean and standard deviation developed over years of testing, the following applies:

$$A : x > \bar{x} + 1.6\sigma$$
$$B : \bar{x} + 0.4\sigma < x \leq \bar{x} + 1.6\sigma$$
$$C : \bar{x} - 0.4\sigma < x \leq \bar{x} + 0.4\sigma$$
$$D : \bar{x} - 1.6\sigma < x \leq \bar{x} - 0.4\sigma$$
$$F : x \leq \bar{x} - 1.6\sigma$$

How many A, C, and D grades are given per 100 students?

4.20 The production of a certain polymer fiber follows a normal distribution with a true mean diameter of 20 μm and a standard deviation of 30 μm. Compute the probability of a measured value greater than 80 μm. Compute the probability of a measured value between 50 and 80 μm.

4.21 An automotive manufacturer removes the friction linings from the clutch plates of drag race cars following test runs. From experience, it expects to find the mean wear to be 209.6 μm with a standard deviation of 52.5 μm. How many clutch plates out of a large set will be expected to show wear of less than 203 μm?

4.22 Determine the mean value of the life of an electric light bulb if

$$p(x) = 0.001e^{-0.001x} \quad x \geq 0$$

and $p(x) = 0$ otherwise. Here x is the life in hours.

4.23 Compare the reduction in the possible range of random error in estimating x' by taking a sample of 16 measurements as opposed to only four measurements. Then compare 100 measurements to 25. Explain "diminishing returns" as it applies to using larger sample sizes to reduce random error in estimating the true mean value.

4.24 A sampling of 409 motor shafts from a very large production batch shows a sample standard deviation in diameter of 0.021 mm with a sample mean of 9.251 mm. Estimate the mean diameter of the entire batch at a 95% level of confidence?

4.25 During the course of a motor test, the motor rpm (revolutions per minute) is measured and recorded at regular intervals as

990   1030   950   1050   1000   980

Calculate the mean value, standard deviation, and best estimate of true value for this data set. Over what interval would 50% of the entire population of motor speed values fall? Test the data set for potential outliers.

4.26 A batch of rivets is tested for shear strength. A sample of 31 rivets shows a mean strength of 924.2 MPa with a standard deviation of 18 MPa. Estimate of the mean shear strength for the batch at 95% probability.

4.27 Three independent sets of data are collected from the population of a variable during similar process operating condition. The statistics are found to be

$$N_1 = 16; \quad \bar{x}_1 = 32; \quad s_{x_1} = 3 \text{ units}$$
$$N_2 = 21; \quad \bar{x}_1 = 30; \quad s_{x_2} = 2 \text{ units}$$
$$N_3 = 9; \quad \bar{x}_1 = 34; \quad s_{x_3} = 6 \text{ units}$$

Neglecting systematic errors and random errors other than the variation in the measured data set, compute an estimate of the pooled mean value of this variable and the range in which the true mean should lie with 95% confidence.

4.28 The following data were collected by measuring the force load acting on a small area of a beam under repeated "fixed" conditions:

	Output (N)		Output (N)
1	923	6	916
2	932	7	927
3	908	8	931
4	932	9	926
5	919	10	923

Determine whether any of these data points should be considered outliers. If so, reject the data point. Estimate the true mean value from this data set assuming that the only error is from variations in the data set.

4.29 An engineer measures the diameter of 20 tubes selected at random from a large shipment. The sample yields $\bar{x} = 47.5$ mm and $s_x = 8.4$ mm. The manufacturer of the tubes claims that $x' = 42.1$ mm. What can the engineer conclude about this claim?

For Problems 4.30 through 4.36, it would be helpful to use a least-squares software package, such as Polynomial_fit, Matlab, or spreadsheet software.

4.30 The following relations exist between material flexural strength and grain size:

Flexural strength x (MPa)	65	63	67	64	68	62	70	66	68	67	69	71
Grain size, y (nm)	68	66	68	65	69	66	68	65	71	67	68	70

Plot the linear relationship and estimate the standard error of the fit.

4.31 Experimental measurements are taken from a physical system that can be modeled as $y = bx^m$. The experimental data is (1.020, 3.270), (1.990, 12.35), (3.000, 28.03), (4.010, 50.04), (4.990, 77.47). Find the coefficients to fit the model using the method of least squares and calculate the standard error of the fit.

4.32 Experimental measurements are taken from a physical system that can be modeled as $y = mx + b$. The experimental data is (1.00, 2.10), (2.00, 2.91), (3.00, 3.89), (4.00, 5.12), (5.00, 6.09). Find the coefficients to fit the model using the method of least squares, calculating the standard error of the fit and the 95% confidence band about the curve fit (Scheffe band).

4.33 In an engineering lab exercise, a student makes 10 independent measurements of a voltage drop across a circuit board resistor:

0.86,	0.83,	0.82,	0.84,	0.95,
0.87,	0.88,	0.89,	0.83,	0.85

Could any value be rejected using Chauvenet's criterion?

4.34 Find the first-order line ($y = a_0 + a_1 x$) that fits the four data pairs (1, 12), (2, 13), (3, 18), (4, 19). Estimate the standard error of the fit. How many degrees of freedom are there in the curve fit? Show the uncertainty bands about the fit.

4.35 A student lab wants to determine the rate of radioactive decay of an isotope by measuring the rate of decay, R, once per hour over a three-hour period. The rate of decay is modeled as $R = R_0 e^{-t/\tau}$ where τ is the time constant of decay. The measurements are (0, 13.8), (1, 7.9), (2, 6.1), (3, 2.9). Apply least-squares regression to the data to obtain an estimate for the constant of decay τ.

4.36 A fan performance test yields the following data pairs for (Q, h): (2000, 5.56), (6000, 5.87), (10000, 4.95), (14000, 4.95), (18000, 3.52), (22000, 1.08). Q is flow rate in m^3/s and h is static head in cm-H_2O. Find the lowest-degree polynomial that best fits the data as $h = f(Q)$. Note: Physically, a second-order polynomial is a reasonable fit. Explain your choice.

4.37 A manufacturer of general aircraft dry vacuum pumps wishes to estimate the mean failure time of its product at 95% confidence. Initially, six pumps are tested to failure with these results (in hours of operation): 1272, 1384, 1543, 1465, 1250, 1319. Estimate the sample mean and the 95% confidence interval of the true mean. How many more data points would be needed to improve the confidence interval to be within ± 50 hours?

4.38 Strength tests made on a batch of cold-drawn steel yield the following:

Strength (MPa)	Occurrences
421–480	4
481–540	8
541–600	12
601–660	6

Test the hypothesis that the data are a random sample that follows a normal distribution at the 5% level of significance.

4.39 Determine the one-sided 95th-percentile of the chi-squared distribution at 16 degrees of freedom. Refer to Figure 4.9 and Table 4.6.

4.40 The manufacturer of an electrical fuse claims that the fuse's link will break with a 10% electrical overload within 12.31 minutes. Independent tests on a random sampling of 20 fuses showed the mean time for link breakage was 10.59 minutes with a standard deviation of 2.61 minutes. Does the sampling support the manufacturer's claim? Test the hypothesis.

4.41 How many independent random measurements of a time-dependent acceleration signal obtained from a vibrating vehicle under fixed operating conditions would lead to a confidence interval about the mean of ± 0.5 g, if the standard deviation of the signal is expected to be 1 g?

4.42   Based on 51 measurements of a time-dependent electrical signal, the standard deviation is 1.52 V. Find the 95% confidence interval in its true mean value. How many more measurements would be required to provide a 95% confidence interval in the true mean to within ± 0.28 V?

4.43   Estimate the probability with which an observed $\chi^2 = 19.0$ will be exceeded if there are 10 degrees of freedom.

4.44   A company produces a computer battery that it claims last for 8 hours on a single charge, with a standard deviation of 20 minutes. A random sampling of seven batteries are tested and found to have a sample standard deviation of 29 minutes. What is the chi-squared value and the predicted level of significance as represented by the test?

4.45   A conductor is insulated using an enameling process. It is known that the number of insulation breaks per meter of wire is 0.07. What is the probability of finding x breaks in a piece of wire 5 m long? Use the Poisson distribution.

4.46   We know that 2% of the screws made by a certain machine are defective, with the defects occurring randomly. The screws are packaged 100 to a box. Estimate the probability that a box will contain x defective screws using (a) the binomial distribution and (b) the Poisson distribution.

4.47   An optical velocity measuring instrument provides an updated signal on the passage of small particles through its focal point. Suppose the average number of particles passing in a given time interval is 4. Estimate the probability of observing x particle passages in a given time. Use the Poisson distribution for $x = 1$ through 10.

4.48   A cantilever is loaded at its tip. Tip deflection is measured. The following results are obtained with 10 repetitions of loading (cm):

5.30	5.73	6.77	5.26	4.33
5.45	6.09	5.64	5.81	5.75

Test for outliers using Chauvenet's criterion. Recompute, as necessary, mean value, standard deviation, and best estimate of true value for this data set. Over what interval would 50% of the entire population of tip deflection values fall?

4.49   A small sample ($N = 7$) of the static coefficient of friction (μ) between two materials is measured with the following data set:

0.0043	0.0050	0.0053	0.0047
0.0031	0.0051	0.0049	

Test the data for outliers. Determine the mean value and its confidence.

4.50   A company packages pretzels into individual bags, each having an advertised mean weight of 454.0 g. A quality control engineer takes regular samples and tests them. From experience, the engineer can expect a population standard deviation of 7.8 g when the machinery is operating correctly. During one such test, a random sample of 25 bags shows the mean weight to be 450.0 g. Is there a problem with the equipment at the 5% level of significance? What would be the maximum deviation between the sample mean weight and the target weight for the machinery to be within specification?

4.51   A transducer is used to indicate the pressure of a vessel when the pressure exceeds 8.16 bar. The test of the system is repeated 20 times, resulting in a mean pressure of 8.23 bar with a standard deviation of 0.201 bar. Determine the reliability of the transducer at $\alpha = 0.05$.

4.52   An engineer takes water samples from 15 lakes in the southern Appalachian mountains region to test for acidity. If the pH is greater than 6, then the lakes are, on average, nonacidic. The engineer finds a mean pH of 6.3 with a sample standard deviation of 0.672. Can the engineer conclude that on average, the lakes are not acidic at the 5% significance level? Set up a single-tailed hypothesis test to justify your answer. (Note: The critical value is 1.761 for $\alpha = 0.05$.)

4.53   A random sample of 51 aircraft rivets is taken from a large batch and tested for shear strength. The mean value was 921.2 lbs with $s_x = 18.1$. The batch is stated to have a mean strength on average of 925.0 lbs. Can we conclude that the batch mean shear strength is 925 lbs at a 95% confidence level?

4.54   Estimate the p-value for $N = 20$, $t = 2.096$ (interpolate as needed).

4.55   Estimate the p-value for $N = 16$, $t = 1.798$ (interpolate the tables as needed to find an approximate value).

4.56 ASTM F558 describes a method for testing the flow performance of vacuum cleaners. To establish performance, it requires that a minimum of three units of a model be tested in a prescribed manner, repeated three times (trials) for each unit. If the spread between the maximum and minimum values in the trials exceeds 6% of the maximum trial value, that unit is to be retested (repeatability limit). The following test data are recorded. Establish the mean and standard deviation for each unit and test that repeatability limits (per code) are met. Performance is in units of air watts (flow rate times suction pressure).

Air Power (W)	Unit 1	Unit 2	Unit 3
Trial 1	293.5	274.6	301.4
Trial 2	290.7	273.6	296.8
Trial 3	276.1	281.8	296.1

4.57 Following the information of Problem 4.56, ASTM 558 requires calculating the 90% confidence interval for each unit tested. If that confidence interval exceeds 5% of the mean value found for that unit, then that unit and its test results must be discarded and a new unit taken from the population and tested. Based on the data set in the preceding problem, must another unit be tested? If not, report the pooled mean for the model.

4.58 To demonstrate the expected frequency of obtaining heads in a coin toss of 10 tries, you are to generate a histogram from repeated trials (1 trial = 10 tosses). In generating the histogram, continue until your expected frequency converges on a value. You might use a spreadsheet or Matlab to conduct a Monte Carlo simulation, recognizing that obtaining either a heads or tails outcome in a single toss is a random event with equal probability.

4.59 Conduct a Monte Carlo simulation to predict the statistical outcome of a planned test to measure drag coefficient, $C_D = D/(0.5\rho U^2 A)$, on an airplane wing model. Experience shows that drag, D, and density, ρ, follow normal distributions, while velocity, U, and area, A, follow rectangular distributions. Expected values are given below. Determine the mean, standard deviation, number of iterations for convergence, and the histogram for C_D.

$$\overline{D} = 5 \text{ N} \qquad s_D = 0.5 \text{ N}$$
$$\overline{\rho} = 1.225 \text{ kg/m}^3 \qquad s_\rho = 0.012 \text{ kg/m}^3$$
$$9 \le U \le 10 \text{ m/s}$$
$$0.48 \le A \le 0.52 \text{ m}^2$$

4.60 Run program *Finite Population*. Describe the evolution of the measured histogram as N increases. Why do the statistical values keep changing? Is there a tendency in these values?

4.61 Run program *Running Histogram*. Study the influence of the number of data points and the number of intervals on the shape of the histogram and magnitude of each interval. Describe this influence. Why are some values "out of range?"

4.62 Run program *Probability Density*. Vary the sample size and number of intervals used to create the density function. Why does it change shape with each new data set?

Chapter 5

Uncertainty Analysis

5.1 INTRODUCTION

Whenever we plan a test or later report a test result, we want to know something about the quality of the result. Uncertainty analysis provides a methodical approach to estimating the quality of the result from a test. This chapter focuses on how to estimate the "± what?" in a test result.

Suppose the competent dart thrower of Chapter 1 tossed several practice rounds of darts at a bullseye. This would give us a good idea of the thrower's tendencies. Then let the thrower toss another round. Without looking, can you guess where the darts will hit? Test measurements that include systematic and random error components are much like this. We can calibrate a measurement system to get a good idea of its behavior and accuracy. However, from the calibration we can only estimate how well any measured value might estimate the exact (or "true") value in a subsequent measurement.

Errors are a property of the measurement. Measurement is the process of assigning a value to a physical variable based on a sampling from the population of that variable. Error causes a difference between the value assigned by measurement and the true value of the population of the variable measured. Measurement errors are introduced from various elements—for example, the individual instrument calibrations, the data set finite statistics, and the approach used. If we knew the exact value for an error, we could just correct for it. But because we do not know the true value of the measured variable but only its measured value, we do not know the exact value of error. Instead, we draw from what we do know about the measurement to estimate a range of probable error. This estimate of the range of probable error is an assigned value called the *uncertainty*. The uncertainty describes an interval about the measured value within which we suspect that the true value must fall at a stated level probability. *Uncertainty analysis* is the process of identifying, quantifying, and combining the estimated values of the errors.

Uncertainty is a property of the result. The outcome of a measurement is a result, and the uncertainty quantifies the quality of that result. Uncertainty analysis provides a powerful design tool for evaluating different measurement systems and methods, designing a test plan, and reporting uncertainty. This chapter presents a systematic approach for identifying, quantifying, and combining the estimates of the errors in a measurement. Although the chapter stresses the methodology of analyses, we emphasize the concomitant need for an equal application of critical thinking and professional judgment in applying the analyses. The quality of an uncertainty analysis depends on the engineer's knowledge of the test, the measured variables, the equipment, and the measurement procedures (1).

Errors are effects, and uncertainties are numbers. Whereas errors are the effects that cause a measured value to differ from the true value, the uncertainty is an assigned numerical value that quantifies the probable range of these errors.

This chapter approaches uncertainty analysis as an evolution of information from test design through final data analysis. Whereas the structure of the analysis remains the same at each step, the number of errors identified and their uncertainty values may change as more information becomes available. In fact, the uncertainty in the result may increase. There is no exact answer to an analysis, just the result from a reasonable approach using honest numbers. This is the nature of an uncertainty analysis.

There are two accepted professional documents on uncertainty analysis. The American National Standards Institute/American Society of Mechanical Engineers (ANSI/ASME) Power Test Codes (PTC) 19.1 Test Uncertainty (2) is the United States engineering test standard, and our approach follows that method. The International Organization for Standardization's "Guide to the Expression of Uncertainty in Measurement" (ISO GUM) (1) is an international metrology standard. The two differ in some terminology and how errors and their uncertainties are cataloged. For example, PTC 19.1 refers to random and systematic errors, terms that classify errors by how they manifest themselves in the measurement. Random and systematic uncertainties are the values assigned to these errors. ISO GUM assigns type A and type B uncertainties to errors, terms that reflect only how the values assigned to errors are estimated. These differences are real but they are not significant to the outcome. Once past the classifications, the two methods are quite similar. The important point is that the end outcome of an uncertainty analysis by either method will yield a similar result!

Upon completion of this chapter, the reader will be able to

- Explain the relation between an error and an uncertainty
- Execute an appropriate uncertainty analysis corresponding to the level and quantity of information available
- Explain the differences between systematic and random errors and how to treat their assigned uncertainties
- Analyze a test system and approach from test design through data presentation to assign and propagate uncertainties
- Propagate uncertainties throughout a test to understand their impact on the final reported result

5.2   MEASUREMENT ERRORS

In the discussion that follows, we discuss how errors affect a result. For this, errors are grouped into two categories: systematic error and random error.

Consider the repeated measurement of a variable under conditions that are expected to produce the same value of the measured variable. The relationship between the true value of the population and the measured data set, containing errors, can be illustrated as in Figure 5.1. The total error in a set of measurements obtained under seemingly fixed conditions can be described by the systematic errors and the random errors in those measurements. The systematic errors shift the sample mean away from the true mean by a fixed amount, and within a sample of many measurements the random errors bring about a distribution of measured values about the sample mean. Even a so-called accurate

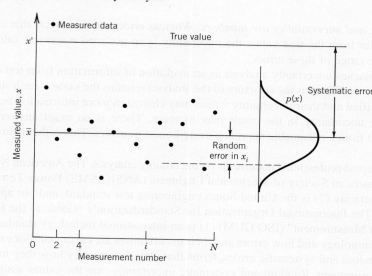

Figure 5.1 Distribution of errors on repeated measurements.

measurement contains small amounts of systematic and random errors. We do not consider the effects of measurement blunders that result in obviously fallacious data—such data should be discarded.

Measurement errors enter during all aspects of a test and obscure our ability to ascertain the information that we desire: the true value of the variable measured. If the result depends on more than one measured variable or test parameter, the errors associated with each entity will propagate through the measurements and affect the result. In Chapter 4, we stated that the best estimate of the true value sought in a measurement is provided by its sample mean value and the uncertainty in that value,

$$x' = \bar{x} \pm u_x \quad (P\%) \tag{4.1}$$

Equation 4.1 still holds, but in Chapter 4 we considered only random uncertainty based on the statistics calculated from a measured data set. In this chapter, we extend this so that the u_x term reflects the uncertainties assigned to all known errors, both systematic and random. Certain assumptions are implicit in an uncertainty analysis:

1. The test objectives are known and the measurement itself is a clearly defined process.

2. Any known corrections for systematic error have been applied to the data set, in which case the systematic uncertainty assigned to that error is the uncertainty of the correction.

3. Except where stated otherwise, we assume a normal distribution of errors and reporting of uncertainties.

4. Unless stated otherwise and for simplicity, the errors are assumed to be independent (uncorrelated) of each other. But errors can be correlated and we discuss how to handle these in Section 5.9.

5. The engineer has some "experience" with the system components.

Regarding item 3, we usually assume that the population of errors is best described by a normal distribution. But this is not always so. For example, when manufacturer specifications cite only "limits of error," the error source population might be better described by a rectangular distribution. This is discussed more in Section 5.7.

In regards to item 5, by "experience" we mean that the engineer is able to assign an uncertainty value based on tests to enable assigning a value, based on past experience of what to expect from a particular measurement, or based on other information, such as manufacturer specifications or the technical literature.

We might begin the design of an engineering test with an idea and some catalogs and end the project after data have been obtained and analyzed. As with any part of the design process, the uncertainty analysis evolves as the design of the measurement system and process matures. In the next several sections, we discuss uncertainty analysis for the following measurement situations: (1) design stage, when tests are being planned but information is limited, (2) advanced stage and single measurement analysis, when additional information about process control is used to improve a design-stage uncertainty estimate, and (3) multiple measurement analysis, when all available test information and test data are combined to assess the uncertainty in the measured test result. The methods we use for situation 3 follow current engineering standards.

General versus Detailed Uncertainty Analysis   In this chapter, we discuss how uncertainty is propagated from individual contributions of error to the uncertainty in the final result. We are particularly interested in how each error affects the result. However, in performing an uncertainty analysis for measurement situations (1) and (2) previously discussed, we often begin with a "general" analysis, in which no distinction is made between random and systematic errors. In a general analysis, we simply assign uncertainty values to errors as indicated by a value u. This is particularly useful when we have little information about the basis of each uncertainty value or its effect on the test outcome or are making a quick analysis perhaps so as to guide further test decisions.

For situation (3), and often for situation (2), we will have sufficient information to distinguish between systematic and random errors, so we can assign uncertainty values separately for each error, enabling a "detailed" analysis. In a detailed analysis, systematic and random uncertainties are designated as values b and s, respectively, until the end result is computed. This is the preferred approach for analyzing test information for product performance assurance and for improving test methods to achieve better results. The added value of using this detailed approach is that by understanding how individual errors affect a result, we can better understand how to interpret the results or improve the test.

5.3   DESIGN-STAGE UNCERTAINTY ANALYSIS

Design-stage uncertainty analysis refers to an analysis performed early in the formulation of a test. It is intended to provide a quick estimate of the minimum uncertainty to be expected based on available information, which in the planning stage is usually just information on the instruments for the measurement technique planned. If this uncertainty value is too large, then alternate approaches will need to be found. So, it is useful for selecting instruments and selecting between alternative measurement techniques. Identifying random and systematic errors separately in this preliminary analysis is unnecessary, so we suggest using a general analysis and designate each uncertainty value in this analysis with the symbol u.

The *zero-order uncertainty* of an instrument, u_0, attempts to estimate the expected variation in the measured values. As a minimum, this variation will be an amount on the level of the instrument resolution (i.e., interpolation error). Assign a numerical value to u_0 of either one-half of the analog instrument resolution[1] or its digital least count. This value will reasonably represent the uncertainty interval on either side of the reading with a probability of 95%. Then,

$$u_0 = \frac{1}{2}\text{resolution} = 1\,\text{LSD} \tag{5.1}$$

where LSD refers to the least significant digit of the readout, its least count. If information exists to better predict the expected variation in a measured value, use it instead.

The second piece of information that is usually available is the manufacturer's statement concerning instrument error. We can assign this stated value as the *instrument uncertainty*, u_c. If no probability level is provided with such information, a 95% level can be assumed. Of course, if additional information exists to better predict measurement uncertainty, use it. Often, experienced users have good idea of what to expect based on their prior use of equipment and techniques. We next look at a way to combine individual uncertainty values into a single design-stage analysis estimate.

Combining Elemental Errors: RSS Method

Each individual measurement error interacts with other errors to affect the overall uncertainty of a measurement. This is called *uncertainty propagation*. Each individual error is called an "elemental error." Errors are often delineated into elemental parts, each part due to an individual contributing factor (recall Table 1.1). For example, the sensitivity error and linearity error of a transducer are two elemental errors, and the numbers associated with these are their uncertainties. These individual uncertainties must be combined in some reasonable manner. An accepted approach of combining uncertainties is termed the *root-sum-squares* (RSS) method.

Consider a measurement of x that is subject to some K elements of error, each of uncertainty u_k, where $k = 1, 2, \ldots, K$. If we know that (1) each error is independent of the other elemental errors and (2) assume that the measured value is equally sensitive to each of the elemental errors, then the uncertainty in the measured variable, u_x, due to these elemental errors can be computed using the RSS *method*:

$$\begin{aligned} u_x &= \sqrt{u_1^2 + u_2^2 + \cdots + u_k^2} \\ &= \sqrt{\sum_{k=1}^{K} u_k^2} \quad (P\%) \end{aligned} \tag{5.2}$$

The RSS method of combining uncertainties is based on the assumption that the square of an uncertainty is a measure of the variance (i.e., σ^2) assigned to an error, and the summation of these variances yields a probable estimate of the total uncertainty. Note that it is imperative to maintain consistency in the units for each uncertainty in Equation 5.2 and to assign each uncertainty term at the same probability level.

In test engineering, it is common to report final uncertainties at a 95% probability level ($P\% = 95\%$), and this is equivalent to assuming the probability covered by two standard deviations. When a probability level equivalent to a spread of one standard deviation is used, this uncertainty is called the "standard" uncertainty (1,2). For a normal distribution, a standard uncertainty equates to a 68% probability level. Whatever level is used, consistency is important.

---

[1] It is possible to assign a value for u_0 that differs from half the scale resolution. Discretion should be used.

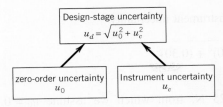

Figure 5.2 Design-stage uncertainty procedure in combining uncertainties.

Design-Stage Uncertainty

The *design-stage uncertainty*, u_d, for an instrument or measurement method is an interval found by combining the instrument uncertainty with the zero-order uncertainty,

$$u_d = \sqrt{u_0^2 + u_c^2} \quad (P\%) \tag{5.3}$$

This procedure for estimating the design-stage uncertainty is outlined in Figure 5.2. The design-stage uncertainty for a test system is arrived at by combining each of the design-stage uncertainties for each component in the system using the RSS method while maintaining consistency of units and confidence levels.

Owing to the limited information used, a design-stage uncertainty estimate is intended only as a guide for selecting equipment and procedures before a test, and is never used for reporting results. *If additional information about other measurement errors is known at the design stage*, then their uncertainties can and should be used and included in Equation 5.3. So Equation 5.3 provides a minimum value for uncertainty. In later sections of this chapter, we move toward more thorough uncertainty analyses.

Example 5.1

A force measuring instrument comes with a certificate of calibration that identifies two instrument errors and assigns each an uncertainty at 95% confidence over its range. Provide an estimate of the instrument design-stage uncertainty. What is the instrument standard uncertainty?

Resolution:	0.25 N
Range:	0 to 200 N
Linearity error:	within 0.20 N over range
Hysteresis error:	within 0.30 N over range

KNOWN Catalog specifications

ASSUMPTIONS Instrument uncertainty at 95% level; normal distribution

FIND u_c, u_d

SOLUTION We follow the procedure outlined in Figure 5.2. An estimate of the instrument uncertainty depends on the uncertainty assigned to each of the contributing elemental errors of linearity, e_1, and hysteresis, e_2, respectively assigned as

$$u_1 = 0.20\,\text{N} \quad (95\%) \qquad u_2 = 0.30\,\text{N} \quad (95\%)$$

Then using Equation 5.2 with $K = 2$ yields the instrument uncertainty

$$u_c = \sqrt{(0.20)^2 + (0.30)^2}$$
$$= 0.36 N$$

The instrument resolution is given as 0.25 N, from which we assume $u_0 = 0.125$ N. From Equation 5.3, the design-stage uncertainty of this instrument when used in a measurement would be

$$u_d = \sqrt{u_0^2 + u_c^2} = \sqrt{(0.125)^2 + (0.36)^2}$$
$$= \pm 0.38 \text{ N} \quad (95\%)$$

We assume that the 95% confidence interval stated in the specifications reflects a coverage factor of two standard deviations of a normal distribution. A standard uncertainty reflects a coverage factor of one standard deviation. Hence the instrument standard uncertainty at a 68% confidence level would be

$$u_c = 0.36/2 = 0.18 \text{ N} \quad (68\%)$$

Example 5.2

A voltmeter is used to measure the electrical output signal from a pressure transducer. The nominal pressure is expected to be about 3 psi($3 \text{ lb/in.}^2 = 0.2$ bar). Estimate the design-stage uncertainty in this combination. The following information is available:

Voltmeter	
Resolution:	10 μV
Accuracy:	within 0.001% of reading
Transducer	
Range:	±5 psi (~ ± 0.35 bar or ~ ± 260 mm Hg)
Sensitivity:	1 V/psi
Input power:	10 VDC ± 1%
Output:	±5 V
Linearity error:	within 2.5 mV/psi over range
Sensitivity error:	within 2 mV/psi over range
Resolution:	Negligible

KNOWN   Instrument specifications

ASSUMPTIONS   Values at 95% probability; normal distribution of errors

FIND u_c for each device and u_d for the measurement system

SOLUTION   The procedure in Figure 5.2 is used for both instruments to estimate the design-stage uncertainty in each. The resulting uncertainties are then combined using the RSS approximation to estimate the system u_d.

The uncertainty in the voltmeter at the design stage is given by Equation 5.3 as

$$(u_d)_E = \sqrt{(u_o)_E^2 + (u_c)_E^2}$$

From the information available,

$$(u_0)_E = 5\,\mu V$$

For a nominal pressure of 3 psi, we expect to measure an output of 3 V. Then,

$$(u_c)_E = (3\,V \times 1.0 \times 10^{-5}) = 30\,\mu V$$

so that the design-stage uncertainty in the voltmeter is

$$(u_d)_E = 30.4\,\mu V$$

The uncertainty in the pressure transducer output at the design stage is also found using Equation 5.2. Assuming that we operate within the input power range specified, the instrument output uncertainty can be estimated by considering the uncertainty in each of the instrument elemental errors of linearity, e_1, and sensitivity, e_2:

$$\begin{aligned} (u_c)_p &= \sqrt{u_1^2 + u_2^2} \\ &= \sqrt{(2.5\,mV/psi \times 3\,psi)^2 + (2\,mV/psi \times 3\,psi)^2} \\ &= 9.61\,mV \end{aligned}$$

Because $(u_0) \approx 0\,V/psi$, the design-stage uncertainty in the transducer in terms of indicated voltage is $(u_d)_p = 9.61\,mV$.

Finally, u_d for the combined system is found by using the RSS method for the design-stage uncertainties of the two devices. The design-stage uncertainty in pressure as indicated by this measurement system is estimated to be

$$\begin{aligned} u_d &= \sqrt{(u_d)_E^2 + (u_d)_p^2} \\ &= \sqrt{(0.030\,mV)^2 + (9.61\,mV)^2} \\ &= \pm 9.6\,mV \quad (95\%) \end{aligned}$$

But because the sensitivity is 1 V/psi, the uncertainty in terms of pressure can be stated as

$$u_d = \pm 0.0096\,psi \quad (95\%)$$

We could round this to ± 0.01 psi (95%). This is equivalent to ± 0.5 mm Hg (95%).

COMMENT   Design-stage uncertainty analysis shows us that the transducer, not the voltmeter, dominates the uncertainty in this setup!

5.4   IDENTIFYING ERROR SOURCES

Design-stage uncertainty is intended to provide a first estimate into the uncertainty of a test. It does not address all the possible errors that influence a measured result. Before proceeding to more detailed analyses, we provide a checklist to help identify common errors. It is not necessary to classify error sources as we do here, but it is helpful.

Suppose we breakdown the measurement process into three distinct stages: calibration, data acquisition (measurement), and data reduction. Errors that enter during each of these steps can be grouped under their respective error source heading: (1) calibration errors, (2) data-acquisition errors, and (3) data-reduction errors. Within each of these three *error source groups*, list the types of errors encountered. Such errors are the elemental errors of the measurement. Later, we will assign uncertainty values to each error. Do not become preoccupied with these groupings. Use them as a guide. If you place an error in an "incorrect" group, it is okay. The important thing is that you recognize each elemental error. The group you place it in does not affect the final uncertainty!

Calibration Errors

Calibration does not eliminate system errors but it provides a way to quantify the uncertainty in the particular pieces of equipment used. *Calibration errors* tend to enter through three sources: (1) the standard or reference value used in the calibration, (2) the instrument or system under calibration, and (3) the calibration process. For example, the laboratory standard used for calibration contains some inherent uncertainty, and this is passed along to the calibrated instrument. Measuring system errors, such as linearity, repeatability, hysteresis, contribute to overall instrument uncertainty. A curve fit of the calibration data contains uncertainty (Section 4.6). All of these effects are built into the calibration data. In Table 5.1, we list the common elemental errors contributing to this error source group.

Data-Acquisition Errors

Errors arising during the act of acquiring data are listed as a data-acquisition error. These errors include sensor and instrument errors not already accounted for by calibration; uncontrolled variables, such as changes or unknowns in measurement system operating conditions; and sensor installation effects on the measured variable. In addition, the inferential statistics of sampled variables and any other temporal and spatial variations that contribute to scatter in the measured data set, all contribute to uncertainty. We list some common elemental errors from this source in Table 5.2.

Table 5.1   Calibration Error Source Group

Element	Error Source[a]
1	Standard or reference value errors
2	Instrument or system errors
3	Calibration process errors
4	Calibration curve fit (or see Table 5.3)
etc.	

[a]Systematic error or random error in each element.

Table 5.2   Data-Acquisition Error Source Group

Element	Error Source[a]
1	Measurement system operating conditions
2	Sensor–transducer stage (instrument error)
3	Signal conditioning stage (instrument error)
4	Output stage (instrument error)
5	Process operating conditions
6	Sensor installation effects
7	Environmental effects
8	Spatial variation error (data set)
9	Temporal variation error (data set)
etc.	

[a]Systematic error or random error in each element.

Note: A total input-to-output measurement system calibration combines elements 2, 3, 4, and possibly 1 within this error source group.

Table 5.3   Data-Reduction Error Source Group

Element	Error Source[a]
1	Curve fit error
2	Truncation error
3	Modeling error
etc.	

[a]Systematic error or random error in each element.

Data-Reduction Errors

Curve fits and correlations of data (Section 4.6) introduce data-reduction errors into test results. Also, truncation errors, interpolation errors, and errors from assumed models or functional relationships affect the quality of a result. We list elemental errors typical of this error source group in Table 5.3.

5.5   SYSTEMATIC AND RANDOM ERRORS AND STANDARD UNCERTAINTIES

Systematic Error

A systematic error[2] remains constant in repeated measurements under fixed operating conditions. A systematic error may cause either a high or a low offset in the estimate of the true value of a measured variable. The estimate of the probable range of a systematic error is assigned a value called the systematic uncertainty. The *systematic standard uncertainty* is represented by b. It can be assigned over an interval, $\pm b$, that has a confidence level of one standard deviation, equivalent to a probability

---

[2] This error was called "bias" error in engineering documents prior to the 1990s.

level of 68% for a normal distribution. The *systematic uncertainty* assigned at some desired confidence level is given by $t_{\nu,p}b$, or simply tb. When stated at a 95% probability level without supporting information, it is defined by the interval, $\pm B$, which we assume is given as

$$B = 2b \tag{5.4}$$

which assigns a value of $t = 2$ or two standard deviations. This t value assumes large degrees of freedom in the assigned systematic uncertainty for which $t_{95} = 1.96$, which is rounded to 2 for convenience (2).

The reader has probably experienced systematic errors in common measurements. For example, improperly using the floating tang[3] on the end of a metal tape measure will offset the measurement, a systematic error. The inability to exactly zero a gauge or device before using it offsets its measured value. The physical act of measurement can offset the measured value from its true value, and small offsets along a measurement chain can demonstrably affect the test outcome.

Consider a home bathroom scale: does it have a systematic error? How might we assign an uncertainty to its indicated weight? Perhaps we notice that it measures differently from the measurement taken in a physician's office or at the gym. By comparing the readings (i.e., a sort of interlaboratory comparison), we can estimate a range of possible systematic error. Or maybe we decide to calibrate the scale using calibrated standard masses, account for local gravitational acceleration, and correct the output, thereby estimating the systematic error of the measurement (i.e., direct calibration against a local standard). Or we could carefully measure body volume displacement in water and compare the results to estimate differences (i.e., concomitant methodology). Or we can use the specification provided by the manufacturer or just use judgment (i.e., experience). But without any of the above, what value would we assign? Would we even suspect a systematic error?

But let us think about this. The insidious aspect of a systematic error has been revealed: It is difficult to recognize. Why suspect a systematic error? Experience teaches us to think through each measurement carefully because systematic error is always present at some magnitude. We see that it is difficult to recognize systematic errors without comparison, so a good test design includes some means of estimating them. Various methodologies can be used: (i) calibration, (ii) concomitant methodology, (iii) interlaboratory comparisons, and (iv) judgment/experience. Calibration can reveal and reduce instrument systematic errors and assist in assigning values to its associated uncertainty. Concomitant methodology uses different methods for estimating the same thing allowing the means to compare the results (3). Last, interlaboratory comparisons, an excellent replication method, introduces different instruments, facilities, or personnel into an otherwise identical measurement procedure. The variations in the results between several facilities can provide a statistical estimate of the systematic uncertainty (2).

In lieu of the above, a judgment value based on past experience may have to be assigned; judgment values are usually understood to be made at the 95% confidence level. For example, the value that first came to mind to you in the bathroom scale example above likely covered a 95% interval.

Note that calibration cannot eliminate systematic error, but it may reduce its uncertainty. Consider the calibration of a temperature transducer against a National Institute of Standards and Technology (NIST) standard certified to be correct to within 0.01 °C. If the calibration data show that

---

[3] The tang is the hook at the end of the tape usually attached by loose rivets.

the transducer output has a systematic offset of 0.2 °C relative to the standard, then we would just correct all the data obtained with this transducer by 0.2 °C. Simple enough! But the standard itself still has an intrinsic systematic uncertainty of 0.01 °C, and this uncertainty plus any uncertainty in the correction value applied remains in the calibrated transducer.

Random Error

When repeated measurements are made under fixed operating conditions, random errors manifest themselves as scatter of the measured data. Hence the effects of these errors are easily observed. Random error[4] is introduced through the repeatability of the measurement system components, calibration, and measurement procedure and technique; by the measured variable's own natural variability as estimated by sample statistics; and by the variations in the process operating and environmental conditions from which the measurements are taken.

The estimate of the probable range of a random error is a value called the random uncertainty. The *random standard uncertainty* is defined by the symbol $s_{\bar{x}}$. It is defined over an interval given by $\pm s_{\bar{x}}$, where

$$s_{\bar{x}} = s_x / \sqrt{N} \tag{5.5}$$

with degrees of freedom $\nu = N - 1$, and assumes the errors are normally distributed.[5] The interval has a confidence level of one standard deviation, equivalent to a probability of 68% for a population of x having a normal distribution. The *random uncertainty* at a desired confidence level is given by $t_{\nu,P} s_{\bar{x}}$, or simply ts, where t is found from Table 4.4.

Other Ways Used to Classify Error and Uncertainty

In the ISO GUM (1), an uncertainty assigned based on a statistical analysis of a data set is called a "Type A" uncertainty. An uncertainty assigned based on any non-statistical approach is called a "Type B" uncertainty. The intent is to inform the ultimate user only to the method used to assign the value. The ISO Gum recommends tagging assigned uncertainties as A or B to keep track. The designations are particularly targeted to manufacturer calibration and instrumentation specifications so that calibration details passed on to users are clear.

The terms *systematic error* and *random error* describe how identified errors affect a test result. In PTC 19.1 (2), the terms systematic uncertainty and random uncertainty are the values assigned to these errors. The ISO GUM does not use these uncertainty terms. An error assigned a value classified as either a Type A or Type B uncertainty could have either a systematic or random effect on a test result (1,2). There is no direct correlation between the different terms and this makes it possible to use both methods simultaneously (2). For example, a statistical analysis of available data used to estimate an average offset within a set of data, i.e., a value used to assign a systematic standard uncertainty, b, would also be tagged as a Type A uncertainty.

The computational community has introduced other terms in describing error and uncertainty in their models. *Aleatory uncertainty* refers to that caused by the natural variability within a system or a parameter. The term is synonymous with random uncertainty, but not with Type A or B. *Epistemic*

---

[4] This error was called a "precision" error in engineering documents prior to the 1990s.
[5] The standard uncertainty when estimated from a rectangular distribution (1) is discussed in Section 5.7.

uncertainty refers to that arising from a lack of knowledge, such as due to deficiencies in the model and simplifications used. Epistemic errors are quantified through parametric variation and sensitivity analyses, probability modeling, or by other means. There is no direct correlation with the other terms described and so it is possible to use the other methods simultaneously.

Each approach has its place (2). Listing uncertainties as Type A and Type B in instrument calibration certification eliminates ambiguity in interpreting specifications. Separating random and systematic effects in reporting test results is a valuable aid for interpreting the results, as well as providing a best practice to improve on the test methods used.

5.6   UNCERTAINTY ANALYSIS: MULTI-VARIABLE ERROR PROPAGATION

Suppose we want to determine how long it would take to fill a small swimming pool using a garden hose. One way is to measure the time required to fill a bucket of known volume to estimate the flow rate from the garden hose. Armed with a measurement of the volume of the pool, we can calculate the time to fill the pool. Clearly, small errors in estimating the flow rate from the garden hose would translate into large differences in the time required to fill the pool. Here we are using measured values, the flow rate and volume, to estimate a result, the time required to fill the pool.

Very often in engineering, a result is determined through a functional relationship between measured values. For example, we just calculated a flow rate above by measuring time, t, and bucket volume, $\forall$, because $Q = f(t, \forall) = \forall/t$. But how do uncertainties in either measured quantity contribute to the uncertainty in flow rate? Is the uncertainty in Q more sensitive to uncertainty in volume, or in time? More generally, how are uncertainties in variables propagated to affect a calculated result? We now explore these questions.

Propagation of Error

A general relationship between some dependent variable y and a measured variable x, that is, $y = f(x)$, is illustrated in Figure 5.3. Now suppose we measure x a number of times at some operating condition so as to establish its sample mean value and the uncertainty u_x. This implies that the true value for x lies somewhere within the interval $\bar{x} \pm u_x$. It is reasonable to assume that the true value of y, which is determined from the measured values of x, falls within the interval defined by

$$\bar{y} \pm \delta y = f(\bar{x} \pm u_x) \tag{5.6}$$

Expanding this as a Taylor series yields

$$\bar{y} \pm \delta y = f(\bar{x}) \pm \left[\left(\frac{dy}{dx} \right)_{x=\bar{x}} u_x + \frac{1}{2} \left(\frac{d^2 y}{dx^2} \right)_{x=\bar{x}} (u_x)^2 + \cdots \right] \tag{5.7}$$

By inspection, the mean value for y must be $f(\bar{x})$ so that the term in brackets estimates $\pm\delta y$. A linear approximation for δy can be made that is valid when u_x is small and that neglects the higher-order terms in Equation 5.7:

$$\delta y \approx \left(\frac{dy}{dx} \right)_{x=\bar{x}} u_{\bar{x}} \tag{5.8}$$

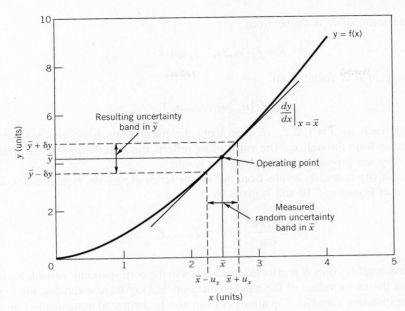

Figure 5.3 Relationship between a measured variable and a resultant calculated using the value of that variable.

The derivative term, $(dy/dx)_{x=\bar{x}}$, defines the slope of a line that passes through the point specified by $\bar{x}$. For small deviations from the value of $\bar{x}$, this slope predicts an acceptable, approximate relationship between u_x and δy. The derivative term is a measure of the sensitivity of y to changes in x. The width of the interval defined by $\pm u_x$ corresponds to $\pm \delta y$, within which we should expect the true value of y to lie. Then the uncertainty in y due to the uncertainty in x is $u_y = \delta y$.

Figure 5.3 illustrates the concept that errors in an independent variable are propagated through to a resultant variable in a predictable way. In general, the uncertainty in x affects the uncertainty in the resultant y as

$$u_y = \left(\frac{dy}{dx}\right)_{x=\bar{x}} u_x \qquad (5.9)$$

Compare the similarities between Equations 5.8 and 5.9 and in Figure 5.3.

This idea can be extended to multivariable relationships. Consider a result R, determined through some functional relationship between independent variables $x_1, x_2, \ldots, x_L$ defined by

$$R = f_1\{x_1, x_2, \ldots, x_L\} \qquad (5.10)$$

where L is the number of independent variables involved. Each variable contains some measure of uncertainty that affects the result. The best estimate of the true mean value R' would be stated as

$$R' = \bar{R} \pm u_R \quad (\text{P}\%) \qquad (5.11)$$

where the sample mean of R is found from

$$\overline{R} = f_1\{\overline{x}_1, \overline{x}_2, \ldots, \overline{x}_L\} \tag{5.12}$$

and the uncertainty in $\overline{R}$ is found from

$$u_R = f_1\{u_{\overline{x}_1}, u_{\overline{x}_2}, \ldots, u_{\overline{x}_L}\} \tag{5.13}$$

In Equation 5.13, each $u_{\overline{x}_i}, i = 1, 2, \ldots, L$ represents the uncertainty associated with the best estimate of x_1 and so forth through x_L. The value of u_R reflects the contributions of the individual uncertainties as they are propagated through to the result.

A general sensitivity index, θ_i, results from the Taylor series expansion, Equation 5.9, and the functional relation of Equation 5.10 and is given by

$$\theta_i = \frac{\partial R}{\partial x_{i_{x=\overline{x}}}} \quad i = 1, 2, \ldots, L \tag{5.14}$$

Each sensitivity index relates how R is affected by changes in the corresponding variable x_i. Each is evaluated at either the mean values of the measurement or, lacking these estimates, the expected values of the corresponding variables. Equation 5.14 can also be estimated numerically using finite differencing methods (4), as discussed next.

The contribution of the uncertainty in x to the result R is estimated by the term $\theta_i u_{\overline{x}_i}$. The most probable estimate of u_R is generally accepted as that value given by the second power relation (5), which is the square root of the sum of the squares (RSS). The propagation of uncertainty in the variables to the result is by

$$u_R = \left[\sum_{i=1}^{L} (\theta_i u_{\overline{x}_i})^2\right]^{1/2} \tag{5.15}$$

Approximating a Sensitivity Index

There are ways to estimate a sensitivity index other than by direct differentiation. One useful numerical approach is to use *dithering*. Here a small perturbation is applied to an input variable of a functional model and the effect on the output variable value noted. The change in output value relative to the change in input value provides the estimate of the sensitivity. The method uses a finite difference method to approximate the derivatives. Dithering is particularly useful when the functional information is built into a numerical or experimental model or built within a spreadsheet or whenever analytical differentiation is cumbersome.

Given that $y = f(x)$, then for $x = x_0$, let $y^+ = f(x + \delta x)$, $y^- = f(x - \delta x)$, such that

$$\left.\frac{\partial y}{\partial x}\right|_{x=x_0} = \frac{y^+ - y^-}{2\delta x} \tag{5.16}$$

The value δx is a small perturbation value. To ensure an accurate estimate of the sensitivity index, the value of δx is successively reduced until the value of $\partial y/\partial x$ converges on a single answer—in other words, no longer changes in value within an acceptable tolerance.

Example 5.3

Suppose we have a calibration model given by $y = x^3 + 2x^2 + 2x + 1$. Then we know that $\partial y/\partial x = 3x^2 + 4x + 2$. Estimate the sensitivity index $\theta = \partial y/\partial x$ about the value $x = 4$ by dithering.

SOLUTION   Analytically, we can solve for $y = f(x)$ for any x as

X	0.000	2.000	4.000	6.000	8.000	10.00
Y	1.00	21.00	105.0	301.0	657.0	1221.0
$\theta = \partial y/\partial x\|_{x_0}$	2.00	22.00	66.00	134.0	226.0	342.0

But we can also use dithering to approximate the sensitivity index about any point x. We do this by successively changing the value of x by an amount δx and calculating the finite difference derivative using Equation 5.16. Using the point $x = 4$ for illustration,

δx	0.4000	0.2000	0.1000	0.0500	0.0100
y+	133.704	118.768	111.741	108.3351	105.6614
y−	80.776	92.352	98.539	101.7349	104.3414
$\theta = \partial y/\partial x\|_{x_0}$	66.16	66.04	66.01	66.00	66.00

The solution converges for $\delta x \leq 0.05$. The approximate value found by dithering $\theta = \partial y/\partial x|_{x=4} = 66.00$ agrees with the value found by exact differentiation.

COMMENT   The procedure of testing against a known result illustrates a verification test that the proposed technique as applied works properly.

In multivariable relationships, such as $R = f_1\{x_1, x_2, \ldots, x_L\}$, the result will be sensitive to changes in the measured variables and test parameters. Each individual sensitivity can be found by dithering the independent variables and test parameters one at a time while holding all other variables constant.

Example 5.4

A measurement model follows $R = Kx^3$, where x is the measured variable and K is a coefficient. Estimate the sensitivities and propagation to uncertainty in the result if $K = 2.201 \text{ kg/mV}$, $u_K = 0.0220 \text{ kg/mV}; \bar{x} = 1.234 \text{ mV}, u_x = 0.0100 \text{ mV}$

KNOWN $R = Kx^3$

$$\bar{x} = 1.234 \text{ mV} \quad u_{\bar{x}} = 0.0100 \text{ mV} \quad (95\%)$$

$$K = 2.201 \text{ kg/mV} \quad u_K = 0.0220 \text{ kg/mV} \quad (95\%)$$

SOLUTION   We evaluate the mean result and its uncertainty as

$$\bar{R} = f(\bar{x}, K) = K\bar{x}^3 = 4.136 \text{ kg} \quad \text{and} \quad u_R = f(u_x, u_K) = \left[\left(\frac{\partial R}{\partial x} u_x \right)^2 + \left(\frac{\partial R}{\partial K} u_K \right)^2 \right]^{\frac{1}{2}}$$

We use dithering to estimate the sensitivities, $\theta_x = \partial R/\partial x$ and $\theta_K = \partial R/\partial K$ about the values $\bar{x} = 1.234$ mV and $K = 2.201$ kg/mV, as

δx	0.100	0.010	0.001	δK	0.100	0.022	0.002
$\delta x+$	5.2250	4.2372	4.1459	$\delta K+$	4.3238	4.1772	4.1396
$\delta x-$	3.2097	4.0361	4.1258	$d K-$	3.9479	4.0945	4.1321
$\theta_x = \partial R/\partial x\vert_{\bar{x},K}$	10.077	10.055	10.055	$\theta_K = \partial R/\partial K\vert_{\bar{x},K}$	1.8791	1.8791	1.8791

So, $\partial R/\partial x = 10.055$ kg/mV and $\partial R/\partial K = 1.8791$ mV. Then,

$$u_R = 0.1087 \text{ kg}$$
$$R' = 4.136 \pm 0.1087 \text{ kg} \quad (95\%)$$

COMMENT We used values about one order above and below the stated uncertainty values for dithering. Alternatively, a value of 1% of the variable value is a good start. In highly nonlinear problems, smaller perturbation values may be needed to get convergence.

The sensitivity index can also be estimated from any information that conveys functional information between two variables, including tabulated data and graphs (such as Figure 5.3), by using finite differencing methods. Suppose we have functional information in tabular or graphical form, such as $y = f(x)$,

x	x_1	x_2	x_3
y	$y(x_1)$	$y(x_2)$	$y(x_3)$

where $x_1 < x_2 < x_3$. Then to find the sensitivity about a point $y(x_2)$, we could use a central finite differencing approximation for the derivative

$$\theta = \frac{\partial y}{\partial x}\bigg|_{x=x_2} = \frac{y(x_3) - y(x_1)}{x_3 - x_1} \tag{5.17}$$

Similarly, one could use forward or backward differencing approximations.

In an experimental setup, sensitivities can be estimated physically by allowing a small change in an input variable or test parameter both above and below a setpoint while noting the change in output value.

Sequential Perturbation

A similar numerical approach can be used to estimate the propagation of uncertainty through to a result directly and is called *sequential perturbation* (6). The method uses a finite difference method to approximate the derivatives:

1. Based on measurements for the independent variables under some fixed operating condition, calculate a result R_o where $R_o = f(x_1, x_2, \ldots, x_L)$. This value fixes the operating point for the numerical approximation (e.g., see Fig. 5.3).

2. Increase the independent variables by their respective uncertainties and recalculate the result based on each of these new values. Call these values R_i^+. That is,

$$
\begin{aligned}
R_1^+ &= f\left(x_1 + u_{x_1}, x_2, \ldots, x_L\right), \\
R_2^+ &= f\left(x_1, x_2 + u_{x_2}, \ldots, x_L\right), \ldots \\
R_L^+ &= f\left(x_1, x_2, \ldots, x_L + u_{x_L}\right),
\end{aligned}
\tag{5.18}
$$

3. In a similar manner, decrease the independent variables by their respective uncertainties and recalculate the result based on each of these new values. Call these values R_i^-.

4. Calculate the differences δR_i^+ and δR_i^- for $i = 1, 2, \ldots, L$

$$
\begin{aligned}
\delta R_i^+ &= R_i^+ - R_o \\
\delta R_i^- &= R_i^- - R_o
\end{aligned}
\tag{5.19}
$$

5. Evaluate the approximation of the uncertainty contribution from each variable,

$$
\delta R_i = \frac{\delta R_i^+ - \delta R_i^-}{2} \approx \theta_i u_i
\tag{5.20}
$$

Then, the uncertainty in the result is

$$
u_R = \left[\sum_{i=1}^{L} (\delta R_i)^2\right]^{1/2}
\tag{5.21}
$$

Equations 5.15 and 5.21 provide two methods for estimating the propagation of uncertainty to a result. In most cases, each equation yields nearly the identical result; the choice of method is left to the user. Any of the methods discussed can be used to estimate the sensitivity index of Equation 5.14 (2).

We point out that sometimes a method may calculate unreasonable estimates of u_R. When this happens the cause can be traced to a sensitivity index that changes rapidly with small changes in the independent variable x_i coupled with a large value of the uncertainty u_{x_i}. This occurs when the operating point is close to an inflection point in the functional relationship. In these situations, reevaluate the cause and extent of the variation in sensitivity index or use a more accurate approximation for the sensitivity, including backward or forward differencing or using the higher-order terms in the Taylor series of Equation 5.7. In subsequent sections, we develop methods to estimate the uncertainty values from available information.

Example 5.5

For a displacement transducer having the calibration curve, $y = KE$. Estimate the uncertainty in displacement y for $E = 5.00$ V, if $K = 10.10$ mm/V with $u_K = \pm 0.10$ mm/V and $u_E = \pm 0.01$ V at 95% confidence.

KNOWN $y = KE$

$$E = 5.00 \text{ V} \qquad u_E = 0.01 \text{ V}$$

$$K = 10.10 \text{ mm/V} \qquad u_K = 0.10 \text{ mm/V}$$

FIND u_y

SOLUTION Based on Equations 5.12 and 5.13, respectively,

$$y = f(E, K) \quad \text{and} \quad u_y = f(u_E, u_K)$$

From Equation 5.15, the uncertainty in the displacement at $y = KE$ is

$$u_y = \left[(\theta_E u_E)^2 + (\theta_K u_K)^2 \right]^{1/2}$$

where the sensitivity indices are evaluated from Equation 5.14 as

$$\theta_E = \frac{\partial y}{\partial E} = K \quad \text{and} \quad \theta_K = \frac{\partial y}{\partial K} = E$$

Or we can write Equation 5.15 as

$$u_y = \left[(K u_E)^2 + (E u_K)^2 \right]^{1/2}$$

The operating point occurs at the nominal or the mean values of $E = 5.00$ V and $y = 50.50$ mm. With $E = 5.00$ V and $K = 10.10$ mm/V and substituting for u_E and u_K, evaluate u_y at its operating point:

$$u_y\big|_{y=50.5} = \left[(0.10)^2 + (0.50)^2 \right]^{1/2} = 0.51 \text{ mm}$$

Alternatively, we can use sequential perturbation. The operating point for the perturbation is again $y = R_o = 50.5$ mm. Using Equations 5.18 through 5.20 gives

i	x_i	R_i^+	R_i^-	δR_i^+	δR_i^-	δR_i
1	E	50.60	50.40	0.10	-0.10	0.10
2	K	51.00	50.00	0.50	-0.50	0.50

Then, using Equation 5.21,

$$u_y\big|_{y=50.5} = \left[(0.10)^2 + (0.50)^2 \right]^{1/2} = 0.51 \text{ mm}$$

The two methods give an identical result. We state the calculated displacement in the form of Equation 5.11 as

$$y' = 50.50 \pm 0.51 \text{ mm} \quad (95\%)$$

Monte Carlo Method

A Monte Carlo simulation provides another effective way to estimate the propagation of the uncertainties in the independent variables to the uncertainty in a result. As presented in Chapter 4, the outcomes of a converged Monte Carlo simulation are the statistics of the predicted population in a result R (i.e., $\bar{x}, s_x$) from which we assign the standard uncertainty in the result as s_x.

Generally, convergence is claimed when the computed standard deviation no longer changes by 1% to 5%. As a more strict test for convergence, we define a numerical tolerance Δ as being half of the least significant digit of the estimated standard uncertainty (7). For example, if $s_x = 0.2$ or 2×10^{-1}, then $\Delta = 0.05$ (i.e., ½ of the 10^{-1} digit), the simulation is converged when s_x no longer changes by 0.05 or more.

5.7   ADVANCED-STAGE UNCERTAINTY ANALYSIS

In designing a measurement system, a pertinent question is how it would affect the result if a particular aspect of the technique or equipment were changed. In design-stage uncertainty analysis, we only considered the errors resulting from a measurement system's resolution and estimated instrument calibration errors. But if additional information is available, we can get a better idea of the uncertainty in a measurement. So an advanced-stage uncertainty analysis permits taking design-stage analysis farther by considering procedural and test control errors that affect the measurement. We consider it a method for a thorough uncertainty analysis when a large data set is not available. This is often the case in the early stages of a test program or for certain tests for which repeating measurements may not be possible. Such an advanced-stage analysis, also known as *single-measurement uncertainty analysis* (3,6), can be used (1) in the advanced design stage of a test to estimate the expected uncertainty beyond the initial design stage estimate and (2) to report the results of a test program that involved measurements over a range of one or more parameters but with no or relatively few repeated measurements of the pertinent variables at each test condition. Essentially, the method assesses different aspects of the main test by quantifying potential errors though various well-focused verification tests.

In this section, the goals are either to estimate the uncertainty in some measured value x or in some general result R through an estimation of the uncertainty in each of the factors that may affect x or R. We present a technique that uses a step-by-step approach for identifying and estimating the uncertainty associated with errors. We seek the combined value of the estimates at each step. We assume that the errors follow a normal distribution—but some errors might be better described by other distributions, and these can be used. For example, calibration errors that specify only a limit of error are well modeled by a rectangular distribution. The standard uncertainty for a rectangular distribution defined over the interval $(m - n)$, where m is the maximum and n is the minimum of the interval (e.g., refer to Table 4.2), is given by

$$u_x = (m - n)/\sqrt{12} \tag{5.22}$$

which covers one standard deviation (or a 58% confidence level). Multiplying by 2 provides for two standard deviations coverage. If we assume that the errors propagate to the result with a normal distribution, then a coverage factor of 2 approximates the 95% confidence level for consistency.

Zero-Order Uncertainty

At zero-order uncertainty, all variables and parameters that affect the outcome of the measurement, including time, are assumed to be fixed except for the physical act of observation itself. For example, suppose we took a photograph of a gauge reading under some condition and showed it to 20 technically capable individuals for them to record their interpretation of the reading. The resulting scatter in those 20 readings would be the result of interpolation error owing to instrument resolution alone, a value that we assign as u_0. The value u_0 estimates the extent of variation expected in the measured value when all influencing effects are controlled except for interpolation error and is found using Equation 5.1. Earlier we placed the value for u_0 as half the instrument resolution or its least significant digit, or a value assigned by discretion. Note that by itself, a zero-order uncertainty is inadequate for the reporting the uncertainty in test results.

Higher-Order Uncertainty

Higher-order uncertainty estimates consider the controllability of the test operating conditions and the variability of all measured variables. For example, at the first-order level, the effect of time as an extraneous variable in the measurement might be considered. That is, what would happen if we started the test, set the operating conditions, and sat back and watched? If a variation in the measured value is observed, then time is a factor in the test, presumably resulting from some extraneous influence affecting process control or simply inherent in the behavior of the variable being measured.

In practice, the uncertainty at this first level would be evaluated for each particular measured variable by operating the test facility at some single operating condition that would be within the range of conditions to be used during the actual tests. A set of data (say, $N \geq 30$) would be obtained under some set operating condition. The first-order uncertainty of our ability to estimate the true value of a measured value could be estimated as

$$u_1 = t_{\nu,P} s_{\bar{x}} \tag{5.23}$$

The uncertainty at u_1 includes the effects of resolution, u_o. So only when $u_1 = u_0$ is time not a factor in the test. In itself, the first-order uncertainty is inadequate for reporting of test results.

With each successive order, another factor identified as affecting the measured value is introduced into the analysis, thus giving a higher but more realistic estimate of the uncertainty. For example, at the second level it might be appropriate to assess spatial variations that affect the outcome, such as when a value from a point measurement is assigned to quantify a larger volume. These are a series of verification tests conducted to assess causality.

Nth-Order Uncertainty

At the Nth-order estimate, instrument calibration characteristics are entered into the scheme through the instrument uncertainty u_c. A practical estimate of the Nth-order uncertainty u_N is given by

$$u_N = \left[u_c^2 + \left(\sum_{i=1}^{N-1} u_i^2 \right) \right]^{1/2} \quad (P\%) \tag{5.24}$$

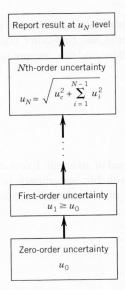

Figure 5.4   Advanced-stage and single-measurement uncertainty procedure in combining uncertainties.

Uncertainty estimates at the Nth order allow for the direct comparison between results of similar tests obtained either using different instruments or at different test facilities. The procedure for a single-measurement analysis is outlined in Figure 5.4.

Note that as a minimum, design-stage analysis includes only the effects found as a result of u_0 and u_c. It is those in-between levels that allow measurement procedure and control effects to be considered in the uncertainty analysis scheme. When done carefully, the Nth-order uncertainty estimate provides a reasonable estimate of the uncertainty value sought in advanced design stage or in single-measurement (one-off or unique tests) analyses.

Example 5.6

As an exercise, obtain and examine an inexpensive dial in-oven thermometer. How would you assess the zero- and first-order uncertainty used to determine the temperature of a kitchen oven using the device?

KNOWN   Dial thermometer

FIND   Estimate u_0 and u_1 in oven temperature

SOLUTION   The zero-order uncertainty would be that contributed by the resolution error of the measurement system only. For example, suppose we have a gauge of this type with a resolution of 10 °C. So from Equation 5.1, we estimate

$$u_0 = 5\,°C$$

At the first order, the uncertainty would be affected by any time variation in the measured value (temperature) and variations in the operating conditions. If we placed the thermometer in the center of an oven, set the oven to some relevant temperature, allowed the oven to preheat to a steady condition, and then recorded the temperature indicated by the thermometer at random intervals, we can estimate the ability of the oven to control its temperature. For J measurements, the first-order uncertainty of the measurement in this oven's mean temperature using this technique and this instrument would be (Equation 5.23),

$$u_1 = t_{J-1,95} s_{\bar{T}}$$

Example 5.7

A stopwatch is to be used to estimate the time between the start and end of an event. Event duration might range from several seconds to 10 minutes. Estimate the probable uncertainty in a time estimate using a hand-operated stopwatch that claims an accuracy of 1 min/month (95%) and a resolution of 0.01 s.

> **KNOWN** $u_0 = 0.005$ s (95%)
>
> $u_c = 60$ s/month (95% assumed)

> **FIND** u_d, u_N

> **SOLUTION** The design-stage uncertainty gives an estimate of the suitability of an instrument for a measurement. At 60 s/month, the instrument accuracy works out to about 0.01 s/10 min of operation. This gives a design-stage uncertainty of
>
> $$u_d = \left(u_o^2 + u_c^2 \right)^{1/2} = \pm 0.01 \text{ s} (95\%)$$
>
> for an event lasting 10 minutes versus ± 0.005 s (95%) for an event lasting 10 s. Note that instrument calibration error controls the longer duration measurement, whereas instrument resolution controls the short duration measurement.
>
> But do instrument resolution and calibration error actually control the uncertainty in this measurement? The design-stage analysis does not include the data-acquisition error involved in the act of physically turning the watch on and off to synchronize with event occurrences. A first-order analysis might be run to estimate the uncertainty that enters through the procedure of using the watch to time an event. Suppose a typical trial run of 30 tries of simply turning a watch on and off against a repeatable visual prompt suggests that the uncertainty in determining the duration of an occurrence is
>
> $$u_1 = t_{\nu, P} s_{\bar{x}} = 0.05 \text{ s}$$
>
> Then, the uncertainty in measuring the duration of an event would then be better estimated by the Nth-order uncertainty of Equation 5.24:
>
> $$u_N = \left(u_1^2 + u_c^2 \right)^{1/2} = \pm 0.05 \text{ s} (95\%)$$
>
> This estimate holds for periods of up to about two hours. Clearly, procedure controls the uncertainty, not the watch. This uncertainty estimate could be further improved by considering how well the operator can synchronize the watch action with the particular event start and finish action.

COMMENT   In the 2014 Olympics, there was much discussion about events that timed outcomes to 0.01 or 0.001 s, the exact precision used being determined by each individual sport's governing body. In all cases, the timing instruments used were accurate to within 10^{-7} s (design-stage). But the overall uncertainty in any measurement remains an outcome of the measurement process, of which the instrument accuracy is only one part, and importantly in this case, the peculiarities associated with the timing of each sport.

Example 5.8

A flow meter can be calibrated by providing a known flow rate through the meter and measuring the meter output. One method of calibration with liquid systems is the use of a catch and time technique whereby a volume of liquid, after passing through the meter, is diverted to a tank for a measured period of time from which the flow rate volume/time, is computed. There are two procedures that can be used to determine the known flow rate Q in, say, ft^3/min:

1. The volume of liquid, $\forall$, caught in known time t can be measured. Suppose we arbitrarily set $t = 6$ s and assume that our available facilities can determine volume (at Nth order) to 0.001 ft^3. Note: The chosen time value depends on how much liquid we can accommodate in the tank.

2. The time t required to collect 1 ft^3 of liquid can be measured.

Suppose we determine an Nth-order uncertainty in time of 0.15 s. In either case the same instruments are used. Determine which method is better to minimize uncertainty over a range of flow rates based on these preliminary estimates.

KNOWN $u_\forall = 0.001 \text{ ft}^3$

$\quad\quad\quad\quad u_t = 0.15 \text{ s}$

$\quad\quad\quad\quad Q = f(\forall, t) = \forall/t$

ASSUMPTIONS   Flow diversion is instantaneous; 95% confidence levels.

FIND   Preferred method

SOLUTION   From the available information, the propagation of probable uncertainty to the result Q is estimated from Equation 5.15:

$$u_Q = \left[\left(\frac{\partial Q}{\partial \forall} u_\forall \right)^2 + \left(\frac{\partial Q}{\partial t} u_t \right)^2 \right]^{1/2}$$

By dividing through by Q, we obtain the relative (fractional percent) uncertainty in flow rate u_Q/Q:

$$\frac{u_Q}{Q} = \left[\left(\frac{u_\forall}{\forall} \right)^2 + \left(\frac{u_t}{t} \right)^2 \right]^{1/2}$$

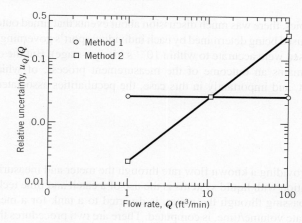

Figure 5.5 Uncertainty plot for the design analysis of Example 5.8.

Representative values of Q are needed to solve this relation. Consider a range of flow rates, say, 1, 10, and 100 ft^3/min; the results are listed in the following table for both methods.

$Q(ft^3/min)$	$t(s)$	$\forall(ft^3)$	$u_\forall/\forall$	u_t/t	$\pm u_Q/Q$
Method 1					
1	6	0.1	0.01	0.025	0.027
10	6	1.0	0.001	0.025	0.025
100	6	10.0	0.0001	0.025	0.025
Method 2					
1	60.0	1.0	0.001	0.003	0.003
10	6.0	1.0	0.001	0.025	0.025
100	0.6	1.0	0.001	0.250	0.250

In method 1, it is clear the uncertainty in time contributes the most to the relative uncertainty in Q, provided that the flow diversion is instantaneous. But in method 2, uncertainty in measuring either time or volume can contribute more to the uncertainty in Q, depending on the time sample length. The results for both methods are compared in Figure 5.5. For the conditions selected and preliminary uncertainty values used, method 2 would be a better procedure for flow rates up to 10 ft^3/min. At higher flow rates, method 1 would be better. However, the minimum uncertainty in method 1 is limited to 2.5%. The engineer may be able to reduce this uncertainty by improvements in the time measurement procedure.

COMMENT These results are without consideration of some other elemental errors that are present in the experimental procedure. For example, the diversion of the flow may not occur instantaneously. A first-order uncertainty estimate could be used to estimate the added uncertainty to flow rate due to the time required to divert the liquid to and from the catch tank. If operator influence is a factor, it should be randomized in actual tests. Accordingly, the above calculations may be used as a guide in procedure selection.

Example 5.9

Repeat Example 5.8, using sequential perturbation for the operating conditions $\forall = 1$ ft^3 and $t = 6$ s.

SOLUTION   The operating point is $R_o = Q = \forall/t = 0.1667$ ft^3/s. Then solving Equations 5.18 to 5.20 gives

i	x_i	R_i^+	R_i^-	δR_i^+	δR_i^+	δR_i
1	$\forall$	0.1668	0.1665	0.000167	−0.000167	0.000167
2	t	0.1626	0.1709	−0.00407	0.00423	0.00415

Applying Equation 5.21 gives the uncertainty about this operating point:

$$u_Q = \left[0.00415^2 + 0.000167^2\right]^{1/2} = \pm 4.15 \times 10^{-3} \text{ ft}^3/\text{s} \quad (95\%)$$

or

$$u_Q/Q = \pm 0.025 \quad (95\%)$$

COMMENT   These last two examples give similar results for the uncertainty aside from insignificant differences as a result of round-off in the computations.

Example 5.10

A temperature sensor comes only with a manufacturer specification that states that its limit of error within its operating range is ± 1.5 °C. Establish a value for the instrument standard uncertainty.

SOLUTION   When faced with vague manufacturer specifications, the engineer needs to makes some assumptions. Specifications of limits of instrument error refer to the extreme values possible. Accordingly, all values of error within the limit band are equally probable (rectangular distribution). The standard uncertainty of a rectangular distribution is $(m - n)/\sqrt{12}$ (see Equation 5.22). Let $m = 1.5$ °C and $n = -1.5$ °C be the bounds of the distribution. The instrument standard uncertainty is

$$u_c = \frac{[1.5 - (-1.5)]}{\sqrt{12}} = \frac{1.5}{\sqrt{3}} = 0.87 \text{ °C} \quad \text{(one standard deviation)}$$

5.8 MULTIPLE-MEASUREMENT UNCERTAINTY ANALYSIS

This section develops a method for estimating the uncertainty in the value assigned to a variable based on a set of measurements obtained under fixed operating conditions. The method parallels the uncertainty standards approved by professional societies in the United States (2,7) and is in harmony with international guidelines (1). The procedures assume that the errors follow a normal probability distribution, although the procedures are actually quite insensitive to deviations away from such behavior (2).

Propagation of Elemental Errors

The procedures for multiple-measurement uncertainty analysis consist of the following steps:

- Identify the elemental errors in the measurement. As an aid, consider the errors in each of the three source groups (calibration, data acquisition, data reduction).
- Estimate the magnitude of systematic and random error in each of the elemental errors.
- Propagate the uncertainties and calculate the uncertainty estimate for the result.

Considerable guidance can be obtained from Tables 5.1 to 5.3 for identifying the elemental errors. In multiple-measurement analysis, it is possible to divide the estimates for elemental errors into random and systematic uncertainties. Random errors identified during calibration are carried forward as systematic errors in a detailed multi measurement analysis. This procedure is called fossilization.

Consider the measurement of variable x, which is subject to elemental random errors each estimated by their random standard uncertainty, $(s_{\bar{x}})_k$, and systematic errors each estimated by their systematic standard uncertainty, $(b_{\bar{x}})_k$. Let subscript k, where $k = 1, 2, \ldots, K$, refer to each of up to K elements of error e_k. A method to estimate the uncertainty in x based on the uncertainties in each of the elemental random and systematic errors is outlined in Figure 5.6 and will now be described.

The *propagation of elemental random uncertainties* owing to the K random errors in measuring x is given by the measurement random standard uncertainty, $s_{\bar{x}}$, as estimated by the RSS method of Equation 5.2,

$$s_{\bar{x}} = \left[(s_{\bar{x}})_1^2 + (s_{\bar{x}})_2^2 + \cdots + (s_{\bar{x}})_K^2 \right]^{1/2} \tag{5.25}$$

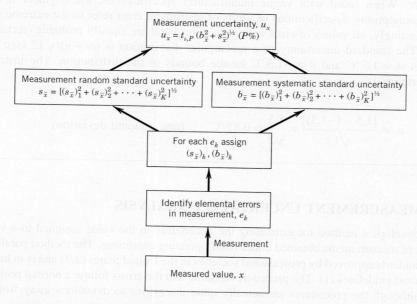

Figure 5.6 Multiple-measurement uncertainty procedure for combining uncertainties.

where each

$$(s_{\bar{x}})_k = s_{x_k} / \sqrt{N_k} \tag{5.26}$$

The *measurement random standard uncertainty* represents a basic measure of the uncertainty due to the known elemental errors affecting the variability of variable x at a one standard deviation confidence level. The degrees of freedom, ν, in the random standard uncertainty, is estimated using the Welch–Satterthwaite formula (2):

$$\nu = \frac{\left(\sum\limits_{k=1}^{K} \left(s_{\bar{x}}^2 \right)_k \right)^2}{\sum\limits_{k=1}^{K} \left(s_{\bar{x}}^4 \right)_k \Big/ \nu_k} \tag{5.27}$$

where k refers to each elemental error with its own $\nu_k = N_k - 1$.

The *propagation of elemental systematic uncertainties* due to K systematic errors in measuring x is treated in a similar manner. The measurement systematic standard uncertainty, $b_{\bar{x}}$, is given by

$$b_{\bar{x}} = \left[(b_{\bar{x}})_1^2 + (b_{\bar{x}})_2^2 + \cdots + (b_{\bar{x}})_K^2 \right]^{1/2} \tag{5.28}$$

The *measurement systematic standard uncertainty*, $b_{\bar{x}}$, represents a basic measure of the uncertainty due to the elemental systematic errors that affect the measurement of variable x at a one standard deviation confidence level.

The *combined standard uncertainty in x* is reported as a combination of the systematic standard uncertainty and random standard uncertainty in x at a one standard deviation confidence level,

$$u_x = \left(b_{\bar{x}}^2 + s_{\bar{x}}^2 \right)^{1/2} \tag{5.29}$$

For a normal distribution, this confidence interval is equivalent to a 68% probability level. A more general form of this equation extends it to include the total uncertainty at other confidence levels through the use of an appropriate t value. This is called the *expanded uncertainty* (1,2) and is found at the corresponding level of confidence ($P\%$) by

$$u_x = t_{\nu,P} \left(b_{\bar{x}}^2 + s_{\bar{x}}^2 \right)^{1/2} \quad (P\%) \tag{5.30}$$

The corresponding degrees of freedom in Equation 5.30 is (2)

$$\nu = \frac{\left(\sum\limits_{k=1}^{K} \left(s_{\bar{x}}^2 \right)_k + \left(b_{\bar{x}}^2 \right)_k \right)^2}{\sum\limits_{k=1}^{K} \left(s_{\bar{x}}^4 \right)_k / \nu_k \right) + \sum\limits_{k=1}^{K} \left(b_{\bar{x}}^4 \right)_k / \nu_k} \tag{5.31}$$

The degrees of freedom in an uncertainty are calculated from measurements or prior information; otherwise the degrees of freedom may be assumed to be large ($\nu > 30$ in Table 4.4). For a 95%

confidence level with large ($\nu > 30$) degrees of freedom, t of 1.96 is assigned in Equation 5.30. This value can be conveniently rounded off to $t = 2$ (i.e., two standard deviations) (2). Reference 1 provides guidance on assigning a value for ν associated with judgment estimates of uncertainty. Note that when the degrees of freedom in the systematic uncertainties are assumed to be large, the effect of the second term in the denominator of Equation 5.31 on ν will be small.

Reference 1 prefers not to assign a probability level to an uncertainty statement but rather to associate a value of the one, two, or three standard deviations, such as we do above for the standard uncertainties. This accommodates different distribution functions and their respective probability coverage in the expanded uncertainty interval. This reflects a difference between the two standards (1,2), but either approach is acceptable.

Example 5.11

Ten repeated measurements of force F are made over time under fixed operating conditions using the force instrument of Example 5.1. The data are listed below. Estimate the random standard uncertainty due to the elemental error in estimating the true mean value of the force from the limited data set.

N	$F[N]$	N	$F[N]$
1	123.2	6	119.8
2	115.6	7	117.5
3	117.1	8	120.6
4	125.7	9	118.8
5	121.1	10	121.9

KNOWN Measured data set $N = 10$

FIND Estimate $s_{\bar{F}}$

SOLUTION The mean value of the force based on this finite data set is computed to be $\bar{F} = 120.13$ N. A random error is associated with assigning the sample mean value as the true value of force because of the small data set. This error enters the measurement during data acquisition (Table 5.2). The random standard uncertainty in this elemental error is computed through the standard deviation of the means, Equation 5.5:

$$s_{\bar{F}} = \frac{s_F}{\sqrt{N}} = \frac{3.04}{\sqrt{10}} = 0.96 \text{ N} \quad \text{with} \quad \nu = 9$$

COMMENT The uncertainty due to data scatter here is Type A and could be classified as arising from a temporal variation error (i.e., variation with time under fixed conditions) as listed in Table 5.2. The resolution error estimated by the zero-order uncertainty u_0 assigned in Example 5.1 was just an initial estimate of variation in a measurement. We now have information about temporal variations in an actual measurement, and we would replace u_0 with this new information at the proper level of confidence. For example, in Example 5.1, the uncertainty estimate is increased to $t_{9,95}s_{\bar{F}} = 2.17$ N (95%).

Example 5.12

The force measuring device of Example 5.1 was used in acquiring the data set of Example 5.11. Estimate the systematic uncertainty due to the known elemental errors in the measurement instrument.

KNOWN $B_1 = 0.20\,N$ (95%)
$B_2 = 0.30\,N$

ASSUMPTIONS   Manufacturer specifications reliable at 95% probability

FIND B (95%)

SOLUTION   Lacking the information used to arrive at the uncertainty values associated with the two element errors, e_1 and e_2, we consider them as Type B uncertainties estimated with large degrees of freedom. We also classify these as systematic uncertainties stated at 95% confidence level. The systematic standard uncertainties are found from Equation 5.4 as $b = B/2$,

$$b_1 = B_1/2 = 0.10\,N \quad b_2 = B_2/2 = 0.15\,N$$

The systematic standard uncertainty in the transducer is

$$b = \left(b_1^2 + b_2^2\right)^{1/2} = 0.18\,N$$

at one standard deviation confidence. The expanded systematic uncertainty with $t_{\nu,95} = 2$ at the 95% confidence level is

$$B = 2\left(b_1^2 + b_2^2\right)^{1/2} = 0.36\,N$$

Elemental errors due to instruments can be considered data-acquisition source errors (Table 5.2). If they were found through calibration by the end-user, we would list them as calibration source errors. In the end, as long as they are accounted for, the error source grouping assigned is not important.

Example 5.13

A laboratory temperature standard is used to calibrate an automated temperature measuring system from sensor through conditioning equipment to its final indicated output value. The measuring system sensor is placed in the identical environment as the standard sensor. The indicated output values from the system are compared over a range of temperatures based on many readings at each of six setpoint temperatures. Estimate a value for the systematic standard uncertainty in the offset correction of the temperature measuring system.

Setpoint (°C)	0.000	20.000	40.000	60.000	80.000	100.000
Indicated Mean (°C)	0.046	20.067	40.072	60.133	80.057	100.116
Offset (°C)	0.046	0.067	0.072	0.133	0.057	0.116

KNOWN   Systems-level calibration at different temperatures

FIND   Systematic standard uncertainty of the temperature system

SOLUTION   The difference between the applied "known" setpoint temperature and the indicated temperature is listed as the offset, which is a systematic error due to the conversion errors within the measuring unit. The mean of the offset found over the full range of temperatures is $< \overline{b} > = \sum_{i=1}^{6} \overline{b}_i = 0.082\,°C$ with standard deviation, $SD = \sqrt{\sum_{i=1}^{6} (\overline{b}_i - < \overline{b} >)^2/5} = 0.034\,°C$. We correct the readings at each setpoint temperature using this mean offset. This reduces the average systematic error but there is still uncertainty in the applied correction. The systematic standard uncertainty in the correction is $b = SD/\sqrt{6} = 0.014$ or

$$b = 0.014\,°C \quad (68\%) \quad (\text{Type A})$$

COMMENT   The standard would likely have an accuracy of no better than two decimal place values. Rounding gives, $b = 0.01\,°C$ (68%), but we would carry the higher number of significant digits throughout any further intermediate calculations. There is also a random component that affects each mean setpoint temperature. This random component is not affected by the offset correction.

Example 5.14

In Examples 4.8 and 4.9, a set of measured data were used to generate a polynomial curve fit equation between variables x and y. If during a test, variable x were to be measured and y computed from x through the polynomial, estimate the random standard uncertainty due to the data-reduction error in the curve fit.

KNOWN   Data set of Example 4.8
Polynomial fit of Example 4.9

ASSUMPTIONS   Data fits curve $y = 1.04x + 0.02$.

FIND   Random standard uncertainty in the curve fit

SOLUTION   The curve fit was found to be given by $y = 1.04x + 0.02$ with a standard error of the fit, $s_{yx} = 0.16$ based on the approximation of Equation 4.38. The random standard uncertainty due to the curve fit is

$$s_{\bar{x}} = s_{yx}/\sqrt{N} = 0.072\,V \quad \text{with} \quad \nu = 3$$

This is a Type A uncertainty.

Example 5.15

The measurement of x contains three elemental random errors from data acquisition. Each elemental error is determined by statistical analysis (thus Type A) and assigned a random standard uncertainty as follows

$$(s_{\bar{x}})_1 = 0.60\,\text{units}, \nu_1 = 29$$
$$(s_{\bar{x}})_2 = 0.80\,\text{units}, \nu_2 = 9$$
$$(s_{\bar{x}})_3 = 1.10\,\text{units}, \nu_3 = 19$$

Assign the random standard uncertainty due to these data-acquisition elemental errors.

KNOWN $(s_{\bar{x}})_k$ with $k = 1, 2, 3$

FIND $s_{\bar{x}}$

SOLUTION   The random standard uncertainty due to data-acquisition errors is

$$s_{\bar{x}} = \left[(s_{\bar{x}})_1^2 + (s_{\bar{x}})_2^2 + (s_{\bar{x}})_3^2\right]^{1/2} = \left(0.60^2 + 0.80^2 + 1.10^2\right)^{1/2} = 1.49 \text{ units}$$

at one standard deviation confidence. This is a Type A uncertainty with degrees of freedom

$$\nu = \frac{\left(\sum\limits_{k=1}^{3} (s_{\bar{x}})_k^2\right)^2}{\sum\limits_{k=1}^{3} (s_{\bar{x}})_k^4/\nu_k} = \frac{\left(0.6^2 + 0.8^2 + 1.1^2\right)^2}{\left(0.6^4/29\right) + \left(0.8^4/9\right) + \left(1.1^4/19\right)} = 38.8 \approx 39$$

Example 5.16

In reporting the results of an experiment to measure stress in a loaded beam, an uncertainty analysis identifies three elemental errors in the stress measurement having the following values of uncertainty:

$$(b_{\bar{\sigma}})_1 = 0.5 \text{ N/cm}^2 \qquad (b_{\bar{\sigma}})_2 = 1.05 \text{ N/cm}^2 \qquad (b_{\bar{\sigma}})_3 = 0 \text{ N/cm}^2$$
$$(s_{\bar{\sigma}})_1 = 4.6 \text{ N/cm}^2, \nu_1 = 14 \quad (s_{\bar{\sigma}})_2 = 10.3 \text{ N/cm}^2, \nu_2 = 37 \quad (s_{\bar{\sigma}})_3 = 1.2 \text{ N/cm}^2, \nu_3 = 8$$

where the degrees of freedom in the systematic uncertainties are very large. If the mean value of the stress in the measurement is $\bar{\sigma} = 223.4 \text{ N/cm}^2$, determine the best estimate of the stress at a 95% confidence level, assuming all errors are accounted for.

KNOWN   Experimental errors with assigned uncertainties

ASSUMPTIONS   All elemental errors ($K = 3$) have been included.

FIND $\bar{\sigma} \pm u_{\sigma}$ (95%)

SOLUTION   We seek values for the statement, $\sigma' = \bar{\sigma} \pm u_{\sigma}(95\%)$, given that $\bar{\sigma} = 223.4 \text{ N/cm}^2$. The measurement random standard uncertainty is

$$s_{\bar{\sigma}} = \left[(s_{\bar{\sigma}})_1^2 + (s_{\bar{\sigma}})_2^2 + (s_{\bar{\sigma}})_3^2\right]^{1/2} = 11.3 \text{ N/cm}^2$$

The measurement systematic standard uncertainty is

$$b_{\bar{\sigma}} = \left[(b_{\bar{\sigma}})_1^2 + (b_{\bar{\sigma}})_2^2 + (b_{\bar{\sigma}})_3^2\right]^{1/2} = 1.16 \text{ N/cm}^2$$

The combined standard uncertainty is

$$u_{\sigma} = \left(b_{\bar{\sigma}}^2 + s_{\bar{\sigma}}^2\right)^{1/2} = 11.4 \text{ N/cm}^2$$

The expanded uncertainty requires the combined degrees of freedom. The degrees of freedom is calculated to be

$$v = \frac{\left(\sum_{k=1}^{K} \left(s_{\bar{x}}^2\right)_k + \left(b_{\bar{x}}^2\right)_k\right)^2}{\sum_{k=1}^{K} \left(s_{\bar{x}}^4\right)_k / v_k + \sum_{k=1}^{K} \left(b_{\bar{x}}^4\right)_k / v_k} = 49$$

where the degrees of freedoms in the systematic uncertainties are assumed to be very large so that the second term in the denominator is essentially zero. Therefore, $t_{49,95} \sim 2$, and the expanded uncertainty is

$$u_\sigma = 2\left[b_{\bar{\sigma}}^2 + s_{\bar{\sigma}}^2\right]^{1/2} = 2\left[\left(1.2\,\text{N/cm}^2\right)^2 + \left(11.3\,\text{N/cm}^2\right)^2\right]^{1/2}$$
$$= 22.7\,\text{N/cm}^2 \quad (95\%)$$

The best estimate of the stress measurement is

$$\sigma' = 223.4 \pm 22.7\,\text{N/cm}^2 \quad (95\%)$$

Propagation of Uncertainty to a Result

Now consider the result R, which is determined through the functional relationship between the measured independent variables x_i, $i = 1, 2, \ldots, L$ as defined by Equation 5.10. Again, L is the number of independent variables involved, and each x_i has an associated measurement systematic standard uncertainty $b_{\bar{x}}$, determined for that variable by Equation 5.28, and a measurement random standard uncertainty $s_{\bar{x}}$, determined using Equation 5.25. The best estimate of the true value R' is given as

$$R' = \bar{R} \pm u_R \quad (P\%) \tag{5.32}$$

where the mean value of the result is determined by

$$\bar{R} = f(\bar{x}_1, \bar{x}_2, \ldots, \bar{x}_L) \tag{5.33}$$

and where the uncertainty in the result u_R is given by

$$u_R = f\left(b_{\bar{x}_1}, b_{\bar{x}_2}, \ldots, b_{\bar{x}_L}; s_{\bar{x}_1}, s_{\bar{x}_2} \ldots, s_{\bar{x}_L}\right) \tag{5.34}$$

where subscripts x_1 through x_L refer to the measurement systematic uncertainties and measurement random uncertainties assigned to the errors in each of these L variables.

The propagation of random uncertainty through the variables to the result gives the random standard uncertainty in the result

$$s_R = \left(\sum_{i=1}^{L} \left[\theta_i s_{\bar{x}_i} \right]^2 \right)^{1/2} \tag{5.35}$$

where θ_i is the sensitivity index as defined by Equation 5.14.

The degrees of freedom in the random uncertainty is estimated by the Welch–Satterthwaite formula:

$$\nu_s = \frac{\left\{ \sum_{i=1}^{L} \left(\theta_i s_{\bar{x}_i} \right)^2 \right\}^2}{\sum_{i=1}^{L} \left\{ \left(\theta_i s_{\bar{x}_i} \right)^4 / \nu_{\bar{x}_i} \right\}} \tag{5.36}$$

By propagation of the systematic standard uncertainties of the variables, the systematic standard uncertainty in the result is

$$b_R = \left(\sum_{i=1}^{L} \left[\theta_i b_{x_i} \right]^2 \right)^{1/2} \tag{5.37}$$

The terms $\theta_i s_{\bar{x}_i}$ and $\theta_i b_{x_i}$ represent the individual contributions of the ith variable to the uncertainty in R. Comparing the magnitudes of each individual contribution identifies the relative importance of the uncertainty terms on the result.

The *combined standard uncertainty in the result*, written as u_R, is

$$u_R = \left[b_R^2 + s_R^2 \right]^{1/2} \tag{5.38}$$

with a confidence level of one standard deviation.

The *expanded uncertainty* in the result is given as

$$u_R = t_{\nu,P} \left[b_R^2 + s_R^2 \right]^{1/2} \quad (P\%) \tag{5.39}$$

The t value is used to provide a reasonable weight to the uncertainty interval to achieve the desired confidence. If the degrees of freedom in each variable is large ($N \geq 30$), then a reasonable approximation is to take $t_{\nu,95} = 2$. When the degrees of freedom in each of the variables, x_i, is not the same, the Welch–Satterthwaite formula is used to estimate the degrees of freedom in the result expressed as

$$\nu_R = \frac{\left(\sum_{i=1}^{L} \left(\theta_i s_{\bar{x}_i} \right)^2 + \left(\theta_i b_{\bar{x}_i} \right)^2 \right)^2}{\sum_{i=1}^{L} \left(\theta_i s_{\bar{x}_i}^{\,4} / \nu_{s_i} \right) + \sum_{i=1}^{L} \left(\theta_i b_{\bar{x}}^{\,4} / \nu_{b_i} \right)} \tag{5.40}$$

When the individual degrees of freedom in each of the systematic uncertainties are very large, as is often the case, the second term in the denominator is essentially zero.

Example 5.17

The density of a gas, ρ, that follows the ideal gas equation of state, $\rho = p/RT$, is estimated through separate measurements of pressure p and temperature T. The gas is housed within a rigid impermeable vessel. The literature accompanying the pressure measurement system states an instrument uncertainty to within $\pm 1\%$ of the reading (95%), and that accompanying the temperature measuring system indicates $\pm 0.6°R$ (95%) uncertainty. Twenty measurements of pressure, $N_p = 20$, and 10 measurements of temperature, $N_T = 10$, are made with the following statistical outcome:

$$\bar{p} = 2253.91 \text{ psfa} \quad s_p = 167.21 \text{ psfa}$$
$$\bar{T} = 560.4°R \quad s_T = 3.0°R$$

where psfa refers to lb/ft^2 absolute. Determine a best estimate of the density. The gas constant is $R = 54.7$ ft-lb/lbm-°R.

KNOWN $\bar{p}, s_p, \bar{T}, s_T$

$$\rho = p/RT; R = 54.7 \text{ ft-lb}/\text{lb}_m\text{-}°R$$

ASSUMPTIONS Gas behaves as an ideal gas

FIND $\rho' = \bar{\rho} \pm u_\rho$ (95%)

SOLUTION The measurement objective is to determine the density of an ideal gas through temperature and pressure measurements. The independent and dependent variables are related through the ideal gas law, $\rho = p/RT$. The mean value of density is

$$\bar{\rho} = \frac{\bar{p}}{R\bar{T}} = 0.0735 \text{ lb}_m/\text{ft}^3$$

The next step must be to identify and estimate the errors and determine how they contribute to the uncertainty in the mean value of density. Since no calibrations are performed and the gas is considered to behave as an ideal gas in an exact manner, the measured values of pressure and temperature are subject only to elemental errors within the data-acquisition error source group (Table 5.2): instrument errors and temporal variation errors.

The tabulated value of the gas constant is not without error. However, estimating the possible error in a tabulated value is sometimes difficult. According to Kestin (8), the systematic uncertainty in the gas constant is on the order of $\pm(0.33$ J/kg-K)/(gas molecular weight) or ± 0.06(ft-lb/lbm-°R)/ (gas molecular weight). Because this yields a small value for a reasonable gas molecular weight, here we assume a zero (negligible) systematic error in the gas constant.

Consider the pressure measurement. The uncertainty assigned to the temporal variation (data scatter) error is based on the variation in the measured data obtained during presumably fixed operating conditions. The instrument error is assigned a systematic uncertainty based on the manufacturer's statement, which is assumed to be stated at 95% confidence:

$$\left(b_{\bar{p}}\right)_1 = \left(B_{\bar{p}}/2\right)_1 = (0.01 \times 2253.91/2) = 11.28 \text{ psfa} \qquad \left(s_{\bar{p}}\right)_1 = 0$$

where the subscript keeps track of the error identity. The temporal variation causes a random uncertainty in establishing the mean value of pressure and is calculated as

$$\left(s_{\bar{p}}\right)_2 = s_{\bar{p}} = \frac{s_p}{\sqrt{N}} = \frac{167.21 \text{ psfa}}{\sqrt{20}} = 37.39 \text{ psfa} \qquad v_{s_p} = 19$$

and assigning no systematic uncertainty to this error gives

$$\left(b_{\bar{p}}\right)_2 = 0$$

In a similar manner, we assign the uncertainty in the data-acquisition source error in temperature. The instrument errors are considered as systematic only and assigned a systematic uncertainty based on the manufacturer's statement as

$$\left(b_{\bar{T}}\right)_1 = \left(b_{\bar{T}}\right)_1/2 = 0.3°\text{R} \qquad \left(s_{\bar{T}}\right)_1 = 0$$

The temporal variation introduces a random uncertainty in establishing the mean value of temperature, and this is calculated as

$$\left(s_{\bar{T}}\right)_2 = s_{\bar{T}} = \frac{s_T}{\sqrt{N}} = \frac{3.0°\text{R}}{\sqrt{10}} = 0.9°\text{R} \qquad v_{s_T} = 9$$
$$\left(b_{\bar{T}}\right)_2 = 0$$

The uncertainties in pressure and temperature owing to the data-acquisition source errors are combined using the RSS method:

$$b_{\bar{p}} = \left[(11.28)^2 + (0)^2\right]^{1/2} = 11.28 \text{ psfa}$$
$$s_{\bar{p}} = \left[(0)^2 + (37.39)^2\right]^{1/2} = 37.39 \text{ psfa}$$

similarly,

$$b_{\bar{T}} = 0.3°\text{R}$$
$$s_{\bar{T}} = 0.9°\text{R}$$

with degrees of freedom in the random standard uncertainties to be

$$\left(v\right)_{s_p} = N_p - 1 = 19$$
$$\left(v\right)_{s_T} = 9$$

The degrees of freedom in the systematic standard uncertainties, v_{b_T} and v_{b_p}, are assumed large.

The systematic and random standard uncertainties propagate through to the result, calculated about the operating point as established by the mean values for temperature and pressure. That is,

$$
\begin{aligned}
s_{\overline{\rho}} &= \left[\left(\frac{\partial \rho}{\partial T}s_{\overline{T}}\right)^2 + \left(\frac{\partial \rho}{\partial p}s_{\overline{p}}\right)^2\right]^{1/2} \\
&= \left[\left(1.3 \times 10^{-4} \times 0.9\right)^2 + \left(3.26 \times 10^{-5} \times 37.39\right)^2\right]^{1/2} \\
&= 0.0012 \, \text{lb}_m/\text{ft}^3
\end{aligned}
$$

and

$$
\begin{aligned}
b_{\overline{\rho}} &= \left[\left(\frac{\partial \rho}{\partial T}b_{\overline{T}}\right)^2 + \left(\frac{\partial \rho}{\partial p}b_{\overline{p}}\right)^2\right]^{1/2} \\
&= \left[\left(1.3 \times 10^{-4} \times 0.3\right)^2 + \left(3.26 \times 10^{-5} \times 11.28\right)^2\right]^{1/2} \\
&= 0.0004 \, \text{lb}_m/\text{ft}^3
\end{aligned}
$$

The degrees of freedom in the density calculation is determined to be

$$
\nu = \frac{\left[\left(\frac{\partial \rho}{\partial T}s_{\overline{T}}\right)^2 + \left(\frac{\partial \rho}{\partial p}s_{\overline{p}}\right)^2 + \left(\frac{\partial \rho}{\partial T}b_{\overline{T}}\right)^2 + \left(\frac{\partial \rho}{\partial T}b_{\overline{p}}\right)^2\right]^2}{\left[\left(\frac{\partial \rho}{\partial p}s_{\overline{p}}\right)^4 \Big/ \nu_{s_p} + \left(\frac{\partial \rho}{\partial T}s_{\overline{T}}\right)^4 \Big/ \nu_{s_T}\right] + \left[\left(\frac{\partial \rho}{\partial T}b_{\overline{T}}\right)^4 \Big/ \nu_{b_T} + \left(\frac{\partial \rho}{\partial p}b_{\overline{p}}\right)^4 \Big/ \nu_{b_p}\right]} = 23
$$

The expanded uncertainty in the mean value of density, using $t_{23,95} = 2.06$, is

$$
\begin{aligned}
u_{\rho} &= t_{23,95}\left[b_{\overline{\rho}}^2 + s_{\overline{\rho}}^2\right]^{1/2} = 2.06 \times \left[0.0004^2 + 0.0012^2\right]^{\frac{1}{2}} \\
&= 0.0026 \, \text{lb}_m/\text{ft}^3 \quad (95\%)
\end{aligned}
$$

The best estimate of the density is reported as

$$
\rho' = 0.0735 \pm 0.0026 \, \text{lb}_m/\text{ft}^3 \quad (95\%)
$$

This value of density has an uncertainty of about 3.4%.

COMMENT (1) We did not consider the uncertainty associated with our assumption of exact ideal gas behavior, a potential modeling systematic error (see Table 5.3) and an epistemic uncertainty (Section 5.5). From experience given these pressures and temperatures, this modeling error should be much smaller than the uncertainties used here. (2) Note how pressure contributes more to either standard uncertainty than does temperature and that the overall systematic uncertainty is small compared to the random uncertainty. The uncertainty in density is best reduced by actions to reduce the effects of the random errors on the pressure measurements, if possible.

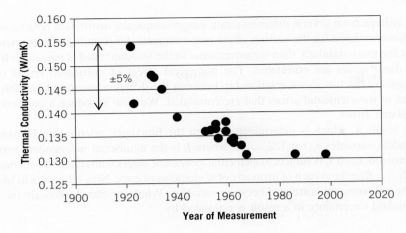

Figure 5.7 The thermal conductivity of liquid toluene at 298.15 K and 0.10 MPa abs (compiled from 11,12).

Example 5.18: Changing Values—Uncertainty in Published Properties

A common feature found in handbooks and textbook appendices is the tabulations of material properties. We use these data in engineering calculations as if they were a matter of fact. However, these data are the result of measurements and so are prone to measurement limitations. Consider the following.

Property reference standards serve to confirm the accuracy of new instrumentation and reproducibility of existing equipment. For example, toluene can be obtained with a very high purity and so serves as a reference standard of choice for calibrating instruments used in measuring the material properties of electrically insulating liquids.

Figure 5.7 shows the thermal conductivity of liquid toluene, k, as officially reported over the last century. Has toluene been changing? Of course not, but measured property values will change over time due to improvements in the methodology used to measure them. Up until the mid-1960s, steady-state thermal methods were used for this measurement. The methods were prone to various systematic errors, particularly those arising from convection and radiation heat transport, which interfere and offset the calculated conductivity to a larger value. Transient techniques are now used and are better able to reduce and to correct for these residual errors. As such, the uncertainty in the reported property data has been reducing with time. Since the 1970s, the reported mean value for thermal conductivity has remained stable to within 0.07%. In 2000, the accepted value was $k(298.15 \text{ K}, 0.10 \text{ MPa}) = 0.13088 \pm 0.00085$ W/m-K(95%). The exact property value varies with operating temperature and its uncertainty varies from 0.6% to 2% over the full usable temperature range. Whenever accuracy is important, be sure to research the source of the information as part of your test plan.

5.9   CORRECTION FOR CORRELATED ERRORS

So far, we have assumed that all of the different elements of error in a test are independent from the others. If two errors are not independent, they are "correlated."

For example, when the same instrument is used to measure different variables, the instrument errors in those measured variables are correlated. So if the same voltmeter was used to measure the bridge deflection voltage from several different strain gauge setups, the instrument errors assigned to the voltmeter are correlated for the various strain measurements. Or if multiple instruments are calibrated against the same standard, then the uncertainty in the standard used is passed on to each instrument, and these errors are correlated. The numerical effect of correlated errors on the uncertainty depends on the functional relationship between the different measured variables and on the magnitudes of the elemental errors that are correlated. We now introduce a correction for treating the correlated errors.

Consider the result R, which is determined through the functional relationship between the measured independent variables $x_i, i = 1, 2, \ldots, L$ where L is the number of independent variables and parameters involved. Each x_i is subject to elemental systematic errors with standard uncertainties, b_k, where $k = 1, 2, \ldots, K$, refer to each of up to any of K elements of error. Now allow that as many as H of these K elemental errors are correlated between variables. When correlated errors are included, the systematic standard uncertainty in a result is estimated by

$$b_R = \left[\sum_{i=1}^{L} \left(\theta_i b_{\bar{x}_i} \right)^2 + 2 \sum_{i=1}^{L-1} \sum_{j=i+1}^{L} \theta_i \theta_j \delta_{ij} b_{\bar{x}_i \bar{x}_j} \right]^{1/2} \tag{5.41}$$

where index j is a counter equal to $i + 1$ and with

$$\theta_i = \frac{\partial R}{\partial x_i}\bigg|_{x=\bar{x}} \tag{5.14}$$

Equation 5.42 introduces the covariance, $b_{\bar{x}_i \bar{x}_j}$, to account for correlated errors, and this is found from

$$b_{\bar{x}_i \bar{x}_j} = \sum_{h=1}^{H} \left(b_{\bar{x}_i} \right)_h \left(b_{\bar{x}_j} \right)_h \tag{5.42}$$

where H is the number of elemental errors that are correlated between variables x_i and x_j and h is a counter for each correlated error. The value δ_{ij} is a covariance coefficient with a value between 0 (no correlation) and 1 (correlated). Normally its value is either 0 or 1. Note that Equation 5.41 reduces to Equation 5.37 when no errors are correlated (i.e., when $H = 0$, or all $\delta_{ij} = 0$). References 2 and 9 discuss treatment of correlated systematic errors in extended detail.

There are situations where random errors can be correlated (10). In those cases, the random standard uncertainty in a result is estimated by

$$s_R = \left[\sum_{i=1}^{L} \left(\theta_i s_{\bar{x}_i} \right)^2 + 2 \sum_{i=1}^{L-1} \sum_{j=i+1}^{L} \theta_i \theta_j r_{\bar{x}_i \bar{x}_j} s_{\bar{x}_i \bar{x}_j} \right]^{1/2} \tag{5.43}$$

where $r_{\bar{x}_i \bar{x}_j}$ is the correlation coefficient between x_i and x_j as given by Equation 4.41 and

$$s_{\bar{x}_i \bar{x}_j} = \sum_{h=1}^{H} \left(s_{\bar{x}_i} \right)_h \left(s_{\bar{x}_j} \right)_h \tag{5.44}$$

In a general uncertainty analysis, the correlated error is handled by the general standard uncertainty equation:

$$u_R = \left[\sum_{i=1}^{L} \left(\theta_i u_{\bar{x}_i} \right)^2 + 2 \sum_{i=1}^{L-1} \sum_{j=i+1}^{L} \theta_i \theta_j \delta_{ij} u_{\bar{x}_i} u_{\bar{x}_j} \right]^{1/2} \tag{5.45}$$

where δ_{ij} is assigned a value between 0 and 1.

Example 5.19

Suppose that a result is a function of three variables, x_1, x_2, x_3. There are four systematic errors associated with both x_1 and x_2 and five associated with x_3. Only the first and second systematic elemental errors associated with the second and third variable (x_2 and x_3) are determined to be correlated because these errors arise from common sources. Express the systematic standard uncertainty in the result and find the covariance term.

   SOLUTION   Equation 5.41 is expanded to the form

$$b_R = \left[\left(\theta_1 b_{\bar{x}_1} \right)^2 + \left(\theta_2 b_{\bar{x}_2} \right)^2 + \left(\theta_3 b_{\bar{x}_3} \right)^2 + 2\theta_1 \theta_2 \delta_{12} b_{\bar{x}_1 \bar{x}_2} + 2\theta_1 \theta_3 \delta_{13} b_{\bar{x}_1 \bar{x}_3} + 2\theta_2 \theta_3 \delta_{23} b_{\bar{x}_2 \bar{x}_3} \right]^{1/2}$$

For the two correlated errors ($H = 2, \delta_{12} = \delta_{13} = 0, \delta_{23} = 1$) associated with variables 2 and 3 ($i = 2, j = 3$), the systematic uncertainty in the result reduces to

$$b_R = \left[\left(\theta_1 b_{\bar{x}_1} \right)^2 + \left(\theta_2 b_{\bar{x}_2} \right)^2 + \left(\theta_3 b_{\bar{x}_3} \right)^2 + 2\theta_2 \theta_3 \delta_{23} b_{\bar{x}_2 \bar{x}_3} \right]^{1/2}$$

where the covariance term is

$$b_{\bar{x}_2 \bar{x}_3} = \sum_{h=1}^{2} \left(b_{\bar{x}_2} \right)_h \left(b_{\bar{x}_3} \right)_h = \left(b_{\bar{x}_2} \right)_1 \left(b_{\bar{x}_3} \right)_1 + \left(b_{\bar{x}_2} \right)_2 \left(b_{\bar{x}_3} \right)_2$$

and subscripts h = 1 and h = 2 refer to the relevant elemental errors.

Example 5.20

Result R is a function of two variables, X and Y, such that $R = X + Y$, and each variable having one elemental error. If $\overline{X} = 10.1$ V with $b_{\bar{x}} = 1.1$ V and $\overline{Y} = 12.2$ V with $b_{\bar{y}} = 0.8$ V, estimate the systematic standard uncertainty in the result if the systematic errors are (1) uncorrelated and (2) correlated.

   SOLUTION   From the stated information, $\overline{R} = 10.1 + 12.2 = 22.3$ V. For the uncorrelated case ($\delta_{XY} = 0$), the systematic standard uncertainty is estimated as

$$b_{R_{\text{unc}}} = \left[\left(\theta_X b_{\bar{X}} \right)^2 + \left(\theta_Y b_{\bar{Y}} \right)^2 \right]^{1/2} = \left[\left(1 \times 1.1 \right)^2 + \left(\left(1 \times 0.8 \right)^2 \right)^2 \right]^{1/2} = 1.36 \text{ V}$$

For the correlated case ($\delta_{XY} = 1$), with $H = 1$ and $L = 2$, the uncertainty is estimated as

$$b_{R_{cor}} = \left[(\theta_X b_{\overline{X}})^2 + (\theta_{\overline{Y}} b_{\overline{Y}})^2 + 2\theta_X \theta_Y \delta_{XY} b_{\overline{XY}} \right]^{1/2}$$

$$= \left[(1 \times 1.1)^2 + (1 \times 0.8)^2 + 2(1)(1)(1)(1.1)(0.8) \right]^{1/2} = 3.12 \text{ V}$$

which is stated at the one-standard deviation confidence level and where

$$b_{\overline{XY}} = \sum_{h=1}^{1} (b_{\overline{X}})_1 (b_{\overline{Y}})_1 = b_{\overline{X}} b_{\overline{Y}}$$

COMMENT In this case, the fact that these systematic errors are correlated has a large impact on the systematic uncertainty. This is not always the case, because it depends on the functional relationship itself. We leave it to the reader to show that if $R = X/Y$, then the correlated error term in this problem would have little effect on the systematic standard uncertainty in the result.

5.10 NONSYMMETRICAL SYSTEMATIC UNCERTAINTY INTERVAL

There are situations in which the error must be bounded on one side or skewed about the measured mean value such that the systematic uncertainty is better described by a nonsymmetrical interval. To develop this case, we assume that the limits of possible systematic error are known, as is the mean of the measured data set $\overline{x}$. Let $\overline{x} + B_{\overline{x}}^+$ and $\overline{x} - B_{\overline{x}}^-$ be the upper and lower limits of the systematic uncertainty relative to the measured mean value (Fig. 5.8), with $B_{\overline{x}}^- = 2b_{\overline{x}}^-$ and $B_{\overline{x}}^+ = 2b_{\overline{x}}^+$ where $b_{\overline{x}}^-$ and $b_{\overline{x}}^+$ are the lower and upper systematic standard uncertainty values. The modeling approach assumes that the errors are symmetric about some mean value of the error distribution but asymmetric about the measured mean value.

If we model the error distribution as a normal distribution, then we assume that the limits defined using $B_{\overline{x}}^-$ to $B_{\overline{x}}^+$ cover 95% of the error distribution. Define the offset between the mean of the error distribution and the mean value of the measurement as

$$q = \frac{(\overline{x} + B_{\overline{x}}^+) + (\overline{x} + B_{\overline{x}}^-)}{2} - \overline{x} = \frac{B_{\overline{x}}^+ - B_{\overline{x}}^-}{2} = b_{\overline{x}}^+ - b_{\overline{x}}^- \tag{5.46}$$

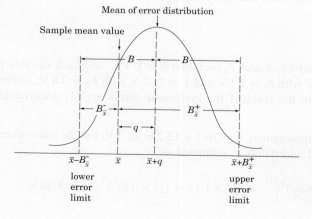

Figure 5.8 Relation between the measured mean value, the mean value of the distribution of errors, and the systematic uncertainties for treating nonsymmetrical uncertainties.

The systematic standard uncertainty has an average width,

$$b_{\bar{x}} = \frac{(\bar{x} + B_{\bar{x}}^+) - (\bar{x} - B_{\bar{x}}^-)}{4} = \frac{(\bar{x} + b_{\bar{x}}^+) - (\bar{x} + b_{\bar{x}}^-)}{2} \tag{5.47}$$

For stating the true value, we use the combined standard uncertainty

$$u_x = \sqrt{b_{\bar{x}}^2 + s_{\bar{x}}^2} \tag{5.48}$$

to achieve the approximate confidence interval, $q \pm t_{\nu,P} u_x$. For large degrees of freedom, we can state

$$x' = \bar{x} + q - t_{\nu,P} u_x, \bar{x} + q + t_{\nu,P} u_x \tag{5.49}$$

for which we choose an appropriate value for t, such as $t = 2$ for 95% confidence.

If we model the error distribution as a rectangular distribution, then we assume that the limits defined using $B_{\bar{x}}^-$ to $B_{\bar{x}}^+$ specify the limits of that distribution. The systematic standard uncertainty is then (from Equation 5.22)

$$b_{\bar{x}} = \frac{(\bar{x} + B_{\bar{x}}^+) - (\bar{x} - B_{\bar{x}}^-)}{\sqrt{12}} \tag{5.50}$$

for use in Equations 5.47 and 5.48. Regardless of the systematic error distribution assumed, we assume that the uncertainties propagate normally.

A normal or rectangular distribution is not always the appropriate model for treating asymmetric systematic uncertainty. References 1 and 2 provide a number of scenarios to treat nonsymmetrical uncertainties including use of other error distributions. As an example, Monsch et al. (13) apply the above approach to estimating the nonsymmetrical systematic uncertainty in the induced drag on an aircraft wing for which the induced drag cannot be less than zero.

5.11 SUMMARY

Uncertainty analysis provides the "± what" to a test result or the anticipated result from a proposed test plan. This chapter discussed the manner in which various errors can enter into a measurement with emphasis on their influences on the test result. Both random errors and systematic errors are considered. Random errors differ on repeated measurements, lead to data scatter, and can be quantified by statistical methods. Systematic errors remain constant in repeated measurements. Errors are quantified by assigning uncertainties. Procedures for uncertainty analysis were introduced both as a means of estimating uncertainty propagation within a measurement and for the propagation of uncertainty among the independent measured variables that are used to determine a result through some functional relationship. We have discussed the role and use of uncertainty analysis in both the design of measurement systems, through selection of equipment and procedures, as well as in the interpretation of measured data. We have presented procedures for estimating uncertainty at various stages of a test. The procedures provide reasonable estimates of the uncertainty to be expected in measurements.

We advise the reader that test uncertainty by its very nature evades exact values. But the methodology presented gives reasonable estimates provided that the engineer has used sound, impartial judgment in its implementation. The important thing is that the uncertainty analysis be performed during the design and development stages of a test, as well as at its completion to assess the stated results. Only then can the engineer use the results with appropriate confidence and professional and ethical responsibility.

REFERENCES

1. International Organization for Standardization (ISO), *Guide to the Expression of Uncertainty in Measurement*, Geneva, Switzerland, 1993.
2. ANSI/ASME Power Test Codes-PTC 19.1, *Test Uncertainty*, American Society of Mechanical Engineers, New York, 2013.
3. Smith, R. E., and S. Wenhofer, From Measurement Uncertainty to Measurement Communications, Credibility and Cost Control in Propulsion Ground Test Facilities, *Journal of Fluids Engineering* **107**: 165–172 1985.
4. Epperson, J. F., *An Introduction to Numerical Methods and Analysis*, Wiley-Interscience, New York, 2007.
5. Kline, S. J., and F. A. McClintock, Describing Uncertainties in Single Sample Experiments, *Mechanical Engineering* **75**: 3–8, 1953.
6. Moffat, R. J., Uncertainty Analysis in the Planning of an Experiment, *Journal of Fluids Engineering* **107**: 173–181, 1985.
7. Taylor, B. N., and C. Kuyatt, *Guidelines for Evaluating and Expressing the Uncertainty of NIST Measurement Results*, NIST Technical Note 1297, Gaithersburg, MD, 1994.
8. Kestin, J., *A Course in Thermodynamics*, revised printing, Hemisphere, New York, 1979.
9. Coleman, H., and W. G. Steele, *Experimentation and Uncertainty Analysis for Engineers*, 2nd ed., Wiley-Interscience, New York, 1999.
10. Dieck, R., *Measurement Uncertainty: Methods and Applications*, 3rd ed., Instrumentation, Systems and Automation Society (ISA), Research Triangle Park, NC, 2002.
11. Ramires, M., C. Nieto de Castro, R. Perkins, Y. Nagasaka, A. Nagashima, M. Assael, W. Wakeham, Standard Reference Data for Thermal Conductivity of Liquids, *Journal of Physical and Chemical Reference Data*, **29** (2): 133–139, 2000.
12. Nieto de Castro, C., S. Li, A. Nagashima, R. Trengrove, W. Wakeham, Standard Reference Data for Thermal Conductivity of Liquids, *Journal of Physical and Chemical Reference Data*, **15** (3): 1073–1086, 1986.
13. Monsch, S., R. S. Figliola, E. Thompson, J. Camberos, Induced Drag for 3D Wing with Volume Integral (Trefftz Plane) Technique, AIAA Paper 2007-1079, AIAA, New York, 2007.

NOMENCLATURE

b	systematic standard uncertainty	m	upper bound (rectangular distribution)
$b_{\bar{x}}$	systematic standard uncertainty in x	n	lower bound (rectangular distribution)
e_k	kth elemental error	p	pressure $(ml^{-1}\,t^{-2})$
$f(\ )$	general functional relation	q	offset factor

s_x	sample standard deviation of x
$s_{\bar{x}}$	random standard uncertainty in x; sample standard deviation of the means
s_{yx}	standard deviation of a fit
$t_{v,95}$	t variable (at 95% probability)
t	time (t); Student's t-variable
u_x	uncertainty of a measured variable
u_d	design-stage uncertainty
u_c	instrument calibration uncertainty
u_0	zero-order uncertainty
u_i	ith-order uncertainty
u_N	Nth-order uncertainty
x	measured variable
x'	true (mean) value of the population of x
$\bar{x}$	sample mean value of x

R	result or resultant value
P	probability
B	systematic uncertainty (at 95% or 2σ probability)
$(P\%)$	confidence level
Q	flow rate $(l^3 t^{-1})$
T	temperature ($°$)
$\forall$	volume (l^{-3})
δ_{ij}	covariance coefficient
θ_i	sensitivity index
ρ	gas density (ml^{-3})
σ	stress $(ml^{-1}t^{-2})$; population standard deviation
v	degrees of freedom
$\langle \rangle$	pooled statistic

PROBLEMS

5.1 What is the difference between an error and an uncertainty? Discuss ways in which uncertainty can be estimated for a measured value.

5.2 Explain what is meant by the terms "true value," "best estimate," "mean value," "uncertainty," and "confidence interval."

5.3 An official Olympics timekeeper once stated, "If you want to claim a result with a very good accuracy to one-hundredth of a second, you have to measure to a thousandth of a second." Explain the possible reasoning.

5.4 A tachometer has an analog display dial graduated in 5-revolutions-per-minute (rpm) increments. The user manual states an accuracy of 1% of reading. Estimate the design-stage uncertainty in the reading at 50, 500, and 5,000 rpm.

5.5 An automobile speedometer is graduated in 5-mph (8-kph) increments and has an accuracy stated as within $\pm 4\%$. Estimate the design-stage uncertainty in indicated speed at 60 mph (90 kph).

5.6 An engineer reads the height of a mercury barometer column as 749.5 mm Hg. The resolution of the vernier scale is 0.1 mm. But the engineer notices that the meniscus of the column makes resolution to 1 mm Hg more likely. Suggest a value for the zero-order uncertainty for the barometer. Would you assign this as a Type A uncertainty, or Type B? Random uncertainty, or systematic?

5.7 The volume of a cylinder is determined by measuring its radius and height. From a design stage analysis, both measurements can be made to an uncertainty of 0.100 mm at one standard deviation. If the radius is 10.00 mm and h is 20.00 mm, express the final result. Discuss how each measurement contributes to the overall uncertainty. Assume all errors are uncorrelated.

5.8 A temperature measurement system is composed of a sensor and a readout device. The readout device has a claimed accuracy of 0.6 °C with a resolution of 0.1 °C. The sensor has an off-the-shelf limit of error of 0.5 °C. Estimate a design-stage uncertainty in the temperature indicated by this combination.

5.9 Two resistors are to be combined to form an equivalent resistance of $1,000\,\Omega$. The resistors are taken from available stock on hand as acquired over the years. Readily available are two common resistors rated at $500 \pm 50\,\Omega$ and two common resistors rated at $2,000\,\Omega \pm 5\%$. What combination of resistors (series or parallel) would provide the smaller uncertainty in an equivalent $1,000\,\Omega$ resistance?

5.10 An equipment catalog boasts that a pressure transducer system comes in $3\frac{1}{2}$-digit (e.g., 19.99) or $4\frac{1}{2}$-digit (e.g., 19.999) displays. The $4\frac{1}{2}$-digit model costs substantially more. Both units are otherwise identical. The specifications list the uncertainties for three elemental errors:

Linearity error:	0.15% FSO
Hysteresis error:	0.20% FSO
Sensitivity error:	0.25% FSO

For a full-scale output (FSO) of 20 kPa, select a readout based on appropriate uncertainty calculations. Explain.

5.11 The shear modulus, G, of an alloy can be determined by measuring the angular twist, θ, resulting from a torque applied to a cylindrical rod made from the alloy. For a rod of radius R and a torque applied at a length L from a fixed end, the modulus is found by $G = 2LT/\pi R^4\theta$. Examine the effect of the relative uncertainty of each measured variable on the shear modulus. If during test planning all of the uncertainties are estimated to be 1%, what is the uncertainty in G?

5.12 An ideal heat engine operates in a cycle and produces work as a result of heat transfer from a thermal reservoir at an elevated temperature T_h and by rejecting energy to a thermal sink at T_c. The efficiency for such an ideal cycle, termed a "Carnot cycle," is

$$\eta = 1 - (T_c/T_h).$$

Determine the required uncertainty in the measurement of temperature to yield an uncertainty in efficiency of 1%. Assume errors are uncorrelated. Use $T_h = 1000$ K and $T_c = 300$ K.

5.13 Heat transfer from a rod of diameter D immersed in a fluid can be described by the Nusselt number, $Nu = hD/k$, where h is the heat-transfer coefficient and k the thermal conductivity of the fluid. If h can be measured to within $\pm7\%$ (95%), estimate the uncertainty in Nu for the nominal value of $h = 150$ W/m²-K. Let $D = 20 \pm 0.5$ mm and $k = 0.6 \pm 2\%$ W/m-K.

5.14 In aircraft design, the pressure coefficient C_p is usually measured during wind tunnel testing of an aircraft component to predict structural loads, as well as lift and drag forces. Let $C_p = (p - p_\infty)/\frac{1}{2}\rho U_\infty^2$ where U_∞ is the airspeed of the test at pressure p_∞ and using a fluid of density ρ. The pressure difference $p - p_\infty$ is measured where p is located at a point on the surface of the test component.. Estimate the uncertainty in C_p under the following conditions: $\Delta p = p - p_\infty = 1000$ N/m²,

$u_{\Delta p} = 15$ N/m²; $\rho = 1.20$ kg/m³, $u_\rho = 0.01$ kg/m³; $U_\infty = 50.0$ m/s, $u_{U_\infty} = 0.21$ m/s. Assume 95% confidence in stated values.

5.15 Estimate the design-stage uncertainty in determining the voltage drop across an electric heating element. The device has a nominal resistance of 30 Ω and a power rating of 500 W. Available is an ohmmeter (accuracy: within 0.5%; resolution: 1 Ω) and ammeter (accuracy: within 0.1%; resolution: 100 mA). Recall $E = IR$.

5.16 Explain the critical difference(s) between a design-stage uncertainty analysis and an advanced-stage uncertainty analysis.

5.17 A displacement transducer has the following specifications:

Linearity error:	$\pm0.25\%$ reading
Drift:	$\pm0.05\%/°C$ reading
Sensitivity error:	$\pm0.25\%$ reading
Excitation:	10-25 V dc
Output:	0-5 V dc
Range:	0-5 cm

The transducer output is to be indicated on a voltmeter having a stated accuracy of $\pm0.1\%$ reading with a resolution of 10 μV. The system is to be used at room temperature, which can vary by ±10 °C. Estimate an uncertainty in a nominal displacement of 2 cm at the design stage. Assume 95% confidence.

5.18 The displacement transducer of Problem 5.17 is used in measuring the displacement of a body impacted by a mass. Twenty measurements are made, which yield

$$\bar{x} = 17.20 \text{ mm} \quad s_x = 1.70 \text{ mm}$$

Determine a best estimate for the mass displacement at 95% probability based on all available information.

5.19 A saline-glycerin blood substitute flows through a compliant tube so as to simulate arterial blood motion. To measure its flow rate, the flow is temporarily deflected and caught in a bucket over a time interval of 10.0 ± 0.20 s. The empty bucket itself has a mass of 201.1 ± 2.0 g. The full bucket has a mass of 403.2 ± 2.0 g. The mass flow rate is determined as the mass (through its measured weight) caught over the time interval. Estimate the mass flow rate and its uncertainty using a general uncertainty analysis. Assume reported uncertainties are standard uncertainties. Assume mass uncertainties are uncorrelated.

5.20 In the previous problem of arterial flow through a tube, suppose that the mass uncertainties contain systematic errors (zero setpoint) that account for 1.5 g of each 2.0 g of uncertainty and these errors are fully correlated. Assume all uncertainties are standard uncertainties. Estimate the mass flow rate and its uncertainty using a general uncertainty analysis.

5.21 Consider determining the spring constant for a linear spring where applied force, F, and resulting elongation or displacement, x, are related as $F = ky$ where k is the spring constant. (a) Suppose the spring constant is measured by hanging weights from the end of the spring and measuring the resulting displacement. For a nominal spring constant of 65 N/m, plot the percent uncertainty in the measured spring constant as a function of the mass in grams hung on the spring for a range of mass from 10 to 100 grams. The uncertainty in the mass is 0.5 grams and the uncertainty in the elongation is 0.02 cm. (b) The following data were collected for a particular spring. Determine the spring constant and its uncertainty.

Mass (g)	10	20	30	40	50	60
Elongation (cm)	0.143	0.295	0.445	0.593	0.725	0.881

5.22 The kinematic viscosity of fresh, pure water is a function of temperature as tabulated below. Determine the viscosity and the uncertainty in determining the viscosity of water if temperature is known to $20.20 \pm 0.20\,°C$

T(°C)	20.00	20.10	20.20	20.30	20.40	20.50
$\nu(m^2/s) \times 10^6$	1.00374	1.00131	0.99888	0.99646	0.99405	0.99165

5.23 The Young's modulus, E, of metals is a function of temperature as shown for copper alloy (C83600), in the table below. Determine the modulus and the uncertainty in the modulus if the temperature is known to vary between $260 \pm 50\,°C$.

Temperature (°C)	149	204	260	316	371
E(GPa)	92.390	91.011	88.943	86.185	82.738

5.24 The drag, D, on a small shape is determined based on available information from $D = C_D \Delta p A$, where C_D is a drag coefficient and A is frontal area. The quantity $\Delta p = \frac{1}{2}\rho U^2$ is the dynamic pressure and it is measured using a pitot-static tube and pressure transducer. Find the drag and its uncertainty using a general analysis based on this information. $\Delta p = 101.6 \pm 1.529\,mm\,H_2O$; $A = 0.01810 \pm 0.00005\,m^2$; $C_D = 1.000 \pm 0.092$. Assume 95% confidence in all values.

5.25 The ideal gas law is a common model for relating the properties of dry air. Find the uncertainty in the density of air $\rho = P/RT$ based on this model where p and T are the absolute pressure and temperature and R is the gas constant. Assume $p = 101,320 \pm 67\,Pa\,abs$; $T = 297 \pm 0.5\,K$; $R = 287.058\,J/kg-K$. Assume 95% confidence levels.

5.26 A capacitance sensor consists of two overlapping plates separated by a small gap filled with either a dielectric material or a vacuum. Capacitance is changed by varying either the area of overlap, A, or changing the thickness of the gap, t, such that $C = c\varepsilon A/t$. For an area of $600 \pm 0.3\,mm^2$ and gap thickness of $0.30 \pm 0.01\,mm$, estimate the uncertainty in capacitance. Let $c\varepsilon = 8.85 \times 10^{-15}\,F/mm$, assume 95% confidence in all values and no correlated errors.

5.27 The Reynolds number is an engineering parameter that relates to the volume flow rate of a fluid through a tube of diameter d as $Re_d = 4Q/\pi\nu d$ where ν is the fluid kinematic viscosity. Determine the Reynolds number and its uncertainty by general analysis given $d = 50.4 \pm 0.1\,mm$, $Q = 0.213 \pm 0.005\,m^3/s$, $\nu = 0.999 \times 10^{-6} \pm 0.005 \times 10^{-6}\,m^2/s$. Which variable contributes the most to the overall uncertainty? Assume uncertainties are at 95% confidence.

5.28 Subsonic aircraft use a pitot tube (impact pressure tube) and a static pressure port to determine aircraft airspeed by $U = \sqrt{2\Delta p/\rho_{air}} = \sqrt{2\rho_{Hg}gH/\rho_{air}}$ where $H = \Delta p/\rho_{Hg}g$ is the differential pressure expressed in either inches Hg or mm Hg. The value of H can be measured to within 1% (95%) uncertainty. Estimate the airspeed and the uncertainty in airspeed at sea level and using International Standard Atmosphere (ISA) conditions of 15 °C and 101320 Pa abs if H is 20 mm Hg. Neglect the uncertainties in densities. Use $\rho_{Hg}(kg/m^3) = 13595 - 2.5T(°C) \pm 0.1$ and treat dry air as an ideal gas, $\rho_{air}(kg/m^3) = 3.4837 \times 10^{-3} \, p(Pa \, abs)/T(K) \pm 0.001$.

5.29 For a thin-walled pressure vessel of diameter D and wall thickness t subjected to an internal pressure p, the tangential stress is given by $\sigma = pD/2t$. During one test, 10 measurements of pressure yielded a mean of $8610 \, lb/ft^2$ with a standard deviation of $273.1 \, lb/ft^2$. Cylinder dimensions are to be based on a set of 10 measurements which yielded: $\overline{D} = 6.2$ in. with $s_D = 0.18$ in. and $\bar{t} = 0.22$ in. with $s_t = 0.04$ in. Estimate the degrees of freedom in the standard uncertainty for stress. Determine the best estimate of stress at 95% confidence. Pressure measurements and dimensions have a systematic uncertainty of 1% of the reading assumed at 95% confidence.

5.30 Suppose a measured normal stress contains three elemental random errors in its calibration with the following values:

$$(s_{\bar{x}})_1 = 0.90 \, N/m^2 \quad (s_{\bar{x}})_2 = 1.10 \, N/m^2 \quad (s_{\bar{x}})_3 = 0.09 \, N/m^2$$
$$v_1 = 21 \qquad\qquad v_2 = 10 \qquad\qquad v_3 = 15$$

Estimate the random standard uncertainty due to calibration errors with its degrees of freedom.

5.31 An experiment to determine force acting at a point on a member contains three elemental errors with resulting uncertainties:

$$(b_{\bar{x}})_1 = 1.0 \, N \quad (b_{\bar{x}})_2 = 2.25 \, N \quad (b_{\bar{x}})_3 = 1.8 \, N$$
$$(s_{\bar{x}})_1 = 0 \qquad\quad (s_{\bar{x}})_2 = 6.1 \, N \quad (s_{\bar{x}})_3 = 4.2 \, N$$
$$\qquad\qquad\qquad v_2 = 17 \qquad\quad v_3 = 19$$

If the mean value for force is estimated to be 200 N, determine the uncertainty in the mean value at 95% confidence. Assume large degrees of freedom in the systematic uncertainties.

5.32 The area of a flat, rectangular parcel of land is computed from the measurement of the length of two adjacent sides, X and Y. Measurements are made using a scaled chain accurate to within 0.5% (95%) over its indicated length. The two sides are measured several times with the following results:

$$\overline{X} = 556 \, m \quad \overline{Y} = 222 \, m$$
$$s_x = 5.3 \, m \quad s_y = 2.1 \, m$$
$$v = 8 \qquad\quad v = 7$$

Estimate the area of the land and state the confidence interval of that measurement at 95%.

5.33 Estimate the random standard uncertainty in the measured value of stress for a data set of 23 measurements if $\bar{\sigma} = 1061$ kPa and $s_\sigma = 22$ kPa. Report the best estimate for stress at 68% confidence. Assume a normal distribution and neglect systematic errors.

5.34 Estimate the uncertainty at 95% confidence in the strength of a metal alloy. Six separate specimens are tested with a mean of 448.1 MPa and standard deviation of 1.23 MPa in the results. A systematic standard uncertainty of 1.48 MPa with $v_b = 19$ is estimated by the operator.

5.35 One use for a GPS rangefinder is on the golf course where it is used to estimate distances between fixed points on the course relative to the player. A typical rangefinder has an instrument systematic uncertainty of 1% and a random uncertainty of 0.5%. Estimate the overall uncertainty of the device at 200 yards (or at 200 meters). Assume values given are at 95% confidence.

5.36 A pressure measuring system outputs a voltage that is proportional to pressure. It is calibrated against a transducer standard (certified error: within ± 0.5 psi) over its 0–100-psi range with the results given below. The voltage is measured with a voltmeter (instrument error: within $\pm 10 \, \mu V$; resolution: $1 \, \mu V$). The engineer intending to use this system estimates that installation effects can cause the indicated pressure to be off by another ± 0.5 psi. Estimate the expanded uncertainty at 95% confidence in using the system based on the known information.

E[mV]:	0.004	0.399	0.771	1.624	2.147	4.121
p[psi]:	0.1	10.2	19.5	40.5	51.2	99.6

5.37 The density of a metal composite is to be determined from the mass of a cylindrical ingot. The volume of the ingot is determined from diameter and length measurements. It is estimated that mass m can be determined to within 0.1 lb$_m$ using an available balance scale; length L can be determined to within 0.05 in. and diameter D to within 0.0005 in. Instrumentation for each variable has a known calibration systematic uncertainty of 1% of its reading. Estimate the design-stage uncertainty in the determination of the density. Which measurement contributes most to the uncertainty in the density? Which measurement method should be improved first if the uncertainty in density is unacceptable? Use the nominal values of $m = 4.5$ lb$_m$, $L = 6$ in., and $D = 4$ in. (*Note:* 1 lb$_m$ = 0.4535 kg; 1 inch = 0.0254 m.)

5.38 A temperature measurement system is calibrated against a standard system certified to an uncertainty of $\pm 0.05\ °C$ at 95%. The system sensor is immersed alongside the standard within a temperature bath so that the two are separated by about 10 mm. The temperature uniformity of the bath is estimated at about $5\ °C/m$. The temperature system sensor is connected to a readout that indicates the temperature in terms of voltage. The following are measured values between the temperature indicated by the standard and the indicated voltage:

$T[°C]$	$E[mV]$	$T[°C]$	$E[mV]$
0.1	0.004	40.5	1.624
10.2	0.399	51.2	2.147
19.5	0.771	99.6	4.121

a. Compute the curve fit for $T = f(E)$ and its standard random uncertainty.

b. Estimate the uncertainty in using the output from the temperature measurement system for temperature measurements.

5.39 The power usage of a strip heater is to be determined by measuring heater resistance and heater voltage drop simultaneously. The resistance is to be measured using an ohmmeter having a resolution of $1\ \Omega$ and uncertainty of 1% of reading, and voltage is to be measured using a voltmeter having a resolution of 1 V and uncertainty of 1% of reading. It is expected that the heater will have a resistance of $100\ \Omega$ and use 100 W of power. Determine the

uncertainty in power determination to be expected with this equipment at the design stage.

5.40 The power usage of a DC strip heater can be determined in either of two ways: (1) Heater resistance and voltage drop can be measured simultaneously and power computed, or (2) heater voltage drop and current can be measured simultaneously and power computed. Instrument manufacturer specifications are listed (assume 95% level):

Instrument	Resolution	Uncertainty (% reading)
Ohmmeter	1 Ω	0.5%
Ammeter	0.5 A	1%
Voltmeter	1 V	0.5%

For loads of 10 W, 1 kW, and 10 kW, determine the best method based on an appropriate uncertainty analysis. Assume nominal values as necessary for resistance, current, and voltage.

5.41 The specific humidity of air is a measure of the amount of water vapor present in a kilogram of dry air. Using a psychrometric chart (available in a thermodynamics text, reference handbook, or online source), determine the sensitivity of the specific humidity of air, γ, to temperature variations, T, for the conditions corresponding to $T = 20\ °C$ and a relative humidity of 50%.

5.42 A thermocouple is a temperature sensor made of two electrical conductors each of different materials. The conductors are electrically connected at two points, called junctions, each junction exposed to a different temperature. The thermocouple generates a voltage relative to the temperature gradient between junctions. Table 8.6 lists values of voltage as a function of temperature relative to one junction maintained at $0\ °C$ for a thermocouple made from joining an iron wire with a constantan (chrome-nickel) wire. Estimate the sensitivity of the wire at $100\ °C$ (one junction) relative to $0\ °C$ (one junction).

5.43 Time variations in a signal require that the signal be measured often to determine its mean value. A preliminary sample is obtained by measuring the signal 50 times. The statistics from this sample show a mean value of 2.112 V with a standard deviation of 0.387 V. Systematic errors are negligible. If it is desired to hold the random uncertainty in the mean value due to data variation to within 0.100 V at 95% confidence, how many measurements of the signal will be required?

5.44   A hand-held infrared thermometer from a well-known manufacturer boasts a measurement range of −30 to 350 °C with an accuracy limit to within ± 1.5 °C at a resolution of 0.1 °C and with repeatability of ± 1 °C. Estimate the uncertainty in a thermometer measurement using this information. Assume 95% confidence in the specifications.

5.45   The pressure in a large vessel is to be maintained at some set pressure for a series of tests. A compressor is to supply air through a regulating valve that is set to open and close at the set pressure. A dial gauge (resolution: 1 psi; uncertainty: 0.5 psi) is used to monitor pressure in the vessel. Thirty trials of pressurizing the vessel to a set pressure of 50 psi are attempted to estimate pressure controllability and a standard deviation in set pressure of 2 psi is found. Estimate the uncertainty to be expected in the vessel set pressure. Compare to the uncertainty in estimating the average set pressure.

5.46   A handbook value for the thermal expansion coefficient of a particular pure metal is $\alpha = 16.52 \times 10^{-6}$ °C^{-1} and states that the possible error in this value should not exceed 0.4×10^{-6} °C^{-1}. Assuming that the population for α follows a rectangular distribution, estimate the standard uncertainty to be assigned.

5.47   The cooling of a thermometer (e.g., Exs. 3.3 and 3.4) can be modeled as a first-order system with $\Gamma = e^{-t/\tau}$. If Γ can be measured to within 2% and time within 1%, determine the uncertainty in τ over the range $0 \leq \Gamma \leq 1$.

5.48   A J-type thermocouple monitors the temperature of air flowing through a duct. Its signal is measured by a thermostat. The air temperature is maintained constant by an electric heater whose power is controlled by the thermostat. To test the control system, 20 measurements of temperature were taken over a reasonable time period during steady operation. This was repeated three times with the following results:

Run	N	$\overline{T}$[°C]	s_T[°C]
1	20	181.0	3.01
2	20	183.1	2.84
3	20	182.1	3.08

The thermocouple itself has an instrument error with uncertainty ± 1 °C (95%) with $\nu = 30$. It has a 90% rise time of 20 ms. Thermocouple insertion errors are estimated to be ± 1.2 °C (95%). What information is found by performing several replications? Identify the elemental errors that affect the system's control of the air temperature. What is the uncertainty in the set temperature?

5.49   The density of air must be known to within $\pm 0.5\%$. If the air temperature at 25 °C can be determined to within ± 1 °C (95%), what uncertainty can be tolerated in the pressure measurement if air behaves as an ideal gas?

5.50   In pneumatic conveying, solid particles such as flour or coal are carried through a duct by a moving air stream. Solids density at any duct location can be measured by passing a laser beam of known intensity I_o through the duct and measuring the light intensity transmitted to the other side, I. A transmission factor is found by

$$T = I/I_o = e^{-KEW} \qquad 0 \leq T \leq 1$$

Here W is the width of the duct, K is the solids density, and E is a factor taken as 2.0 ± 0.04 (95%) for spheroid particles. Determine how u_K/K is related to the relative uncertainties of the other variables. If the transmission factor and duct width can be measured to within $\pm 1\%$, can solids density be measured to within 5%? 10%? Discuss your answer, remembering that the transmission factor varies from 0 to 1.

5.51   A step test is run to determine the time constant of a first-order instrument (see Chapter 3). If the error fraction $\Gamma(t)$ can be estimated to within $\pm 2\%$ (95%) and time t can be estimated in seconds to within $\pm 0.5\%$ (95%), plot u_τ/τ versus $\Gamma(t)$ over its range, $0 \leq \Gamma(t) \leq 1$.

5.52   The acceleration of a cart down a plane inclined at an angle α to horizontal can be determined by measuring the change in speed of the cart at two points, separated by a distance s, along the inclined plane. Suppose two photocells are fixed at the two points along the plane. Each photocell measures the time for the cart, which has a length L, to pass it. If $L = 5 \pm 0.5$ cm, $s = 100 \pm 0.2$ cm, $t_1 = 0.054 \pm 0.001$ s, and $t_2 = 0.031 \pm 0.001$ s, all (95%), estimate the uncertainty in acceleration:

$$a = \frac{L^2}{2s}\left(\frac{1}{t_2^2} - \frac{1}{t_1^2}\right)$$

Compare as a concomitant check $a = g \sin \alpha$ using values for $\alpha = 30°$ and $90°$. What uncertainty in α is needed to make this the better method?

5.53 A particular flow meter allows the volumetric flow rate Q to be calculated from a measured pressure drop Δp. Theory predicts $Q \propto \sqrt{\Delta p}$, and a calibration provides the following data:

$Q(\text{m}^3/\text{min})$	10	20	30	40
$\Delta p(\text{Pa})$	1,000	4,271	8,900	16,023

What uncertainty in the measurement of Δp is required to yield an uncertainty of 0.25% in Q over the range of flow rates from 10 to 40 m³/min?

5.54 A steel cantilever beam is fixed at one end and free to move at the other. A load (F) of 980 N with a systematic error (95%) of 10 N is applied at the free end. The beam geometry, as determined by a metric ruler having 1-mm increments, is 100 mm long (L), 30 mm wide (w), and 10 mm thick (t). Estimate the maximum stress at the fixed end and its uncertainty. The maximum stress is given by

$$\sigma = \frac{Mc}{I} = \frac{FL(t/2)}{wt^3/12}.$$

5.55 The heat flux in a reaction is estimated by $Q = 5(T_1 - T_2)$ kJ/s. Two thermocouples are used to measure T_1 and T_2. Each thermocouple has a systematic uncertainty of 0.2 °C (95%). Based on a large sample of measurements, $\overline{T}_1 = 180\,°C$, $\overline{T}_2 = 90\,°C$, and the random standard uncertainty in each temperature measurement is 0.1 °C. Assume large degrees of freedom. Compare the uncertainty in heat flux if the thermocouple systematic errors are (a) uncorrelated and (b) correlated. Explain.

5.56 A comparative test uses the relationship $R = p_2/p_1$, where pressure p_2 is measured to be 54.7 MPa and pressure p_1 is measured to be 42.0 MPa. Each pressure measurement has a single systematic error of 0.5 MPa. Compare the systematic uncertainty in R if these errors are (a) correlated and (b) uncorrelated. Explain.

5.57 A sensitive material is to be contained within a glovebox. Codes require that the transparent panels that make up the walls of the glovebox withstand the impact of a 22-kg mass falling at 8 m/s. An impact test is performed where the 22 kg mass is dropped a known variable height onto a panel. The mass motion is sensed by photocells from which an impact velocity is estimated. The mean kinetic energy for failure was 717 N-m, with a standard deviation of 60.7 N-m based on 61 failures. The systematic uncertainty in mass is 0.001 kg (95%). The velocity at impact is 8 m/s, with a systematic uncertainty in velocity estimated as 0.27 m/s (95%). Estimate the combined standard uncertainty and the expanded uncertainty at 95% in the kinetic energy for failure. Assume a normal distribution of errors.

5.58 A geometric stress concentration factor, K_t, is used to relate the actual maximum stress to a well-defined nominal stress in a structural member. The maximum stress is given by $\sigma_{max} = K_t \sigma_o$, where σ_{max} represents the maximum stress and σ_o represents the nominal stress. The nominal stress is most often stated at the minimum cross section.

Consider the axial loading shown in Figure 5.9, where the structural member is in axial tension and experiences a stress concentration as a result of a transverse hole. In this geometry, $\sigma_o = F/A$, where area $A = (w - d)t$. If $d = 0.5\,w$, then $K_t = 2.2$. Suppose $F = 10,000 \pm 500$ N, $w = 1.5 \pm 0.02$ cm, $t = 0.5 \pm 0.02$ cm, and the uncertainty in the value for d is 3%. Neglecting the uncertainty in stress concentration factor, determine the uncertainty in the maximum stress experienced by this part.

5.59 The result of more than 60 temperature measurements over time during fixed conditions in a furnace

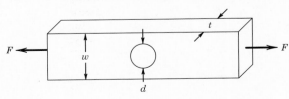

Figure 5.9 Structural member discussed in Problem 5.58.

determines $\overline{T} = 624.7\,°C$ with $s_{\overline{T}} = 2.4\,°C$. The engineer suspects that owing to a radiation systematic error, the measured temperature may be higher by up to $10\,°C$ at a 95% confidence level but not lower, so the lower bound of error is $0\,°C$ relative to the measured mean value. Determine the statement for the true mean temperature with its confidence interval assuming a normal distribution for the systematic error. Then repeat, assuming the systematic errors follow a rectangular distribution with upper bound $10\,°C$ and lower bound $0\,°C$ relative to the measured mean value.

5.60 Estimate the uncertainty at 95% confidence in drag coefficient as predicted by the Monte Carlo simulation using the information from Problem 4.59. Consider only the random uncertainties in the simulation.

5.61 The calibration certificate for a laboratory standard resistor lists its resistance as $10.000742\,\Omega\ \pm$ $130\,\mu\Omega$ (95%). The certificate lists a systematic standard uncertainty of $10\,\mu\Omega$. Estimate the random standard uncertainty in the resistor.

5.62 In a mechanical loading test, the operator is to load a specimen to a value of 100 N. The control is such that this value can be achieved with a possible error of no more than 0.5 N. Estimate a value for the uncertainty in loading at the 95% level using (1) a zero-order estimate based on instrument resolution and (2) an estimate based on the information provided assuming a rectangular distribution. Comment on the results.

5.63 In Problem 5.9, we assumed that the errors in the known resistor values were uncorrelated. Suppose that the resistors are certified by the manufacturer to have specifications based on a common calibration. Repeat Problem 5.9 by assuming that the errors are correlated systematic errors.

Chapter 6

Analog Electrical Devices and Measurements

6.1 INTRODUCTION

This chapter provides an introduction to basic electrical analog devices used with analog signals or to display signals in an analog form. Information is often transferred between stages of a measurement system as an analog electrical signal. This signal typically originates from the measurement of a physical variable using a fundamental electromagnetic or electrical phenomenon and then propagates from stage to stage of the measurement system (see Fig. 1.1). Accordingly, we discuss some signal conditioning and output methods.

Although analog devices have been supplanted by their digital equivalents in many applications, they are still widely used and remain engrained in engineered devices. Analog technology forms the basis for power management in cell phones and for innovative cardiac monitoring and is abundant in modern flat panel displays. Also, an analog output format is often ergonomically superior in monitoring, as evidenced by modern car speedometer dials and dial wristwatches. Within a signal chain, it is common to find digital and analog electrical devices being used together and so requiring special attention to signal conditioning methods. An understanding of analog device function provides insight, as well as a historical reference point, into the advantages and disadvantages of digital counterparts; such issues are discussed in this chapter.

Upon completion of this chapter, the reader will be able to

- Understand the principles behind common analog voltage and current measuring devices
- Understand the operation of balanced and unbalanced resistance bridge circuits
- Define, identify, and minimize loading errors
- Understand the basic principles involved in signal conditioning, especially filtering and amplification
- Apply proper grounding and shielding techniques in measuring system hookups

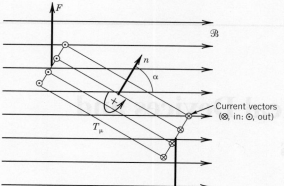

Figure 6.1 Forces and resulting torque on a current loop in a magnetic field.

6.2   ANALOG DEVICES: CURRENT MEASUREMENTS

Direct Current

One way to measure a DC electrical current is to use an analog device that responds to the force exerted on a current-carrying conductor in a magnetic field. Because an electric current is a series of moving charges, a magnetic field exerts a force on a current-carrying conductor. This force can be used as a measure of the flow of current in a conductor by moving a pointer on a display.

A current loop in a magnetic field experiences a torque if the loop is not aligned with the magnetic field, as illustrated in Figure 6.1. The torque[1] on a loop composed of N turns is given by

$$T_\mu = NIAB \sin \alpha \tag{6.1}$$

where

A = cross-sectional area defined by the perimeter of the current loop

B = magnetic field strength(magnitude)

I = current

α = angle between the normal to the cross-sectional area of the current loop and the magnetic field

One approach to measure current is in the very common D'Arsonval movement shown in Figure 6.2. In this arrangement, the uniform radial magnetic field and torsional spring result in a steady angular deflection of the coil that corresponds to the existing current through the coil. The coil and fixed permanent magnet are arranged in the normal direction to the current loop—that is, $\alpha = 90$ degrees. If an analog dial indicates current, voltage, or resistance, then it is likely to employ this mechanism.

Most devices that use the D'Arsonval movement employ a pointer whose deflection increases with the magnitude of current applied. This mode of operation is called the *deflection mode*. A typical

---

[1] In vector form, the torque on a current loop can be written $\mathbf{T}_\mu = \mu \times \mathbf{B}$, where μ is the magnetic dipole moment.

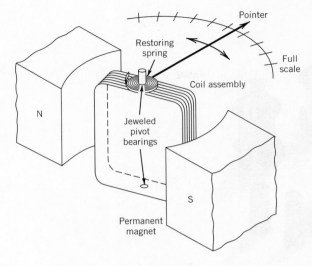

Figure 6.2 Basic D'Arsonval meter movement.

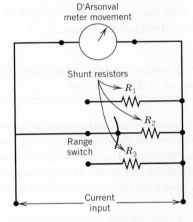

Figure 6.3 Simple multirange ammeter (with make-before-break selector switch). Shunt resistors determine meter range.

circuit for an analog current measuring device, an ammeter, is shown in Figure 6.3. Here the deflection of the pointer indicates the magnitude of the current flow.

Example 6.1: Case Study: Galvanometer Rotary Actuator

The operating principle of the D'Arsonval movement can be employed to create a high-precision, high performance rotary actuator, termed a galvanometer, or "galvo." One area of application is the fast and precise positioning of mirrors for the deflection of laser beams. The system consists of a "motor" section based on moving magnet technology coupled with a high-precision position sensor. These are employed to rotate mirrors in scanning systems, and have step times in the 100 μsec range. Applications include laser scanning for engraving, medical applications including ophthalmology, and laser light shows! Figure 6.4 shows a few galvanometer actuators with appropriate mirrors used for laser scanning.

Figure 6.4 Scanning galvanometer heads with mirrors. (Courtesy of SCANLAB AG-Puchheim Germany.)

COMMENT In this example the term galvanometer is used to describe an actuator. The term galvanometer is also used to describe a sensor-transducer, as is discussed next.

The sensor-transducer called a *galvanometer* is a measuring device used to detect a current flow in a circuit. It is a highly sensitive D'Arsonval movement calibrated about zero current; the indicator normally indicates zero but can deflect to the plus or minus direction. The pointer is used to adjust a circuit to a zero current state. This mode of operation is called the *null mode*. The highest sensitivity of commercial galvanometers is approximately 0.1 μA/division for a galvanometer with a mechanical pointer. That's quite sensitive.

Errors inherent to the D'Arsonval movement include hysteresis and repeatability errors due to mechanical friction in the pointer-bearing movement, and linearity errors in the spring that provides the restoring force for equilibrium. Also, in developing a torque, the D'Arsonval movement must extract energy from the current flowing through it. This draining of energy from the signal being measured changes the measured signal, and such an effect is called a *loading error*. Loading errors occur in all instruments that operate in deflection mode. A quantitative analysis of loading errors is provided later.

Example 6.2

A galvanometer sensor-transducer consists of N turns of a conductor wound about a core of length l and radius r that is situated perpendicular to a magnetic field of uniform flux density **B**. A DC current passes through the conductor due to an applied potential, $E_i(t)$. The output of the device is the rotation of the core and pointer, θ, as shown in Figure 6.5. Develop a lumped parameter model relating pointer rotation and current.

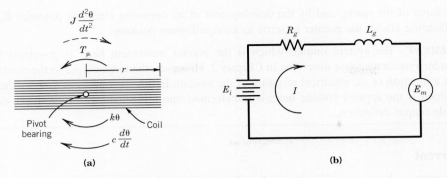

Figure 6.5 Circuit and free-body diagram for Example 6.1.

FIND   A dynamic model relating θ and I

SOLUTION   The galvanometer is a rotational system consisting of a torsional spring, a pointer, and core that are free to deflect; the rotation is subject to frictional damping. The mechanical portion of the device consists of a coil having moment of inertia J, bearings that provide damping from friction, with a damping coefficient c, and a torsional spring of stiffness k. The electrical portion of the device consists of a coil with a total inductance L_g and a resistance R_g.

We apply Newton's second law with the mechanical free-body diagram shown in Figure 6.4a.

$$J\frac{d^2\theta}{dt^2} + c\frac{d\theta}{dt} + k\theta = T_\mu \tag{6.2}$$

As a current passes through the coil, a force is developed that produces a torque on the coil:

$$T_\mu = 2NBlrI \tag{6.3}$$

This torque, which tends to rotate the coil, is opposed by an electromotive force E_m, due to the Hall effect,

$$E_m = \left(r\frac{d\theta}{dt} \times \mathbf{B}\right)i\hat{k} = 2NBr\frac{d\theta}{dt} \tag{6.4}$$

which produces a current in the direction opposite to that produced by E_i. Applying Kirchhoff's law to the electrical free body of Figure 6.5b gives

$$L_g\frac{dI}{dt} + R_gI = E_i - E_m \tag{6.5}$$

Equations 6.3 and 6.4 define the coupling relations between the mechanical Equation 6.2 and the electrical Equation 6.5. From these, we see that the current due to potential E brings about developed torque T_μ that moves the galvanometer pointer, and that this motion is opposed by the mechanical restoring force of the spring and by the development of an opposing electrical potential E_m. The system damping allows the pointer to settle to an equilibrium position.

COMMENT The system output, which is the pointer movement here, is governed by the second-order system response described in Chapter 3. However, the torque input to the mechanical system is a function of the electrical response of the system. In this measurement system, the input signal, which is the applied voltage here, is transformed into a mechanical torque to provide the measurable output deflection.

Alternating Current

An AC current can be measured in any number of ways. One technique, found in common deflection meters, uses diodes to form a rectifier that converts the time-dependent AC current into a DC current. This current then can be measured with a calibrated D'Arsonval movement meter as previously described. This is the same technique used in those ubiquitous small transformers used to convert AC wall current into a DC current at a set voltage to power or to charge electronic devices. An *electrodynamometer* is essentially a D'Arsonval movement modified for use with AC current by replacing the permanent magnet with an electromagnet in series with the current coil. These AC meters have upper limits on the frequency of the alternating current that they can effectively measure; most common instruments are calibrated for use with standard line frequency (60 Hz).

An accurate measuring solution for large AC current is the Hall effect probe. This is a probe clamped over the current-carrying wire (conductor) to measure its unknown current flow. To understand its use, let us mention two phenomena. The first is the *Hall effect*, which is a voltage that is developed from any current-carrying conductor placed perpendicular to a magnetic field. For a known current, the magnitude of this voltage directly depends on the magnitude of the magnetic field. The second is that a current passing through a wire generates a magnetic field. So in practice, a Hall-effect probe is realized by coupling these two processes concurrently: Use the unknown current within a wire to generate a magnetic field that develops a measurable voltage across a Hall-effect sensor.

The Hall-effect sensor is a thin conducting semiconductor wafer driven by a known current; this current is unrelated to the unknown current being measured and is provided by a separate source, such as a battery. The Hall-effect probe is an iron-core ring that is placed around the wire carrying the current to be measured. This ring is used to concentrate the magnetic field, and a Hall-effect sensor is attached to the iron core so as to be aligned parallel to the wire carrying the current being measured.

The companion Labview program *Basic DC-RMS Measurement* allows the user to manipulate DC and AC signal levels so as to measure AC and DC magnitudes. The program builds from discussions in Sections 2.3–2.4.

6.3 ANALOG DEVICES: VOLTAGE MEASUREMENTS

Often we are interested in measuring either static or dynamic voltage signals. Depending on the source, the magnitude of these signals may range over several orders of magnitude throughout the measured signal chain. The frequency content of dynamic voltage signals is often of interest as well. As such, a wide variety of measurement systems have been developed for voltage measurement of static and dynamic signals. This section discusses several convenient and common methods to indicate voltage in measurement systems.

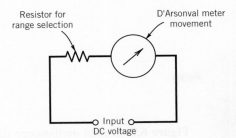

Figure 6.6   A DC voltmeter circuit.

Analog Voltage Meters

A DC voltage can be measured in through the analog circuit shown in Figure 6.6, where a D'Arsonval movement is used in series with a resistor. Although fundamentally sensitive to current flow, the D'Arsonval movement can be calibrated in terms of voltage by using an appropriate known fixed resistor and through Ohm's law relating it to the measured current. This basic circuit is employed in the construction of analog dial voltmeters and volt-ohmmeters (VOMs), which for many years served as the common measurement device for current, voltage, and resistance.

An AC voltage can be measured through rectification or through the use of an electromagnet, either in an electrodynamometer or with a movable iron vane. These instruments are sensitive to the *rms* (root-mean-square) value of a simple periodic AC current, and can be calibrated in terms of voltage; shunt resistors can be used to establish the appropriate scale. The circuit shown in Figure 6.6 can also be used to measure an AC voltage if the input voltage is rectified prior to input to this circuit. A common AC meter indicates a true rms value for a simple periodic signal only, but a *true* rms AC voltmeter performs the signal integration (e.g., Eq. 2.4) required to accurately determine the rms value in a signal-conditioning stage and indicates true signal rms regardless of waveform.

Oscilloscope

The *oscilloscope* is a practical graphical display device providing an analog representation of a measured signal. It is used to measure and to visually display voltage magnitude versus time for dynamic signals over a wide range of frequencies with a signal bandwidth extending commonly into the megahertz range and, with some units, into the 50 GHz range and beyond (1,2). A useful diagnostic tool, the oscilloscope provides a visual output of signal magnitude, frequency, and a delineation of the DC and AC components, as well as signal quality attributes, such as distortion and clipping. The visual image provides a direct means to detect the superposition of noise and interference on a measured signal, something nonvisual metering devices cannot do. In addition to signal versus time, a typical unit can also display two or more signals [$X(t)$ and $Y(t)$], perform addition and subtraction of signals ($X + Y, X - Y$), and display amplitude versus amplitude (XY) plots and other features. Some oscilloscopes have significant internal storage so as to mimic a data-logging signal recorder. Others have internal fast Fourier transform (FFT) circuitry to provide for direct spectral analysis of a signal (see Chapter 2).

A digital oscilloscope, such as shown in Figure 6.7, provides an analog representation of the measured signal. But it does so by first sampling the signal to convert it into a digital form and then reconstructing the signal on a display as an analog output. In this way, the digital oscilloscope

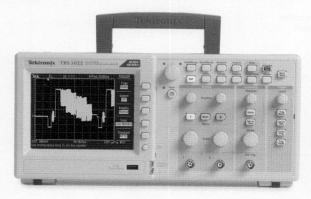

Figure 6.7   Digital oscilloscope illustrating oscilloscope output. (Copyright © Tektronix. Printed with permission. All rights reserved.)

measures and stores voltage in a digital manner but then displays it in an analog format as a sequence of measured points, as shown in Figure 6.7. These common lab devices can be found as small and very portable packages and are often a virtual software component of a 12- to 18-bit analog-to-digital converter data acquisition system. State of the art units include the capability to do signal analysis using application specific software, and some units can run third-party software as well. Analog-to-digital sampling techniques used to measure voltages are discussed in Chapter 7.

Example 6.3

We can specify the requirements for an oscilloscope based on the intended signal to be measured. For example, in the original USB 1.1 protocol for data transmission, a single frame of data lasts for 1 ms with data transmitted serially at 12 Mbps (million bits per second). We can simplify this as trying to capture a 12 MHz square wave for 1 ms on the oscilloscope screen to base our requirements.

From the discussions of Chapter 2, we know that we need at least five harmonics to reconstruct a square wave with any reasonable fidelity, so the sampling rate required to reconstruct a 12 MHz square wave is at least five times the fundamental frequency or at least 60 MHz (i.e., 60×10^6 samples/s). So that sets the lowest value on required oscilloscope response. The storage capacity of the digital oscilloscope required to capture one frame of the USB data signal is 60×10^6 samples/s $\times 0.001$ s or 60,000 samples at 60 MHz. For comparison of potential oscilloscope needs, the more common USB 2.0 transmits at a nominal 480 Mbps. In 2013 a successor standard was issued, USB 3.1, which sets transfer rates up to 10 Gbps and is called SuperSpeed+. Color coding and the letters SS on connectors identify hardware compatible with this standard.

Potentiometer

The *potentiometer*[2] is a device used to measure DC voltages that are in the microvolt to millivolt range. Equivalent to a balance scale, a potentiometer balances an unknown input voltage against a known internal voltage until both sides are equal. A potentiometer is a null balance instrument in that it drives the loading error to essentially zero. Potentiometers have been supplanted by digital

---

[2] The term "potentiometer" is used in several different ways: a sliding contact precision variable resistor, the divider circuit in Figure 6.9, and a high-precision circuit used as a voltage measuring instrument.

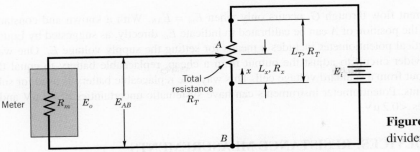

Figure 6.8 Voltage divider circuit.

voltmeters, which are deflection mode devices but have very high input impedances so as to keep loading errors small, even at low-level voltages.

Voltage Divider Circuit

A general purpose component found in the potentiometer circuit is the *voltage divider circuit* shown in Figure 6.8. The point labeled A in Figure 6.8 represents a sliding contact, which makes an electrical connection with the resistor R. The resistance between points A and B in the circuit is a linear function of the distance from A to B. So the output voltage sensed by the meter is given by

$$E_o = \frac{L_x}{L_T} E_i = \frac{R_x}{R_T} E_i \tag{6.6}$$

which holds so long as the internal resistance of the meter, R_m, is very large relative to R_T.

Potentiometer Instruments

A simple potentiometer measuring instrument can be derived from the voltage divider circuit as shown in Figure 6.9. In this arrangement a current-sensing device, such as a galvanometer, is used to detect current flow; any current flow through the galvanometer, G, would be a result of an imbalance between the measured voltage, E_m, and the voltage imposed across points A to B, E_{AB}. The voltage E_{AB} is adjusted by moving the sliding contact A; if this sliding contact is calibrated in terms of a known supply voltage, E_i, the circuit may be used to measure voltage by creating a balanced condition, indicated by a zero current flow through the galvanometer. A null balance, corresponding

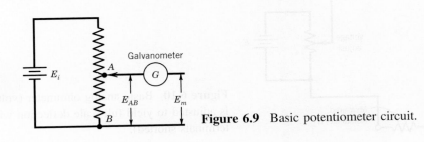

Figure 6.9 Basic potentiometer circuit.

to zero current flow through G, occurs only when $E_m = E_{AB}$. With a known and constant supply voltage E_i, the position of A can be calibrated to indicate E_m directly, as suggested by Equation 6.6.

A practical potentiometer includes a means for setting the supply voltage E_i. One way uses a separate divider circuit to adjust the output from a cheap, replaceable battery to equal the fixed, known output from a standard voltage cell. This way, the replaceable battery is used for subsequent measurements. Potentiometer instruments can have systematic uncertainties of $<1\,\mu V$ and random uncertainties $<0.2\,\mu V$.

6.4 ANALOG DEVICES: RESISTANCE MEASUREMENTS

Resistance measurements may range from simple continuity checks, to determining changes in resistance on the order of $10^{-6}\,\Omega$, to determining absolute resistance ranging from 10^{-5} to $10^{15}\,\Omega$. As a result of the tremendous range of resistance values that have practical application, numerous measurement techniques that are appropriate for specific ranges of resistance or resistance changes have been developed. Some of these measurement systems and circuits provide necessary protection for delicate instruments, allow compensation for changes in ambient conditions, or accommodate changes in transducer reference points. In addition, the working principle of many transducers is a change in resistance relative to a change in the measured variable. We will discuss the measurement of resistance using basic techniques of voltage and current measurement in conjunction with Ohm's law.

Ohmmeter Circuits

A simple way to measure resistance is by imposing a voltage across the unknown resistance and measuring the resulting current flow where $R = E/I$. Clearly, from Ohm's law, the value of resistance can be determined in a circuit such as in Figure 6.10, which is the basis of a common analog ohmmeter. Practical analog ohmmeters use circuits similar to those shown in Figure 6.11, which use shunt resistors and a D'Arsonval mechanism for measuring a wide range of resistance while limiting the flow of current through the meter movement. In this technique, the lower limit on the measured

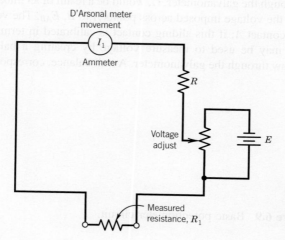

Figure 6.10 Basic analog ohmmeter (voltage is adjusted to yield full-scale deflection with terminals shorted).

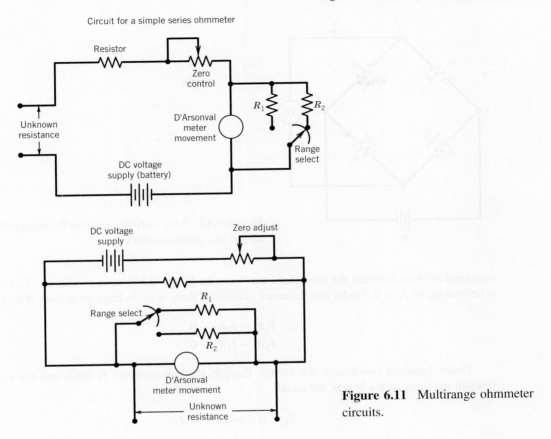

Circuit for a simple series ohmmeter

Figure 6.11 Multirange ohmmeter circuits.

resistance R_1 is determined by the upper limit of current flow through it, that is, I_1. A practical limit to the maximum current flow through a resistance is imposed by the ability of the resistance element to dissipate the power generated by the flow of current (I^2R heating). For example, the limiting ability of a thin metallic conductor to dissipate heat is the principle on which fuses are based. At too large a value of I_1, they melt. The same is true for delicate, small resistance sensors!

Bridge Circuits

A variety of bridge circuits have been devised for measuring capacitance, inductance, and, most often, resistance. A purely resistive bridge, called a *Wheatstone bridge*, provides a means for accurately measuring resistance, and for detecting very small changes in resistance. Figure 6.12 shows the basic arrangement for a bridge circuit, where R_1 is a sensor that experiences a change in resistance associated with a change in some physical variable. A DC voltage is applied as an input across nodes A to D, and the bridge forms a parallel circuit arrangement between these two nodes. The currents flowing through the resistors R_1 to R_4 are I_1 to I_4, respectively. Under the condition that the current flow through the galvanometer, I_g, is zero, the bridge is in a balanced condition (alternatively, a current-sensing digital voltmeter might be used in place of the galvanometer). A specific

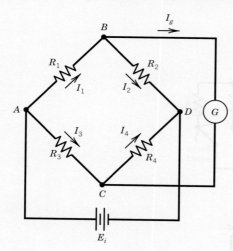

Figure 6.12 Basic current-sensitive Wheatstone bridge circuit (G, galvanometer).

relationship exists between the resistances that form the bridge at balanced conditions. To find this relationship, let $I_g = 0$. Under this balanced condition, there is no voltage drop from B to C and

$$I_1 R_1 - I_3 R_3 = 0$$
$$I_2 R_2 - I_4 R_4 = 0 \qquad (6.7)$$

Under balanced conditions, the current through the galvanometer is zero, and the currents through the arms of the bridge are equal:

$$I_1 = I_2 \quad \text{and} \quad I_3 = I_4 \qquad (6.8)$$

Solving Equation 6.7 simultaneously with the condition stated in Equation 6.8 yields the necessary relationship among the resistances for a balanced bridge:

$$\frac{R_2}{R_1} = \frac{R_4}{R_3} \qquad (6.9)$$

Resistance and resistance change can be measured in one of two ways with this bridge circuit. If the resistor R_1 varies with changes in the measured physical variable, one of the other arms of the bridge can be adjusted to null the circuit and determine resistance. Another method uses a voltage measuring device to measure the voltage unbalance in the bridge as an indication of the change in resistance. We will both these methods further.

Null Method

Consider the circuit shown in Figure 6.12, where R_2 is an adjustable variable resistance. If the resistance R_1 changes due to a change in the measured variable, the resistance R_2 can be adjusted to compensate so that the bridge is once again balanced. In this null method of operation, the resistance R_2 must be a calibrated variable resistor, such that adjustments to R_2 directly indicate the value of R_1.

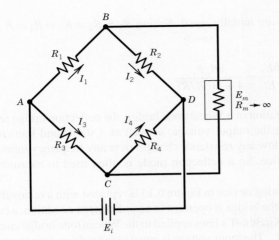

Figure 6.13 Voltage-sensitive Wheatstone bridge.

The balancing operation may be accomplished either manually or automatically through a closed-loop controller circuit. An advantage of the null method is that the applied input voltage need not be known, and changes in the input voltage do not affect the accuracy of the measurement. In addition, the current detector or controller need only detect if there is a flow of current, not measure its value.

Deflection Method

In an unbalanced condition, the magnitude of the current or voltage drop for the meter portion of the bridge circuit is a direct indication of the change in resistance of one or more of the arms of the bridge. Consider first the case in which the voltage drop from node B to node C in the basic bridge is measured by a meter with infinite internal impedance so that there is no current flow through the meter, as shown in Figure 6.13. Knowing the conditions for a balanced bridge given in Equation 6.8, the voltage drop from B to C can be determined, since the current I_1 must equal the current I_2, as

$$E_o = I_1 R_1 - I_3 R_3 \tag{6.10}$$

Under these conditions, substituting Equations 6.7 to 6.9 into Equation 6.10 yields

$$E_o = E_i \left(\frac{R_1}{R_1 + R_2} - \frac{R_3}{R_3 + R_4} \right) \tag{6.11}$$

The bridge is usually initially balanced at some reference condition. Any transducer resistance change resulting from a change in the measured variable would then cause a deflection in the bridge voltage away from the balanced condition. Assume that from an initially balanced condition where $E_o = 0$, a change in R_1 occurs to some new value, $R_1' = R_1 + \delta R$. The output from the bridge becomes

$$E_o + \delta E_o = E_i \left(\frac{R_1'}{R_1' + R_2} - \frac{R_3}{R_3 + R_4} \right) = E_i \frac{R_1' R_4 - R_3 R_2}{(R_1' + R_2)(R_3 + R_4)} \tag{6.12}$$

In many designs, the bridge resistances are initially equal. Setting $R_1 = R_2 = R_3 = R_4 = R$ allows Equation 6.12 to be reduced to

$$\frac{\delta E_o}{E_i} = \frac{\delta R/R}{4 + 2(\delta R/R)} \tag{6.13}$$

In contrast to the null method of operation of a Wheatstone bridge, the deflection bridge requires a meter capable of accurately indicating the output voltage, as well as a stable and known input voltage. But the bridge output should follow any resistance changes over any frequency input, up to the frequency limit of the detection device. So a deflection mode is often used to measure time-varying signals.

If the high-impedance voltage measuring device in Figure 6.13 is replaced with a relatively low-impedance current measuring device and the bridge is operated in an unbalanced condition, a current-sensitive bridge circuit results. Consider Kirchhoff's laws applied to the Wheatstone bridge circuit for a current-sensing device with resistance R_g. The input voltage is equal to the voltage drop in each arm of the bridge,

$$E_i = I_1 R_1 + I_2 R_2 \tag{6.14}$$

but $I_2 = I_1 - I_g$, which gives

$$E_i = I_1(R_1 + R_2) - I_g R_2 \tag{6.15}$$

If we consider the voltage drops in the path through R_1, R_g, and R_3, the total voltage drop must be zero:

$$I_1 R_1 + I_g R_g - I_3 R_3 = 0 \tag{6.16}$$

For the circuit formed by R_g, R_4, and R_2,

$$I_g R_g + I_4 R_4 - I_2 R_2 = 0 \tag{6.17}$$

or with $I_2 = I_1 - I_g$ and $I_4 = I_3 + I_g$,

$$I_g R_g + (I_3 + I_g)R_4 - (I_1 - I_g)R_2 = 0 \tag{6.18}$$

Equations 6.16 to 6.18 form a set of three simultaneous equations in the three unknowns I_1, I_g, and I_3. Solving these three equations for I_g yields

$$I_g = \frac{E_i(R_1 R_4 - R_2 R_3)}{R_1 R_4(R_2 + R_3) + R_2 R_3(R_1 + R_4) + R_g(R_1 + R_2)(R_3 + R_4)} \tag{6.19}$$

and $E_o = I_g R_g$. Then the change in resistance of R_1 can be found in terms of the bridge deflection voltage, E_o, by

$$\frac{\delta R}{R_1} = \frac{(R_3/R_1)\left[E_o/E_i + R_2/(R_2 + R_4)\right]}{1 - E_o/E_i - R_2/(R_2 + R_4)} - 1 \tag{6.20}$$

Consider the case when all of the resistances in the bridge are initially equal to R, and subsequently R_1 changes by an amount δR. The current through the meter is given by

$$I_g = E_i \frac{\delta R/R}{4(R + R_g)} \tag{6.21}$$

and the output voltage is given by $E_o = I_g R_g$,

$$E_o = E_i \frac{\delta R/R}{4(1 + R/R_g)} \tag{6.22}$$

The bridge impedance can affect the output from a constant voltage source having an internal resistance R_s. The effective bridge resistance, based on a Thévenin equivalent circuit analysis, is given by

$$R_B = \frac{R_1 R_3}{R_1 + R_3} + \frac{R_2 R_4}{R_2 + R_4} \tag{6.23}$$

such that for a power supply of voltage E_s

$$E_i = \frac{E_s R_B}{R_s + R_B} \tag{6.24}$$

In a similar manner, the bridge impedance can affect the voltage indicated by the voltage-measuring device. For a voltage-measuring device of internal impedance R_g, the actual bridge deflection voltage, relative to the indicated voltage, E_m, is

$$E_o = \frac{E_m}{R_g} \left(\frac{R_1 R_2}{R_1 + R_2} + \frac{R_3 R_4}{R_3 + R_4} + R_g \right) \tag{6.25}$$

The difference between the measured voltage E_m and the actual voltage E_o is a loading error, in this case due to the bridge impedance load. Loading errors are discussed next.

Example 6.4

A certain temperature sensor experiences a change in electrical resistance with temperature according to the equation

$$R = R_o[1 + \alpha(T - T_o)] \tag{6.26}$$

where

R = sensor resistance (Ω)
R_o = sensor resistance at the reference temperature, T_o (Ω)
T = temperature (°C)
T_o = reference temperature (0 °C)
A = the constant 0.00395 °C^{-1}

This temperature sensor is connected in a Wheatstone bridge such as the one shown in Figure 6.12, where the sensor occupies the R_1 location, and R_2 is a calibrated variable resistance. The bridge is operated using the null method. The fixed resistances R_3 and R_4 are each equal to 500 Ω. If the temperature sensor has a resistance of 100 Ω at 0 °C, determine the value of R_2 that would balance the bridge at 0 °C.

   KNOWN $R_1 = 100 \, \Omega$

$R_3 = R_4 = 500 \, \Omega$

   FIND R_2 for null balance condition

   SOLUTION From Equation 6.9, a balanced condition for this bridge would be achieved when $R_2 = R_1 R_4 / R_3$ or $R_2 = 100 \, \Omega$. Notice that to be a useful circuit, R_2 must be adjustable and provide an indication of its resistance value at any setting.

Example 6.5

Consider a deflection bridge, which initially has all four arms of the bridge equal to 100 Ω, with the temperature sensor described in Example 6.4 again as R_1. The input or supply voltage to the bridge is 10 V. If the temperature of R_1 is changed such that the bridge output is 0.569 V, what is the temperature, T_1, of the sensor? How much current, I_1, flows through the sensor, and how much power, P_1, must it dissipate?

   KNOWN $E_i = 10 \, \text{V}$
   Initial (reference) state: $R_1 = R_2 = R_3 = R_4 = 100 \, \Omega$

$$E_o = 0 \, \text{V}$$

   Deflection state: $E_o = 0.569 \, \text{V}$

ASSUMPTION   Voltmeter has infinite input impedance but source has negligible impedance $(E_s = E_i)$

FIND T_1, I_1, P_1

SOLUTION   The change in the sensor resistance can be found from Equation 6.13 as

$$\frac{\delta E_o}{E_i} = \frac{\delta R/R}{4 + 2(\delta R/R)} \Rightarrow \frac{0.569}{10} = \frac{\delta R}{400 + 2\delta R}$$

$$\delta R = 25.67\,\Omega$$

This gives a total sensor resistance of $R_1 + \delta R = 125.7\,\Omega$, which equates to a sensor temperature of $T_1 = 65\,°C$.

To determine the current flow through the sensor, consider first the balanced case where all the resistances are equal to $100\,\Omega$. The equivalent bridge resistance R_B is simply $100\,\Omega$ and the total current flow from the supply, $E_i/R_B = 100$ mA. Thus, at the initially balanced condition, the current flow through each arm of the bridge and through the sensor is 50 mA. If the sensor resistance changes to $125.67\,\Omega$, the current will be reduced. If the output voltage is measured by a high-impedance device such that the current flow to the meter is negligible, the current flow through the sensor is given by

$$I_1 = E_i \frac{1}{(R_1 + \delta R) + R_2} \tag{6.27}$$

This current flow is then 44.3 mA.

The power, $P_1 = I_1^2(R_1 + \delta R)$, that must be dissipated from the sensor is 0.25 W, which, depending on the surface area and local heat transfer conditions, may cause a change in the temperature of the sensor.

COMMENT   The current flow through the sensor results in a sensor temperature higher than would occur with zero current flow due to I^2R heating. This is a loading error that offsets the indicated temperature, a systematic error. There is a trade-off between the increased sensitivity, dE_o/dR_1, and the correspondingly increased current associated with E_i. The input voltage must be chosen appropriately for a given application.

6.5   LOADING ERRORS AND IMPEDANCE MATCHING

In an ideal sense, an instrument or measurement system should not in itself affect the variable being measured. Any such effect will alter the variable and be considered as a "loading" that the measurement system exerts on the measured variable. A *loading error* is the difference between the value of the measurand and the indicated value brought on by the act of measurement. Loading effects can be of any form: mechanical, electrical, or optical. When the insertion of a sensor into a process somehow changes the physical variable being measured, that is a loading error. A loading error can occur anywhere along the signal path of a measurement system. If the output from one system stage is in any way affected by the subsequent stage, then the signal is affected by *interstage loading error*. Good measurement system design minimizes all loading errors.

To illustrate this idea, consider measuring the temperature of a volume of a high-temperature liquid using a glass bulb thermometer. Some finite quantity of energy must flow from the liquid to the thermometer bulb to achieve thermal equilibrium between the bulb and the liquid (i.e., the thermometer may cool down or heat up the liquid). As a result of this energy flow, the liquid temperature is changed, and the measured value does not correspond to the original liquid temperature sought. This measurement has introduced a loading error. If the relative energy exchange is small, the loading error can be kept small or negligible.

Loading Errors for Voltage-Dividing Circuit

Consider the voltage-divider circuit shown in Figure 6.8 for the case where R_m is finite. Under these conditions, the circuit can be represented by the equivalent circuit shown in Figure 6.14. As the sliding contact at point A moves, it divides the full-scale deflection resistance R into R_1 and R_2, such that $R_1 + R_2 = R_T$. The resistances R_m and R_1 form a parallel loop, yielding an equivalent resistance, R_L, given by

$$R_L = \frac{R_1 R_m}{R_1 + R_m} \tag{6.28}$$

The equivalent resistance for the entire circuit, as seen from the voltage source, is R_{eq}:

$$R_{eq} = R_2 + R_L = R_2 + \frac{R_1 R_m}{R_1 + R_m} \tag{6.29}$$

The current flow from the voltage source is then

$$I = \frac{E_i}{R_{eq}} = \frac{E_i}{R_2 + R_1 R_m/(R_1 + R_m)} \tag{6.30}$$

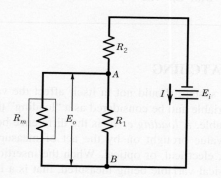

Figure 6.14 Instruments in parallel to signal path form an equivalent voltage-dividing circuit.

and the output voltage is given by

$$E_o = E_i - IR_2 \tag{6.31}$$

This result can be expressed as

$$\frac{E_o}{E_i} = \frac{1}{1 + (R_2/R_1)(R_1/R_m + 1)} \tag{6.32}$$

The limit of this expression as R_m tends to infinity is

$$\frac{E_o}{E_i} = \frac{R_1}{R_1 + R_2} \tag{6.33}$$

which is just Equation 6.6. Also, as R_2 approaches zero, the output voltage approaches the supply voltage, as expected. Expressing the value found from Equation 6.6 as the true value $(E_o/E_i)'$, the loading error e_I may be given by

$$e_I = E_i \left[\left(\frac{E_o}{E_i} \right) - \left(\frac{E_o}{E_i} \right)' \right] \tag{6.34}$$

Here the loading error goes to zero as $R_m \to \infty$. Hence a large meter resistance reduces loading error in voltage measurements.

Interstage Loading Errors

Consider a common situation in the measurement chain in which the output voltage signal from one system device provides the input signal to the following device. This following device could be either a measuring device or an interstage device, such as a signal conditioning device like an amplifier. An open circuit voltage E_1 will be present at the output terminal of device 1 with its equivalent

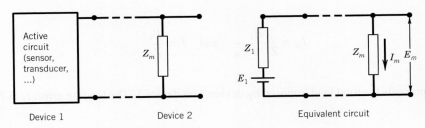

Figure 6.15 Equivalent circuit formed by parallel connections.

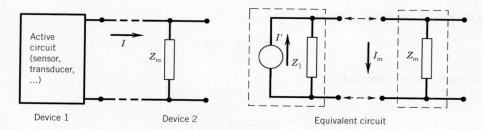

Figure 6.16 Instruments in series with signal path.

impedance Z_1. Device 2 is then attached across the terminals. As shown in Figure 6.15, the Thévenin equivalent circuit consists of a voltage, E_1, with internal series impedance Z_1. Connecting device 2 across the terminals of device 1 is equivalent to placing an impedance Z_m across the terminals. This causes a current, I, to flow in the loop formed by the two terminals, and Z_m acts as a load on the first device. The voltage sensed by device 2 is

$$E_m = IZ_m = E_1 \frac{1}{1 + Z_1/Z_m} \tag{6.35}$$

Consequently, the voltage sensed by device 2 differs from the original voltage owing to the interstage connection. This is a loading error of magnitude, $e_I = E_m - E_1$, or

$$e_I = E_1 \left(\frac{1}{1 + Z_1/Z_m} - 1 \right) \tag{6.36}$$

For the measured voltage between stages to nearly equal the actual voltage (i.e., to maximize the voltage potential between stages), it is required that $Z_m \gg Z_1$ such that $e_I \to 0$. The interstage or measuring device should have a high input impedance to minimize loading error.

When the signal is current-driven, such as a 4-20 mA current loop, the maximum current transfer between devices 1 and 2 is desirable. Consider the circuit shown in Figure 6.16. The current through the loop indicated by the device 2 is given by I_m, whereas when device 2 is not present in the circuit, the short circuit value is I'. That is,

$$I_m = \frac{E_1}{Z_1 + Z_m} \quad \text{and} \quad I' = \frac{E_1}{Z_1} \tag{6.37}$$

where Z_m represents the total load imposed by device 2 on device 1. The loading error, $e_I = I_m - I'$, is

$$e_I = E_1 \frac{-Z_m}{Z_1^2 + Z_1 Z_m} \tag{6.38}$$

From Equation 6.38 or 6.37, it is clear that to reduce current-driven loading error, then $Z_m \ll Z_1$, such that $e_I \to 0$.

When a 4-20 mA current loop is used to drive the signal, often a resistor of impedance Z_m is placed across the output terminals and voltage measured. The measured voltage, E_m, is

$$E_m = I' Z_m \left(\frac{Z_1}{Z_1 + Z_m} \right) \tag{6.39}$$

The term in parentheses in Equation 6.39 shows the loading influence, which is minimized as $Z_m \ll Z_1$.

Example 6.6

For the Wheatstone bridge shown in Figure 6.17, find the open circuit output voltage (when $R_m \to \infty$) if the four resistances change by

$$\delta R_1 = +40\,\Omega \quad \delta R_2 = -40\,\Omega \quad \delta R_3 = +40\,\Omega \quad \delta R_4 = -40\,\Omega$$

KNOWN   Measurement device resistance given by R_m

ASSUMPTIONS $R_m \to \infty$ Negligible source impedance

FIND E_o

SOLUTION   From Equation 6.11,

$$
\begin{aligned}
E_o &= E_i \left(\frac{R_1}{R_1 + R_2} - \frac{R_3}{R_3 + R_4} \right) \\
&= 50\,\text{V} \left(\frac{5040\,\Omega}{5040\,\Omega + 4960\,\Omega} - \frac{8040\,\Omega}{8040\,\Omega + 7960\,\Omega} \right) = +0.0750\,\text{V}
\end{aligned}
$$

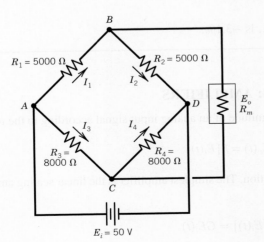

$R_1 = 5000\,\Omega$

$R_2 = 5000\,\Omega$

$R_3 = 8000\,\Omega$

$R_4 = 8000\,\Omega$

E_o
R_m

$E_i = 50\,\text{V}$

Figure 6.17   Bridge circuit for Example 6.6.

Example 6.7

Now consider the case where a meter with internal impedance $R_m = 200\,\text{k}\Omega$ is used in Example 6.6. What is the output voltage with this particular meter in the circuit? Estimate the loading error.

KNOWN $R_m = 200,000\,\Omega$

FIND E_o, e_I

SOLUTION   Because the output voltage E_o is equal to $I_m R_m$, where I_m is found from Equation 6.19,

$$E_o = R_g I_m = \frac{E_i R_m (R_1 R_4 - R_2 R_3)}{R_1 R_4 (R_2 + R_3) + R_2 R_3 (R_1 + R_4) + R_m (R_1 + R_2)(R_3 + R_4)}$$

which for $R_m = 200,000\,\Omega$, gives $E_o = 0.0726\,\text{V}$.

The loading error e_I is the difference in the two output voltages between infinite meter impedance (Ex. 6.6) and finite meter impedance. This gives $e_I = 0.0726 - 0.0750 = -2.4\,\text{mV}$ with the negative sign reflecting a lower measured value under load.

The bridge output may also be expressed as a ratio of the output voltage with $R_m \to \infty$, E_o', and the output voltage with a finite meter resistance E_o:

$$\frac{E_o}{E_o'} = \frac{1}{1 + R_e/R_g}$$

where $R_e = \dfrac{R_1 R_2}{R_1 + R_2} + \dfrac{R_3 R_4}{R_3 + R_4}$, which yields for the present example

$$\frac{E_o}{E_o'} = \frac{1}{1 + 6500/200,000} = 0.969$$

The percent loading error, $100 \times \left(\dfrac{E_o}{E_o'} - 1\right)$, is -3.1%.

6.6   ANALOG SIGNAL CONDITIONING: AMPLIFIERS

An amplifier is a device that scales the magnitude of an analog input signal according to the relation

$$E_o(t) = h\{E_i(t)\}$$

where $h\{E_i(t)\}$ defines a mathematical function. The simplest amplifier is the linear scaling amplifier in which

$$h\{E_i(t)\} = GE_i(t) \tag{6.40}$$

where the gain G is a constant that may be any positive or negative value. Many other types of operation are possible, including the "base x" logarithmic amplifier in which

$$h\{E_i(t)\} = G \log_x(E_i(t)) \tag{6.41}$$

Amplifiers have a finite frequency response and limited input voltage range.

The most widely used type of amplifier is the solid-state operational amplifier. This device is characterized by a high input impedance ($Z_i > 10^7 \, \Omega$), a low output impedance ($Z_o < 100 \, \Omega$), and a high internal gain ($A_o \approx 10^5$ to 10^6). As shown in the general diagram of Figure 6.18a, an operational amplifier has two input ports, a noninverting and an inverting input, and one output port. The signal at the output port is in phase with a signal passed through the noninverting input port but is 180 degrees out of phase with a signal passed through the inverting input port. The amplifier requires dual-polarity DC excitation power ranging from ± 5 V to ± 15 V. In addition, two DC voltage offset null (bias) input ports provide a means to zero out any output offset signal at zero input; usually a variable 10 kΩ resistor is placed across these inputs to adjust offset null.

As an example, the pin connection layout of a common operational amplifier circuit, the type 741, is shown in Figure 6.18b in its eight-pin, dual-in-line ceramic package form. This is the familiar rectangular black integrated circuit package seen on circuit boards. Each pin port is numbered and labeled as to function. An internal schematic diagram is shown in Figure 6.18c with the corresponding pin connections labeled. As shown, each input port (i.e., 2 and 3) is attached to the base of an npn transistor.

The high internal open loop gain, A_o, of an operational amplifier is given as

$$E_o = A_o\big(E_{i_2}(t) - E_{i_1}(t)\big) \tag{6.42}$$

The magnitude of A_o, flat at low frequencies, falls off rapidly at high frequencies, but this intrinsic gain curve is overcome by using external input and feedback resistors that set the circuit gain G and circuit response. Some possible amplifier configurations using an operational amplifier are shown in Figure 6.19.

Because the amplifier has a very high internal gain and negligible current draw, resistors R_1 and R_2 are used to form a feedback loop and control the overall amplifier circuit gain, called the *closed loop gain*, G. The noninverting linear scaling amplifier circuit of Figure 6.19a has a closed loop gain of

$$G = \frac{E_o(t)}{E_i(t)} = \frac{R_1 + R_2}{R_2} \tag{6.43}$$

Resistor R_s does not affect the gain but is used to balance out potential variation problems at small currents. Its value is selected such that $R_s \approx R_1 R_2 / (R_1 + R_2)$. The inverting linear scaling amplifier circuit of Figure 6.19b provides a gain of

$$G = \frac{E_o(t)}{E_i(t)} = \frac{R_2}{R_1} \tag{6.44}$$

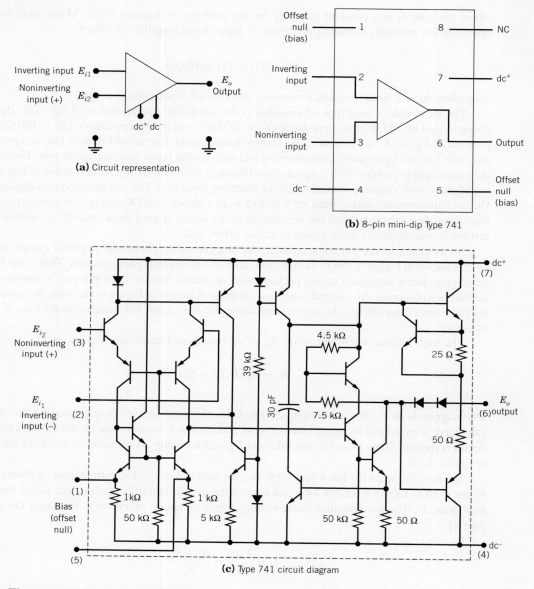

(a) Circuit representation

(b) 8–pin mini-dip Type 741

(c) Type 741 circuit diagram

Figure 6.18   Operational amplifier.

The Labview programs *Opamp*, *Inverting-Opamp*, and *Noninverting_Opamp* allow the user to vary resistance values and input signal amplitude to manipulate the output signal.

By using both inputs, the arrangement forms a differential amplifier, Figure 6.19c, in which

$$E_o(t) = \left(E_{i_2}(t) - E_{i_1}(t)\right)(R_2/R_1) \tag{6.45}$$

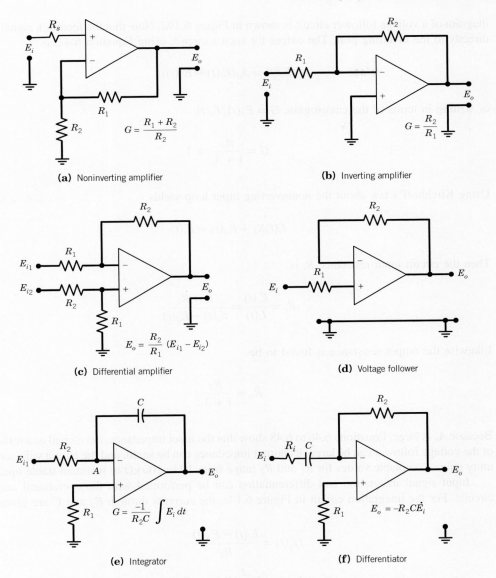

(a) Noninverting amplifier

(b) Inverting amplifier

(c) Differential amplifier

(d) Voltage follower

(e) Integrator

(f) Differentiator

Figure 6.19   Some common amplifier circuits using operational amplifiers.

The differential amplifier circuit is effective as a voltage comparator for many instrument applications (see Section 6.7).

A voltage follower circuit is commonly used to isolate an impedance load from other stages of a measurement system, such as might be needed with a high output-impedance transducer. A schematic diagram of a voltage follower circuit is shown in Figure 6.19d. Note that the feedback signal is sent directly to the inverting port. The output for such a circuit, using Equation 6.42, is

$$E_o(t) = A_o(E_i(t) - E_o(t))$$

or writing in terms of the circuit gain, $G = E_o(t)/E_i(t)$,

$$G = \frac{A_o}{1 + A_o} \approx 1 \tag{6.46}$$

Using Kirchhoff's law about the noninverting input loop yields

$$I_i(t)R_1 + E_o(t) = E_i(t)$$

Then the circuit input resistance R_i is

$$R_i = \frac{E_i(t)}{I_i(t)} = \frac{E_i(t)R_1}{E_i(t) - E_o(t)} \tag{6.47}$$

Likewise the output resistance is found to be

$$R_o = \frac{R_2}{1 + A_o} \tag{6.48}$$

Because A_o is large, Equations 6.46 to 6.48 show that the input impedance, developed as a resistance, of the voltage follower can be large, its output impedance can be small, and the circuit will have near unity gain. Acceptable values for R_1 and R_2 range from 10 to 100 kΩ to maintain stable operation.

Input signal integration and differentiation can be performed with the operational amplifier circuits. For the integration circuit in Figure 6.19e, the currents through R_2 and C are given by

$$I_{R_2}(t) = \frac{E_i(t) - E_A(t)}{R_2}$$

$$I_C(t) = C\frac{d}{dt}(E_o(t) - E_A(t))$$

Summing currents at node A yields the integrator circuit operation:

$$E_o(t) = -\frac{1}{R_2 C}\int E_i(t)dt \tag{6.49}$$

As a special note, if the input voltage is a constant positive DC voltage, then the output signal will be a negative linear ramp, $E_0 = -E_i t/RC$, with t in seconds, a feature used in ramp converters and integrating digital voltmeters (see Section 7.5 in Chapter 7). Integrator circuits are relatively unaffected by high-frequency noise, because integration averages out noise.

Following a similar analysis, the differentiator circuit shown in Figure 6.19f performs the operation

$$E_o(t) = -R_2 C \frac{dE_i(t)}{dt} \tag{6.50}$$

Differentiator circuits amplify high-frequency noise in signals to the point of masking the true signal. Adding resistor R_i limits high-frequency gain to a -3 dB cutoff frequency of $f_c = 1/2\pi R_i C$. In noisy environments, a passive RC differentiator may simply perform better.

6.7   ANALOG SIGNAL CONDITIONING: SPECIAL-PURPOSE CIRCUITS

Analog Voltage Comparator

A voltage comparator provides an output that is proportional to the difference between two input voltages. As shown in Figure 6.20a, a basic comparator consists of an operational amplifier operating in a high-gain differential mode. In this case, any difference between inputs E_{i_1} and E_{i_2} is amplified at the large open-loop gain A_o. Accordingly, the output from the comparator saturates when $E_{i_1} - E_{i_2}$ is either slightly positive or negative, the actual value set by the threshold voltage E_T. The saturation output value E_{sat} is nearly the supply voltage E_s. For example, a 741 op-amp driven at ± 15 V might saturate at ± 13.5 V. The value for E_T can be adjusted by the amplifier bias (offset null) voltage E_{bias}. The comparator output is given by

$$
\begin{aligned}
E_o &= A_o\left(E_{i_1} - E_{i_2}\right) & \text{for} & \quad \left|E_{i_1} - E_{i_2}\right| \le E_T \\
&= +E_{\text{sat}} & \text{for} & \quad E_{i_1} - E_{i_2} > E_T \\
&= -E_{\text{sat}} & \text{for} & \quad E_{i_1} - E_{i_2} < -E_T
\end{aligned}
\tag{6.51}
$$

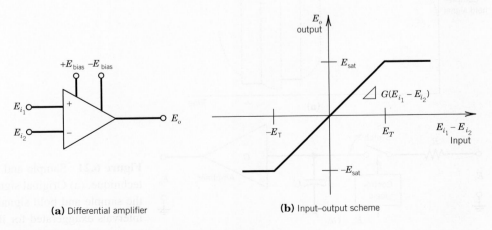

(a) Differential amplifier   **(b)** Input–output scheme

Figure 6.20 Analog voltage comparator.

This input–output relation is shown in Figure 6.20b. For a ± 15 V supply and a gain of about 200,000, the comparator might saturate with a voltage difference of only ~68 μV.

Often E_{i_2} is a known reference voltage. This allows the comparator output to be used for control circuits to decide whether E_{i_1} is less than or greater than E_{i_2}; a positive difference gives a positive output. One frequent use of the comparator is in an analog-to-digital converter (see Chapter 7). A zener-diode connected between the + input and the output provides a TTL (transistor-transistor logic) compatible output signal for digital system use.

Sample-and-Hold Circuit

The sample-and-hold circuit (SHC) is used to take a narrow-band measurement of a time-changing signal and to hold that measured value until reset. It is widely used in data acquisition systems when using analog-to-digital converters. The circuit tracks the signal until it is triggered to hold it at a fixed value while it is actually measured. This is illustrated in Figure 6.21a, in which the track-and-hold logic provides the appropriate trigger.

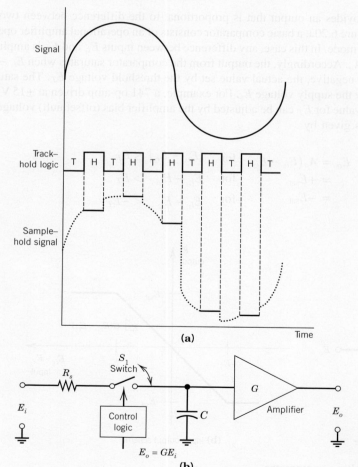

Figure 6.21 Sample and hold technique. (a) Original signal and the sample and hold signal (time intervals exaggerated for illustrative purposes), (b) Circuit.

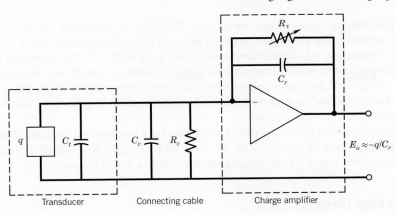

Figure 6.22 Charge amplifier circuit shown connected to a transducer.

$E_o \approx -q/C_r$

The basic circuit for sample and hold is shown in Figure 6.21b. The switch S_1 is a fast analog device. When the switch is closed, the "hold" capacitor C is charged through the source resistor R_s. When the capacitor is charged, the switch is opened. The amplifier presents a very high input impedance and very low current, which, together with a very low leakage capacitor, allows for a sufficiently long hold time. The typical SHC is noninverting with a unit gain ($G = 1$).

Charge Amplifier

A *charge amplifier* is used to convert a high-impedance charge q into an output voltage E_o. The circuit consists of a high-gain, inverting voltage operational amplifier such as shown in Figure 6.22. These circuits are commonly used with transducers that utilize piezoelectric crystals or capacitance sensors. A piezoelectric crystal develops a time-dependent surface charge under varying mechanical load.

The circuit output voltage is determined by

$$E_o = -q/\left[C_r + (C_T/A_o)\right] \tag{6.52}$$

with $C_T = C_t + C_c + C_r$ where C_t, C_c, and C_r represent the transducer, cable, and feedback capacitances, respectively, R_c is the cable and transducer resistances, and A_o is the amplifier open-loop gain. Because the open-loop gain of operational amplifiers is very large, Equation 6.52 can often be simplified to

$$E_o \approx -q/C_r \tag{6.53}$$

A variable resistance R_τ and fixed feedback capacitor are used to average out low-frequency fluctuations in the signal.

4–20 mA Current Loop

Low-level voltage signals below about 100 mV are quite vulnerable to noise. Examples of transducers having low-level signals include strain gauges, pressure gauges, and thermocouples.

One means of transmitting such signals over long distances is by signal boosting through amplification. But amplification also boosts noise. A practical alternative method that is well suited to an industrial environment is a *4–20 mA current loop* (read as "4 to 20"). In this approach, the low-level voltage is converted into a standard current loop signal of between 4 and 20 mA, the lower value for the minimum voltage and the higher value for the maximum voltage in the range. The 4–20 mA current loop can be transmitted over several hundred meters without degradation. The 4–20 mA output is a common option for transducers.

At the output display end, a receiver converts the current back to a low-level voltage. This can be as simple as a single resistor in parallel with the loop. For example, a 4–20 mA signal can be converted back to a 1–5 V signal by terminating the loop with a 250 Ω resistor.

Multivibrator and Flip-Flop Circuits

Multivibrator and *flip-flop* circuits are analog circuits that also form the basis of digital signals. They are useful for system control and triggering of events. The *astable multivibrator* is a switching circuit that toggles on and off continuously between a high- and low-voltage state in response to an applied time-dependent input voltage. It is used to generate a square waveform on a prompt, where typically amplitudes vary between 0 V (low) and 5 V (high) in a manner known as a *TTL signal*. Because the circuit continuously switches between high and low state, it is astable. The heart of this circuit, as shown in Figure 6.23, is the two transistors T_1 and T_2, which conduct alternately; the circuit is symmetric around the transistors. This generates a square wave signal of fixed period and amplitude at the output as shown in Figure 6.24a. The working mechanism is straightforward. The two transistors change state as the currents through capacitors C_1 and C_2 increase and decrease due to the applied input. For example, when T_2 turns on and its collector switches from V_c toward 0, the base of T_1 is driven negative, which turns T_1 off. While C_2 discharges, C_1 charges. As the voltage across C_1 increases, the current across it decreases. But so long as the current through C_1 is large enough, T_2 remains on. It eventually falls to a value that turns T_2 off. This causes C_2 to charge, which turns T_1 on. The period of the resulting square wave is proportional to $R_1 C_1$.

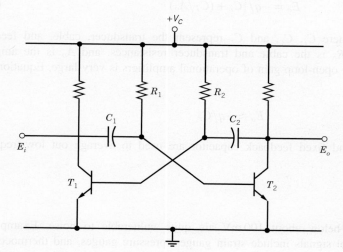

Figure 6.23 Basic multivibrator circuit.

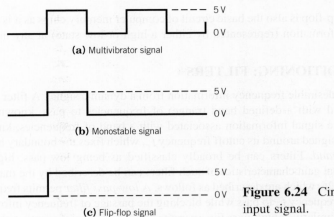

(a) Multivibrator signal

(b) Monostable signal

(c) Flip-flop signal

Figure 6.24 Circuit output response to an applied input signal.

A useful variation of the multivibrator circuit is the *monostable*. In this arrangement T_2 stays on until a positive external pulse or change in voltage is applied to the input. At that moment, T_2 turns off for a brief period of time, providing a jump in the output voltage. The monostable circuit cycles just once, awaiting another pulse as shown in Figure 6.24b. Hence, it is often referred to as a *one-shot*. Because it fires just once and resets itself, the monostable is an effective trigger. Labview program *Monostable* illustrates the function of this important circuit.

Another variation of this circuit is the *flip-flop* or *bistable multivibrator* in Figure 6.25. This circuit is an effective electronic switch. Its operation is analogous to that of a light switch; it is either in an "on" (high) or an "off" (low) state. In practice, transistors T_1 and T_2 change state every time a pulse is applied. If T_1 is on with T_2 off, a pulse turns T_2 off, turning T_1 on, producing the output shown in Figure 6.24c. So the flip-flop output level changes from low to high voltage or high to low

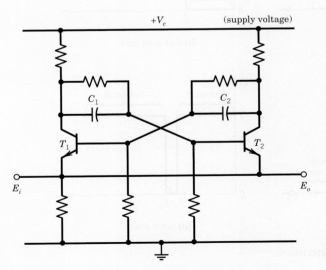

Figure 6.25 Basic flip-flop circuit.

voltage on command. The flip-flop is also the basic circuit of computer memory chips as it is capable of holding a single bit of information (represented by either a high or low state) at any instant.

6.8   ANALOG SIGNAL CONDITIONING: FILTERS

A *filter* is used to remove undesirable frequency information from a dynamic signal. A filter permits signal information associated with a defined band (range) of frequencies to pass, known as the *passband*, while blocking the signal information associated with a band of frequencies, known as the *stopband*. The filter is designed around its cutoff frequency f_c, which fixes the boundary between the *passband* and the *stopband*. Filters can be broadly classified as being low-pass, high-pass, bandpass, and notch. The ideal gain characteristics of such filters can be described by the magnitude ratio plots shown in Figure 6.26, which are described as follows. A *low-pass filter* permits frequencies below the prescribed cutoff frequency f_c to pass while blocking the passage of frequency information above the cutoff frequency. Similarly, a *high-pass filter* permits only frequency information above the cutoff frequency to pass. A *bandpass filter* combines features of both the low- and high-pass filters. It is described by a low cutoff frequency f_{c_1} and a high cutoff frequency f_{c_2} to define a band of frequency information that is permitted to pass through the filter. A *notch filter* permits the passage of all frequency information except those within a narrow frequency band. An intensive treatment of filters for analog and digital signals can be found in many specialized texts (3–8).

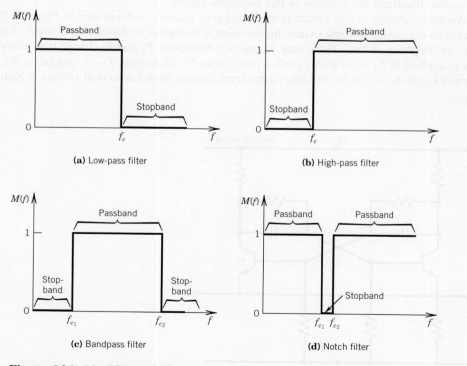

Figure 6.26   Ideal filter characteristics.

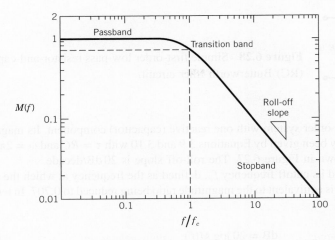

Figure 6.27   Magnitude ratio for a low-pass Butterworth filter.

Filters work by performing a well-defined mathematical operation on the input signal as specified by their transfer function. The transfer function is defined by the position and values of elements in the filter design. *Passive analog filter circuits* consist of combinations of resistors, capacitors, and inductors. *Active filters* incorporate operational amplifiers into the circuit, as well.

The sharp cutoff of the ideal filter cannot be realized in a practical filter. As an example, a plot of the magnitude ratio for a real low-pass filter is shown in Figure 6.27. All filter response curves contain a transition band over which the magnitude ratio decreases with frequency from the passband to the stopband. This rate of transition is known as the filter *roll-off*, usually specified in units of dB/decade. In addition, the filter introduces a phase shift between its input and output signal.

The design of real filters is focused on certain desirable response features. The magnitude and phase characteristics of a real filter can be optimized to meet one of the following: (1) maximum magnitude flatness over the passband, (2) a linear phase response over the passband, or (3) a sharp transition from the passband to stopband with steep roll-off. No one filter can meet all three characteristics, but we can design for any one of them.

For example, a relatively flat magnitude ratio over its passband with a moderately steep initial roll-off and acceptable phase response is a characteristic of a *Butterworth* filter response. On the other hand, a very linear phase shift over its passband but with a relatively gradual initial roll-off is a characteristic of a *Bessel* filter response. The frequency-dependent behavior of low-pass, bandpass, and high-pass filters can be explored in the LabView programs *Butterworth filters* and *Bessel filters*.

Butterworth Filter Design

A Butterworth filter is optimized to achieve maximum flatness in magnitude ratio over the passband. A simple passive low-pass Butterworth filter can be constructed using the resistor-and-capacitor (RC) circuit of Figure 6.28. By applying Kirchhoff's law about the input loop, we derive the model relating the input voltage E_i to the output voltage E_o:

$$RC\dot{E}_o(t) + E_o(t) = E_i(t) \qquad (6.54)$$

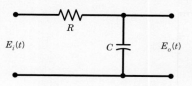

Figure 6.28 Simple first-order low-pass resistor-and-capacitor (RC) Butterworth filter circuit.

This real filter model is a first-order system with one reactive (capacitor) component. Its magnitude and phase response has already been given by Equations 3.9 and 3.10 with $\tau = RC$ and $\omega = 2\pi f$, and its magnitude response is shown in Figure 6.27. The roll-off slope is 20 dB/decade.

A filter is designed around its cutoff frequency f_c, defined as the frequency at which the signal power is reduced by half. This is equivalent to the magnitude ratio being reduced to 0.707. In terms of the decibel (dB),

$$dB = 20 \log M(f) \tag{3.11}$$

f_c occurs at -3 dB—that is, the frequency at which the signal is attenuated by 3 dB. For the filter of Figure 6.28, this requires that $\tau = RC = 1/(2\pi f_c)$.

Improved Butterworth Filter Designs

With its simplicity, the RC filter begins to roll off well before the cutoff frequency; the roll-off is not very sharp, and the phase shift response (recall Fig. 3.13) is not linear, leading to potential signal distortion. But the flatness in the passband can be extended and the roll-off slope of a filter improved by staging multiple filters in series, called *cascading filters*. This is done by adding reactive elements, such as by alternating inductors (L) and capacitors (C), to the circuit as shown in the resulting LC ladder circuit of Figure 6.29. The resistances R_s and R_L refer to the circuit source and load impedances, respectively. A kth-order or k-stage filter contains k reactive elements in stages. The magnitude ratio and phase shift for a k-stage low-pass Butterworth filter are given by

$$M(f) = \frac{1}{\left[1 + (f/f_c)^{2k}\right]^{1/2}} \tag{6.55a}$$

$$\Phi(f) = \sum_{i=1}^{k} \Phi_i(k) \tag{6.55b}$$

where f/f_c is the normalized frequency. The general normalized magnitude response curves for k-stage Butterworth filters are shown in Figure 6.30. The roll-off slope is $20 \times k$ (dB/decade) or $6 \times k$

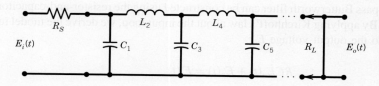

Figure 6.29 Ladder circuit for multistage (cascading) low-pass LC filter to achieve a higher-order response.

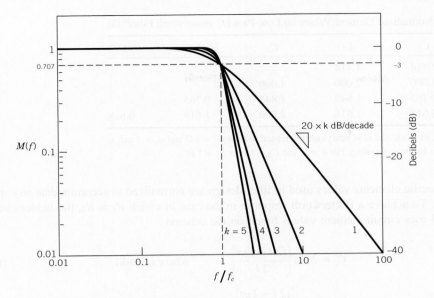

Figure 6.30   Magnitude characteristics for Butterworth low-pass filters of various stages.

(dB/octave). Frequency-dependent signal attenuation is found from the dynamic error, $\delta(f)$, or directly in decibels by

$$A(f/f_c) = \text{Attenuation(dB)} = 10 \log \left[1 + (f/f_c)^{2k} \right] \qquad (6.56)$$

Equation 6.56 is useful for specifying the required order of a filter based on desired response needs. This is emphasized in Figure 6.31, showing how the stages affect attenuation and roll-off slope versus normalized frequency.

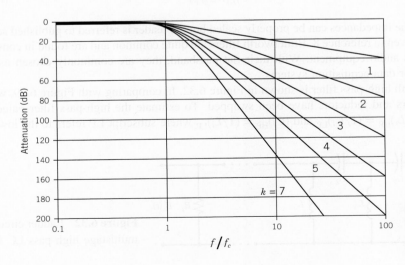

Figure 6.31 Attenuation with various stages of Butterworth filters.

Table 6.1 Normalized Element Values for Low-Pass LC Butterworth Filter[a] (8)

k	C_1	L_2	C_3	L_4	C_5
2	1.414	1.414			
3	1.000	2.000	1.000		
4	0.765	1.848	1.848	0.765	
5	0.618	1.618	2.000	1.618	0.618

[a]Values for C_i in farads and L_i in henrys are referenced to $R_s = R_L = 1\ \Omega$ and $\omega_c = 1$ rad/s. See discussion for proper scaling. For $k = 1$, use $C_1 = 1$ F or $L_1 = 1$ H.

The specific elements values used in filter design are normalized to accommodate any specific application. To achieve a Butterworth response in the case in which $R_L = R_S$, the ladder circuit of Figure 6.29 uses circuit element values based on the scheme

$$C_i = 2 \sin\left[\frac{(2\,i - 1)\pi}{2\,k}\right] \qquad \text{where } i \text{ is odd} \tag{6.57a}$$

$$L_i = 2 \sin\left[\frac{(2\,i - 1)\pi}{2\,k}\right] \qquad \text{where } i \text{ is even} \tag{6.57b}$$

where $i = 1, 2, \ldots, k$. Specific values are provided in Table 6.1 for elements C_i and L_i for a two-through five-stage Butterworth filter associated with Figure 6.29 (3,8). The lead reactive element is assumed to be a capacitor; if the lead reactive element is switched to an inductor (such that L_i is odd), then the table values assigned to C and L switch. The values in Table 6.1 are normalized with $R_s = R_L = 1\ \Omega$ and $\omega_c = 2\pi f_c = 1$ rad/s. For other values, L_i and C_i are scaled by

$$L = L_i R / 2\pi f_c \tag{6.58a}$$

$$C = C_i / (R 2\pi f_c) \tag{6.58b}$$

When $R_L \neq R_S$ the impedances can be properly scaled but the reader is referred to published scaling tables, such as given in reference 8. Butterworth filters are quite common and are found in consumer products such as audio equipment. With their flat passband, they are commonly chosen as anti-aliasing filters for data acquisition systems.

A Butterworth high-pass filter is shown in Figure 6.32. In comparing with Figure 6.29, we see that the capacitors and inductors have been swapped. To estimate the high-pass filter values for Figure 6.32, let $(L_i)_{HP} = (1/C_i)_{LP}$ and $(C_i)_{HP} = (1/L_i)_{LP}$ where subscript LP refers to the low-pass

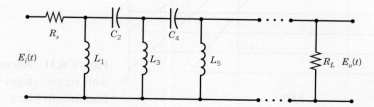

Figure 6.32 Ladder circuit for multistage high-pass LC filter.

values from Table 6.1 and HP refers to the high-pass filter values to be used with Figure 6.32. Reflecting this, the scaling values for a high-pass filter are

$$L = R/2\pi f_c C_i \tag{6.59a}$$

$$C = 1/2\pi f_c R L_i \tag{6.59b}$$

The magnitude ratio for a high-pass filter is given by

$$M(f) = \frac{1}{\left[1 + (f_c/f)^{2k}\right]^{1/2}} \tag{6.60}$$

The phase shift properties described by Equation 6.55b apply.

Example 6.8

Design a one-stage Butterworth RC low-pass filter with a cutoff frequency of 100 Hz at −3 dB if the source and load impedances are 50 Ω. Calculate the expected dynamic error and attenuation at 192 Hz in the realized filter.

KNOWN $f_c = 100$ Hz

$\quad\quad\quad\quad\quad k = 1$

FIND C and δ

SOLUTION   A single-stage low-pass Butterworth RC filter circuit would be similar to that of Figure 6.28, which is just a first-order system with time constant $\tau = RC$. With the relation $\omega = 2\pi f$, the magnitude ratio for this circuit is given by

$$M(f) = \frac{1}{\left[1 + (\omega\tau)^2\right]^{1/2}} = \frac{1}{\left[1 + (2\pi f\tau)^2\right]^{1/2}} = \frac{1}{\left[1 + (f/f_c)^2\right]^{1/2}}$$

Setting $M(f) = 0.707 = -3$ dB with $f = f_c = 100$ Hz gives

$$\tau = 1/2\pi f_c = RC = 0.0016 \text{ s}$$

With $R = 50$ Ω, we need a capacitor of $C = 32$ μF.

Alternately, we could use Figure 6.29 and Table 6.1 with $k = 1$ for which the normalized value is $C = 1$ F. This value is scaled to $R = 50$ Ω and $f_c = 100$ Hz by

$$C = C_1/(R2\pi f_c) = (1 \text{ F})/(50 \text{ Ω})(2\pi)(100 \text{ Hz}) = 32 \text{ μF}$$

A commercially available capacitor size is 33 µF. Using this size in our realized circuit, the cutoff frequency shifts to 96 Hz. We use $f_c = 96$ Hz below.

At $f = 192$ Hz, the dynamic error, $\delta(f) = M(f) - 1$ is

$$\delta(192\,\text{Hz}) = M(192\,\text{Hz}) - 1 = -0.55$$

meaning that the input signal frequency content at 192 Hz is reduced by 55%. The attenuation at the normalized frequency of $f/f_c = 2$ is given by equation 6.56 as

$$A(2) = 10\log\left[1 + (2)^2\right] = 46\,\text{dB}$$

Example 6.9

How many filter stages are required to design a low-pass Butterworth filter with a cutoff frequency of 3 kHz if the filter must provide attenuation of at least 60 dB at 30 kHz? If $R_s = R_L = 50\,\Omega$, find the filter element values.

KNOWN $f_c = 3$ kHz; $f = 30$ kHz

FIND k to achieve $A \geq 60$ dB at 30 kHz

SOLUTION   The normalized frequency is $f/f_c = 30{,}000/3000 = 10$.

$$A(f/f_c = 10) = A(10) = 10\log\left[1 + (f/f_c)^{2k}\right]$$

We can solve directly for k with $A(10) = 60$ dB to get $k = 3$. Or, using Figure 6.31, we find that $A(10) \geq 60$ dB for $k \geq 3$.

Using the ladder circuit of Figure 6.29 with $R_s = R_L = 1\,\Omega$ and $\omega_c = 2\pi f_c = 1$ rad/s, Table 6.1 gives $C_1 = C_3 = 1$ F and $L_2 = 2$ H. Scaling these to $f_c = 3$ kHz and $R_s = R_L = 50\,\Omega$, the element values are

$$L = L_2 R/2\pi f_c = 5.3\,\text{mH}$$

$$C = C_1/(R2\pi f_c) = C_3/(R2\pi f_c) = 106\,\mu\text{F}$$

Example 6.10

In a two-speaker (two-way) system, the low-pass circuit passes only low-frequency signal content to the woofer, while the high-pass circuit passes high-frequency content to the wide-frequency speaker.

The Linkwitz–Riley high-pass or low-pass filter demonstrates a practical use of Butterworth filters for loudspeaker designs. The 4th order ($k = 4$) design is a cult standard achieving a 24 dB/octave slope. As shown in Figure 6.33 with component design values listed below, the fourth-order

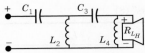
Fourth-order high-pass filter

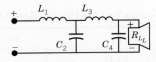
Fourth-order low-pass filter

Figure 6.33 Linkwitz–Riley fourth-order Butterworth filters.

design uses two each of inductors and capacitors in conjunction with the natural impedance of the loudspeaker driver to achieve a high-pass or low-pass design, or the two can be coupled as a band-pass filter.

High-Pass Circuit[a]	Low-Pass Circuit[a]
$C_1(\text{farad}) = 0.0844/(R_{L_H} \times f_c)$	$C_2(\text{farad}) = 0.2533/(R_{L_L} \times f_c)$
$C_3(\text{farad}) = 0.1688/(R_{L_H} \times f_c)$	$C_4(\text{farad}) = 0.0563/(R_{L_L} \times f_c)$
$L_2(\text{henry}) = 0.1000 \times R_{L_H}/f_c$	$L_1(\text{henry}) = 0.3000 \times R_{L_L}/f_c$
$L_4(\text{henry}) = 0.04501 \times R_{L_H}/f_c$	$L_3(\text{henry}) = 0.1500 \times R_{L_L}/f_c$

[a] f_c refers to the cutoff frequency for that particular circuit; R_{L_H} and R_{L_L} refer to the impedances of the high and low frequency speakers, respectively.

Design a filter for a 120-Hz cut off frequency to be used with a speaker system composed of a woofer and a wide frequency speaker. Assume $8\,\Omega$ drivers.

KNOWN $f_c = 120$ Hz; $R_{L_H} = 8\,\Omega$; $R_{L_L} = 8\,\Omega$

FIND C and L for the circuit

SOLUTION Because the filter circuits have identical qualities when using the same f_c, the output power to each driver is halved at f_c, so the total output sum to each of the two drivers remains at near 0 dB at the crossover frequency. This provides for a desirable nearly flat filter response across the listening frequency range. Applying the information with $f_c = 120$ Hz, we get

High-Pass Circuit	Low-Pass Circuit
$C_1(\mu\text{F}) = 87.9$	$C_2(\mu\text{F}) = 263.9$
$C_3(\mu\text{F}) = 175.8$	$C_4(\mu\text{F}) = 58.7$
$L_2(\text{mH}) = 6.7$	$L_1(\text{mH}) = 20.0$
$L_4(\text{mH}) = 30.0$	$L_3(\text{mH}) = 10.0$

Commercial values for each element that are close to these calculated would be used in the circuit of Figure 6.33 to achieve a cutoff as close to 120 Hz as possible.

Bessel Filter Design

A Bessel filter will sacrifice a flat gain over its passband with a gradual initial roll-off in order to maximize a linear phase response over portions of the passband. A linear phase response closely

Table 6.2   Normalized Element Values for Low-Pass LC Bessel Filters[a] (8)

k	C_1	L_2	C_3	L_4	C_5
2	0.576	2.148			
3	0.337	0.971	2.203		
4	0.233	0.673	1.082	2.240	
5	0.174	0.507	0.804	1.111	2.258

[a]Values for C_i in farads and L_i in henrys are referenced to $R_s = R_L = 1\,\Omega$ and $\omega_c = 1$ rad/s. See discussion for proper scaling. For $k = 1$, use $C_1 = 2$ F or $L_1 = 2$ H.

resembles a time delay, reducing distortion. Thus, the attractiveness of a Bessel filter is the ability to pass wideband signals with a minimum of distortion. The filter is widely used for different applications, including audio design where crossover phase shift can have an effect on sound quality.

A k-stage low-pass Bessel filter has the transfer function

$$KG(s) = \frac{a_o}{a_o + a_1 s + \cdots + a_k s^k} \tag{6.61}$$

For design purposes, this can be rewritten as

$$KG(s) = \frac{a_o}{D_k(s)} \tag{6.62}$$

where $D_k(s) = (2k - 1)D_{k-1}(s) + s^2 D_{k-2}(s)$, $D_o(s) = 1$ and $D_1(s) = s + 1$.

A k-stage LC low-pass Bessel filter can be designed based on the filter circuit shown in Figure 6.29. Table 6.2 lists the normalized corresponding values for elements L_i and C_i to achieve a two- through five-stage Bessel filter corresponding to $\omega_c = 1$ rad with $R_s = R_L = 1\,\Omega$ (6,7). For other values, L_i and C_i are found from Equations 6.58a and 6.58b. Similarly, a high-pass filter with Bessel characteristics and topology similar to Figure 6.32 could be scaled using Table 6.2 and values found from Equations 6.59a and 6.59b.

Active Filters

An active filter capitalizes on the high-frequency gain characteristics of the operational amplifier to form an effective analog filter. A low-pass active filter is shown in Figure 6.34a using the type 741 operational amplifier. This is a first-order, inverting single-stage, low-pass Butterworth filter. It has a low-pass cutoff frequency given by

$$f_c = \frac{1}{2\pi R_2 C_2} \tag{6.63}$$

The static sensitivity (or gain) of the filter is given by $K = R_2/R_1$. The filter retains the Butterworth characteristics described by Equation 6.55a.

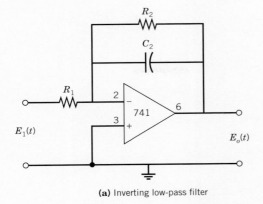

(a) Inverting low-pass filter

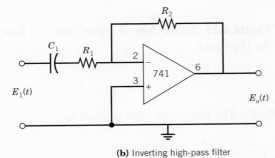

(b) Inverting high-pass filter

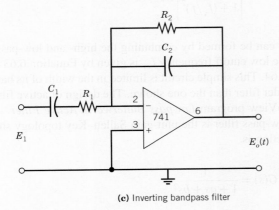

(c) Inverting bandpass filter

Figure 6.34 Common basic active filters.
(a) Inverting low-pass. (b) Inverting high-pass.
(c) Inverting bandpass.

A first-order, inverting single-stage, high-pass Butterworth active filter is shown in Figure 6.34b, using the 741 operational amplifier. The filter has a high-pass cutoff frequency of

$$f_c = \frac{1}{2\pi R_1 C_1} \tag{6.64}$$

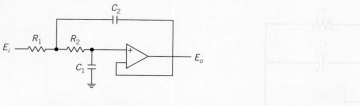

(a) Low-Pass Filter

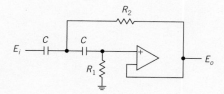

(b) High-Pass Filter

Figure 6.35 Sallen–Key unit-gain filter. (a) Low-pass. (b) High-pass.

with static sensitivity (or gain) of $K = R_2/R_1$. The magnitude ratio is given by

$$M(f) = \frac{1}{\left[1 + (f_c/f)^2\right]^{1/2}} \tag{6.65}$$

An active inverting bandpass filter can be formed by combining the high- and low-pass filters above and is shown in Figure 6.34c. The low cutoff frequency f_{c_1} is given by Equation 6.63 and the high cutoff frequency f_{c_2} by Equation 6.64. This simple circuit is limited in the width of its bandpass. A narrow bandpass requires a higher order filter than the one shown. The design of active filters can be studied with the accompanying LabView program *Lowpass Butterworth Active Filter*.

A commonly used second-order low-pass filter is the unit-gain Sallen–Key topology shown in Figure 6.35a. The generic transfer function is

$$G(s) = \frac{1}{1 + as + bs^2} \tag{6.66}$$

with $a = \omega_c C_1(R_1 - R_2)$ and $b = \omega_c^2 C_1 C_2 R_1 R_2$ where $\omega_c = 2\pi f_c$. The values for a and b enabling Butterworth or Bessel filter characteristics are listed in Table 6.3. As capacitors come in fewer stock sizes than resistors, usually the capacitor sizes are first selected with $C_2 = nC_1$, where n is a multiple number, such that $C_2 \geq 4bC_1/a^2$, and the resistors are then found by

$$R_{1,2} = \frac{aC_2 \mp \sqrt{(aC_2)^2 - 4bC_1C_2}}{4\pi f_c C_1 C_2} \tag{6.67}$$

Table 6.3   Parameters for Second-Order Sallen–Key Filter Design

Parameter	Bessel	Butterworth
a	1.3617	1.4142
b	0.618	1
Q	$1/\sqrt{3}$	$1/\sqrt{2}$

The filter cutoff frequency is set by

$$f_c = \frac{1}{2\pi\sqrt{R_1 R_2 C_1 C_2}} \tag{6.68}$$

The unit-gain Sallen–Key high-pass filter is shown in Figure 6.35b. The design switches the resistors and capacitors from the low-pass filter. For high-pass filters, we can simplify the design and set $C_1 = C_2 = C$. The high-pass filter transfer function is

$$G(s) = \frac{1}{1 + a/s + b/s^2} \tag{6.69}$$

where $a = 2/\omega_c C R_1$ and $b = 1/\omega_c^2 C^2 R_1 R_2$. Values for a and b to meet Butterworth or Bessel filter characteristics are given in Table 6.3 with the resistor values set by $R_1 = 1/\pi f_c a C$ and $R_2 = a/4\pi f_c b C$.

At some frequency, the reactive elements in a filter circuit resonate, leading to a peak in the response behavior. This is behavior predicted by its Q-factor, an inverse damping ratio-like parameter. For $Q = 1/2$ response is critically damped, for $Q < 1/2$ response is overdamped, and for $Q > 1/2$ response is underdamped. A higher Q-factor generally allows for sharper roll-off and is used in place of using a higher number of stages in a design. Like its mechanical system analog $\left(Q = \sqrt{mk}/\zeta\right)$, an underdamped filter shows ringing and has a resonance behavior. The Q-factors for Butterworth and Bessel filters are shown in Table 6.3, where

$$Q = \sqrt{b}/a \tag{6.70}$$

6.9   GROUNDS, SHIELDING, AND CONNECTING WIRES

The type of connecting wires used between electrical devices can have a significant impact on the noise level of the signal. Low-level signals of <100 mV are particularly susceptible to errors induced by noise. Some simple rules help to keep noise levels low: (1) Keep the connecting wires as short as possible, (2) keep signal wires away from noise sources, (3) use a wire shield and proper ground, and (4) twist wire pairs along their lengths.

Ground and Ground Loops

The voltage at the end of a wire that is connected to a rod driven far into the soil would likely be at the same voltage level as the earth—a reference datum called zero or *earth ground*. A *ground* is simply a return path to earth. Now suppose that wire is connected from the rod through an electrical box and then through various building routings to the ground plug of an outlet. Would the ground potential at the outlet still be at zero? Probably not. The network of wires that form the return path to earth would likely act as antennae and pick up some voltage potential relative to earth ground. Any instrument grounded at the outlet would be referenced back to this voltage potential, not to earth ground. The point is that an electrical ground does not represent an absolute value. Ground values vary between ground points because the ground returns pass through different equipment or building wiring on their return path to earth. Thus, if a signal is grounded at two points, say at a power source ground and then at an earth ground, the two grounds could be at different voltage levels. The difference between the two ground point voltages is called the *common-mode voltage* (cmv). This can lead to problems, as discussed next.

Ground loops are caused by connecting a signal circuit to two or more grounds that are at different voltage potentials. A ground wire of finite resistance usually carries some current, and so it develops a potential. Thus two separate and different grounds, even two in close proximity, can be at different potential levels. So when these ground points are connected into a circuit the potential difference itself drives a small unwanted current. The result is a ground loop—an electrical interference superimposed onto the signal. A ground loop can manifest itself in various forms, such as a sinusoidal signal or simply a voltage offset.

Ensure that a system, including its sensors and all electrical components, has only one ground point. Figure 6.36 shows one proper connection between a grounded source, such as a transducer, and a measuring device. In this case, the ground is at the transducer. Note that the common (−) line at the measuring device is not grounded because it is grounded at the source. Thus it is referred to as being isolated or as a floating ground device. Alternately, the ground can be at the measuring device rather than at the source. The key point is to ground the circuit and all equipment back to only one point, the common ground point. For stationary devices, an excellent ground is created by driving a conducting rod well into the earth for use as the common ground point. Incidentally, many devices are grounded

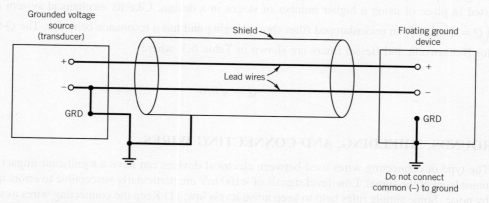

Figure 6.36 Signal grounding arrangements: grounded source with shield and floating signal at the measuring device. It is important to ground the shield at one end only.

through their AC power lines by means of a third prong. Unless the device power supply is isolated from the circuit, this can set up a ground loop relative to other circuit grounds. To create a floating ground, it is necessary to break this ground connection at this point.

Shields

Long wires act as antennas and pick up stray signals from nearby electrical fields. The most common problem is AC line noise. Electrical shields are effective against such noise. A *shield* is a piece of metal foil or wire braid wrapped around the signal wires and connected to ground. The shield intercepts external electrical fields, returning them to ground. A shield ground loop is prevented by grounding the shield to only one point, usually the signal ground at the transducer. Figure 6.36 shows such a shield-to-ground arrangement for a grounded voltage source connected to a floating ground measuring device.

A common source of electrical fields is an AC power supply transformer. A capacitance coupling between the 60- or 50-Hz power supply wires and the signal wires is set up. For example, a 1-pF capacitance superimposes a 40-mV interference on a 1-mV signal.

A different source of noise is from a magnetic field. When a lead wire moves within a magnetic field, a voltage is induced in the signal path. The same effect occurs with a lead wire in proximity to an operating motor. The best prevention is to separate the signal lead wires from such sources. Also, twisting the lead wires together tends to cancel any induced voltage, as the currents through the two wires are in opposite directions; the more twists per meter, the better. A final recourse is a magnetic shield made from a material having a high ferromagnetic permeability.

Connecting Wires

There are several types of wire available. *Single cable* refers to a single length of common wire or wire strand. The conductor is coated for electrical insulation. The wire is readily available and cheap, but should not be used for low-level signal (millivolt level) connections or for connections of more than a few wires. Flat cable is similar but consists of multiple conductors arranged in parallel strips, usually in groups of 4, 9, or 25. Flat cable is commonly used for short connections between adjacent electrical boards, but in such applications the signals are on the order of 1 V or more or are TTL signals. Neither of these two types of wire offers any shielding.

Twisted pairs of wires consist of two electrically insulated conductors twisted about each other along their lengths. Twisted pairs are widely used to interconnect transducers and equipment. The intertwining of the wires offers some immunity to noise with that effectiveness increasing with the number of twists per meter. Cables containing several twisted pairs are available; for example, CAT5 network cable (such as Internet cable) consists of four twisted pairs of wires. Shielded twisted pairs wrap the twisted pairs within a metallic foil shield. CAT6 network cable consists of four twisted pairs with a shield separating its wire pairs. Shielded twisted pairs are one of the best choices for applications requiring low-level signals or high data transfer rates.

Coaxial cable consists of an electrically insulated inner single conductor surrounded within an outer conductor made of stranded wire and a metal foil shield. In general, current flows in one direction along the inner wire and the opposite direction along the outer wire. Any electromagnetic fields generated will cancel. Coaxial cable is often the wire of choice for low-level, high-frequency signals. Signals can be sent over very long distances with little loss. A variation of coaxial cable is *triaxial cable*, which contains two inner conductors. It is used in applications as with twisted pairs but offers superior noise immunity.

Optical cable is widely used to transmit low-level signals over long distances. This cable may contain one or more fiber-optic ribbons within a polystyrene shell. A transmitter converts the low-level voltage signal into infrared light. The light is transmitted through the cable to a receiver, which converts it back to a low-level voltage signal. The cable is virtually noise-free from magnetic fields and harsh environments and is incredibly light and compact compared to its metal wire counterpart.

6.10   SUMMARY

This chapter has focused on classic but basic analog electrical measurement and signal conditioning devices. Devices that are sensitive to current, resistance, and voltage were presented. Signal conditioning devices that can modify or condition a signal include amplifiers, current loops, and filters, and these were also presented. Because all of these devices are widely used, often in combination and often masked within more complicated circuits, the reader should strive to become familiar with the workings of each.

The important issue of loading error was presented. Loading errors, which are due to the act of measurement or to the presence of a sensor, can only be minimized by proper choice of sensor and technique. Loading errors that occur between the connecting stages of a measurement system can be minimized by proper impedance matching. Such impedance matching was discussed, including using the versatile voltage follower circuit for this purpose.

REFERENCES

1. Hickman, I., *Oscilloscopes*, 5th ed., Newnes, Oxford, UK, 2000.
2. Bleuler, E., and R. O. Haxby, *Methods of Experimental Physics*, 2nd ed., Vol. 2, Academic, New York, NY, 1975.
3. Lacanette, K., *A Basic Introduction to Filters: Active, Passive, and Switched-Capacitor Application Note 779*, National Semiconductor, Santa Clara, CA, 1991.
4. Stanley, W. D., G. R. Dougherty, and R. Dougherty, *Digital Signal Processing*, 2nd ed., Reston (a Prentice Hall company), Reston, VA, 1984.
5. DeFatta, D. J., J. Lucas, and W. Hodgkiss, *Digital Signal Processing*, Wiley, New York, 1988.
6. Lam, H. Y., *Analog and Digital Filters: Design and Realization*, Prentice-Hall, Englewood Cliffs, NJ, 1979.
7. Niewizdomski, S., *Filter Handbook—A Practical Design Guide*, CRC Press, Boca Raton, FL, 1989.
8. Bowick, C., C. Ajluni, and J. Blyler, *RF Circuit Design*, 2nd ed., Newnes, Oxford, UK, 2007.

NOMENCLATURE

c	damping coefficient	f_c	filter cutoff frequency (t^{-1})
e	error	f_m	maximum analog signal frequency (t^{-1})
f	frequency (t^{-1})	$h\{E_i(t)\}$	function

k	cascaded filter stage order
$\hat{k}$	unit vector aligned with current flow
l	length (l)
$\hat{n}$	unit vector normal to current loop
q	charge (C)
r	radius, vector (l)
t	time (t)
u_R	uncertainty in resistance R
A	cross-sectional area of a current-carrying loop; attenuation
A_o	operational amplifier open-loop gain
E	voltage (V)
E_1	open circuit potential (V)
E_i	input voltage (V)
E_m	indicated output voltage (V)
E_o	output voltage (V)
F	force (mlt^{-2})
G	amplifier closed-loop gain
$G(s)$	transfer function
I	electric current $(C - t^{-1})$

I_e	effective current $(C - t^{-1})$
I_g	galvanometer or current device current $(C - t^{-1})$
K	static sensitivity
L	inductance (H)
$M(f)$	magnitude ratio at frequency f
N	number of turns in a current-carrying loop
R	resistance (Ω)
R_B	effective bridge resistance (Ω)
R_g	galvanometer or detector resistance (Ω)
R_{eq}	equivalent resistance (Ω)
R_m	meter resistance (Ω)
δR	change in resistance (Ω)
T_μ	torque on a current carrying loop in a magnetic field $(ml^2 t^{-2})$
α	angle
τ	time constant (t)
$\Phi(f)$	phase shift at frequency f
θ	angle

PROBLEMS

6.1 Determine the maximum torque on a current loop having 20 turns, a cross-sectional area of 1 in.^{2}, and experiencing a current of 20 mA. The magnetic field strength is 0.4 Wb/m^2.

6.2 Plot the torque on a current loop as a function of the angle between the normal to the cross-sectional area of the current loop and the magnetic field for

$$N = 100$$
$$A = 5 \times 10^{-4}\ \text{m}^2$$
$$I = 100\ \text{mA}$$
$$B = 1\ \text{Wb/m}^2$$

6.3 A 10-V voltage is applied across a Wheatstone bridge of Figure 6.12 to measure the value of a variable resistance sensor R_1. The resistance in arms 2 and 4 are fixed at $R_2 = R_4 = 250\ \Omega$. Variable resistance arm 3 is adjusted to $R_3 = 300\ \Omega$ to achieve a condition of zero current through the galvanometer. What is the value of R_1? Should we be concerned about loading error? Explain or estimate.

6.4 Determine the loading error as a percentage of the output for the voltage dividing circuit of Figure 6.14, if $R_T = R_1 + R_2$ and $R_1 = kR_T$. The parameters of the circuit are

$$R_T = 500\ \Omega \quad E_i = 10\ \text{V} \quad R_m = 10,000\ \Omega \quad k = 0.5$$

What is the loading error if expressed as a percentage of the full-scale output? Show that the two answers for the loading error are equal when expressed in volts.

6.5 Consider the Wheatstone bridge shown in Figure 6.12. Suppose

$$R_3 = R_4 = 200\ \Omega$$
$$R_2 = \text{variable calibrated resistor}$$
$$R_1 = \text{transducer resistance} = 40x + 100$$

a. When $x = 0$, what is the value of R_2 required to balance the bridge?

b. If the bridge is operated in a balanced condition in order to measure x, determine the relationship between R_2 and x.

6.6   For the voltage-dividing circuit of Figure 6.14, develop and plot the family of solutions for loading error versus $r = R_1/(R_1 + R_2)$ as a function of $(R_1 + R_2)/R_m$. Under what conditions will the loading error in measuring the open circuit potential E_o be less than 7% of the input voltage?

6.7   For the Wheatstone bridge shown in Figure 6.12, R_1 is a sensor whose resistance is related to a measured variable x by the equation $R_1 = 20x^2$. If $R_3 = R_4 = 100\,\Omega$ and the bridge is balanced when $R_2 = 46\,\Omega$, determine x.

6.8   A force sensor has as its output a change in resistance. The sensor forms one leg (R_1) of a basic Wheatstone bridge. The sensor resistance with no force load is $500\,\Omega$, and its static sensitivity is $0.5\,\Omega/N$. Each arm of the bridge is initially $500\,\Omega$

    a.   Determine the bridge output for applied loads of 100, 200, and 350 N. The bridge is operated as a deflection bridge, with an input voltage of 10 V.

    b.   Determine the current flow through the sensor.

    c.   Repeat (a) and (b) with $R_m = 10\,k\Omega$ and $R_s = 600\,\Omega$.

6.9   Consider a deflection bridge arrangement as shown in Figure 6.13. Initially all four arms of the bridge are equal to $120\,\Omega$. The resistance R_1 is a sensor, whose resistance changes with input. The supply voltage to the bridge is 9 V dc. If the resistance R_1 changes such that the bridge output is 0.5 V, determine the value of R_1 and the current flow through R_1.

6.10   A reactance bridge arrangement replaces the resistor in a Wheatstone bridge with a capacitor or inductor. Such a reactance bridge is then excited by an AC voltage. Consider the bridge arrangement shown in Figure 6.37. Show that the balance equations for this bridge are given by $C_2 = C_1R_1/R_2$.

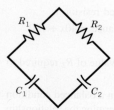

Figure 6.37   Bridge circuit for Problem 6.10.

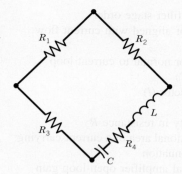

Figure 6.38   Bridge circuit for Problem 6.11.

6.11   Circuits containing inductance- and capacitance-type elements exhibit varying impedance depending on the frequency of the input voltage. Consider the bridge circuit of Figure 6.38. For a capacitor and an inductor connected in a series arrangement, the impedance is a function of frequency, such that a minimum impedance occurs at the resonance frequency $f = 1/2\pi\sqrt{LC}$ where f is the frequency (Hz), L is the inductance (H), and C is the capacitance (F). Design a bridge circuit that could be used to calibrate a frequency source at 500 Hz.

6.12   A Wheatstone bridge initially has resistances equal to $R_1 = 200\,\Omega$, $R_2 = 400\,\Omega$, $R_3 = 500\,\Omega$, and $R_4 = 600\,\Omega$. For an input voltage of 5 V, determine the output voltage at this condition. If R_1 changes to $250\,\Omega$, what is the bridge output?

6.13   Construct a plot of the voltage output of a Wheatstone bridge having all resistances initially equal to $500\,\Omega$, with a voltage input of 10 V for the following cases:

    a. R_1 changes over the range 500–$1{,}000\,\Omega$.

    b. R_1 and R_2 change equally, but in opposite directions, over the range 500–$600\,\Omega$.

    c. R_1 and R_3 change equally over the range 500–$600\,\Omega$.

Discuss the possible implications of these plots for using bridge circuits for measurements with single and multiple transducers connected as arms of the bridge.

6.14   Consider Equation 6.11 that expresses the bridge output voltage in terms of the bridge resistances. Assume that the bridge is in an initially balanced condition so that the output voltage is zero. Then, assuming that the resistances change but that the changes are small enough that we can model the output voltage as

$$dE_o = \sum_{i=1}^{4} \frac{\partial E_o}{\partial R_i} dR_i$$

show that

$$dE_o = E_i \left[\frac{R_2 dR_1 - R_1 dR_2}{(R_1 + R_2)^2} + \frac{R_3 dR_4 - R_4 dR_3}{(R_3 + R_4)^2} \right]$$

6.15 Consider a Wheatstone bridge circuit that has all resistances equal to $100\,\Omega$. One arm of the bridge is a sensor that cannot sustain a power dissipation of more than 0.25 W. What is the maximum value of E_i that can be employed for this bridge?

6.16 Consider the simple potentiometer circuit shown in Figure 6.9 with reference to Figure 6.8. Perform a design-stage uncertainty analysis to determine the minimum uncertainty in measuring a voltage. The following information is available concerning the circuit (assume 95% confidence):

$$E_i = 10 \pm 0.1 \text{ V} \quad R_T = 100 \pm 1\,\Omega \quad R_g = 100\,\Omega$$

$$R_x = \text{reading} \pm 2\%$$

where R_g is the internal resistance of a galvanometer. The null condition of the galvanometer may be assumed to have negligible error. The uncertainty associated with R_x is associated with the reading obtained from the location of the sliding contact. Estimate the uncertainty in the measured value of voltage at nominal values of 2 and 8 V.

6.17 The temperature sensor of Example 6.4 is connected to a Wheatstone bridge using $R_3 = R_4 = 100\,\Omega$. The sensor is immersed into boiling water at 1 atm abs. pressure. What is the resistance of balancing resistor R_2 to balance the bridge? $R_{sensor}(0\,^\circ\text{C}) = 100\,\Omega$.

6.18 A differential pressure transducer transmits its signal using a 4–20-mA current loop, as shown in Figure 6.39. The short-circuit output impedance is $1\,\text{M}\Omega$. The input pressure range is 0 to 20 kPa. To measure this current, the loop is terminated in a $250\,\Omega$ shunt resistor. Accordingly, the voltage across the shunt resistor should vary with pressure between 1 to 5 V. Estimate the loading effect that

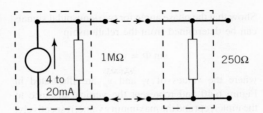

Figure 6.39 Circuit diagram for Problem 6.18.

the shunt has on the voltage measured. Suppose the connecting cables increase the load impedance by $250\,\Omega$. Recalculate the effect.

Problems 6.19 through 6.22 relate to the comparison of two sinusoidal input signals using a dual-trace oscilloscope. A schematic diagram of the measurement system is shown in Figure 6.40, in which one of the signals is assumed to be a reference standard signal, having a known frequency and amplitude. The oscilloscope trace (Input signal versus Reference signal or X versus Y) for these inputs is called a Lissajous diagram. The problems can also be worked in the laboratory.

6.19 Develop the characteristic shape of the Lissajous diagrams for two sinusoidal inputs having the same amplitude, but with the following phase relationships:

a. in phase

b. ± 90 degrees out of phase

c. 180 degrees out of phase

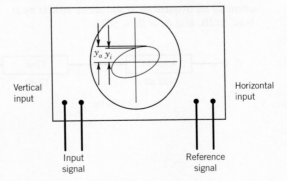

Figure 6.40 Dual-trace oscilloscope for measuring signal characteristics by using Lissajous diagrams.

6.20   Show that the phase angle for two sinusoidal signals can be determined from the relationship

$$\sin \Phi = y_i / y_a$$

where the values of y_i and y_a are illustrated in Figure 6.40 and represent the vertical distance to the y intercept and the maximum y value, respectively.

6.21   Draw a schematic diagram of a measurement system to determine the phase lag resulting from an electronic circuit. Your schematic diagram should include a reference signal, the electronic circuit, and the dual-trace oscilloscope. Discuss the expected result and how a quantitative estimate of the phase lag can be determined.

6.22   Construct Lissajous diagrams for sinusoidal inputs to a dual trace oscilloscope having the following horizontal signal to vertical signal frequency ratios: (a) 1:1, (b) 1:2, (c) 2:1, (d) 1:3, (e) 2:3, (f) 5:2.

These plots can be easily developed using spreadsheet software and plotting a sufficient number of cycles of each of the input signals.

6.23   As a data acquisition anti-alias filter, an RC Butterworth filter is designed with a cutoff frequency of 100 Hz using a single-stage topology. Determine the attenuation of the filtered analog signal at 10, 50, 75, and 200 Hz.

6.24   A three-stage LC Bessel filter with $f_c = 100$ Hz is used to filter an analog signal. Determine the attenuation of the filtered analog signal at 10, 50, 75, and 200 Hz.

6.25   Design a cascading LC low-pass filter with maximally flat magnitude response. Use a passband of 0 to 5 kHz with 5 kHz cutoff frequency and filter to attenuate all frequencies at and above 10 kHz by at least 30 dB. Use $R_s = R_L = 50 \, \Omega$.

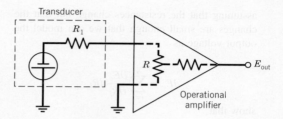

Figure 6.41   pH transducer circuit for Problem 6.28.

6.26   An electrical displacement transducer has an output impedance of $Z = 500 \, \Omega$. Its voltage is input to a voltage measuring device that has an input impedance of $Z = 100 \, \text{k}\Omega$. Estimate the ratio of true voltage to the voltage measured by the measuring device.

6.27   Consider a transducer connected to a voltage measuring device. Plot the ratio E_m / E_1 as the ratio of output impedance of the transducer to input impedance of the measuring device varies from 1:1 to 1:10,000 (just look at each order of magnitude). Comment on the effect that input impedance has on measuring a voltage. What value of input impedance would be acceptable in a quality instrument?

6.28   The pH meter of Figure 6.41 consists of a transducer of glass containing paired electrodes. This transducer can produce voltage potentials up to 1 V with an internal output impedance of up to $10^9 \, \Omega$. If the signal is to be conditioned, estimate the minimum input impedance required to keep loading error below 0.1% for the op-amp shown.

6.29   Consider the circuit of Figure 6.42, which consists of an amplifier providing 32-dB gain, followed by a filter, which causes an attenuation of 12 dB at the

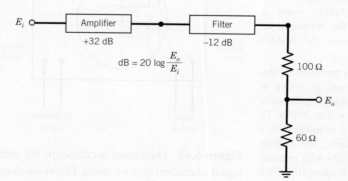

Figure 6.42   Multistage signal path for Problem 6.29.

frequency of interest, followed by a voltage divider. Find E_o/E_i across the 60-Ω resistor.

6.30 What internal impedance is needed for the bridge circuit of Figure 6.13 for the loading error to be under 1% if the bridge resistances change by

$$\delta R_1 = +40\,\Omega \quad \delta R_2 = -40\,\Omega \quad \delta R_3 = +40\,\Omega$$

$$\delta R_4 = -40\,\Omega$$

6.31 The input to a subwoofer loudspeaker is to pass through a passive low-pass Butterworth filter having a cutoff frequency of 100 Hz. Specify a filter that meets the following specifications: At 50 Hz, the signal magnitude ratio should be at least at 0.95; at 200 Hz, the magnitude ratio should be no more than 0.01. The sensor and load resistances are 10 Ω. You will need to specify the number of even stages and the values for the components.

6.32 For the application in Problem 6.31, repeat using a Linkwitz–Riley topology to specify the reactive element values in both a fourth-order low-pass filter and high-pass filter for 10 Ω speakers.

6.33 A high-pass Butterworth filter with cutoff frequency of 5,000 Hz is used as a crossover to a high-frequency loudspeaker. Specify a filter such that at 2,500 Hz, the attenuation should be at least −20 dB. The source and load resistances are 10 Ω. You need to specify the number of stages and the values for the components.

6.34 Design an active-RC low-pass first-order Butterworth filter for a cutoff frequency of 10 kHz, and a passband gain of 20. Use a 741 op-amp and 0.1-μF capacitors for your design. The program *Lowpass Butterworth Active Filter* can be used.

6.35 Design an active-RC first-order high-pass filter for a cutoff frequency of 10 kHz and passband gain of 10. Use a 741 op-amp.

6.36 Use a laboratory oscilloscope with a function generator (or use the LabView program *Two_Channel_Oscilloscope*, which is provided as a separate source code file in the companion software download folder) to explore the workings of an actual oscilloscope. Vary between channel A and B.

a. Characterize each signal by its waveform (i.e., square or sine). Determine the amplitude and the period of both signals that you apply.

b. Vary the oscilloscope signal gain (volts/division). Explain the effect on the displayed signals.

c. Vary the oscilloscope time base (ms/div). Explain the effect on the displayed signals.

d. Make sure channel B is active. With the trigger source set to channel B, vary the trigger level dial. Explain the function of the trigger level. Why do the waveforms disappear at some settings of the trigger?

6.37 Use the LabView program *Butterworth_Filters* to explore the behavior of low-pass, bandpass, and high-pass Butterworth filters chosen for their flat amplitude passband. In this program, the signal $y(t) = 2 \sin 2\pi f t$ is passed through the filter, and it is the filtered signal $y^*(t) = B \sin[2\pi f t + \varphi(f)]$ that is displayed. The single-tone results show the effect at the input frequency f. The amplitude spectrum results show the effect over the full-frequency band of interest. Describe the amplitude behavior of the filtered signal as the single input frequency is increased. Pay particular attention near the filter cutoff frequencies, which are set at 300 and 500 Hz.

6.38 Use the LabView program *Bessel_Filters* to explore the behavior of low-pass, bandpass, and high-pass Bessel filters. Using the information from Problem 6.37, describe the amplitude behavior of the filtered signal as a single-input frequency is increased from 1 to 1,000 Hz.

6.39 The Labview program *Filtering_Noise* demonstrates the effect of using a low-pass Butterworth filter to treat a signal containing high-frequency noise. Set the signal frequency at 5 Hz. Set the cutoff frequency at 10 Hz. Discuss the behavior of the filter as the number of stages is increased. Then incrementally decrease the cutoff frequency and discuss the behavior. Tie your discussion back to filter roll-off. What happens when the cutoff frequency is less than the signal frequency? Why?

6.40 Labview program *Monostable Circuit* provides a simulation of a monostable integrated circuit based on the type 555 op-amp. The trigger controls the simulation. The user can vary the "on" time and values for R and C. Discuss the output vs. time results for a simulation. Vary R and C to create a 1-s square wave.

6.41 Labview program *741 Opamp* simulates op-amp gain characteristics. Determine if it is noninverting or inverting and explain. For an input voltage of 1 V, find a resistor combination for a gain of 5. What is this op-amp's maximum output? Repeat for a gain of 0.5. Labview programs *Inverting_Opamp* and *Non-inverting_Opamp* are similar and are also available to complete this problem.

6.42   Consider a Wheatstone bridge that has been balanced within the sensitivity of the meter such that I_g is smaller than the smallest current detectable. This current flow due to the meter resolution is a loading error with an associated (systematic) uncertainty in the measured resistance, u_R. Using a basic analysis of the circuit with a current flow through the galvanometer, show that

$$\frac{u_R}{R_1} = \frac{I_g(R_1 + R_g)}{E_i}$$

where R_g is the internal resistance of the galvanometer. Alternatively, if a current-sensing digital voltmeter is used, then the input bias current from this meter remains and a small offset voltage is present as a loading error.

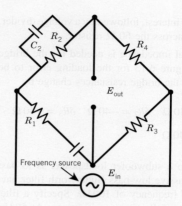

Figure 6.43   Wien Bridge Circuit.

6.43   Consider the bridge circuit shown in Figure 6.43. This circuit is termed a Wien Bridge Circuit, and can be used to measure frequency. Show that under balanced conditions the frequency in Hz is given by

$$f = \frac{1}{2\pi\sqrt{R_1 R_2 C_1 C_2}}$$

Chapter 7

Sampling, Digital Devices, and Data Acquisition

7.1 INTRODUCTION

Integrating analog electrical transducers with digital devices is cost-effective and commonplace. Digital microprocessors are central to most controllers and data acquisition systems today. There are many advantages to digital data acquisition and control, including the efficient handling and rapid processing of large amounts of data and varying degrees of artificial intelligence. But there are fundamental differences between analog and digital systems that impose some limitations and liabilities upon the engineer. As pointed out in Chapter 2, the most important difference is that analog signals are continuous in both amplitude and time, whereas digital signals are discrete (non-continuous) in both amplitude and time. It is not immediately obvious how such a discrete signal, which is a series of numbers, can be used to represent the continuous behavior of an analog process variable.

This chapter begins with an introduction to the fundamentals of *sampling*, the process by which continuous signals are made discrete. The major pitfalls are explored. Criteria are presented that circumvent the loss or the misinterpretation of signal information while undergoing the sampling process. We show how a discrete series of data can actually contain all the information available in a continuous signal or at least provide a very good approximation.

The discussion moves on to the devices most often involved in analog and digital systems. Analog devices interface with digital devices through an analog-to-digital (A/D) converter. The reverse process of a digital device interfacing with an analog device occurs through a digital-to-analog (D/A) converter. A digital device interfaces with another digital device through a digital input–output (I/O) port. These interfaces are the major components of computer-based data acquisition systems. Necessary components and the basic layout of these systems are introduced, and standard methods for communication between digital devices are presented. Digital image acquisition and processing are introduced because of their increasing importance in a wide variety of applications ranging from quality assurance inspection to high-speed imaging.

Upon completion of this chapter, the reader will be able to

- Describe analog, discrete time, and digital signals
- Properly choose a sample rate for data acquisition to eliminate aliasing
- Clearly describe the functioning of A/D and D/A converters

- Define and calculate quantization errors
- Understand how instruments interface with data acquisition systems
- Perform basic image processing on digital images

7.2 SAMPLING CONCEPTS

Consider an analog signal and its discrete time series representation in Figure 7.1. The information contained in the analog and discrete representations may appear to be quite different. However, the important analog signal information concerning amplitude and frequency can be well represented by such a discrete series—just how well represented depends on

1. The frequency content of the analog signal
2. The size of the time increment between each discrete number
3. The total sample period of the measurement

Chapter 2 discussed how a continuous dynamic signal could be represented by a Fourier series. The discrete Fourier transform (DFT) was also introduced as a method for reconstructing a dynamic signal from a discrete set of data. The DFT conveys all the information needed to reconstruct the Fourier series of a continuous dynamic signal from a representative discrete time series. Hence Fourier analysis provides certain guidelines for sampling continuous data. This section discusses the concept of converting a continuous analog signal into an equivalent discrete time series. Extended discussions on this subject of signal analysis can be found in many texts (1–3).

Sample Rate

Just how frequently must a time-dependent signal be measured to determine its frequency content? Consider Figure 7.2a, in which the magnitude variation of a 10-Hz sine wave is plotted versus time

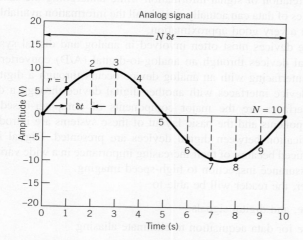

Figure 7.1 Analog and discrete representations of a time-varying signal.

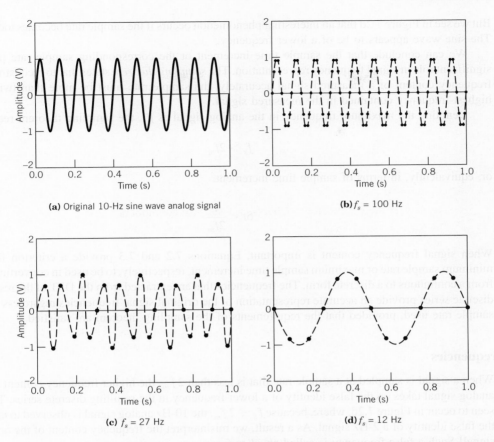

Figure 7.2 The effect of sample rate on signal frequency and amplitude interpretation.

over time period t_f. Suppose this sine wave is measured repeatedly at successive sample time increments δt. This corresponds to measuring the signal with a sample frequency (in samples/second) or sample rate (in Hz) of

$$f_s = 1/\delta t \qquad (7.1)$$

For this discussion, we assume that the signal measurement occurs at a constant sample rate. For each measurement, the amplitude of the sine wave is converted into a number. For comparison, in Figures 7.2b–d the resulting series versus time plots are given when the signal is measured using sample time increments (or the equivalent sample rates) of (b) 0.010 second ($f_s = 100$ Hz), (c) 0.037 second ($f_s = 27$ Hz), and (d) 0.083 second ($f_s = 12$ Hz). We can see that the sample rate has a significant effect on our perception and reconstruction of a continuous analog signal in the time domain. As sample rate decreases, the amount of information per unit time describing the signal decreases. In Figures 7.2b,c we can still discern the 10-Hz frequency content of the original signal.

But we see in Figure 7.2d that an interesting phenomenon occurs if the sample rate becomes too slow: The sine wave appears to be of a lower frequency.

We can conclude that the sample time increment or the corresponding sample rate plays a significant role in signal frequency representation. The *sampling theorem* states that to reconstruct the frequency content of a measured signal accurately, the sample rate must be more than twice the highest frequency contained in the measured signal.

Denoting the maximum frequency in the analog signal as f_m, the sampling theorem requires

$$f_s > 2f_m \tag{7.2}$$

or, equivalently, in terms of sample time increment,

$$\delta t < \frac{1}{2f_m} \tag{7.3}$$

When signal frequency content is important, Equations 7.2 and 7.3 provide a criterion for the minimum sample rate or maximum sample time increment, respectively, to be used in converting data from a continuous to a discrete form. The frequencies that are extracted from the DFT of the resulting discrete series provide an accurate representation of the original signal frequencies regardless of the sample rate used, provided that the requirements of the sampling theorem are satisfied.

Alias Frequencies

When a signal is sampled at a sample rate that is less than $2f_m$, the higher frequency content of the analog signal takes on the false identity of a lower frequency in the resulting discrete series. This is seen to occur in Figure 7.2d, where, because $f_s < 2f_m$, the 10-Hz analog signal is observed to take on the false identity of a 2-Hz signal. As a result, we misinterpret the frequency content of the original signal! Such a false frequency is called an *alias frequency*.

The alias phenomenon is an inherent consequence of a discrete sampling process. To illustrate this, consider a simple periodic signal:

$$y(t) = C \sin[2\pi f t + \phi(f)] \tag{7.4}$$

Suppose $y(t)$ is measured with a sample time increment of δt, so that its discrete time signal is given by

$$\{y(r\delta t)\} = C \sin[2\pi f r \delta t + \phi(f)] \quad r = 0, 1, \ldots, N - 1 \tag{7.5}$$

Now using the identity $\sin x = \sin(x + 2\pi q)$, where q is any integer, we rewrite $\{y(r\delta t)\}$ as

$$\begin{aligned} C \sin[2\pi f r \delta t + \phi(f)] &= C \sin[2\pi f r \delta t + 2\pi q + \phi(f)] \\ &= C \sin\left[2\pi\left(f + \frac{m}{\delta t}\right) r \delta t + \phi(f)\right] \end{aligned} \tag{7.6}$$

where $m = 0, 1, 2, \ldots$ (and hence, $mr = q$ is an integer). This shows that for any value of δt, the frequencies of f and $f + m/\delta t$ are indistinguishable in a sampled discrete series. Hence, all frequencies given by $f + m/\delta t$ are the alias frequencies of f. However, by adherence to the sampling theorem criterion of either Equation 7.2 or 7.3, all $m \geq 1$ are eliminated from the sampled signal, and thus this ambiguity between frequencies is avoided.

This same discussion applies equally to all time-dependent waveforms. This is shown by examining the general Fourier series that can be used to represent such signals. A discrete series such as

$$\{y(r\delta t)\} = \sum_{n=1}^{\infty} C_n \sin \left[2\pi n f r \delta t + \phi_n(f) \right] \tag{7.7}$$

for $r = 0, 1, \ldots, N - 1$ can be rewritten as

$$\{y(r\delta t)\} = \sum_{n=1}^{\infty} C_n \sin \left[2\pi \left(nf + \frac{nm}{\delta t} \right) r \delta t + \phi_n(f) \right] \tag{7.8}$$

and so it displays the same aliasing phenomenon shown in Equation 7.6.

In general, the Nyquist frequency defined by

$$f_N = \frac{f_s}{2} = \frac{1}{2\delta t} \tag{7.9}$$

represents a folding point for the aliasing phenomenon. All frequency content in the analog signal that is at frequencies above f_N appears as alias frequencies of less than f_N in the sampled signal. That is, such frequencies are folded back and superimposed on the signal as lower frequencies. The aliasing phenomenon occurs in spatial sampling [i.e., $y(x)$ at sampling intervals δx] as well, and the above discussion applies equally.

The alias frequency f_a, can be predicted from the folding diagram of Figure 7.3. Here the original input frequency axis is folded back over itself at the folding point of f_N and again for each of its

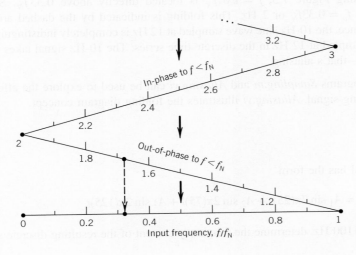

Figure 7.3 The folding diagram for alias frequencies.

harmonics mf_N, where $m = 1, 2, \ldots$ For example, and as noted by the solid arrows in Figure 7.3, the frequencies $f = 0.5\,f_N, 1.5\,f_N, 2.5\,f_N, \ldots$ that may be present in the original input signal all appear as the frequency $0.5\,f_N$ in the discrete series $y(r\delta t)$. Use of the folding diagram is illustrated further in Example 7.1.

How does one avoid this alias phenomenon when sampling a signal of unknown frequency content? The preferred option is to choose a sample rate based on the maximum frequency of interest while adhering to the criterion of Equation 7.2 and to pass the signal through a low-pass filter prior to sampling. Based on Equation 7.9, the filter should be set to remove signal content at and above f_N. This type of filter is called an *anti-aliasing filter*. Another option is to assign a sample rate so large that the measured signal does not have significant amplitude content above f_N. In this way, you choose f_s high enough such that the amplitudes of any frequency content above f_N are small compared to those below f_N. With this option, the resulting effects of aliasing are minimized but not eliminated.

Example 7.1

A 10-Hz sine wave is sampled at 12 Hz. Compute the maximum frequency that can be represented in the resulting discrete signal. Compute the alias frequency.

> **KNOWN** $f = 10\,\text{Hz}$
> $f_s = 12\,\text{Hz}$

ASSUMPTION Constant sample rate

FIND f_N and f_a

SOLUTION The Nyquist frequency f_N sets the maximum frequency that can be represented in a resulting data set. Using Equation 7.9 with $f_s = 12\,\text{Hz}$ gives the Nyquist frequency $f_N = 6\,\text{Hz}$. This is the maximum frequency that can be sampled correctly. So the measured frequency is in error.

All frequency content in the analog signal above f_N appears as an alias frequency f_a between 0 and f_N. Because $f/f_N \approx 1.67$, f is folded back and appears in the sampled signal as a frequency between 0 and f_N. Reading Figure 7.3, $f = 1.67 f_N$ is located directly above $0.33 f_N$. So f is folded back to appear as $f_a = 0.33 f_N$ or 2 Hz. This folding is indicated by the dashed arrow in Figure 7.3. As a consequence, the 10-Hz sine wave sampled at 12 Hz is completely indistinguishable from a 2-Hz sine wave sampled at 12 Hz in the discrete time series. The 10-Hz signal takes on the identity of a 2-Hz signal—that's aliasing.

COMMENT The programs *Sampling.m* and *Aliasing.vi* can be used to explore the effects of sample rate on the resulting signal. *Aliasing.vi* illustrates the folding diagram concept.

Example 7.2

A complex periodic signal has the form

$$y(t) = A_1 \sin 2\pi(25)t + A_2 \sin 2\pi(75)t + A_3 \sin 2\pi(125)t$$

If the signal is sampled at 100 Hz, determine the frequency content of the resulting discrete series.

KNOWN $f_s = 100\ \text{Hz}$
$f_1 = 25\ \text{Hz}\ f_2 = 75\ \text{Hz}\ f_3 = 125\ \text{Hz}$

ASSUMPTION   Constant sample rate

FIND   The alias frequencies f_{a_1}, f_{a_2}, and f_{a_3} and discrete set $\{y(r\delta t)\}$

SOLUTION   From Equation 7.9 for a sample rate of 100 Hz, the Nyquist frequency is 50 Hz. All frequency content in the analog signal above f_N takes on an alias frequency between 0 and f_N in the sampled data series. With $f_1 = 0.5 f_N, f_2 = 1.5 f_N$, and $f_3 = 2.5 f_N$, we can use Figure 7.3 to determine the respective alias frequencies.

i	f_i	f_{a_i}
1	25 Hz	25 Hz
2	75 Hz	25 Hz
3	125 Hz	25 Hz

In the resulting series, $f_{a_1} = f_{a_2} = f_{a_3} = 25\ \text{Hz}$. Because of the aliasing phenomenon, the 75- and 125-Hz components would be completely indistinguishable from the 25-Hz component. The discrete series would be described by

$$y(r\delta t) = (B_1 + B_2 + B_3)\sin 2\pi(25)r\delta f \quad r = 0, 1, 2, \ldots$$

where $B_1 = A_1, B_2 = -A_2$, and $B_3 = A_3$. Note from Figure 7.3 that when the original frequency component is out of phase with the alias frequency, the corresponding amplitude is negative.

COMMENT   Example 7.2 illustrates the potential for misrepresenting a signal with improper sampling. With signals for which the frequency content is not known before their measurement, sample frequencies should be increased according to the following scheme:

- Sample the input signal at increasing sample rates using a fixed total sample time, and examine the time plots for each signal. Look for changes in the shape of the waveform (Fig. 7.2).
- Compute the amplitude spectrum for each signal at increasing sample rates and compare the resulting frequency content.
- Always use (anti-aliasing) analog filters set at the Nyquist frequency.

Amplitude Ambiguity

For simple and complex periodic waveforms, the DFT of the sampled discrete time signal remains unaltered by a change in the total sample period $N\delta t$ provided that (1) the total sample period remains an integer multiple of the fundamental period T_1 of the measured continuous waveform (that is, $mT_1 = N\delta t$ where m is an integer) and (2) the sample frequency meets the sampling theorem criterion. If both criteria are met, the amplitudes associated with each frequency in the periodic signal, the spectral amplitudes, are accurately represented by the DFT. This means that an original periodic waveform can be completely reconstructed from a discrete time series regardless of the sample time increment used. The total sample period defines the frequency resolution of the DFT:

$$\delta f = \frac{1}{N\delta t} = \frac{f_s}{N} \tag{7.10}$$

The frequency resolution plays a crucial role in the reconstruction of the signal amplitudes, as will be noted.

An important difficulty arises when $N\delta t$ is not coincident with an integer multiple of the fundamental period of $y(t)$: The resulting DFT cannot *exactly* represent the spectral amplitudes of the sampled continuous waveform. This is exaggerated when $N\delta t$ represents only a relatively few fundamental periods of the signal. The problem is brought on by the truncation of one complete cycle of the signal (Fig. 7.4) and from spectral resolution, because the associated fundamental frequency and its harmonics are not coincident with a discrete frequency ($k\delta f$) of the DFT. However, this error decreases either as the value of $N\delta t$ more closely approximates an exact integer multiple of T_1 or as f_s becomes very large relative to f_m.

This situation is illustrated in Figure 7.4, which compares the amplitude spectrum resulting from sampling the signal $y(t) = 10 \cos 628t$ over different sample periods. Two sample periods of 0.0256 and 0.1024 second were used with $\delta t = 0.1$ ms, and a third sample period of 0.08 s with $\delta t = 0.3125$ ms.[1] These sample periods provide for DFT frequency resolution δf of about 39, 9.8, and 12.5 Hz, respectively.

Leakage

The two spectra shown in Figures 7.4a,b display a spike near the correct 100 Hz with surrounding noise spikes, known as *leakage*, at adjacent frequencies. Note that the original signal $y(t)$ cannot be exactly reconstructed from either of these spectra. The spectrum in Figure 7.4c has been constructed using a longer sample time increment (i.e., lower sample rate) than that of the spectra in Figures 7.4a and with fewer data points than the spectrum of Figure 7.4b. Yet the spectrum of Figure 7.4c provides an exact representation of $y(t)$. The N and δt combination used has reduced leakage to zero, which maximizes the amplitude at 100 Hz. Its sample period corresponds to exactly eight periods of $y(t)$, and the frequency of 100 Hz corresponds to the center frequency of the eighth frequency interval of the DFT. As seen in Figure 7.4, the loss of accuracy in a DFT representation occurs in the form of amplitude "leakage" to adjacent frequencies. To the DFT, the truncated segment of the sampled signal appears as an aperiodic signal. The DFT returns the correct spectral amplitudes for both the periodic and aperiodic signal portions. But as a result, the spectral amplitudes for any truncated segment are superimposed onto portions of the spectrum adjacent to the correct frequency. Recall how the amplitude varied in Figure 7.2c. This is how leakage affects the time domain reconstruction. By varying the sample period or its equivalent, the DFT resolution, one can minimize leakage and control the accuracy of the spectral amplitudes.

If $y(t)$ is an aperiodic or nondeterministic waveform, there may not be a fundamental period. In such situations, one controls the accuracy of the spectral amplitudes by varying the DFT resolution δf to minimize leakage. As δf tends toward zero, leakage decreases.

---

[1] Many DFT (including fast Fourier transform [FFT]) algorithms require that $N = 2^M$, where M is an integer. This affects the selection of δt.

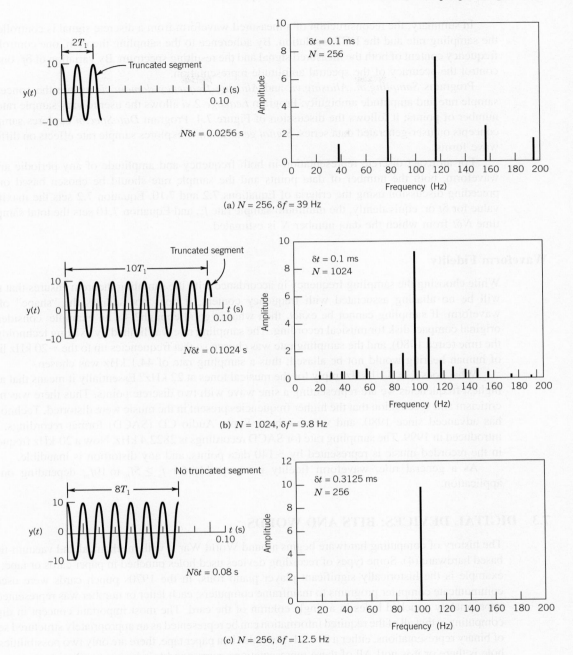

Figure 7.4 Amplitude spectra for $y(t)$. **(a)** $\delta f = 39$ Hz. **(b)** $\delta f = 9.8$ Hz. **(c)** $\delta f = 12.5$ Hz.

In summary, the reconstruction of a measured waveform from a discrete signal is controlled by the sampling rate and the DFT resolution. By adherence to the sampling theorem, one controls the frequency content of both the measured signal and the resulting spectrum. By variation of δf, one can control the accuracy of the spectral amplitude representation.

Programs *Sampling.m*, *Aliasing.vi*, and *Aliasing frequency domain.vi* explore the concept of sample rate and amplitude ambiguity. Program *Leakage.2.vi* allows the user to vary sample rate and number of points; it follows the discussion of Figure 7.4. Program *DataSpec.m* explores sampling concepts on user-generated data series. *Signal generation.vi* explores sample rate effects on different wave forms.

For an exact discrete representation in both frequency and amplitude of any periodic analog waveform, both the number of data points and the sample rate should be chosen based on the preceding discussion using the criteria of Equations 7.2 and 7.10. Equation 7.2 sets the maximum value for δt or, equivalently, the minimum sample rate f_s, and Equation 7.10 sets the total sampling time $N\delta t$ from which the data number N is estimated.

Waveform Fidelity

While choosing the sampling frequency in accordance with the sampling theorem ensures that there will be no aliasing associated with frequency content, it does not address the "shape" of the waveform. If sampling cannot be exact, then what is good enough? As an example, consider the original compact disk for musical recording. The sampling rate was limited by existing technology at the time (circa 1980), and the sampling rate was chosen so that frequencies up to the $\sim$20 kHz limits of human hearing would not be aliased; thus a sampling rate of 44.1 kHz was chosen.

What does $f_s = 44.1$ kHz imply for the musical tones at 22 kHz? Essentially it means that at the highest frequencies we are representing a sine wave with two discrete points. Thus there was much criticism and speculation that the higher frequencies present in the music were distorted. Technology has advanced since 1980, and we now have Super Audio CD (SACD) format recordings, first introduced in 1999. The sampling rate for SACD recordings is 2822.4 kHz. Now a 20-kHz frequency in the recorded music is represented by $\sim$140 data points, and any distortion is inaudible.

As a general rule, waveform fidelity is acceptable for $f_s \geq 5f_m$ to $10f_m$ depending on the application.

7.3 DIGITAL DEVICES: BITS AND WORDS

The history of computing hardware begins around World War II with mechanical and vacuum-tube–based hardware (4). Some types of recording devices used holes punched in paper cards or tape; one example is the historically significant player piano rolls. In the 1970s punch cards were used to communicate computer programs to mainframe computers; each letter or number was represented by appropriately punched holes in a single column of the card. The most important concept in digital computing is that all of the required information can be represented as an appropriately structured series of binary representations, either a 0 or a 1. For holes in a paper tape, there are only two possibilities—a hole is there or it is not! All of these representations correspond to a binary numbering system.

Digital systems use some variation of a binary numbering system both to represent and transmit signal information. Binary systems use the binary digit or *bit* as the smallest unit of information. A bit is a single digit, either a 1 or a 0. Bits are like electrical switches and are used to convey both logical and numerical information. From a logic standpoint, the 1 and 0 are represented by the on and off switch settings. By appropriate action a bit can be reset to either on or off, thereby permitting control and logic actions. By combining bits it is possible to define integer numbers greater than 1 or 0. A numerical *word* is an ordered sequence of bits, with a *byte* being a specific sequence usually consisting of 8 bits. Computer memory is usually byte-addressed. The memory location where numerical information is stored is known as a *register*, with each register assigned its own *address*.

A combination of M bits can be arranged to represent 2^M different words. For example, a combination of 2 bits can represent 2^2, or four, possible combinations of bit arrangements: 00, 01, 10, or 11. We can alternatively reset this 2-bit word to produce these four different arrangements that represent the decimal: base 10, integer numbers 0, 1, 2, or 3, respectively. So an 8-bit word can represent the numbers 0 through 255; a 16-bit word can represent 0 through 65,535.

The numerical value for a word is computed by moving by bit from right to left. From the right side, each successive bit increases the value of the word by a unit, a 2, a 4, an 8, and so forth through the progression of 2^M, provided that the bit is in its on (value of 1) position; otherwise, the particular bit increases the value of the word by zero. A weighting scheme of an M-bit word is given as follows:

Bit $M-1$	. . .	Bit 3	Bit 2	Bit1	Bit 0
2^{M-1}	. . .	2^3	2^2	2^1	2^0
Decimal value	. . .	8	4	2	1

Using this scheme, bit $M-1$ is known as the most significant bit (MSB), because its contribution to the numerical level of the word is the largest relative to the other bits. Bit 0 is known as the least significant bit (LSB). In hexadecimal (base 16), a sequence of four bits is used to represent a single digit so that the 2^4 different digits (0 through 9 plus A through F) form the alphabet for creating a word.

Several binary codes are in common use. In 4-bit straight binary, 0000 is equivalent to analog 0, and 1111 is equivalent to decimal 15 or hexadecimal F. Straight binary is considered to be a unipolar code, because all numbers are of like sign. Bipolar codes allow for sign changes. *Offset binary* is a bipolar code with 0000 equal to -7 and 1111 equal to $+7$. The comparison between these two codes and several others is shown in Table 7.1. In offset binary, there are two zero levels, with true zero being centered between them. Note how the MSB is used as a sign bit in a bipolar code.

The *ones-* and *twos-complement binary codes* shown in Table 7.1 are both bipolar. They differ in the positioning of the zero, with the twos-complement code containing one more negative level (-8) than positive ($+7$). Twos-complement code is widely used in digital computers. The bipolar AVPS code uses the MSB as the sign bit but uses the lower bits in a straight binary representation; for example, 1100 and 0100 represent -4 and $+4$, respectively.

A code often used for digital decimal readouts and for communication between digital instruments is binary coded decimal (BCD). In this code, each individual digit of a decimal number is represented separately by its equivalent value coded in straight binary. For example, the three digits of the base 10 number 532_{10} are represented by the BCD number 0101 0011 0010 (i.e., binary 5, binary 3, binary 2). In this code each binary number need span but a single decade—that is, have values from 0 to 9. Hence, in digital readouts, when the limits of the lowest decade are exceeded— that is, 9 goes to 10—the code just carries into the next higher decade.

Table 7.1   Binary Codes (Example: 4-Bit Words)

Bits	(Hex) Straight	Offset	Twos Complement	Ones Complement	AVPS[a]
0000	0	−7	+0	+0	−0
0001	1	−6	+1	+1	−1
0010	2	−5	+2	+2	−2
0011	3	−4	+3	+3	−3
0100	4	−3	+4	+4	−4
0101	5	−2	+5	+5	−5
0110	6	−1	+6	+6	−6
0111	7	−0	+7	+7	−7
1000	8	+0	−8	−7	+0
1001	9	+1	−7	−6	+1
1010	A	+2	−6	−5	+2
1011	B	+3	−5	−4	+3
1100	C	+4	−4	−3	+4
1101	D	+5	−3	−2	+5
1110	E	+6	−2	−1	+6
1111	F	+7	−1	−0	+7

[a]AVPS, absolute-value plus-sign code.

Example 7.3

A 4-bit register contains the binary word 0101. Convert this to its decimal equivalent assuming a straight binary code.

KNOWN   4-bit register

FIND   Convert to base 10

ASSUMPTION   Straight binary code

SOLUTION   The content of the 4-bit register is the binary word 0101. This represents

$$0 \times 2^3 + 1 \times 2^2 + 0 \times 2^1 + 1 \times 2^0 = 5$$

The equivalent of 0101 in decimal is 5.

7.4   TRANSMITTING DIGITAL NUMBERS: HIGH AND LOW SIGNALS

Electrical devices transmit binary code by using differing voltage levels. Because a bit can have a value of 1 or 0, the presence of a particular voltage (call it HIGH) at a junction could be used to represent a 1-bit value, whereas a different voltage (LOW) could represent a 0. Most simply, this

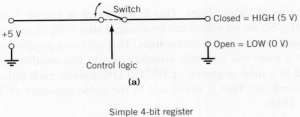

(a)

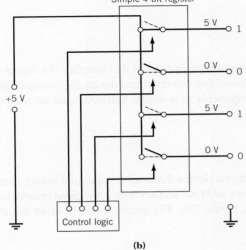

(b)

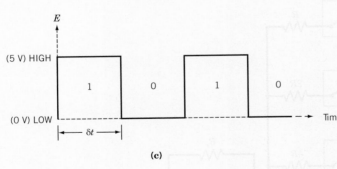

(c)

Figure 7.5 Methods for transmitting digital information. **(a)** Simple on/off (1 or 0) switch. **(b)** Four-bit register transmitting 1010 in parallel. **(c)** Serial transmission of 1010.

voltage can be affected by use of an open or closed switch using a flip-flop (see Chapter 6) such as depicted in Figure 7.5a.

For example, a signal method common with many digital devices is a +5 V form of TTL (true transistor logic), which uses a nominal +5 V HIGH/0 V LOW scheme for representing the value of a bit. To avoid any ambiguity from voltage fluctuations and provide for more precision, a voltage level between +2 and +5.5 V is taken as HIGH and a bit value of 1, whereas a voltage between −0.6 and +0.8 V is taken as LOW or a 0-bit value. But this scheme is not unique.

Binary numbers can be formed through a combination of HIGH and LOW voltages. Several switches can be grouped in parallel to form a register. Such an *M*-bit register forms a number based on the value of the voltages at its M output lines. This is illustrated in Figure 7.5b for a 4-bit register forming the number 1010. So the bit values can be changed simply by changing the opened or closed state of the switches that connect to the output lines. This defines a parallel form in which all of the bits needed to form a word are available simultaneously. But another form is serial, whereby the bits are separated in a time sequence of HIGH/LOW pulses, each pulse lasting for only one predefined time duration, δt. This is illustrated in the pulse sequence of Figure 7.5c, which also forms the number 1010.

7.5 VOLTAGE MEASUREMENTS

Digital measurement devices consist of several components (5) that interface the digital device with the analog world. In particular, the digital-to-analog converter and the analog-to-digital converter are discussed next. They form the major components of a digital voltmeter and an analog-to-digital/digital-to-analog data acquisition system.

Digital-to-Analog Converter

A digital-to-analog converter is an M-bit digital device that converts a digital binary number into an analog voltage (5). One possible scheme uses an M-bit register with a weighted resistor ladder circuit and operational amplifier, as depicted in Figure 7.6. The circuit consists of M binary weighted

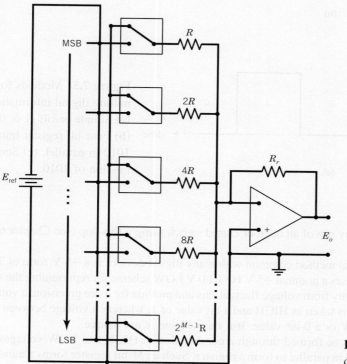

Figure 7.6 Digital-to-analog converter.

resistors having a common summing point. The resistor associated with the register MSB has a value of R. At each successive bit the resistor value is doubled, so that the resistor associated with the LSB has a value of ($2^{M-1}R$). The circuit output is a current, I, given by

$$I = E_{ref} \sum_{m=1}^{M} \frac{c_m}{2^{m-1}R} \qquad (7.11)$$

The values for c_m are either 0 or 1 depending on the associated mth bit value of the register that controls the switch setting. The output voltage from the amplifier is

$$E_o = IR_r \qquad (7.12)$$

Note that this is equivalent to an operation in which an M-bit D/A converter would compare the magnitude of the actual input binary number, X, contained within its register to the largest possible number 2^M. This ratio determines E_o from

$$E_o = \frac{X}{2^M} \qquad (7.13)$$

The D/A converter will have both digital and analog specifications, the latter expressed in terms of its full-scale analog voltage range output (E_{FSR}). Typical values for E_{FSR} are 0 to 10 V (unipolar) and ±5 V (bipolar) and for M are 8, 12, 16, and 18 bits.

Analog-to-Digital Converter

An analog-to-digital converter converts an analog voltage value into a binary number through a process called *quantization*. The conversion is discrete, taking place one number at a time.

The A/D converter is a hybrid device having both an analog side and a digital side. The analog side is specified in terms of a full-scale voltage range, E_{FSR}. The E_{FSR} defines the voltage span over which the device operates. The digital side is specified in terms of the number of bits of its register. An M-bit A/D converter outputs an M-bit binary number. It can represent 2^M different binary numbers. For example, a typical 8-bit A/D converter with an $E_{FSR} = 10$ V would be able to represent analog voltages in the range between 0 and 10 V (unipolar) or the range between ±5 V (bipolar) with $2^8 = 256$ different binary values. Principal considerations in selecting a type of A/D converter include resolution, voltage range, and conversion speed (5).

Resolution

The A/D converter resolution is defined in terms of the smallest voltage increment that causes a bit change. Resolution is specified in volts per bit and is determined by

$$Q = E_{FSR}/2^M \qquad (7.14)$$

Primary sources of error intrinsic to any A/D converter are the resolution and associated quantization error, saturation error, and conversion error.

Quantization Error

An A/D converter has a finite resolution, and any input voltage that falls between two adjacent output codes results in an error. This error is referred to as the *quantization error* e_Q. It behaves as noise imposed on the digital signal.

Consider the basic analog input to digital output relationship for a 0- to 4-V, 2-bit A/D converter as represented in Figure 7.7. This scheme is common in digital readout devices. From Equation 7.14, the resolution of this device is $Q = 1$ V/bit. For such a converter, an input voltage of 0.0 V would result in the same binary output of 00 as would an input of 0.9 V. Yet an input of 1.1 V results in an output of 01. Clearly, the output error is directly related to the resolution. In this encoding scheme, e_Q is bounded between 0 and 1 LSB above E_i, so we estimate $e_Q = Q$.

Now, consider the scheme most commonly used in the A/D converters within commercial data acquisition systems. While the resolution is not changed, this scheme makes the quantization error symmetric about the input voltage. To do this, the analog voltage is shifted internally within the A/D converter by a bias voltage, E_{bias}, of an amount equivalent to ½ LSB. This shift is transparent to the user. The effect of such a shift is shown on the second lower axis in Figure 7.7. This now makes e_Q bounded by $\pm 1/2$ LSB about E_i. So here the quantization error is $e_Q = \pm 1/2Q$. Regardless of the scheme used, the span of the quantization error remains 1 LSB, and its effect is most significant at smaller measured voltages. In an uncertainty analysis we would set the resolution uncertainty as $u_Q = e_Q$.

The resolution of an A/D converter is sometimes specified in terms of signal-to-noise ratio (SNR). The SNR relates the power of the signal, given by Ohm's law as E^2/R, to the power that can be resolved by quantization, given by $E^2/(R \times 2^M)$. The SNR is just the ratio of these values. Defined in terms of the decibel (dB), this gives

$$SNR(\text{dB}) = 20 \log 2^M \tag{7.15}$$

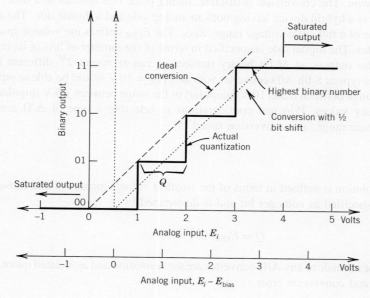

Figure 7.7 Binary quantization and saturation.

Table 7.2   Conversion Resolution

Bits M	Q^a (V/bit)	SNR (dB)
2	2.50	12
4	0.625	24
8	0.0390	48
12	0.00244	72
16	0.153 (10^{-3})	96
18	0.0381 (10^{-3})	108

^{a}Assumes $E_{FSR} = 10$ V.

The effect of bit number on resolution and SNR is detailed in Table 7.2. The program *Bits of Resolution.vi* explores the influence of a combination of M, E_{FSR}, and Q on a measured signal.

Saturation Error

The analog range of an A/D converter limits the minimum and maximum analog voltage. If either limit is exceeded, the A/D converter output saturates and does not change with a subsequent increase in input level. As noted in Figure 7.7, an input to the 0- to 4-V, 2-bit A/D converter above 4 V results in an output of binary 11 and below 0 V of binary 00. A saturation error is defined by the difference between the input analog signal level and the equivalent digital value assigned by the A/D converter. Saturation error can be avoided by conditioning signals to remain within the limits of the A/D converter.

Conversion Error

An A/D converter is not immune to elemental errors arising during the conversion process that lead to misrepresenting the input value. As with any device, the A/D errors can be delineated into hysteresis, linearity, sensitivity, zero, and repeatability errors. The extent of such errors depends on the particular method of A/D conversion. Factors that contribute to conversion error include A/D converter settling time, signal noise during the analog sampling, temperature effects, and excitation power fluctuations (5).

Linearity errors result from the ideal assumption that an M-bit A/D converter resolves the analog input range into 2^{M-1} equal steps of width Q. In practice, the steps may not be exactly equal, which causes a nonlinearity in the ideal conversion line drawn in Figure 7.7. This error is specified in terms of bits.

Example 7.4

Compute the relative effect of quantization error (e_Q/E_i) in the quantization of a 100-mV and a 1-V analog signal using an 8-bit and a 12-bit A/D converter, both having a full-scale range of 0 to 10 V.

KNOWN $E_i = 100$ mV and 1 V

$M = 8$ and 12

$E_{FSR} = 0 - 10$ V

FIND e_Q/E_i, where e_Q is the quantization error

SOLUTION The resolutions for the 8-bit and 12-bit converters can be estimated from Equation 7.14 as

$$Q_8 = \frac{E_{FSR}}{2^8} = \frac{10}{256} = 39 \text{ mV}$$

$$Q_{12} = \frac{E_{FSR}}{2^{12}} = \frac{10}{4096} = 2.4 \text{ mV}$$

Assume that the A/D converter is designed so that the absolute quantization error is given by $\pm\frac{1}{2} Q$. The relative effect of the quantization error can be computed by e_Q/E_i. The results are tabulated as follows:

E_i	M	$e_Q = \pm 1/2\, Q$	$100 \times e_Q/E_i$
100 mV	8	± 19.5 mV	19.5%
100 mV	12	± 1.22 mV	1.22%
1 V	8	± 19.5 mV	1.95%
1 V	12	± 1.22 mV	0.122%

From a relative point of view, the quantization error is much greater at lower voltage levels.

Example 7.5

The A/D converter with the specifications listed below is planned to be used in an environment in which the A/D converter temperature may change by ± 10 °C. Estimate the contributions of conversion and quantization errors to the uncertainty in the digital representation of an analog voltage by the converter.

	Analog-to-Digital Converter
E_{FSR}	0–10 V
M	12 bits
Linearity error	± 3 bits
Temperature drift error	1 bit/5 °C

KNOWN 12-bit resolution (see Ex. 7.4)

FIND $(u_c)_E$ measured

SOLUTION   We can estimate a design-stage uncertainty as a combination of uncertainty arising from quantization errors u_Q and from conversion errors u_c:

$$(u_d)_E = \sqrt{u_o^2 + u_c^2}$$

The resolution of a 12-bit A/D converter with full-scale range of 0 to 10 V is $Q = 2.4$ mV (see Ex. 7.4), so that the quantization error provides an estimate for the zero-order uncertainty as

$$u_o = \frac{1}{2}Q = 1.2 \text{ mV}$$

Now the conversion error is affected by two elements:

$$\text{Linearity uncertainty} = u_2 = 3 \text{ bits} \times 2.4 \text{ mV}$$

$$= 7.2 \text{ mV}$$

$$\text{Temperature uncertainty} = u_3 = \frac{1 \text{ bit}}{5\,^\circ\text{C}} \times 10\,^\circ\text{C} \times 2.4 \text{ mV}$$

$$= 4.8 \text{ mV}$$

An estimate of the uncertainty resulting from conversion errors is found using the RSS method:

$$u_c = \sqrt{u_2^2 + u_3^2} = \sqrt{(7.2 \text{ mV})^2 + (4.8 \text{ mV})^2} = 8.6 \text{ mV}$$

The combined uncertainty in the digital representation of an analog value due to these uncertainties is an interval described as

$$(u_d)_E = \sqrt{(1.2 \text{ mV})^2 + (8.6 \text{ mV})^2}$$

$$= \pm 8.7 \text{ mV (95\% assumed)}$$

The effects of the conversion errors dominate the uncertainty. A go-no go decision regarding this planned system can be made based on this result.

Successive Approximation Converters

We next discuss several common methods for converting voltage signals into binary words. More detailed discussions and additional methods for A/D conversion are presented in specialized texts (e.g., 5).

The most common type of A/D converter uses the *successive approximation* technique. This technique uses a trial-and-error approach for converting the input voltage. Basically, the successive approximation A/D converter guesses successive binary values as it narrows in on the appropriate binary representation for the input voltage. As depicted in Figure 7.8, this A/D converter uses an

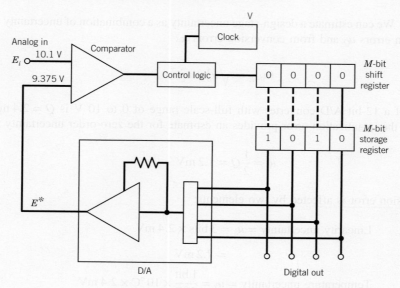

Figure 7.8 Successive approximation A/D converter. Four-bit scheme is shown with register = 1010 with $E_i = 10.1$ V.

M-bit register to generate a trial binary number, a D/A converter to convert the register contents into an analog voltage, and a voltage comparator (see Section 6.7 in Chapter 6) to compare the input voltage to the internally generated voltage. The conversion sequence is as follows:

1. The MSB is set to 1. All other bits are set to 0. This produces a value of E^* at the D/A converter output equivalent to the register setting. If $E^* > E_i$, the comparator goes LOW, causing the MSB to be reset to 0. If $E^* < E_i$, the MSB is kept HIGH at 1.

2. The next highest bit (MSB-1) is set to 1. Again, if $E^* > E_i$, it is reset to 0; otherwise its value remains 1.

3. The process continues through to the LSB. The final register value gives the quantization of E_i.

The process requires a time of one clock tick per bit.

An example of this sequence is shown in Table 7.3 for an input voltage of $E_i = 10.1$ V and using the 0 to 15 V, 4-bit successive approximation A/D converter of Figure 7.8. For this case, the converter has a resolution of 0.9375 V. The final register count of 1010 or its equivalent of 9.375 V is the output from the A/D converter. This measured value differs from the actual input voltage of 10.1 V as a result of the quantization error. This error can be reduced by increasing the bit size of the register.

The successive approximation converter is typically used when conversion speed at a reasonable cost is important. The number of steps required to perform a conversion equals the number of bits in the A/D converter register. With one clock tick per step, a 12-bit A/D converter operating with a 1-MHz clock would require a maximum time of 12 μs per conversion. But this also reveals the trade-off between increasing the number of bits to lower quantization error and the resulting increase in conversion time. Faster clocks enable higher sample rates. Common maximum sample rates are on the order of 100 kHz to 1 MHz using 12, 16, or 24 bits.

Table 7.3 Example of Successive Approximation Conversion Sequence for a 4-Bit Converter

Sequence	Register	E^*	E_i	Comparator
Initial status	0000	0	10.1	
MSB set to 1	1000	7.5	10.1	High
Leave at 1	1000	7.5		
Next highest bit set to 1	1100	11.25	10.1	Low
Reset to 0	1000	7.5		
Next highest bit set to 1	1010	9.375	10.1	High
Leave at 1	1010	9.375		
LSB set to 1	1011	10.3125	10.1	Low
Reset to 0	1010	9.375		

Sources of conversion error originate in the accuracies of the D/A converter and the comparator. Noise is the principal weakness of this type of converter, particularly at the decision points for the higher-order bits. The successive approximation process requires that the voltage remain constant during the conversion process. Because of this, a sample-and-hold circuit (SHC), as introduced in Chapter 6, is used ahead of the converter input to measure and to hold the input voltage value constant throughout the duration of the conversion. The SHC also minimizes noise during the conversion.

Ramp (Integrating) Converters

Low-level (<1-mV) measurements often rely on ramp converters for their low-noise features. These integrating analog-to-digital converters use the voltage level of a linear reference ramp signal to discern the voltage level of the analog input signal and convert it to its binary equivalent. Principal components, as shown in Figure 7.9, consist of an analog comparator, ramp function generator, and counter and M-bit register. The reference signal, initially at zero, is increased at set time steps, within which the ramp level is compared to the input voltage level. This comparison is continued until the two are equal.

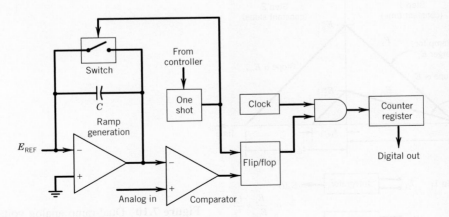

Figure 7.9 Ramp A/D converter.

The usual method for generating the ramp signal is to use a capacitor of known fixed capacitance C, which is charged from a constant current source of amperage, I. Because the charge is related to current and time by

$$q = \int_0^t I \, dt \tag{7.16}$$

the reference ramp voltage is linearly related to the elapsed time by

$$E_{\text{ref}} = \frac{q}{C} = \text{constant} \times t \tag{7.17}$$

Time is integrated by a counter that increases the register value by 1 bit at each time step. Time step size depends on the value of 2^M. When the input voltage and ramp voltage magnitudes cross during a time step, the comparator output goes to zero, which flips a flip-flop halting the process. The register count value then indicates the digital binary equivalent of the input voltage.

The accuracy of this single ramp operation is limited by the accuracy of the timing clock and the known values and constancy of the capacitor and the charging current. Increased accuracy can be achieved using a dual ramp converter, sometimes referred to as a dual-slope integrator, in which the measurement is accomplished in two steps, as illustrated in Figure 7.10. In the first step, the input voltage E_i is applied to an integrator for a fixed time period t_1. The integrator output voltage increases in time with a slope proportional to the input voltage. At the end of the time interval, the output of the integrator has reached the level E^*. The second step in the process involves the application of a fixed reference voltage E_{ref} to the integrator. This step occurs immediately at the end of the fixed time interval in the first step, with the integrator voltage at exactly the same level established by the input. The reference voltage has the opposite polarity to the input voltage and thus reduces the output of the integrator at a known rate. The time required for the output to return to zero is a direct measure of the

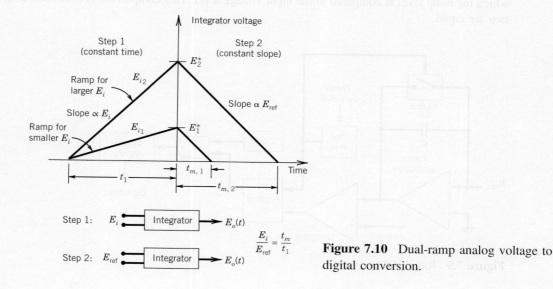

Figure 7.10   Dual-ramp analog voltage to digital conversion.

input voltage. The time intervals are measured using a digital counter, which accumulates pulses from a stable oscillator. The input voltage is given by the relationship

$$E_i/E_{\text{ref}} = t_m/t_1 \tag{7.18}$$

where t_m is the time required for step two. Dual-ramp converter accuracy depends on the stability of the counter during conversion and requires a very accurate reference voltage.

Ramp converters are inexpensive but relatively slow devices. However, their integration process tends to average out any noise in the input signal. This is the feature that makes them so attractive for measuring low-level signals. The maximum conversion time using a ramp converter is estimated as

$$\text{Maximum conversion time} = 2^M/\text{Clock speed} \tag{7.19}$$

For a dual-ramp converter, this time is doubled. If one assumes that a normal distribution of input values are applied over the full range, the average conversion time can be taken to be one-half of the maximum value. For a 12-bit register with a 1-MHz clock and required to count the full 2^{12} pulses, a dual-ramp converter would require $\sim$8 ms for a single conversion, corresponding to a sample rate of 125 Hz.

Parallel Converters

A very fast A/D converter is the parallel or flash converter depicted in Figure 7.11. These converters are common to high-end stand-alone digital oscilloscopes and spectral analyzers. An M-bit parallel

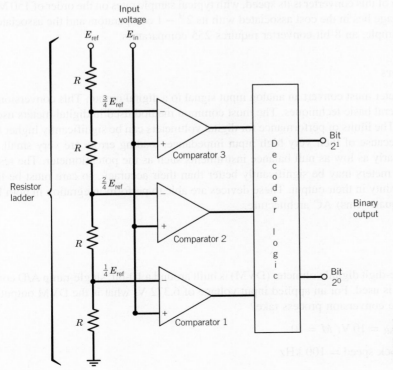

Figure 7.11 Parallel or flash A/D converter. Two-bit scheme is shown.

Table 7.4 Logic Scheme of a 2-Bit Parallel A/D Converter

Comparator			
1	2	3	Binary Output
HIGH	HIGH	HIGH	11
LOW	HIGH	HIGH	10
LOW	LOW	HIGH	01
LOW	LOW	LOW	00

converter uses $2^M - 1$ separate voltage comparators to compare a reference voltage to the applied input voltage. As indicated in Figure 7.11, the reference voltage applied to each successive comparator is increased by the equivalent of 1 LSB by using a voltage-dividing resistor ladder. If the input voltage is less than its reference voltage, a comparator will go LOW; otherwise it will go HIGH.

Consider the 2-bit converter shown in Figure 7.11. If $E_{in} \geq \frac{1}{2}E_{ref}$ but $E_{in} < \frac{3}{4}E_{ref}$, then comparators 1 and 2 will be HIGH but comparator 3 will be LOW. In this manner, there can only be $2^M = 2^2$ different HIGH/LOW combinations from $2^2 - 1$ comparators, as noted in Table 7.4 which shows how these combinations correspond to the 2^2 possible values of a 2-bit binary output. Logic circuits transfer this information to a register.

Because all the comparators act simultaneously, the conversion occurs in a single clock count. Thus an attraction of this converter is its speed, with typical sample rates on the order of 150 MHz at 8 bits. Its disadvantage lies in the cost associated with its $2^M - 1$ comparators and the associated logic circuitry. For example, an 8-bit converter requires 255 comparators.

Digital Voltmeters

The digital voltmeter must convert an analog input signal to a digital output. This conversion can be done through several basic techniques. The most common method used in digital meters uses dual-ramp converters. The limits of performance for digital voltmeters can be significantly higher than for analog meters. Because of their very high input impedance, loading errors are very small at low-voltage levels, nearly as low as null balance instruments such as the potentiometer. The resolution (1 LSB) of these meters may be significantly better than their accuracy, so care must be taken in estimating uncertainty in their output. These devices are able to perform integration in both DC and true root-mean-square (rms) AC architecture.

Example 7.6

A 0- to 10-V, three-digit digital voltmeter (DVM) is built around a 10-bit, single-ramp A/D converter. A 100-kHz clock is used. For an applied input voltage of 6.372 V, what is the DVM output value? How long will the conversion process take?

KNOWN $E_{FSR} = 10 \text{ V}$; $M = 10$

Clock speed = 100 kHz

Three significant digit readout (x.xx)

Input voltage, $E_i = 6.372$ V

FIND   Digital output display value

Conversion time

SOLUTION   The resolution of the A/D converter is

$$Q = E_{\text{FSR}}/2^{\text{M}} = 10\text{ V}/1024 = 9.77\text{ mV/bit}$$

Because each ramp step increases the counter value by 1 bit, the ramp converter has a slope

$$\text{slope} = Q \times 1\text{ bit/step} = 9.77\text{ mV/step}$$

The integer number of steps required for conversion is

$$\text{steps required} = \frac{E_i}{\text{slope}} = \frac{6.372\text{ V}}{0.00977\text{ V/step}} = 652.49\text{ or }653\text{ steps}$$

so the indicated output voltage is

$$E_0 = \text{slope} \times \text{number of steps}$$
$$= 9.77\text{ mV/step} \times 653\text{ steps} = 6.3769\text{ V} \approx 6.38\text{ V}$$

A three-digit display rounds E_0 to 6.38 V. The difference between the input and output values is attributed to quantization error. With a clock speed of 100 kHz, each ramp step requires 10 μs/step, and conversion time = 653 steps × 10 μs/step = 6530 μs.

COMMENT   A similar dual-ramp converter would require twice as many steps and twice the conversion time.

7.6   DATA ACQUISITION SYSTEMS

A *data acquisition system* is the portion of a measurement system that quantifies and stores data. There are many ways to do this. The engineer who reads a transducer dial, associates a number with the dial position, and records the number in a log book performs all the tasks germane to a data acquisition system.

This section focuses on microprocessor-based data acquisition systems, which are used to perform automated data quantification and storage in engineering testing. The needed microprocessor technology can be found in many devices from specialized control boards to personal mobile devices to laptop, desktop and dedicated computers. We use the generic term "computer" to refer to any of these devices.

Figure 7.12 shows how a data acquisition system (DAS) might fit into the general measurement scheme between an analog measurement and its subsequent data reduction. A typical analog signal

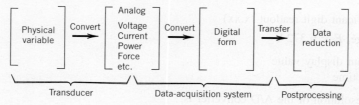

Figure 7.12 Typical signal and measurement scheme.

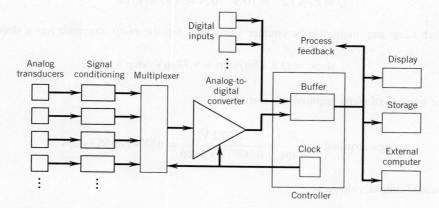

Figure 7.13 Analog signal flow scheme using an automated data acquisition system.

flow scheme is shown in Figure 7.13 for multiple input signals to a single microprocessor-based/controller DAS.

Dedicated microprocessor systems can continuously perform their programming instructions to measure, store, interpret and provide process control without any intervention. Such microprocessors use I/O ports to interface with other devices to measure and to output instructions. Programming allows for such operations as which sensors to measure and when and how often, and data reduction. Programming can allow for decision-making and feedback to control process variables. A cost-effective microprocessor system may be derived from a programmable logic controller (PLC).

Computer-based laboratory data acquisition systems combine a data acquisition module with both the microprocessor and interface capability of a computer. The interface between external instruments and the computer is done through the dedicated module, which connects to the computer via an external bus port, an I/O board, or wireless port. The computer also provides access to the Internet, an increasingly useful method for transmitting data between locations.

7.7 DATA ACQUISITION SYSTEM COMPONENTS

Analog Signal Conditioning: Filters and Amplification

Analog signals usually require some type of signal conditioning to properly interface with a digital system. Filters and amplifiers are the most common components used.

Filters

Analog filters are used to control the frequency content of the signal being sampled. These were discussed in Chapter 6. Anti-alias analog filters remove signal information above the Nyquist frequency before sampling. Not all data acquisition boards contain analog filters, so these necessary components may need to be added.

Digital filters are effective for digital signal analysis. They are not used to prevent aliasing. One digital filtering scheme involves taking the Fourier transform of the sampled signal, multiplying the signal amplitude in the targeted frequency domain to attain the desired frequency response (i.e., type of filter and desired filter settings), and transforming the signal back into time domain using the inverse Fourier transform (1,2).

A simpler digital filter is the moving average or smoothing filter, which is used for removing noise or showing trends of discrete information. Essentially, this filter replaces a current data point value with an average based on a series of successive data point values. A center-weighted moving averaging scheme takes the form

$$y_i^* = (y_{i-n} + \cdots + y_{i-1} + y_i + y_{i+1} + \cdots + y_{i+n})/(2n + 1) \tag{7.20}$$

where y_i^* is the averaged value that is calculated and used in place of y_i, and $(2n + 1)$ equals the number of successive values used for the averaging. For example, if $y_4 = 3, y_5 = 4, y_6 = 2$, then a three-term average $(2n + 1 = 3, n = 1)$ of y_5 $(i = 5)$ produces $y_5^* = 3$.

In a similar manner, an $(n + 1)$-term forward-moving averaging smoothing scheme takes the form

$$y_i^* = (y_i + y_{i+1} + \cdots + y_{i+n})/(n + 1) \tag{7.21}$$

and a backward-moving averaging smoothing scheme takes the form

$$y_i^* = (y_{i-n} + \cdots + y_{i-1} + y_i)/(n + 1) \tag{7.22}$$

Light filtering might use three terms, whereas heavy filtering might use 10 or more terms. This filtering scheme is easily accomplished within a spreadsheet or programming package.

Amplifiers

All data acquisition systems are input range limited; that is, there is a minimum value of a signal that they can resolve and a maximum value that initiates the onset of saturation. Thus, some transducer signals need amplification or attenuation before conversion. Most data acquisition systems contain onboard instrumentation amplifiers as part of the signal conditioning stage, as depicted in Figure 7.13, with selectable gains ranging from less than to greater than unity. Gain is varied either by logic switches set by software or by resistor jumpers, which effectively reset resistor ratios across op-amplifiers. Section 6.6 in Chapter 6 discusses amplifiers.

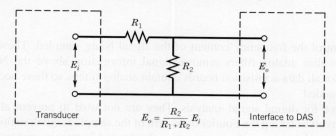

Figure 7.14 Voltage divider circuit for signal amplitude attenuation.

Although instrument amplifiers offer good output impedance characteristics, voltages can also be attenuated using a voltage divider. The output voltage from the divider circuit of Figure 7.14 is determined by

$$E_o = E_i \frac{R_2}{R_1 + R_2} \tag{7.23}$$

For example, a 0- to 50-V signal can be measured by a 0- to 10-V A/D converter using $R_1 = 40 \, k\Omega$ and $R_2 = 10 \, k\Omega$.

When only the dynamic content of time-dependent signals is important, amplification may require a strategy. For example, suppose the mean value of a voltage signal is large but the dynamic content is small, as with the 5-Hz signal

$$y(t) = 2 + 0.1 \sin 10 \, \pi t (V) \tag{7.24}$$

Setting the amplifier gain at or more than $G = 2.5$ to improve the resolution of the dynamic content would saturate a ± 5-V A/D converter. In such situations, you can remove the mean component from the signal before amplification either by (1) using an offset nulling circuit to zero the mean value as discussed next or, similarly, by (2) adding a mean voltage of equal but opposite sign to the signal, such as by adding -2 V in this example or (3) passing the signal through a very low frequency high-pass filter (also known as *AC coupling*).

Shunt Resistor Circuits

An A/D converter requires a voltage signal at its input. It is straightforward to convert current signals into voltage signals using a shunt resistor. The circuit in Figure 7.15 provides a voltage

$$E_o = IR_{\text{shunt}} \tag{7.25}$$

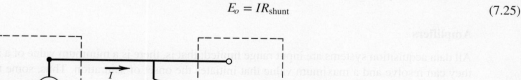

Figure 7.15 Simple shunt resistor circuit.

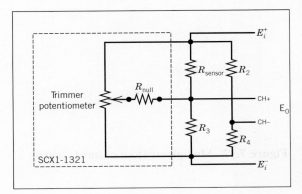

Figure 7.16 Circuit for applying a null offset voltage. (Courtesy of National Instruments.)

for signal current I. For a transducer using a standard 4- to 20-mA current loop, a 250-Ω shunt would convert this current output to a 1- to 5-V signal.

Offset Nulling Circuit

Offset nulling is used to subtract a small constant voltage from a signal, such as to zero out a transducer output signal. Hence it is an important tool for reducing uncertainty arising from zero error while keeping loading error to a minimum. The technique uses a trim potentiometer (R_{null}) in conjunction with a bridge circuit, such as shown in Figure 7.16, to produce and to apply a null voltage, E_{null}, to the circuit. The available null voltage is

$$E_{null} = \pm \left[\frac{E_i}{2} - \frac{E_i R_3 (R_{null} + R_{sensor})}{R_{null} R_{sensor} + R_3 (R_{null} + R_{sensor})} \right] \tag{7.26}$$

The corrected signal voltage is $E_0 = E_i - E_{null}$. The technique is analogous to nulling (zeroing) the output indicator on a mechanical scale.

Components for Acquiring Data

Multiplexer

When multiple input signal lines are connected by a common throughput line to a single A/D converter, a multiplexer is used to switch between the connections and the A/D converter one at a time. The technique is quite common in commercial data acquisition modules. The approach is illustrated by the multiplexer depicted in Figure 7.17, which uses parallel flip-flops (see Chapter 6) to open and close the connection paths sequentially. The switching rate is determined by the conversion timing control logic.

A/D Converters

High-speed data acquisition modules employ successive approximation A/D converters with conversion rates typically up to the 100-kHz to 10-MHz range or parallel A/D converters for rates

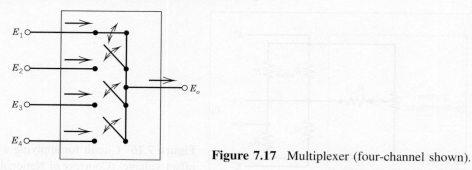

Figure 7.17 Multiplexer (four-channel shown).

up to and over 10 GHz. Low-level voltage measurements require the high noise rejection of the dual-ramp A/D converter. Here the trade-off is in speed, with maximum conversion rates of up to 100 Hz more common.

D/A Converters

A digital-to-analog (D/A) converter permits a DAS to convert digital numbers into analog voltages, which might be used for process control, such as to change a process variable, activate a device, or to drive a sensor positioning motor. The digital-to-analog signal is initiated by software or a controller.

Digital Input/Output

Digital signals are composed of discrete states, either high or low. Digital input/output (I/O) connection lines may be used to communicate between instruments (see Section 7.9), control relays (to power equipment on or off), or indicate the state or status of a device. A common way to transmit digital information is the use of a 5-V TTL signal (e.g., see Fig. 7.5).

Digital I/O signals may be as a single state (HIGH or LOW; 5 V or 0 V) or as a series of pulses of HIGH/LOW states. The single-state signal might be used to operate a switch or relay, or to signal an alarm. A series of pulses transmits a data series. Gate-pulse counting entails counting pulses that occur over a specified period of time. This enables frequency determination (number of pulses/unit time) and counting/timing applications. Pulse stepping, sending a predetermined number of pulses in a series, is used to drive stepper motors and servos. Several I/O ports can be grouped to send parallel information.

Closed-Loop Controller

In closed-loop control, the controller is used to compare the state of a process, as determined through the value of a measured variable, with the value of a set condition and to take appropriate action to reduce the difference in value of the two. This difference in value is called the error signal. This error can change as part of the dynamics of the process, which causes the measured variable to change value, or because the set condition itself is changed. The controller action is to adjust the process so as to keep the error within a desired range.

When integrated within a data acquisition system, the control process consists of this sequence: sample, compare, decide, correct. An example of a digital control loop is depicted in Figure 7.18.

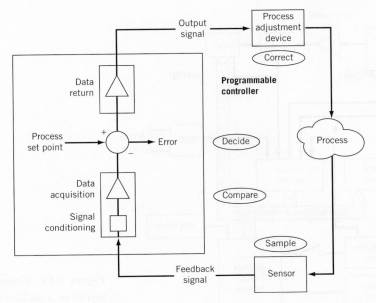

Figure 7.18 Closed-loop control concept built around a data acquisition–based programmable controller.

Here the controller receives input about the process through a sensor that monitors the measured variable. The controller measures the sensor signal by sampling through an A/D converter or some digital input. The controller compares the measured value to the set value, computes the error signal, executes calculations through its control algorithm to decide on the correction needed, and acts to correct the process. In this example, we anticipate that the corrective elements are analog based, and so the controller sends the necessary signal for appropriate corrective action at the output of a D/A converter. The controller repeats this procedure at each cycle. A more extensive treatment of feedback control is provided in Chapter 12.

7.8   ANALOG INPUT-OUTPUT COMMUNICATION

Data Acquisition Modules

Data acquisition systems (DAS) consist of a module that supports interfacing with analog and digital external transducers and devices and a microprocessor/computer that controls the process and stores/retrieves information from memory. The most common arrangement uses an external module connected by a wired bus to a tablet, mobile device, laptop, or stationary computer. A plug-in computer board is also common, particularly in industrial environments. Fully self-contained standalone modules exist with an onboard microprocessor and memory. Some modules can interface with other devices using wireless communication, such as Bluetooth or the wireless network. With a wireless or hardwired Internet connection, it is possible to acquire or send data and transmit information to different locations on command.

Consider the generic layout of a data acquisition system shown in Figure 7.19. Field wiring from/to transducers or other external devices is usually made through a mechanical screw terminal that attaches directly or indirectly to a DAS module. The module allows for multichannel analog and

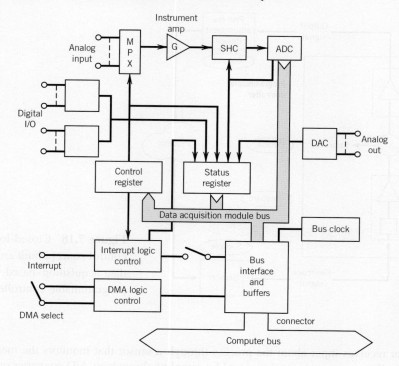

Figure 7.19 Generic layout of a multipurpose data acquisition module interfaced with a computer.

digital input and output. The layout shows the signal path through each of the various module components, which passes onto the module's internal bus. The module interfaces with the controlling computer bus through some communication interface (e.g., a USB port or a Bluetooth connection), labeled here as a "connector." Standalone systems simply port the DAS bus to the onboard microprocessor bus.

A commercially available multichannel high-speed data acquisition module is shown in Figure 7.20. It uses a 16-channel multiplexer and instrument amplifier enabling 8 differential or 16-single-ended analog input channels. Input impedance exceeds 10 GΩ. The 16-bit successive approximation A/D converter is capable of sample rates of up to 250 kHz. Input signals may be unipolar (e.g., 0–10 V) or bipolar (e.g., ±5 V). For the board shown, this gives a unit gain input resolution ($E_{FSR} = 10$ V) of $Q = 152.6\ \mu$V/bit. The selectable gain onboard amplifier extends the analog input range from ±10V down to ± 0.2 V. The module also contains two 16-bit analog output channels driven by digital-to-analog converters. There are four digital input/output channels and two 32-bit counters for instrument control applications and external triggering. This USB-powered device is comparable in size to a pocket-sized camera, a small size that can still accommodate a row of screw terminal wire connectors to attach transducers and devices.

Data acquisition technology is rapidly changing to bring new capabilities to the marketplace. A full-featured, self-contained DAS module is shown in Figure 7.21. This particular system is powered by an onboard dual core microprocessor with an operating system enabling standalone programmable real-time data processing and control. Data acquisition and control capabilities are retained: 10 single-ended and differential-ended analog inputs, six analog outputs, 40 digital input/output, and

Figure 7.20 Photograph of a multi-purpose USB-based data acquisition module. (Photograph of NI USB-621x; Courtesy of National Instruments.)

Figure 7.21 Example of a self-contained, portable multipurpose data acquisition system. (Photograph of NI myRIO; Courtesy of National Instruments.)

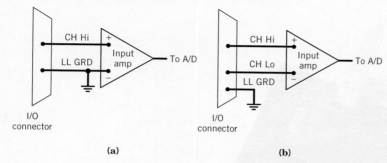

Figure 7.22 **(a)** Single-ended connection. **(b)** Differential-ended connection.

three-axis accelerometer along with the needed I/O connector terminals. It can interface with other devices, such as tablets and external computers, either through a USB port or its onboard WiFi communication. Aside from the portability offered by the small package, the design is also attractive for data acquisition needs in moving (e.g., vehicles, robotics) and rotating systems, because there is no need for connection cables to computers and power sources.

Single- and Differential-Ended Connections

Analog signal input connections to a DAS board may be single- or differential-ended. *Single-ended connections* use only one signal line (+ or HIGH) that is measured relative to ground (GRD), as shown in Figure 7.22. The return line (− or LOW) and ground are connected together. There is a common external ground point, usually through the DAS board. Multiple single-ended connections to a DAS board are shown in Figure 7.23.

Single-ended connecting wires should never be long (>1 m), as they are susceptible to EMI (electromagnetic interference) noise, particularly at high speed transfer rates. Because of noise, the

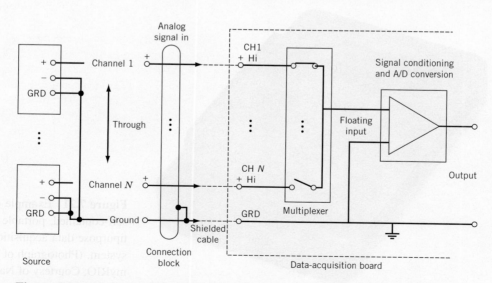

Figure 7.23 Multiple single-ended analog connections to a data acquisition board.

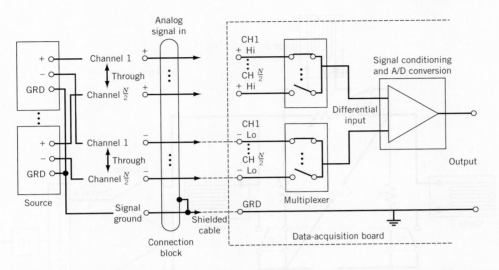

Figure 7.24   Multiple differential-ended analog connections to a data acquisition board.

signal should be large compared to the DAS resolution. Single-ended connections are suitable only when all the analog signals can be made relative to the common ground point.

Why the concern over ground points? Electrical grounds are not all at the same voltage value (see Ground and Ground Loops, in Chapter 6). So if a signal is grounded at two different points, such as at its source and at the DAS board, the grounds could be at different voltage levels. When a grounded source is wired as a single-ended connection, the difference between the source ground voltage and the board ground voltage gives rise to a *common-mode voltage* (CMV). The CMV combines with the input signal superimposing interference and noise on the signal. This effect is referred to as a ground loop. The measured signal is unreliable.

Differential-ended connections allow the voltage difference between two distinct input signals to be measured. Here the signal input (+ or HIGH) line is paired with a signal return (− or LOW) line, which is isolated from ground (GRD), as shown in Figure 7.22. By using twisted pairs, the effects of noise are greatly reduced. A differential-ended connection to a DAS board is shown in Figure 7.24. The measured voltage is the voltage difference between these two (+ and −) lines for each channel.

Differential-ended connections are the preferred way to measure low-level signals. However, for low-level measurements, a 10-kΩ to 100-kΩ resistor should be connected between the signal return (− or LOW) line and ground (GRD) at the DAS board. The differential-ended connection is less prone to common-mode voltage errors. But if the measured voltage exceeds the board's CMV range limit specification, it can be damaged.

Special Signal Conditioning Modules

Signal conditioning modules exist for different transducer applications (5). They connect between the transducer and the data acquisition module. For example, resistance bridge modules allow the direct interfacing of strain gauges or other resistance sensors through an on-board Wheatstone bridge, such as depicted in Figure 7.25. There are temperature modules to allow for electronic thermocouple cold

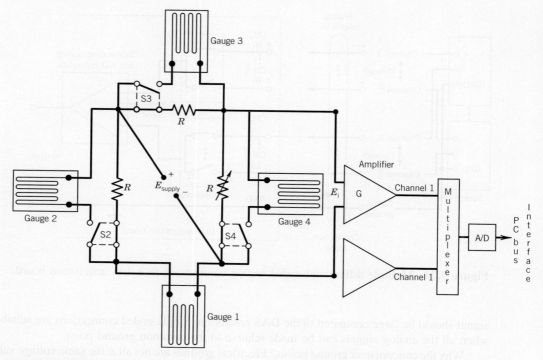

Figure 7.25 A strain-gauge interface. Gauges are connected by wires to the interface.

junction compensation that can provide for signal linearization for reasonably accurate (down to 0.5 °C) temperature measurements.

Data Acquisition Triggering

With DAS boards, data acquisition or equipment control can be triggered by software command, external pulse, or on-board clock. Software-controlled triggers can send TTL-level pulses to external devices for synchronization, activation, and control.

7.9 DIGITAL INPUT–OUTPUT COMMUNICATION

Certain standards exist for the manner in which digital information is communicated between digital devices. Serial communication methods transmit data bit by bit. Parallel communication methods transmit data in simultaneous groups of bits—for example, byte by byte. Both methods use a *handshake*, an interface protocol that initiates, allows, and terminates the data transfer between devices and TTL-level signals. Most lab equipment can be equipped to communicate by at least one of these means, and standards have been defined to codify communications between devices of different manufacturers. Standards continuously evolve, so the present discussion focuses on concepts and application of representative standards.

Data Transmission

Serial communication implies that data are sent in successive streams of information 1 bit at a time. The value of each bit is represented by an analog voltage pulse with a 1 and 0 distinguished by two equal voltages of opposite polarity in the 3- to 25-V range. Communication rates are measured in baud, which refers to the number of signal pulses per second.

A typical asynchronous transmission is 10 serial bits composed of a start bit followed by a 7- or 8-bit data stream, either one or no parity bit, and terminated by 1 or 2 stop bits. The start and stop bits form the serial "handshake." *Asynchronous transmission* means that information may be sent at random intervals. So the start and stop bits are signals used to initiate and to end the transmission of each byte of data transmitted, hence the analogy of a handshake. The start bit allows for synchronization of the clocks of the two communicating devices. The parity bit allows for limited error checking. *Parity* involves counting the number of 1s in a byte. In 1 byte of data, there will be an even or odd number of bits with a value of 1. An additional bit added to each byte to make the number of 1 bits a predetermined even or odd number is called a parity bit. The receiving device counts the number of transmitted bits, checking for the predetermined even (even parity) or odd (odd parity) number. In *synchronous transmission*, the two devices initialize and synchronize communication with each other. Even when there are no data to send, the devices transmit characters just to maintain synchronization. Stop and start bits are then unnecessary, allowing for higher data transfer rates.

Devices that employ data buffers, a region of RAM that serves as a data holding area, use a software handshaking protocol such as XON/XOFF. With this, the receiving device transmits an XOFF signal to halt transmission as the buffer nears full and an XON signal when it has emptied and is again ready to receive.

Universal Serial Bus

The Universal Serial Bus (USB) permits peripheral expansion for up to 127 devices at low- to high-speed serial data transfer rates. The original USB (USB 1.0/1.1) supports transfer rates from 1.5 Mbs up to 12 Mbs, and the high-speed USB (USB 2.0) supports rates up to 480 Mbs. The USB 3.x standard boasts data rates of up to 10 Gbit/sec. Notably, the maximum current that a USB 3.0 device can draw is 900 mA, an 80% increase over USB 2.0. This bus connects USB devices to a single computer host through a USB root hub. The USB physical interconnect is a tiered star topology, as illustrated in Figure 7.26a. In this setup, the root hub permits one to four attachments, which can be a combination of USB peripheral devices and additional USB hubs. Each successive hub can in turn support up to four devices or hubs. Cable length between a device and hub or between two hubs is limited to 5 m for USB 2.0 and 3 m for USB 3.0. The connecting cable (Fig. 7.26b) is a four-line wire consisting of a hub power line, two signal lines (+ and −), and a ground (GRD) line.

Bluetooth Communications

Bluetooth is a standardized specification permitting wireless communication between devices over short distances (< 10 m). The specification is based on the use of radio waves to transmit information, allowing signals to pass through enclosures, walls, and clothes. Devices are independent of aiming position. As such, the specification overcomes problems inherent with popular infrared port communication devices—for example, those used on common electronic remote controls, which require line-of-sight interaction. However, Bluetooth is not intended or written to be a wireless LAN

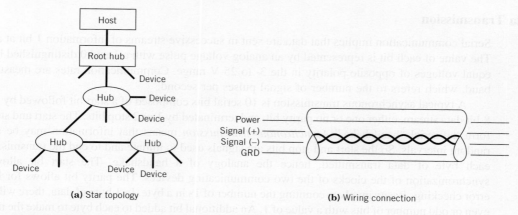

Figure 7.26 General purpose interface bus (GPIB) assignments to a 25-pin connector.

network protocol, such as is IEEE 802.11. Instead, it does permit communication between peripheral devices that use their own protocol by setting up a personal area network (PAN). Up to eight devices can communicate simultaneously over distances of up to about 10 m of separation within the 2.402- and 2.480-GHz radio spectrum on a PAN. Bluetooth 4.0 can manage data transfer rates up to 24 Mbps.

Other Serial Communications: RS-232C

The RS-232C protocol, initially set up to translate signals between telephone lines and computers via a modem (modulator–demodulator), remains a well-used interface for communication between a computer and any serial device. Basically, this protocol allows two-way communication using two single-ended signal (+) wires, noted as TRANSMIT and RECEIVE, between data communications equipment (DCE), such as an instrument, and data terminal equipment (DTE), such as the data acquisition system or computer. These two signals are analogous to a telephone's mouthpiece and earpiece signals. A signal GROUND wire allows signal return (−) paths. Many lab computers have an RS-232C–compatible I/O port. The popularity of this interface comes from the wide range of equipment that can use it. Either a 9-pin or 25-pin connector can be used.

Parallel Communications

The *general purpose interface bus* (*GPIB*) is a classic example of a high-speed parallel interface. Originally developed by Hewlett-Packard, it is sometimes referred to as the HP-IB. The GPIB is usually operated under the IEEE-488 communication standard. In 2004 the international and IEEE standards were combined and the current standard is designated IEC-60488-1. This 8-bit parallel bus allows for the control of other devices through a central controller, usually a lab computer, and it allows devices to receive/transmit information from/to the controller. This standard is well defined to interface communication between computers and printers and scientific instrumentation. The standard for the GPIB operates from a 16-wire bus. A 25-pin connector is standard. Bit-parallel, byte-serial communication at data rates up to 8 Mbytes/s is possible.

Example 7.7

A strain transducer has a static sensitivity of 2.5 V/unit strain (2.5 μV/με) and requires a supply voltage of 5 VDC. It is to be connected to a DAS having a ±5 V, 12-bit A/D converter and its signal measured at 1000 Hz. The transducer signal is to be amplified and filtered. For an expected measurement range of 1 to 500 με, specify appropriate values for amplifier gain, filter type, and cutoff frequency, and show a signal flow diagram for the connections.

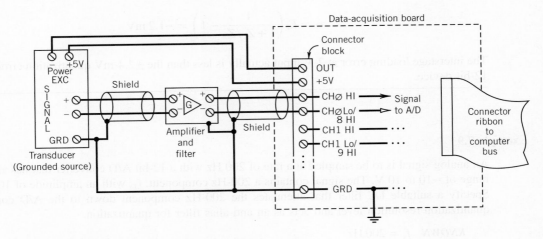

Figure 7.27 Line connections for Example 7.7.

SOLUTION The signal flow diagram is shown in Figure 7.27. Power is drawn off the data acquisition board and routed to the transducer. Transducer signal wires are shielded and routed through the amplifier, filter, and connected to the data acquisition board connector block using twisted pairs to channel 0, as shown. The differential-ended connection at the board reduces noise.

The amplifier gain is determined by considering the minimum and maximum signal magnitudes expected, and the quantization error of the DAS. The nominal signal magnitude ranges from 2.5 μV to 1.25 mV. Using an amplifier gain of $G = 1000$ would boost these signal levels to 2.5 mV to 1.25 V. This lower value is on the order of the quantization error of the 12-bit converter. Using a gain of $G = 3000$ would raise the low end of the signal out of quantization noise while keeping the high end out of saturation achieving an input signal range of 7.5 mV to 3.75 V.

With a sample rate of $f_s = 1000$ Hz, an anti-alias, low-pass Butterworth filter with 12 dB/octave roll-off and a cutoff frequency set at $f_c = f_N = 500$ Hz would meet the task.

Example 7.8

The output from an analog device (nominal output impedance of 600 Ω) is the input to the 12-bit A/D converter (nominal input impedance of 1 MΩ) of a data acquisition system. For a 2-V signal, will interstage loading be a problem?

KNOWN $Z_1 = 600\ \Omega$

$Z_m = 1\ \text{M}\Omega$

$E_1 = 2\ \text{V}$

FIND e_l

SOLUTION   The loading error is given by $e_l = E_m - E_1$, where E_1 is the true voltage and E_m is the measured voltage. From Equation 6.39,

$$e_l = E_1 \left(\frac{1}{1 + Z_1/Z_m} - 1 \right) = -1.2\ \text{mV}$$

The interstage loading error at 2-V input actually is less than the ± 2.4-mV quantization error of the 12-bit device.

Example 7.9

An analog signal is to be sampled at a rate of 200 Hz with a 12-bit A/D converter that has an input range of -10 to 10 V. The signal contains a 200-Hz component, f_1, with an amplitude of 100 mV. Specify a suitable LC filter that attenuates the 200-Hz component down to the A/D converter quantization resolution level and acts as an anti-alias filter for quantization.

KNOWN $f_s = 200\ \text{Hz}$

$f_1 = 200\ \text{Hz}$

$A_1 = 100\ \text{mV}$

$E_{\text{FSR}} = 10\ \text{V}$

$M = 12$

FIND   Specify a suitable filter

SOLUTION   For $f_s = 200$ Hz, the Nyquist frequency f_N is 100 Hz. So an appropriate design for an anti-alias filter might have the properties of $f_c = 100$ Hz and $M(100\ \text{Hz}) = -3$ dB and a maximally flat passband (Butterworth). The A/D converter quantization resolution level is

$$Q = \frac{E_{\text{FSR}}}{2^M} = \frac{20\ \text{V}}{4096} = 4.88\ \text{mV/bit}$$

For a 100-mV signal at 200 Hz, this requires a k-stage, low-pass Butterworth filter with an attenuation of

$$M(200\ \text{Hz}) = \frac{4.88\ \text{mV}}{100\ \text{mV}} = 0.0488(\text{or} - 26\ \text{dB})$$

$$= \left[1 + (f/f_c)^{2k} \right]^{-1/2} = \left[1 + (200/100)^{2k} \right]^{-1/2}$$

so $k = 4.3 \approx 5$. Appropriate values for L and C can be set from Table 6.1 and Figure 6.29 and scaled.

Example 7.10: Digital Signal Processors

We enjoy our personal mobile devices to store audio (e.g., music, books) and use them for playback as we go about our normal activities. At some point, this audio information was converted from an analog to a digital signal by passing it through an A/D converter and subsequently storing it in digital form, often compressed to minimize memory needs. This would be the form of a file that you might download from a website.

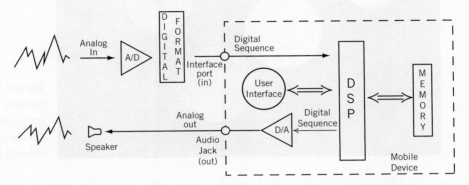

Figure 7.28   An internal digital signal processor (DSP) is common within personal mobile devices to handle digital data processing, conditioning, and handling.

Digital signal processors (DSP) are devices that process digital information (5). They use simple mathematical algorithms to manipulate information quickly. A common mobile audio device (schematic in Figure 7.28) uses a DSP to encode incoming digital information, such as from a download, and save it to the audio device's memory. In storing the information, the user interacts directly with the DSP through a user interface, assigning it a filename or organizing the order of the files. To play back some particular information, the user again interacts with the DSP so that the DSP pulls the correct digital file from memory and decodes, expands, and converts it back to its original analog signal by passing it through a D/A converter. The now analog signal passes through an amplifier before the signal is coupled with the headset earpiece or external speakers.

A DSP is programmable and works with any type of physical signal that has been digitized. More and more, we interact with DSPs without realizing it. For example, a cell phone uses a DSP to provide echo cancellation during conversations, and your television or video player uses a DSP to expand the movie you download whose file had been previously compressed from the original signal to manage resource usage. Furthermore, that movie's audio format is decoded by the DSPs within both the video player and audio amplifier to create an impressive home theater sound.

7.10   DIGITAL IMAGE ACQUISITION AND PROCESSING

Digital images are widely used in applications as varied as inspection, medical imaging, quality assurance, and sorting. This section serves as a brief introduction to acquiring and processing digital images.

Figure 7.29 Characteristics of a digital grayscale image. (Photo courtesy of Donald E. Beasley.)

Image Acquisition

Consider the characteristics of a digital grayscale image, such as the one in Figure 7.29. A digital image is composed of an $m \times n$ array of pixels, as illustrated in Figure 7.30. Here we have zoomed in on the edge of one of the coins in Figure 7.29, and we see that the image is composed of pixels having various levels of gray shading. The grayscale images in Figures 7.29 and 7.30 are 8-bit digital images; each pixel is assigned a number between 0 and 256, with zero corresponding to pure black.

The device that enables digital imaging is termed a charge-coupled device (CCD).[2] A CCD image sensor is essentially an array of capacitors that gain charge when exposed to light. Each

Figure 7.30 Digital image of an m × n pixel array. (Photo courtesy of Donald E. Beasley.)

---

[2] The CCD was invented in 1969 at AT&T Bell Labs by Willard Boyle and George Smith (*Bell System Technical Journal* 49(4), 1970). In 2009 they shared the Nobel Prize in physics for this invention.

capacitor represents one pixel in the resulting digital image. A digital camera may have a physical or an electronic shutter, but in either case the CCD array is active for an exposure time exactly as film cameras function. Once the exposure is over, the CCD accumulated charge is converted to a digital representation of the image. The resolution of the image is determined by the number of pixels. Common cameras may have resolutions that range from 0.3 million pixels (MP) to 21 MP, but higher resolutions are available.

In many applications for digital imaging, the images are transferred directly from the camera to a computer, with image acquisition rates as fast as 200,000 frames per second. The term framegrabber is used to describe the computer hardware that acquires the images from a CCD camera. Most framegrabber hardware is in the form of a plug-in board or an external module for a computer. As with any photography, very short exposure times require high-intensity lighting.

Image Processing

A basic treatment of the fundamentals and applications of digital image processing would easily require an entire book. So our purpose here is to describe one basic issue as illustrative of the impact of digital image processing. Let's consider again Figure 7.29, which shows five coins. Two tasks that might be reasonable to ask an imaging system to accomplish are to locate the coins and determine their diameter, sensing tasks with parallels in automated manufacturing.

Both these tasks can be accomplished if we can locate the outer edges of the coins. In principle, an edge occurs where the gradient of the pixel intensity is a maximum. So, some numerical scheme for finding the gradient is required. In a two-dimensional image, the gradient has both magnitude and direction. For the two-dimensional array that comprises a grayscale image, two gradients are calculated, one in the x-direction and one in the y-direction (horizontal and vertical in the image). Using these two numerical estimates of the gradient, a magnitude and direction can be determined. We will explore two gradient-based methods, *Sobel* and *Canny*, as they are implemented in Matlab. The basic differences in these two methods are the degree of smoothing and the criteria for defining an edge. The *Canny* method smoothes the image to suppress noise and uses two threshold values in determining the location of an edge. Gradient values that are below the low threshold result in a pixel being assigned as a non-edge. Pixels where gradient values are above the high threshold are set as edges pixels. Gradient values between the two thresholds are examined to see if adjacent pixels are an edge, and if there is a direct path to an edge pixel that pixel is included as an edge.

The implementation of edge detection in Matlab is a very straightforward process for grayscale images. The function *imread* creates the $m \times n$ matrix of grayscale values. The function *edge* identifies the edges with the method selected. The function *imshow* views the image.

Figure 7.31 shows the results of the two edge detection methods for our image of coins. Clearly the grayscale image contains information that causes the edge detection methods to find many edges that have nothing to do with our coins. One method to improve the chances of finding the edges of the coins is to subject the image to a process called *thresholding* whereby all pixel values above a certain level are set to pure white and all below to pure black. Let's threshold our image at level 127. The resulting image is shown in Figure 7.32. Applying the same edge detection methods used on the grayscale image gives the image shown in Figure 7.33. This is a step closer to finding the edges of the coins and sizing their diameters.

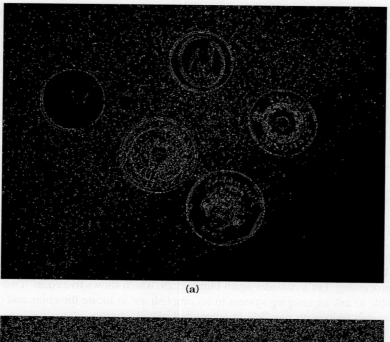

(a)

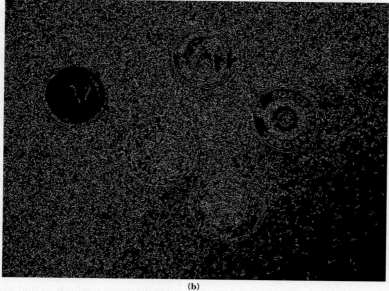

(b)

Figure 7.31 Edge detection methods. **(a)** Sobel method of edge detection. **(b)** Canny method of edge detection. (Photo courtesy of Donald E. Beasley.)

Figure 7.32   Improved effect of using threshold values to eliminate noise. (Photo courtesy of Donald E. Beasley.)

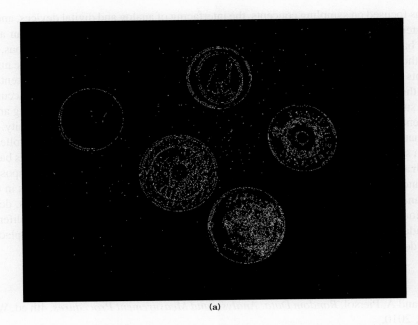

(a)

Figure 7.33   Effects of using threshold values on a grayscale image. **(a)** Sobel method of edge detection. **(b)** Canny method of edge detection. (Photo courtesy of Donald E. Beasley.)

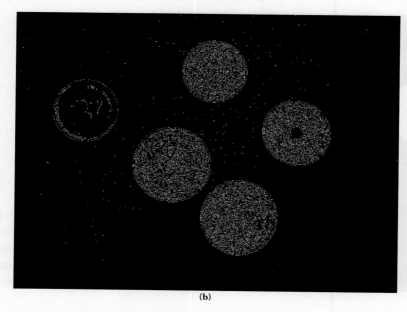

(b)

Figure 7.33   (*Continued*)

7.11   SUMMARY

This chapter has focused on sampling concepts, the interfacing of analog and digital devices, and data acquisition systems. Despite the advantages of digital systems, they must interact with an analog world, and the back-and-forth exchange between an analog and digital signal has limitations. With this in mind, a thorough discussion has been provided on the selection of sample rate and the number of measurements required to reconstruct a continuous process variable from a discrete representation. Equations 7.2 through 7.10 explain the criteria by which a periodic waveform can be accurately represented by such a discrete time series. The limitations resulting from improper sampling and the discrete representation include frequency alias and leakage, as well as amplitude ambiguity.

The fact that practically any electrical instrument can be interfaced with a microcontroller or a data acquisition system is significant. The working mechanics of A/D and D/A converters is basic to interfacing such analog and digital devices with data acquisition systems. The limitations imposed by the resolution and range of these devices mandate certain signal conditioning requirements in terms of signal gain and filtering prior to acquisition. These require analog amplifiers and filters, devices discussed previously. Communication between digital devices surrounds us. But their proliferation means that standards for communication must be set. These standards are evaluated and replaced as newer devices demand faster data transfer.

REFERENCES

1. Bendat, J., and A. Piersol, *Random Data: Analysis and Measurement Procedures*, 4th ed, Wiley, New York, 2010.
2. Proakis, J. G., and D. K. Manolakis, *Digital Signal Processing*, 4th ed., Prentice-Hall, Pearson Education, Upper Saddle River, NJ, 2007.

3. Lyons, R., *Understanding Digital Signal Processing*, 3rd ed., Pearson Education, Upper Saddle River, NJ, 2010.

4. Swedlin, E. G., and D. L. Ferro, *Computers: The Life Story of a Technology*, Johns Hopkins University Press, Baltimore, MD, 2007.

5. Di Paolo, M., *Data-Acquisition Systems: From Fundamentals to Applied Design*, Springer, New York, 2013.

NOMENCLATURE

e	error		E_o	output voltage (V)
e_Q	quantization error		E_{FSR}	full-scale analog voltage range
f	frequency (t^{-1})		G	amplifier gain
f_a	alias frequency (t^{-1})		I	electric current (A)
f_c	filter cutoff frequency (t^{-1})		M	number of bits
f_m	maximum analog signal frequency (t^{-1})		$M(f)$	magnitude ratio at frequency f
f_N	Nyquist frequency (t^{-1})		N	data set size
f_s	sample rate frequency (t^{-1})		$N\delta t$	total digital sample period (t)
k	cascaded filter stage number		R	resistance (Ω)
t	time (t)		Q	A/D converter resolution (V)
$y(t)$	analog signal y_i,		δf	frequency resolution of DFT (t^{-1})
$y(r\delta t)$	discrete values of a signal, $y(t)$		δR	change in resistance (Ω)
$\{y(r\delta t)\}$	complete discrete time signal of $y(t)$		δt	sample time increment (t)
A	amplifier open loop gain; also, constant		τ	time constant (t)
E	voltage (V)		$\phi(f)$	phase shift at frequency f
E_i	input voltage (V)			

PROBLEMS

For many of these problems, using spreadsheet software or the accompanying software will facilitate solution.

7.1 Convert the analog voltage $E(t) = 2 \sin 4\pi t$ mV into a discrete time signal. Specifically, use sample time increments of (a) 1/8 s, (b) 1/5 s, (c) 1/3 s, and (d) 1/21 s, and use a data set of 128 points. Plot 4 s of each series as a function of time. Discuss apparent differences between the discrete representations of the analog signal.

7.2 For the analog voltage $E(t) = 2 \sin 4\pi t$ mV sampled at time increments of (a) 1/8 s, (b) 1/5 s, (c) 1/3 s, and (d) 1/21 s, compute the DFT for each. Use a data set of 128 points. Discuss apparent differences.

7.3 An analog signal has the form $E(t) = 4 \sin 8\pi t$ (V). Compute the DFT at sample rates of 4 Hz and 16 Hz. Use a data set of 256 points. Discuss and compare your results.

7.4 Determine the alias frequency that results from sampling f_1 at sample rate f_s:

$$f_1 = 60 \text{ Hz}; f_s = 90 \text{ Hz}$$
$$f_1 = 1.2 \text{ kHz}; f_s = 2 \text{ kHz}$$
$$f_1 = 10 \text{ Hz}; f_s = 6 \text{ Hz}$$
$$f_1 = 16 \text{ Hz}; f_s = 8 \text{ Hz}$$

7.5 An experimental analysis of the natural oscillations in a particular structure shows the dominant frequencies of interest appear at below 200 Hz. However, frequency information also exists at 350, 450, and 750 Hz. If the signal is sampled at 400 Hz, how will the information in the aliased frequencies appear in the sampled data. How can one prevent this information from being superposed onto the desired signal?

7.6 A particular data acquisition system is used to convert the analog signal $E(t) = (\sin 2\pi t + 2 \sin 8\pi t)$ V into a discrete time signal using a sample rate of 16 Hz. Build the discrete time signal, and from that, use the Fourier transform to reconstruct the Fourier series.

7.7 Consider the continuous signal found in Example 2.3. What would be an appropriate sample rate and sample period to use in sampling this signal if the resulting discrete series must have a size of 2^M, where M is an integer and the signal is to be filtered at and above 2 Hz?

7.8 A golf cart engine operates between 1,000 to 3,000 rpm. Engine vibrations are transmitted at these frequencies to the cart chassis. Accelerometer measurements are taken with the engine operating at 3,000 rpm. Estimate a proper sample size and sample rate to acquire at least 100 cycles of the signal and to minimize spectral leakage. What is the frequency resolution of your sampled signal for spectral analysis? What is the Nyquist frequency? Show your resulting amplitude spectrum (assume a unit amplitude for the time-based signal).

7.9 Convert the analog signal $E(t) = (4 + 2 \sin 4\pi t + 3 \sin 16\pi t)$ V into a discrete time signal using a sample rate of 32 Hz. Build the discrete time signal and its amplitude and phase spectra. Then try at $f_s = 16$ Hz and at $f_s = 40$ Hz. Discuss results.

7.10 Let $y(t) = \sin 100\pi t$. Discuss and show the effects of leakage by acquiring 128 data points of this signal using $f_s = 250$ Hz and 200 Hz and examining the amplitude spectrum. Use an amplitude of 1 unit.

7.11 The time-dependent pressure and flow rate are measured downstream of an artificial heart valve situated within an experimental mock circulation of the human body. The signals are sampled at a rate of 160 Hz. The valve opens and closes with the heart rate at 75 beats per minute. Blocks of data are recorded at 128 samples per block. What is the frequency resolution of the sampled data? How many signal (heart beat) cycles are included in each block of 128 samples?

7.12 Convert the following straight binary numbers to positive integer base 10 numbers:
a. 1010
b. 11111
c. 10111011
d. 1100001

7.13 Convert (a) 1100111.1101 (binary) into a base 10 number; (b) 4B2F into straight binary; (c) 278.632 (base 10) into straight binary.

7.14 Convert the following decimal (base 10) numbers into bipolar binary numbers using a twos complement code:
a. 10
b. −10
c. −247
d. 1013

7.15 List some possible sources of uncertainty in the dual-slope procedure for A/D conversion. Derive a relationship between the uncertainty in the digital result and the slope of the integration process.

7.16 Compute the resolution and SNR for an M-bit A/D converter having a full-scale range of ±5 V. Let M be 4, 8, 12, and 16.

7.17 A 12-bit A/D converter having an $E_{FSR} = 5$ V has a relative accuracy of 0.03% FS (full-scale). Estimate its quantization error in volts. What is the total possible error expected in volts? What value of relative uncertainty might be used for this device?

7.18 A 16-bit A/D converter has a full-scale range of +10 V. What is the resolution of this A/D in volts? If this A/D were used to measure temperature using a sensor having a sensitivity of 0.1 mV/°C, what is the resolution in °C?

7.19 Determine the minimum number of bits required for an A/D converter having an analog range of 0 to 5 V if the minimum resolution that can be tolerated is 0.61 mV.

7.20 An 8-bit single-ramp A/D converter with $E_{FSR} = 10$ V uses a 2.5-MHz clock and a comparator having a threshold voltage of 1 mV. Estimate the following:
a. The binary output when the input voltage is $E = 6.000$ V and when $E = 6.035$ V
b. The actual conversion time for the 6.000-V input and average conversion times
c. The resolution of the converter

7.21 Compare the maximum conversion times of a 10-bit successive approximation A/D converter to a dual-slope ramp converter if both use a 1-MHz clock and $E_{FSR} = 10$ V.

7.22 An 8-bit D/A converter shows an output of 3.58 V when straight binary 10110011 is applied. What is the output voltage when 01100100 is applied?

7.23   During a test, the exact output from a load cell is an analog voltage of 3.150 V. This output signal is measured by a 0 to 5V, 8-bit successive approximation A/D converter. Show the comparator conversion sequence and register values as used by the A/D converter to achieve a final quantized value. Determine the actual measured voltage value and the register value corresponding to the measured voltage.

7.24   A 0- to 10-V, 4-bit successive approximation A/D converter is used to measure an input voltage of 4.9 V.

    a. Determine the binary representation and its analog approximation of the input signal. Explain your answer in terms of the quantization error of the A/D converter.

    b. If we wanted to ensure that the analog approximation be within 2.5 mV of the actual input voltage, estimate the number of bits required of an A/D converter.

7.25   Discuss the trade-offs between resolution and conversion rate for successive approximation, ramp, and parallel converters.

7.26   A dual-slope A/D converter has 12-bit resolution and uses a 10-kHz internal clock. Estimate the conversion time required for an output code equivalent to 2011_{10}.

7.27   How long does it take an 8-bit single ramp A/D converter using a 1-MHz clock to convert the number 173_{10}?

7.28   A successive approximation A/D converter has a full-scale output of 0 to 10 V and uses an 8-bit register. An input of 6.185 V is applied. Go through the comparator conversion sequence as used by the A/D converter to arrive at the measured value. Determine the actual measured voltage value and the register value corresponding to the measured voltage.

7.29   A 12-bit A/D converter has a full-scale output of 0 to 10 V. A voltage measurement is performed, and the register value is 010101001000. What is the value of the measured voltage?

7.30   A pressure transducer has a range of ±100 mm Hg and a sensitivity of 50 μV/mm Hg/V (excitation). The excitation voltage is set to 10 VDC. The transducer output signal is passed through a nulling amplifier circuit with a gain of G before being measured on a bipolar, ±5 V, 16-bit data acquisition system. For a dynamic signal varying between ±100 mm Hg, assign an appropriate gain G and estimate the output voltage range as measured by the DAS. What is the resolution in pressure (per bit) measured?

7.31   The voltage from a 0- to 5-kg strain gauge balance scale is expected to vary from 0 to 3.50 mV. The signal is to be recorded using a 12-bit A/D converter having a ±5-V range with the weight displayed on a computer monitor. Suggest an appropriate amplifier gain for this situation.

7.32   A strain gauge based balance scale has an input range of 0 to 5 kg and a corresponding output range of 0 to 3.5 mV. The output signal from the scale is to be measured using a 0 to 10V, 16 bit A/D converter. Determine the value of gain required so that 1% of the full-scale output signal of the balance scale will be within 1 bit of the resolution of the converter.

7.33   An aircraft wing oscillates under wind gusts. The oscillations are expected to be at about 2 Hz. Wing-mounted strain-gauge sensors are connected to a ±5-V, 12-bit A/D converter and data acquisition system to measure this. For each test block, 10 seconds of data are sampled.

    a. Suggest an appropriate sample rate. Explain.

    b. If the signal oscillates steadily with an amplitude of 2 V, express the signal as a Fourier series.

    c. Based on (a) and (b), sketch a plot of the expected amplitude spectrum. What is the frequency spacing on the abscissa? What is its Nyquist frequency?

7.34   A bipolar, ±5 V, 12-bit successive approximation A/D converter is used in a data acquisition system. The analog input signal to the A/D converter is first passed through an amplifier of gain, G. Determine the effect on the quantization error (given as $\frac{1}{2}$ Q) and the measurable analog voltage range if the gain is set to $G = 1, 5, 10, 50, 100$.

7.35   Select an appropriate sample rate and data number to acquire with minimal leakage the first five terms of a square wave signal having a fundamental period of 10 seconds. Select an appropriate cutoff frequency for an anti-alias filter. Hint: Approximate the square wave as a Fourier series.

7.36 A triangle wave with a period of 2 s can be expressed by the Fourier series

$$y(t) = \sum \left[2D_1(1 - \cos n\pi)/n\pi \right] \cos \pi nt$$
$$n = 1, 2, \ldots$$

Specify an appropriate sample rate and data number to acquire the first seven terms with minimal leakage. Select an appropriate cutoff frequency for an anti-alias filter.

7.37 Using Fourier transform software (or equivalent software), generate the amplitude spectrum for the square wave of Problem 7.35.

7.38 Using Fourier transform software (or equivalent software), generate the amplitude spectrum for the triangle wave of Problem 7.36. Use $D_1 = 1$ V.

7.39 A single-stage low-pass RC Butterworth filter with $f_c = 100$ Hz is used to filter an analog signal. Determine the attenuation of the filtered analog signal at 10, 50, 75, and 200 Hz.

7.40 A three-stage LC Bessel filter with $f_c = 100$ Hz is used to filter an analog signal. Determine the attenuation of the filtered analog signal at 10, 50, 75, and 200 Hz.

7.41 Design a cascading LC Butterworth low-pass filter that has a magnitude ratio flat to within 3 dB from 0 to 5 kHz but with an attenuation of at least 30 dB for all frequencies at and above 10 kHz.

7.42 Choose an appropriate cascading low-pass filter to remove a 500-Hz component contained within an analog signal that is to be passed through an 8-bit A/D converter having 10-V range and 200-Hz sample rate. Attenuate the component to within the A/D converter quantization error.

7.43 The voltage output from a J-type thermocouple referenced to 0 °C is to be used to measure temperatures from 50 to 70 °C. The output voltages vary linearly over this range from 2.585 to 3.649 mV.

a. If the thermocouple voltage is input to a 12-bit A/D converter having a ±5-V range, estimate the percent quantization error in the digital value.

b. If the analog signal can be first passed through an amplifier circuit, compute the amplifier gain required to reduce the quantization error to 5% or less.

c. If the ratio of signal-to-noise level (SNR) in the analog signal is 40 dB, compute the magnitude of the noise after amplification. Discuss the results of 7.40b in light of this.

7.44 Specify an appropriate ±5-V M-bit A/D converter (8- or 12-bit), sample rate (up to 100 Hz) and signal conditioning to convert these analog signals into digital series. Estimate the quantization error and dynamic error resulting from the system specified:

a. $E(t) = 2 \sin 20\pi t$ V

b. $E(t) = 1.5 \sin \pi t + 20 \sin 32\pi t - 3 \sin(60\pi t + \pi/4)$ V

c. $P(t) = -10 \sin 4\pi t + 5 \sin 8\pi t$ psi; $K = 0.4$ V/psi

7.45 The following signal is to be sampled using a 12-bit, ±5-V data acquisition board

$$y(t) = 4 \sin 8\pi t + 2 \sin 20\pi t + 3 \sin 42\pi t$$

Select an appropriate sample rate and sample size that provide minimal spectral leakage.

7.46 A strain-gauge sensor is used with a bridge circuit and connected to a DAS as indicated in Figure 7.16. Estimate the range of offset nulling voltage available if the bridge excitation is 3.333 V, sensor and bridge resistors are each at a nominal value of 120 Ω, and the adjustable trim potentiometer is rated at 39 kΩ.

7.47 Design a low-pass Butterworth filter around a 10 Hz cutoff (−3 dB) frequency. The filter is to pass 95% of signal magnitude at 5 Hz but no more than 10% at 20 Hz. Source and load impedances are 10 Ω.

7.48 A two-stage LC Butterworth filter with $f_c = 100$ Hz is used as an anti-alias filter for an analog signal. Determine the signal attenuation experienced at 10, 50, 75, and 200 Hz.

The following problems make use of the accompanying software.

7.49 In the discussion of Figure 7.4, we point out the effects of sample rate and total sample period on the reconstructed time signal and its amplitude spectrum. Use the program *Leakage.2* to duplicate Figure 7.4. Then develop a similar example (signal frequency, sample rate, and sample period) in which fewer points and a slower sample rate lead to a better reconstruction in both time and frequency domains. Incidentally, for a fixed sample rate, you can increment N and watch the acquired waveform develop such that the leakage decreases to zero as the acquired signal reaches an exact integer period of the waveform. For a fixed N, the same can be shown for changes in sample rate.

7.50 Using program *Aliasing*, solve Example 7.1 to find the alias frequency. Observe the plot of the original and the acquired signal, as well as the amplitude spectrum. Decrement the signal frequency 1 Hz at a time until it is within the region where there no longer is aliasing. Based on these observations, discuss how the acquired time signal changes and how this is related to aliasing.

7.51 Use program *Aliasing* to understand the folding diagram of Figure 7.3. For a sample rate of 20 Hz, vary the signal frequency over its full range. Determine the frequencies corresponding to f_N, $2f_N$, $3f_N$, and $4f_N$ on Figure 7.3. Determine the alias frequencies corresponding to $1.6f_N$, $2.1f_N$, $2.6f_N$, and $3.2f_N$.

7.52 Using program *Signal generation*, examine the rule that when exact discrete representations are not possible, a sample rate of at least 5 to 10 times the maximum signal frequency gives adequate approximation. Select a sine wave of 2 Hz and

discuss the acquired waveform as the sample rate is increased incrementally from a low to a high value. At what sample rate does the signal look like a sine wave? Compare with the corresponding frequency and amplitude content from the amplitude spectrum. Write a short discussion of your observations and conclusions.

7.53 Program *Leakage.2* samples a single frequency signal at a user-defined sample rate and period. Describe how sample period corresponds to the length of the signal measured and how this affects leakage in the amplitude spectrum. Does frequency resolution matter?

7.54 The image file with the companion software, *coins.jpg*, is the original color photograph of the coins used in Figures 7.29 through 7.33. The image file *graycoins.jpg* is the corresponding grayscale image. Using the Matlab commands IMREAD, IMSHOW, IM2BW, and EDGE, reproduce the results in Figures 7.29 through 7.33.

7.55 Using an image you download from the Internet or acquire with your own camera, create a grayscale image and employ edge detection to process the image. Explore the effects of the gradient threshold value on the quality of the edge detection. Create a binary image and repeat the process.

Chapter **8**

Temperature Measurements

8.1 INTRODUCTION

Temperature is one of the most commonly used and measured engineering variables. Although much of our lives is affected by the diurnal and seasonal variations in ambient temperature, the fundamental scientific definition of temperature and a scale for the measurement of temperature are not commonly understood. This chapter explores the establishment of a practical temperature scale and common methods of temperature measurement. In addition, errors associated with the design and installation of temperature sensors are discussed.

Upon completion of this chapter, the reader will be able to

- Describe the primary standards for temperature
- State the role of fixed point calibration and the necessity for an interpolation method in establishing a temperature standard
- State the physical principle underlying electrical resistance thermometry
- Employ standard relationships to determine temperature from resistance devices
- Analyze thermoelectric circuits designed to measure temperature
- State the principles employed in radiation temperature measurements
- Estimate the impact of loading errors in temperature measurement

Historical Background

Guillaume Amontons (1663–1705), a French scientist, was one of the first to explore the thermo-dynamic nature of temperature. His efforts examined the behavior of a constant volume of air that was subject to temperature changes. The modern liquid-in-glass bulb thermometer traces its origin to Galileo (1565–1642), who attempted to use the volumetric expansion of liquids in tubes as a relative measure of temperature. Unfortunately, this open tube device was sensitive to both barometric pressure and temperature changes. A major advance in temperature measurement occurred in 1630 as a result of a seemingly unrelated event: the development of the technology to manufacture capillary glass tubes. These tubes were then used with water and alcohol in a thermometric device resembling a bulb thermometer, and these devices eventually led to the development of a practical temperature-measuring instrument.

A temperature scale proposed by Gabriel D. Fahrenheit, a German physicist (1686–1736), in 1715 attempted to incorporate body temperature as the median point on a scale having 180 divisions between the freezing point and the boiling point of water. Fahrenheit also successfully used mercury as the liquid in a bulb thermometer, making significant improvements over the attempts of Ismael Boulliau in 1659. In 1742, the Swedish astronomer Anders Celsius[1] (1701–1744) described a temperature scale that divided the interval between the boiling and freezing points of water at 1 atm pressure into 100 equal parts. The boiling point of water was fixed as 0, and the freezing point of water as 100. Shortly after Celsius's death, Carolus Linnaeus (1707–1778) reversed the scale so that the 0 point corresponded to the freezing point of water at 1 atm. Even though this scale may not have been originated by Celsius (1), in 1948 the change from degrees centigrade to degrees Celsius was officially adopted.

As stated by H. A. Klein in *The Science of Measurement: A Historical Survey* (2),

From the original thermoscopes of Galileo and some of his contemporaries, the measurement of temperature has pursued paths of increasing ingenuity, sophistication and complexity. Yet temperature remains in its innermost essence the average molecular or atomic energy of the least bits making up matter, in their endless dance. Matter without motion is unthinkable. Temperature is the most meaningful physical variable for dealing with the effects of those infinitesimal, incessant internal motions of matter.

8.2 TEMPERATURE STANDARDS AND DEFINITION

Temperature can be loosely described as the property of an object that describes its hotness or coldness, concepts that are clearly relative. Our experiences indicate that heat transfer tends to equalize temperature, or more precisely, systems that are in thermal communication eventually have equal temperatures. The *zeroth law of thermodynamics* states that two systems in thermal equilibrium with a third system are in thermal equilibrium with each other. Thermal equilibrium implies that no heat transfer occurs between the systems, defining the equality of temperature. Although the zeroth law of thermodynamics essentially provides the definition of the equality of temperature, it provides no means for defining a temperature scale.

A temperature scale provides for three essential aspects of temperature measurement: (1) the definition of the size of the degree, (2) fixed reference points for establishing known temperatures, and (3) a means for interpolating between these fixed temperature points. These provisions are consistent with the requirements for any standard, as described in Chapter 1.

Fixed Point Temperatures and Interpolation

To begin, consider the definition of the triple point of water as having a value of 0.01 for our temperature scale, as is done for the Celsius scale (0.01 °C). This provides for an arbitrary starting, or fixed, point for a temperature scale; in fact, the numeric value assigned to this temperature could be anything. For example, on the Fahrenheit temperature scale it has a value very close to 32. Fixed points are typically defined by melting or boiling point temperatures, or the triple point, of a pure substance. The point at which pure water boils at one standard atmosphere pressure is an easily

---

[1] It is interesting to note that in addition to his work in thermometry, Celsius published significant papers on the aurora borealis and the falling level of the Baltic Sea.

reproducible fixed point temperature. For our purposes, let's assign this fixed point a numerical value of 100.

The next problem is to define the size of the degree. Because we have two fixed points on our temperature scale, we can see that the degree is 1/100th of the temperature difference between the ice point and the boiling point of water at atmospheric pressure.

Conceptually, this defines a workable scale for the measurement of temperature, but as yet we have made no provision for interpolating between the two fixed-point temperatures.

Interpolation

The calibration of a temperature measurement device entails not only the establishment of fixed temperature points, but also the indication of any temperature between fixed points. The operation of a liquid-in-glass thermometer is based on the thermal expansion of a liquid, often mercury or alcohol, contained in a glass capillary. The level of the liquid is read as an indication of the temperature. Imagine that we submerged the thermometer in water at the ice point, made a mark on the glass at the height of the column of liquid, and labeled it 0 °C, as illustrated in Figure 8.1. Next we submerged the thermometer in boiling water and again marked the level of the liquid, this time labeling it 100 °C.

Using reproducible fixed temperature points, we have calibrated our thermometer at two points, but we want to be able to measure temperatures other than these two fixed points. How can we determine the appropriate place on the thermometer to mark, say, 50 °C?

The process of establishing 50 °C without a fixed-point calibration is called interpolation. The simplest option would be to divide the distance on the thermometer between the marks representing 0 and 100 into equally spaced degree divisions. This places 50 °C as shown in Figure 8.1. What assumption is implicit in this method of interpolation? It is obvious that we do not have enough information to appropriately divide the interval between 0 and 100 on the thermometer into degrees. A theory of the behavior of the liquid in the thermometer or many fixed points for calibration is necessary to resolve our dilemma.

Even by the late eighteenth century, there was no standard for interpolating between fixed points on the temperature scale; the result was that different thermometers indicated different temperatures away from fixed points, sometimes with surprisingly large errors.

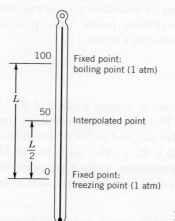

Figure 8.1 Calibration and interpolation for a liquid-in-glass thermometer.

Temperature Scales and Standards

It is necessary to reconcile an arbitrary temperature scale with the idea of absolute temperature. Thermodynamics defines a temperature scale that has an absolute reference and defines an absolute zero for temperature. For example, this absolute temperature governs the energy behavior of an ideal gas and is used in the ideal gas equation of state. The behavior of real gases at very low pressure may be used as a temperature standard to define a practical measure of temperature that approximates the thermodynamic temperature. The unit of degrees Celsius (°C) is a practical scale related to the absolute Kelvin scale as °C = K − 273.15.

The modern engineering definition of the temperature scale is provided by a standard called the International Temperature Scale of 1990 (ITS-90) (3). This standard establishes fixed points for temperature, and provides standard procedures and devices for interpolating between fixed points. It establishes the Kelvin (K) as the unit for the fundamental increment in temperature. Temperatures established according to ITS-90 do not deviate from the thermodynamic temperature scale by more than the uncertainty in the thermodynamic temperature at the time of adoption of ITS-90. The primary fixed points from ITS-90 are shown in Table 8.1. In addition to these fixed points, other fixed points of secondary importance are available in ITS-90.

Standards for Interpolation

Along with the fixed temperature points established by ITS-90 (3), a standard for interpolation between these fixed points is necessary. Standards for acceptable thermometers and interpolating equations are provided in ITS-90. For temperatures ranging from 13.8033 to 1,234.93 K, ITS-90 establishes a platinum resistance thermometer as the standard interpolating instrument, and establishes interpolating equations that relate temperature to resistance. Above 1,234.93 K, temperature is defined in terms of blackbody radiation, without specifying an instrument for interpolation.

Table 8.1   Temperature Fixed Points as Defined by ITS-90

Defining State	Temperature[a]	
	K	°C
Triple point of hydrogen	13.8033	−259.3467
Liquid–vapor equilibrium for hydrogen at 25/76 atm	≈17	≈−256.15
Liquid–vapor equilibrium for hydrogen at 1 atm	≈20.3	≈−252.87
Triple point of neon	24.5561	−248.5939
Triple point of oxygen	54.3584	−218.7916
Triple point of argon	83.8058	−189.3442
Triple point of water	273.16	0.01
Solid–liquid equilibrium for gallium at 1 atm	302.9146	29.7646
Solid–liquid equilibrium for tin at 1 atm	505.078	231.928
Solid–liquid equilibrium for zinc at 1 atm	692.677	419.527
Solid–liquid equilibrium for silver at 1 atm	1,234.93	961.78
Solid–liquid equilibrium for gold at 1 atm	1,337.33	1,064.18
Solid–liquid equilibrium for copper at 1 atm	1,357.77	1,084.62

[a]significant digits shown are as provided in ITS-90.

In summary, temperature measurement, a practical temperature scale, and standards for fixed points and interpolation have evolved over a period of about two centuries. Present standards for fixed-point temperatures and interpolation allow for practical and accurate measurements of temperature. In the United States, the National Institute of Standards and Technology (NIST) provides for a means to obtain accurately calibrated platinum resistance thermometers for use as secondary standards in the calibration of a temperature measuring system to any practical level of uncertainty.

8.3 THERMOMETRY BASED ON THERMAL EXPANSION

Most materials exhibit a change in size with changes in temperature. Since this physical phenomenon is well defined and repeatable, it is useful for temperature measurement. The liquid-in-glass thermometer and the bimetallic thermometer are based on this phenomenon.

Liquid-in-Glass Thermometers

A liquid-in-glass thermometer measures temperature by virtue of the thermal expansion of a liquid. The construction of a liquid-in-glass thermometer is shown in Figure 8.2. The liquid is contained in a glass structure that consists of a bulb and a stem. The bulb serves as a reservoir and provides sufficient fluid for the total volume change of the fluid to cause a detectable rise of the liquid in the stem of the thermometer. The stem contains a capillary tube, and the difference in thermal expansion between the liquid and the glass produces a detectable change in the level of the liquid in the glass capillary. Principles and practices of temperature measurement using liquid-in-glass thermometers are described elsewhere (4).

During calibration, such a thermometer is subject to one of three measuring environments:

1. For a *complete immersion thermometer*, the entire thermometer is immersed in the calibrating temperature environment or fluid.

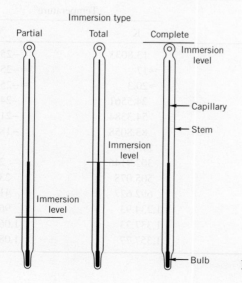

Figure 8.2 Liquid-in-glass thermometers.

2. For a *total immersion thermometer*, the thermometer is immersed in the calibrating temperature environment up to the liquid level in the capillary.

3. For a *partial immersion thermometer*, the thermometer is immersed to a predetermined level in the calibrating environment.

For the most accurate temperature measurements, the thermometer should be immersed in the same manner in use as it was during calibration. Liquid-in-glass thermometers have limited engineering applications but do provide reliable, accurate temperature measurement. Thus they may be used as a local standard for calibration of other temperature sensors.

Bimetallic Thermometers

The physical phenomenon employed in a bimetallic temperature sensor is the differential thermal expansion of two metals. Figure 8.3 shows the construction and response of bimetallic sensors to a change in temperature. The sensor is constructed by bonding two strips of different metals, A and B. The resulting bimetallic strip may be in a variety of shapes, depending on the particular application. Consider the simple linear construction shown in Figure 8.3. At the assembly temperature, T_1, the bimetallic strip is straight, but for temperatures other than T_1 the strip has a curvature. The physical basis for the relationship between the radius of curvature and temperature is given as

$$r_c \propto \frac{d}{\left[(C_\alpha)_A - (C_\alpha)_B \right](T_2 - T_1)} \tag{8.1}$$

where r_c is the radius of curvature, C_α is the material thermal expansion coefficient, T is temperature, and d is sensor thickness.

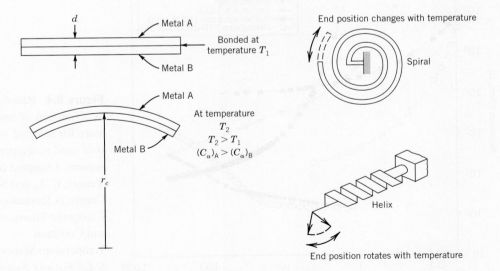

Figure 8.3 Expansion thermometry using bimetallic materials: strip, spiral, and helix forms.

Bimetallic strips employ one metal having a high coefficient of thermal expansion with another having a low coefficient, providing increased sensitivity. Invar is often used as one of the metals, because for this material $C_\alpha = 1.7 \times 10^{-8}$ m/m °C, as compared to typical values for other metals, such as steels, which range from approximately 2×10^{-5} to 20×10^{-5} m/m °C.

The bimetallic sensor is the primary element in most dial thermometers and many thermostats. The geometries shown in Figure 8.3 serve to provide the desired deflection in the bimetallic strip for a given application. Dial thermometers using a bimetallic strip as their sensing element typically provide temperature measurements with uncertainties of ± 1 °C.

8.4 ELECTRICAL RESISTANCE THERMOMETRY

As a result of the physical nature of the conduction of electricity, electrical resistance of a conductor or semiconductor varies with temperature. Using this behavior as the basis for temperature measurement is extremely simple in principle and leads to two basic classes of resistance thermometers: resistance temperature detectors (conductors) and thermistors (semiconductors). Resistance temperature detectors (RTDs) may be formed from a solid metal wire that exhibits a change in electrical resistance with temperature. Depending on the material selected, the resistance may increase or decrease with temperature. A thermistor may have a positive temperature coefficient (PTC) or a negative temperature coefficient (NTC). Figure 8.4 shows resistance as a function of temperature for a variety of conductor and semiconductor materials used to measure temperature. The PTC materials are metals or alloys and the NTC materials are semiconductors. Cryogenic temperatures are included in this figure, and germanium is clearly an excellent choice for low temperature measurement because of its large sensitivity.

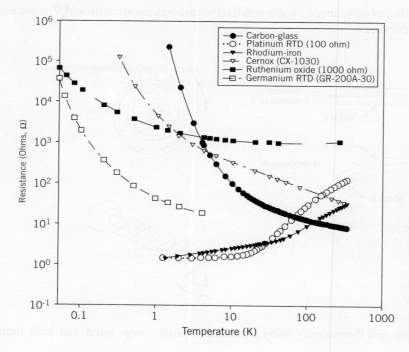

Figure 8.4 Resistance as a function of temperature for selected materials used as temperature sensors. (Adapted from Yeager, C. J., and S. S. Courts, A Review of Cryogenic Thermometry and Common Temperature Sensors, *IEEE Sensors Journal*, 1(4), 2001.)

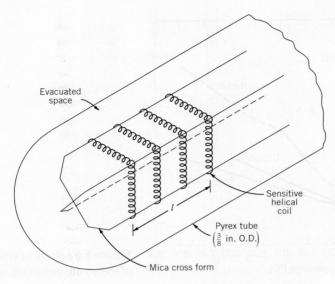

Figure 8.5   Construction of a platinum RTD. (From Benedict, R. P., *Fundamentals of Temperature, Pressure, and Flow Measurements*, 3rd ed. Copyright © 1984 by John Wiley and Sons, New York.)

Resistance Temperature Detectors

In the case of a resistance temperature detector (RTD),[2] the sensor is generally constructed by mounting a metal wire on an insulating support structure to eliminate mechanical strains and by encasing the wire to prevent changes in resistance resulting from influences from the sensor's environment, such as corrosion. Figure 8.5 shows such a typical RTD construction.

Mechanical strain significantly changes a conductor's resistance and must be eliminated if accurate temperature measurements are to be made. Such mechanical stresses and resulting strains can be created by thermal expansion. Thus provision for strain-free expansion of the conductor as its temperature changes is essential in the construction of an RTD. The support structure also expands as the temperature of the RTD increases, and the construction allows for strain-free differential expansion.

The physical basis for the relationship between resistance and temperature is the temperature dependence of the resistivity, ρ_e, of a material. The resistance of a conductor of length l and cross-sectional area A_c may be expressed in terms of the resistivity ρ_e as

$$R = \frac{\rho_e l}{A_c} \tag{8.2}$$

The relationship between the resistance of a metal conductor and its temperature may also be expressed by the polynomial expansion:

$$R = R_0\left[1 + \alpha(T - T_0) + \beta(T - T_0)^2 + \cdots\right] \tag{8.3}$$

---

[2] The term RTD in this context refers to metallic PTC resistance sensors.

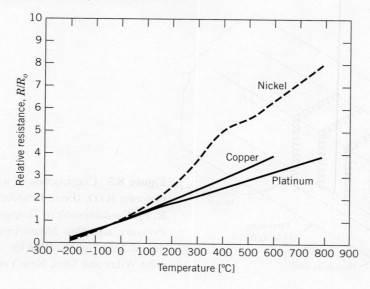

Figure 8.6 Relative resistance of three pure metals (R_0 at 0°C).

where R_0 is a reference resistance measured at a reference temperature T_0. The coefficients α, β, ... are material constants. Figure 8.6 shows the relative relation between resistance and temperature for three common metals. This figure provides evidence that the relationship between temperature and resistance over specific small temperature ranges is linear. This approximation can be expressed as

$$R = R_0[1 + \alpha(T - T_0)] \tag{8.4}$$

where α is termed the temperature coefficient of resistivity. For example, for platinum conductors the linear approximation is accurate to within an uncertainty of $\pm 0.3\%$ over the range 0–200 °C and $\pm 1.2\%$ over the range 200–800 °C. Table 8.2 lists a number of temperature coefficients of resistivity α for materials at 20 °C.

Table 8.2 Temperature Coefficient of Resistivity for Selected Materials at 20 °C

Substance	α [°C^{-1}]
Aluminum (Al)	0.00429
Carbon (C)	−0.0007
Copper (Cu)	0.0043
Gold (Au)	0.004
Iron (Fe)	0.00651
Lead (Pb)	0.0042
Nickel (Ni)	0.0067
Nichrome	0.00017
Platinum (Pt)	0.003927
Tungsten (W)	0.0048

Platinum Resistance Temperature Device (RTD)

Platinum is the most common material chosen for the construction of RTDs. The principle of operation is quite simple: platinum exhibits a predictable and reproducible change in electrical resistance with temperature, which can be calibrated and interpolated to a high degree of accuracy. The linear approximation for the relationship between temperature and resistance is valid over a wide temperature range, and platinum is highly stable. To be suitable for use as a secondary temperature standard, a platinum resistance thermometer should have a value of α not less than $0.003925\,°C^{-1}$. This minimum value is an indication of the purity of the platinum. In general, RTDs may be used for the measurement of temperatures ranging from cryogenic to approximately 650 °C. By properly constructing an RTD and correctly measuring its resistance, an uncertainty in temperature measurement as low as $\pm0.005\,°C$ is possible. Because of this potential for low uncertainties and the predictable and stable behavior of platinum, the platinum RTD is widely used as a local standard

Resistance Temperature Device Resistance Measurement

The resistance of an RTD may be measured by a number of means, and the choice of an appropriate resistance measuring device must be made based on the required level of uncertainty in the final temperature measurement. Conventional ohmmeters cause a small current to flow during resistance measurements, creating self-heating in the RTD. An appreciable temperature change of the sensor may be caused by this current, in effect causing a loading error. This is an important consideration for RTDs.

Bridge circuits, as described in Chapter 6, are used to measure the resistance of RTDs to provide low uncertainties in measured resistance values. Wheatstone bridge circuits are commonly used for these measurements. However, the basic Wheatstone bridge circuit does not compensate for the resistance of the leads in measuring the resistance of an RTD, which is a major source of error in electrical resistance thermometers. When greater accuracies are required, three-wire and four-wire bridge circuits are used. Figure 8.7a shows a three-wire Callendar-Griffiths bridge circuit. The lead

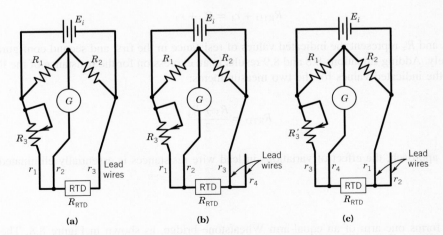

Figure 8.7   Bridge circuits. (**a**) Callender-Griffiths 3-wire bridge; (**b**) and (**c**) Mueller 4-wire bridge. An average of the readings in (**b**) and (**c**) eliminates the effect of lead wire resistances.

wires numbered 1, 2, and 3 have resistances $r_1, r_2,$ and r_3, respectively. At balanced conditions neglecting lead wire effects,

$$\frac{R_1}{R_2} = \frac{R_3}{R_{RTD}}$$

(8.5)

but with the lead wire resistances included in the circuit analysis,

$$\frac{R_1}{R_2} = \frac{R_3 + r_1}{R_{RTD} + r_3}$$

(8.6)

and with $R_1 = R_2$, the resistance of the RTD, R_{RTD}, can be found as

$$R_{RTD} = R_3 + r_1 - r_3$$

(8.7)

If $r_1 = r_3$, the effect of these lead wires is eliminated from the determination of the RTD resistance by this bridge circuit. Note that the resistance of lead wire 2 does not contribute to any error in the measurement at balanced conditions, because no current flows through the galvanometer G.

The four-wire Mueller bridge, as shown in Figure 8.7b,c, provides increased compensation for lead-wire resistances compared to the Callendar-Griffiths bridge and is used with four-wire RTDs. The four-wire Mueller bridge is typically used when low uncertainties are desired, as in cases where the RTD is used as a laboratory standard. A circuit analysis of the bridge circuit in the first measurement configuration, Figure 8.7b, yields

$$R_{RTD} + r_3 = R_3 + r_1$$

(8.8)

and in the second measurement configuration, Figure 8.7c, yields

$$R_{RTD} + r_1 = R_3' + r_3$$

(8.9)

where R_3 and R_3' represent the indicated values of resistance in the first and second configurations, respectively. Adding Equations 8.8 and 8.9 results in an expression for the resistance of the RTD in terms of the indicated values for the two measurements:

$$R_{RTD} = \frac{R_3 + R_3'}{2}$$

(8.10)

With this approach, the effect of variations in lead wire resistances is essentially eliminated.

Example 8.1

An RTD forms one arm of an equal-arm Wheatstone bridge, as shown in Figure 8.8. The fixed resistances, R_2 and R_3 are equal to 25 Ω. The RTD has a resistance of 25 Ω at a temperature of 0 °C and is used to measure a temperature that is steady in time.

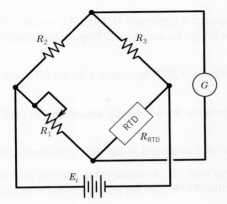

Figure 8.8   RTD Wheatstone bridge arrangement.

The resistance of the RTD over a small temperature range may be expressed, as in Equation 8.4:

$$R_{RTD} = R_0[1 + \alpha(T - T_0)]$$

Suppose the coefficient of resistance for this RTD is $0.003925\,°C^{-1}$. A temperature measurement is made by placing the RTD in the measuring environment and balancing the bridge by adjusting R_1. The value of R_1 required to balance the bridge is $37.36\,\Omega$. Determine the temperature of the RTD.

KNOWN $R(0°C) = 25\,\Omega$

$\quad\quad\quad\quad \alpha = 0.003925°\,C^{-1}$

$\quad\quad\quad\quad R_1 = 37.36\,\Omega$ (when bridge is balanced)

FIND   The temperature of the RTD.

SOLUTION   The resistance of the RTD is measured by balancing the bridge; recall that in a balanced condition

$$R_{RTD} = R_1 \frac{R_3}{R_2}$$

The resistance of the RTD is measured to be $37.36\,\Omega$. With $R_0 = 25\,\Omega$ at $T = 0\,°C$ and $\alpha = 0.003925\,°C^{-1}$, Equation 8.4 becomes

$$37.36\,\Omega = 25(1 + \alpha T)\Omega$$

The temperature of the RTD is $126\,°C$.

Example 8.2

Consider the bridge circuit and RTD of Example 8.1. To select or design a bridge circuit for measuring the resistance of the RTD in this example, the required uncertainty in temperature would be specified. If the required uncertainty in the measured temperature is $\leq 0.5\,°C$, would a 1% total uncertainty in each of the resistors that make up the bridge be acceptable? Neglect the effects of lead wire resistances for this example.

KNOWN A required uncertainty in temperature of $\pm 0.5\,°C$, measured with the RTD and bridge circuit of Example 8.1

FIND The uncertainty in the measured temperature for a 1% total uncertainty in each of the resistors that make up the bridge circuit

ASSUMPTION All uncertainties are provided and evaluated at the 95% confidence level

SOLUTION Perform a design-stage uncertainty analysis. Assuming at the design stage that the total uncertainty in the resistances is 1%, then with initial values of the resistances in the bridge equal to $25\,\Omega$, the design-stage uncertainties are set at

$$u_{R_1} = u_{R_2} = u_{R_3} = (0.01)(25) = 0.25\,\Omega$$

The root-sum-squares (RSS) method is used to estimate the propagation of uncertainty in each resistor to the uncertainty in determining the RTD resistance, u_{RTD}, by

$$u_{\text{RTD}} = \sqrt{\left[\frac{\partial R}{\partial R_1}u_{R_1}\right]^2 + \left[\frac{\partial R}{\partial R_2}u_{R_2}\right]^2 \left[\frac{\partial R}{\partial R_3}u_{R_3}\right]^2}$$

where

$$R = R_{\text{RTD}} = \frac{R_1 R_3}{R_2}$$

and we assume that the uncertainties are not correlated. Then the design-stage uncertainty in the resistance of the RTD is

$$u_{\text{RTD}} = \sqrt{\left[\frac{R_3}{R_2}u_{R_1}\right]^2 + \left[\frac{-R_1 R_3}{R_2^2}u_{R_2}\right]^2 + \left[\frac{R_1}{R_2}u_{R_3}\right]^2}$$

$$u_{\text{RTD}} = \sqrt{(1 \times 0.25)^2 + (1 \times -0.25)^2 + (1 \times 0.25)^2} = 0.433\,\Omega$$

To determine the uncertainty in temperature, we know

$$R = R_{\text{RTD}} = R_0[1 + \alpha(T - T_0)]$$

and

$$u_T = \sqrt{\left(\frac{\partial T}{\partial R}u_{\text{RTD}}\right)^2}$$

Setting $T_0 = 0\,°\text{C}$ with $R_0 = 25\,\Omega$, and neglecting uncertainties in T_0, α, and R_0, we have

$$\frac{\partial T}{\partial R} = \frac{1}{\alpha R_0}$$

$$\frac{1}{\alpha R_0} = \frac{1}{(0.003925°\,\text{C}^{-1})(25\,\Omega)}$$

Then the design-stage uncertainty in temperature is

$$u_T = u_{\text{RTD}}\left(\frac{\partial T}{\partial R}\right) = \frac{0.433\,\Omega}{0.098\,\Omega/°\text{C}} = 4.4°\text{C}$$

The desired uncertainty in temperature is not achieved with the specified levels of uncertainty in the pertinent variables.

COMMENT   Uncertainty analysis, in this case, would have prevented performing a measurement that would not provide acceptable results.

Example 8.3

Suppose the total uncertainty in the bridge resistances of Example 8.1 was reduced to 0.1%. Would the required level of uncertainty in temperature be achieved?

KNOWN   The uncertainty in each of the resistors in the bridge circuit for temperature measurement from Example 8.1 is $\pm 0.1\%$

FIND   The resulting uncertainty in temperature

SOLUTION   The uncertainty analysis from the previous example may be directly applied, with the uncertainty values for the resistances appropriately reduced. The uncertainties for the resistances are reduced from 0.25 to 0.025, yielding

$$u_{\text{RTD}} = \sqrt{(1 \times 0.025)^2 + (1 \times -0.025)^2 + (1 \times 0.025)^2} = 0.0433\,\Omega$$

and the resulting 95% uncertainty interval in temperature is $\pm 0.44\,°\text{C}$, which satisfies the design constraint.

COMMENT   This result provides confidence that the effect of the resistors' uncertainties will not cause the uncertainty in temperature to exceed the target value. However, the uncertainty in temperature also depends on other aspects of the measurement system. The design-stage uncertainty analysis performed in this example may be viewed as ensuring that the factors considered do not produce a higher than acceptable uncertainty level.

Practical Considerations

The transient thermal response of typical commercial RTDs is generally quite slow compared with other temperature sensors, and for transient measurements bridge circuits must be operated in a deflection mode or use auto-balancing circuits. For these reasons, RTDs are not generally chosen for transient temperature measurements. A notable exception is the use of very small platinum wires and films for temperature measurements in noncorrosive flowing gases. In this application, wires having diameters on the order of 10 μm can have frequency responses higher than any other temperature sensor because of their extremely low thermal capacitance. Obviously, the smallest impact would destroy this sensor. Other resistance sensors in the form of thin metallic films provide fast transient response temperature measurements, often in conjunction with anemometry or heat flux measurements. Such metallic films are constructed by depositing a film, commonly of platinum, onto a substrate and coating the film with a ceramic glass for mechanical protection (5). Typical film thickness ranges from 1 to 2 μm, with a 10-μm protective coating. Continuous exposure at temperatures of 600 °C is possible with this construction. Some practical uses for film sensors include temperature control circuits for heating systems and cooking devices and surface temperature monitoring on electronic components subject to overheating. Uncertainty levels range from about ±0.1 to 2 °C.

Thermistors

Thermistors (from *thermally* sensitive *resistors*) are ceramic-like semiconductor devices. The most common thermistors are NTC, and the resistance of these thermistors decreases rapidly with temperature, in contrast to the small increases of resistance with temperature for RTDs.

An accurate functional relationship between resistance and temperature for a thermistor is generally assumed to be of the form

$$R = R_0 e^{\beta(1/T - 1/T_0)} \tag{8.11}$$

The parameter β ranges from 3,500 to 4,600 K, depending on the material, temperature, and individual construction for each sensor, and thus must be determined for each thermistor. Figure 8.9 shows the variation of resistance with temperature for two common thermistor materials; the ordinate

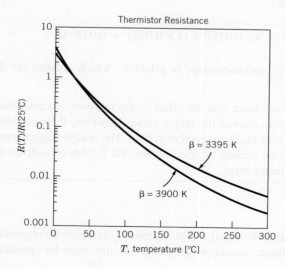

Figure 8.9 Representative thermistor resistance variations with temperature.

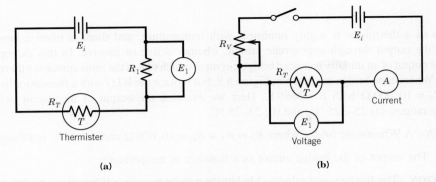

Figure 8.10   Circuits for determining β for thermistors. **(a)** Voltage divider method: $R_T = R_1(E_i/E_1 - 1)$. Note: Both R_1 and E_i must be known values. The value of R_1 may be varied to achieve appropriate values of thermistor current. **(b)** Volt-ammeter method. Note: Both current and voltage are measured.

is the ratio of the resistance to the resistance at 25 °C. Thermistors exhibit large resistance changes with temperature in comparison to a typical RTD as indicated by comparison of Figures 8.6 and 8.9. Equation 8.11 is not accurate over a wide range of temperature unless β is taken to be a function of temperature; typically the value of β specified by a manufacturer for a sensor is assumed to be constant over a limited temperature range. A simple calibration is possible for determining β as a function of temperature, as illustrated in the circuits shown in Figure 8.10. Other circuits and a more complete discussion of measuring β may be found in the Electronic Industries Association standard Thermistor Definitions and Test Methods (6).

Thermistors are generally used when high sensitivity, ruggedness, or fast response times are required (7). Thermistors are often encapsulated in glass and thus can be used in corrosive or abrasive environments. The resistance characteristics of the semiconductor material may change at elevated temperatures, and some aging of a thermistor occurs at temperatures above 200 °C. The high resistance of a thermistor, compared to that of an RTD, eliminates the problems of lead wire resistance compensation.

A commonly reported specification of a thermistor is the zero-power resistance and dissipation constant. The zero-power resistance of a thermistor is the resistance value of the thermistor with no flow of electric current. The zero power resistance should be measured such that a decrease in the current flow to the thermistor results in not more than a 0.1% change in resistance. The dissipation constant for a thermistor is defined at a given ambient temperature as

$$\delta = \frac{P}{T - T_\infty} \tag{8.12}$$

where

δ = dissipation constant

P = power supplied to the thermistor

T, T_∞ = thermistor and ambient temperatures, respectively

Example 8.4

The output of a thermistor is highly nonlinear with temperature, and there is often a benefit to linearizing the output through appropriate circuit, whether active or passive. In this example we examine the output of an initially balanced bridge circuit in which one of the arms contains a thermistor. Consider a Wheatstone bridge as shown in Figure 8.8, but replace the RTD with a thermistor having a value of $R_0 = 10,000\,\Omega$ with $\beta = 3680$ K. Here we examine the output of the circuit over two temperature ranges: (a) 25–325 °C, and (b) 25–75 °C.

KNOWN A Wheatstone bridge where $R_2 = R_3 = R_4 = 10,000\,\Omega$ and where R_1 is a thermistor.

FIND The output of the bridge circuit as a function of temperature.

SOLUTION The fundamental relationship between resistances in a Wheatstone bridge and the normalized output voltage is provided in Equation 6.11:

$$\frac{E_o}{E_i} = \left(\frac{R_1}{R_1 + R_2} - \frac{R_3}{R_3 + R_4} \right) \tag{6.11}$$

and the resistance of the thermistor is

$$R = R_o e^{\beta(1/T - 1/T_o)}$$

Substituting in Equation 6.11 for R_1 yields

$$\frac{E_o}{E_i} = \left(\frac{R_o e^{\beta(1/T - 1/T_o)}}{R_o e^{\beta(1/T - 1/T_o)} + R_2} - \frac{R_3}{R_3 + R_4} \right)$$

Figure 8.11a is a plot of this function over the range 25 to 325 °C. Clearly the sensitivity of the circuit to changes in temperature greatly diminishes as the temperature increases above 100 °C, with an asymptotic value of 0.5. Figure 8.11b shows the behavior over the restricted range 25 to 75 °C; a

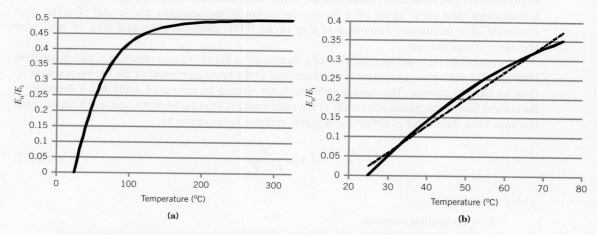

Figure 8.11 Normalized bridge output voltage as a function of temperature with a thermistor as the temperature sensor: **(a)** 25 to 325°C, **(b)** 25 to 75°C.

linear curve fit is also shown for comparison. Over this range of temperature, assuming a linear relationship between normalized output and temperature would be suitable for many applications provided that the accompanying linearity error is acceptable.

Example 8.5

The material constant β is to be determined for a particular thermistor using the circuit shown in Figure 8.10a. The thermistor has a resistance of 60 kΩ at 25 °C. The reference resistor in the circuit, R_1, has a resistance of 130.5 kΩ. The dissipation constant δ is 0.09 mW/°C. The voltage source used for the measurement is constant at 1.564 V. The thermistor is to be used at temperatures ranging from 100 to 150 °C. Determine the value of β.

KNOWN The temperature range of interest is from 100 to 150 °C.

$$R_0 = 60,000\ \Omega, \quad T_0 = 25\ °C$$
$$E_i = 1.564\ V, \quad \delta = 0.09\ mW/°C, \quad R_1 = 130.5\ k\Omega$$

FIND The value of β over the temperature range from 100 to 150 °C.

SOLUTION The voltage drop across the fixed resistor is measured for three known values of thermistor temperature. The thermistor temperature is controlled and determined by placing the thermistor in a laboratory oven and measuring the temperature of the oven. For each measured voltage across the reference resistor, the thermistor resistance R_T is determined from

$$R_T = R_1 \left(\frac{E_i}{E_1} - 1 \right)$$

The results of these measurements are as follows:

Temperature (°C)	E_1 Voltage (V)	$R_T(\Omega)$
100	1.501	5,477.4
125	1.531	2,812.9
150	1.545	1,604.9

Equation 8.11 can be expressed in the form of a linear equation as

$$\ln \frac{R_T}{R_0} = \beta \left(\frac{1}{T} - \frac{1}{T_0} \right) \tag{8.13}$$

Applying this equation to the measured data, with $R_0 = 60,000\ \Omega$, the three data points above yield the following:

$\ln(R_T/R_0)$	$(1/T - 1/T_0)\left[K^{-1}\right]$	$\beta(K)$
-2.394	-6.75×10^{-4}	3,546.7
-3.060	-8.43×10^{-4}	3,629.9
-3.621	-9.92×10^{-4}	3,650.2

COMMENT   These results are for constant β and are based on the behavior described by Equation 8.11 over the temperature range from T_0 to the temperature T. The significance of the measured differences in β is examined further.

The measured values of β in Example 8.5 are different at each value of temperature. If β were truly a temperature-independent constant and these measurements had negligible uncertainty, then all three measurements would yield the same value of β. The variation in β may result from a physical effect of temperature or be attributable to the uncertainty in the measured values.

Are the measured differences significant? If so, what value of β best represents the behavior of the thermistor over this temperature range? To perform the necessary uncertainty analysis, additional information must be provided concerning the instruments and procedures used in the measurement.

Example 8.6

Perform an uncertainty analysis to determine the uncertainty in each measured value of β in Example 8.5, and evaluate a single best estimate of β for this temperature range. The measurement of β involves the measurement of voltages, temperatures, and resistances. For temperature, there is a random error associated with spatial and temporal variations in the oven temperature with a random standard uncertainty of $s_{\overline{T}} = 0.19°\,C$ for 20 measurements. In addition, based on a manufacturer's specification, there is a known measurement systematic uncertainty for temperature of $0.36\,°C$ (95%) in the thermocouple.

The systematic errors in measuring resistance and voltage are negligible, and estimates of the instrument repeatability, which are based on the manufacturer's specifications in the measured values and assumed to be at a 95% confidence level, are assigned systematic uncertainties of 1.5% for resistance and 0.002 V for the voltage.

KNOWN   Standard deviation of the means for oven temperature, $s_{\overline{T}} = 0.19\,°C$, $N = 20$. The remaining errors are assigned systematic uncertainties at 95% confidence assuming large degrees of freedom such that $B_x = 2b_x$:

$$B_T = 2b_T = 0.36°\,C$$
$$B_R/R = (2b_R)/R = 1.5\%$$
$$B_E = 2b_E = 0.002\,V$$

FIND   The uncertainty in β at each measured temperature, and a best estimate for β over the measured temperature range

SOLUTION   Consider the problem of providing a single best estimate of β. One method of estimation might be to average the three measured values. This results in a value of 3,609 K. However, because the relationship between $\ln(R_T/R_0)$ and $(1/T - 1/T_0)$ is expected to be linear, a least-squares fit can be performed on the three data points, and include the point $(0, 0)$. The resulting value of β is 3,638 K. Is this difference significant? Which value best represents the behavior of the thermistor? To answer these questions, an uncertainty analysis is performed for β.

For each measured value,

$$\beta = \frac{\ln(R_T/R_0)}{1/T - 1/T_0}$$

Uncertainties in voltage, temperature, and resistance are propagated into the resulting value of β for each measurement.

Consider first the sensitivity indices θ_i for each of the variables $R_T, R_0, T,$ and T_0. These may be tabulated by computing the appropriate partial derivatives of β, evaluated at each of the three temperatures, as follows:

$T\,(°C)$	$\theta_{R_T}(K/\Omega)$	$\theta_{R_0}(K/\Omega)$	θ_T	θ_{T_0}
100	−0.270	0.0247	−37.77	59.17
125	−0.422	0.0198	−27.18	48.48
150	−0.628	0.0168	−20.57	41.45

The determination of the uncertainty in β, u_β requires the uncertainty in the measured value of resistance for the thermistor, u_{R_T}. But R_T is determined from the expression

$$R_T = R_1 \left[(E_i/E_1) - 1 \right]$$

and thus requires an analysis of the uncertainty in the resulting value of R_T from measured values of $R_1, E_i,$ and E_1. All errors in R_T are treated as uncorrelated systematic errors, yielding

$$b_{R_T} = \sqrt{\left[\frac{\partial R_T}{\partial R_1}b_{R_1}\right]^2 + \left[\frac{\partial R_T}{\partial E_i}b_{E_i}\right]^2 + \left[\frac{\partial R_T}{\partial E_1}b_{E_1}\right]^2} = \sqrt{\left[\theta_{R_1}b_{R_1}\right]^2 + \left[\theta_{E_i}b_{E_i}\right]^2 + \left[\theta_{E_1}b_{E_1}\right]^2}$$

To arrive at a representative value, we compute b_{R_T} at 125 °C. The systematic standard uncertainty in R_1 is 0.75% of 130.5 kΩ, or 978 Ω, and in R_0 is 450 Ω. The systematic standard uncertainties in E_i and E_1 are each 0.001 V, and $\theta_{R_1} = 0.022$, $\theta_{E_1} = 87076$, and $\theta_{E_1} = 87076$. This gives $b_{R_T} = 123.7\ \Omega$.

An uncertainty for β is determined for each of the measured temperatures. The propagation of the measurement systematic errors for temperature and resistance is found as

$$b_\beta = \sqrt{\left[\theta_T b_T\right]^2 + \left[\theta_{T_0} b_{T_0}\right]^2 + \left[\theta_{R_T} b_{R_T}\right]^2 + \left[\theta_{R_0} b_{R_0}\right]^2}$$

where

$$b_T = 0.18°\,C \quad b_{R_T} = 123.7\ \Omega$$
$$b_{T_0} = 0.18°\,C \quad b_{R_0} = 450\ \Omega$$

The random standard uncertainty for β contains contributions only from the statistically determined oven temperature characteristics and is found from

$$s_{\bar{\beta}} = \sqrt{\left(\theta_T s_{\bar{T}}\right)^2 + \left(\theta_{T_0} s_{\bar{T}_0}\right)^2}$$

where both $s_{\bar{T}}$ and $s_{\bar{T}_0}$ are 0.19, as determined with $\nu = 19$.

Table 8.3 Uncertainties in β.

| T [°C] | Uncertainty | | |
	Random $s_{\bar{\beta}}$ [K]	Systematic b_β [K]	Total $\pm u_\beta$ 95%[K]
100	13.3	37.4	79.4
125	10.6	53.9	109.9
150	8.8	78.4	157.8

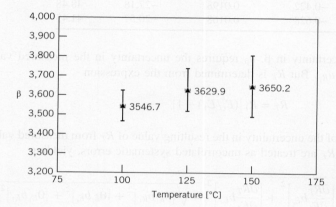

Figure 8.12 Measured values of β and associated uncertainties for three temperatures. Each data point represents $\bar{\beta} \pm u_\beta$.

The resulting values of uncertainty in β are found from

$$u_\beta = t_{\nu,95}\sqrt{b_\beta^2 + s_{\bar{\beta}}^2}$$

where ν_β is sufficiently large (see Eq. 5.39) so that $t_{\nu,95} \approx 2$. At each temperature the uncertainty in β is determined as shown in Table 8.3. The effect of increases in the sensitivity indices θ_i on the total uncertainty is to cause increased uncertainty in β as the temperature increases.

The original results of the measured values of β must now be reexamined. The results from Table 8.3 are plotted as a function of temperature in Figure 8.12, with 95% uncertainty limits on each data point. Clearly, there is no justification for assuming that the measured values indicate a trend of changes with temperature, and it would be appropriate to use either the average value of β or the value determined from the linear least-squares curve fit.

8.5 THERMOELECTRIC TEMPERATURE MEASUREMENT

A common method of measuring and controlling temperature uses an electrical circuit called a thermocouple. A *thermocouple* consists of two electrical conductors that are made of dissimilar metals and have at least one electrical connection. This electrical connection is referred to as a *junction*. A thermocouple junction may be created by welding, by soldering, or by any method that provides good electrical contact between the two conductors, such as twisting the wires around one another. The output of a thermocouple circuit is a voltage, and there is a definite relationship between

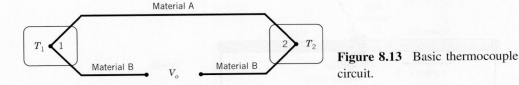

Figure 8.13   Basic thermocouple circuit.

this voltage and the temperatures of the junctions that make up the thermocouple circuit. We will examine the causes of this voltage and develop the basis for using thermocouples to make engineering measurements of temperature.

Consider the thermocouple circuit shown in Figure 8.13. The junction labeled 1 is at a temperature T_1 and the junction labeled 2 is at a temperature T_2. This thermocouple circuit measures the difference between T_1 and T_2. If T_1 and T_2 are not equal, a finite open-circuit voltage, V_o, is measured. The magnitude of the potential depends on the difference in the temperatures and the particular metals used in the thermocouple circuit.

A thermocouple junction is the source of a voltage that gives rise to the total voltage difference in a thermocouple circuit. It is the basis for temperature measurement using thermocouples. The circuit shown in Figure 8.13 is the most common form of a thermocouple circuit used for measuring temperature.

It is our goal to understand the origin of thermoelectric phenomena and the requirements for providing accurate temperature measurements using thermocouples. In an electrical conductor that is subject to a temperature gradient, there will be both a flow of thermal energy and a flow of electricity. These phenomena are closely tied to the behavior of the free electrons in a metal; it is no coincidence that good electrical conductors are, in general, good thermal conductors. The characteristic behavior of these free electrons in an electrical circuit composed of dissimilar metals results in a useful relationship between temperature and thermocouple voltage. There are three basic phenomena that can occur in a thermocouple circuit: (1) the *Seebeck effect*, (2) the *Peltier effect*, and (3) the *Thomson effect*.

The open-circuit voltage, V_o, generated by a thermocouple circuit is the result of the Seebeck effect only. Under measurement conditions with no loading errors the measured voltage would equal the open-circuit voltage.

Seebeck Effect

The Seebeck effect, named for Thomas Johann Seebeck (1770–1831), refers to the generation of a voltage difference in an open thermocouple circuit due to a difference in temperature between junctions in the circuit. The Seebeck effect refers to the case when there is no current flow in the circuit. There is a fixed, reproducible relationship between the voltage and the junction temperatures T_1 and T_2(Fig. 8.13). This relationship is expressed by the Seebeck coefficient, α_{AB}, defined as

$$\alpha_{AB} = \left[\frac{\partial (V)}{\partial T} \right]_{\text{open circuit}}$$ (8.14)

where A and B refer to the two materials that comprise the thermocouple. Since the Seebeck coefficient specifies the rate of change of voltage with temperature for the materials A and B, it is equal to the static sensitivity of the open-circuit thermocouple.

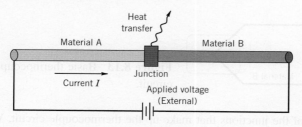

Figure 8.14 Peltier effect due to current flow across a junction of dissimilar metals.

Peltier Effect

A familiar concept is that of I^2R or joule heating in a conductor through which an electrical current flows. Consider the two conductors having a common junction, shown in Figure 8.14, through which an electrical current I flows due to an externally applied voltage. For any portion of either of the conductors, the energy removal rate required to maintain a constant temperature is I^2R, where R is the resistance to a current flow and is determined by the resistivity and size of the conductor. However, at the junction of the two dissimilar metals, the removal of a quantity of energy different than I^2R is required to maintain a constant temperature. The difference in I^2R and the amount of energy generated by the current flowing through the junction is due to the Peltier effect. The Peltier effect is a thermodynamically reversible conversion of energy as a current flows across the junction, in contrast to the irreversible dissipation of energy associated with I^2R losses. The Peltier heat is the quantity of heat in addition to the quantity I^2R that must be removed from the junction to maintain the junction at a constant temperature. This amount of energy is proportional to the current flowing through the junction; the proportionality constant is the Peltier coefficient π_{AB}, and the heat transfer required to maintain a constant temperature is

$$Q_\pi = \pi_{AB}I \tag{8.15}$$

caused by the Peltier effect alone. This behavior was discovered by Jean Charles Athanase Peltier (1785–1845) during experiments using Seebeck's thermocouple. He observed that passing a current through a thermocouple circuit having two junctions, as in Figure 8.13, raised the temperature at one junction while lowering the temperature at the other junction. This effect forms the basis of a device known as a Peltier refrigerator, which provides cooling without moving parts. An assessment of the prospects of thermoelectric cooling technologies is provided in (8).

Thomson Effect

In addition to the Seebeck effect and the Peltier effect, there is a third phenomenon that occurs in thermoelectric circuits. Consider the conductor shown in Figure 8.15, which is subject to a longitudinal temperature gradient and also to a voltage difference, such that there is a flow of current and heat in the conductor. Again, to maintain a constant temperature in the conductor it is found that a quantity of energy different than the joule heat, I^2R, must be removed from the conductor. First noted by William Thomson (1824–1907, known as Lord Kelvin from 1892) in 1851, this energy is expressed in terms of the Thomson coefficient, σ, as

$$Q_\sigma = \sigma I(T_1 - T_2) \tag{8.16}$$

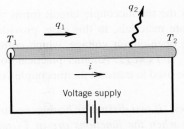

q_1 Energy flow as a result of a temperature gradient
q_2 Heat transfer to maintain constant temperature

Figure 8.15 Thomson effect due to simultaneous flows of current and heat.

For a thermocouple circuit, all three of these effects may be present and may contribute to the overall voltage difference in the circuit.

Fundamental Thermocouple Laws

The basic thermocouple circuit shown in Figure 8.16 can be used to measure the difference between the two temperatures T_1 and T_2. For practical temperature measurements, one of these junctions becomes a reference junction and is maintained at some known, constant reference temperature—say, T_2. The other junction then becomes the measuring junction, and the voltage existing in the circuit provides a direct indication of the temperature of the measuring junction T_1.

The use of thermocouple circuits to measure temperature is based on observed behaviors of carefully controlled thermocouple materials and circuits. The following laws provide the basis necessary for temperature measurement with thermocouples:

1. *Law of homogeneous materials: A thermoelectric current cannot be sustained in a circuit of a single homogeneous material by the application of heat alone, regardless of how it might vary in cross section.* Simply stated, this law requires that at least two materials be used to construct a thermocouple circuit for the purpose of measuring temperature. It is interesting to note that a current may occur in an inhomogeneous wire that is nonuniformly heated, but this is neither useful nor desirable in a thermocouple.

2. *Law of intermediate materials: The algebraic sum of the voltage differences in a circuit composed of any number of dissimilar materials is zero if all of the circuit is at a uniform temperature.* This law allows a material other than the thermocouple materials to be inserted into a thermocouple circuit without changing the output voltage of the circuit. As an example, consider the thermocouple circuit shown in Figure 8.16, where the junctions of the measuring device are made of copper and material B is an alloy (not pure copper). The electrical

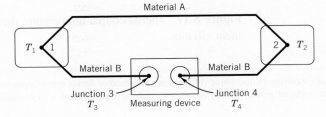

Figure 8.16 Typical thermocouple measuring circuit.

connection between the measuring device and the thermocouple circuit forms yet another thermocouple junction. The law of intermediate materials, in this case, provides that the measured voltage will be unchanged from the open-circuit voltage, which corresponds to the temperature difference between T_1 and T_2, if $T_3 = T_4$. Another practical consequence of this law is that copper extension wires may be used to transmit thermocouple voltages to a measuring device.

3. ***Law of successive or intermediate temperatures:*** *If two dissimilar homogeneous materials that form a thermocouple circuit produce V_1 when the junctions are at T_1 and T_2 and produce V_2 when the junctions are at T_2 and T_3, the voltage generated when the junctions are at T_1 and T_3 will be $V_1 + V_2$. This law allows a thermocouple calibrated for one reference temperature—say, T_2—to be used at another reference temperature, such as T_3, to determine temperature T_1.*

Basic Temperature Measurement with Thermocouples

Let's first examine a historically significant method of using a thermocouple circuit to measure temperature. Figure 8.17 shows two basic thermocouple circuits using a chromel–constantan thermocouple and a reference temperature. In Figure 8.17a, the thermocouple wires are connected directly to a voltmeter[3] to measure the voltage. In Figure 8.17b, copper extension wires are employed, creating two reference junctions. The law of intermediate materials ensures that neither the voltmeter

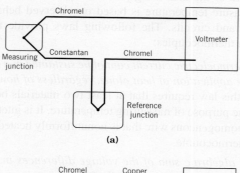

(a)

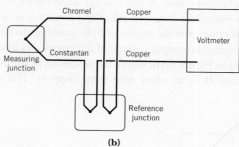

(b)

Figure 8.17 Thermocouple temperature measurement circuits.

---

[3] Historically the most accurate method of measuring thermocouple voltage was a null-balance device termed a potentiometer. However, modern nano-voltmeters have replaced potentiometers for accurate voltage measurement with essentially zero loading error.

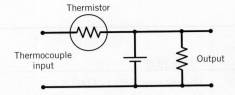

Thermistor

Thermocouple input

Output

Figure 8.18   Basic thermistor circuit for thermocouple reference junction compensation.

nor the extension wires will change the output voltage of the circuit so long as the two connecting junctions at the voltmeter and the two at the reference temperature experience no temperature difference. All that is required to be able to measure temperature with this circuit is to know the relationship between the output voltage and the temperature of the measuring junction, for the particular reference temperature. One method of determining this relationship is to calibrate the thermocouple. However, we shall see that for reasonable levels of uncertainty for temperature measurement, standard materials and procedures allow thermocouples to be accurate temperature measuring devices without the necessity of calibration.

Reference Junction

The provisions for a reference junction should provide a temperature that is accurately known, stable, and reproducible. A reference junction temperature could be provided by the ice point, 0 °C, because of the relative ease with which it could historically be obtained. An ice bath can provide the reference junction temperature. An ice bath is typically made by filling a vacuum flask, or Dewar, with finely crushed ice, and adding just enough water to create a transparent slush. Ice baths can be constructed to provide a reference junction temperature to an uncertainty within ±0.01 °C.

Modern reference junction temperatures are created electronically. Termed *electronic reference junction compensation*, these dedicated circuits provide a convenient means of the measurement of temperature without the necessity to construct an ice bath. Numerous manufacturers produce commercial temperature measuring devices with built-in reference junction compensation, and many digital data acquisition cards include reference junction compensation. The electronics generally rely on a thermistor, a temperature-sensitive integrated circuit, or an RTD to determine the reference junction temperature, as shown for a thermistor in Figure 8.18. Uncertainties for the reference junction temperature in this case are on the order of ±0.1 °C, with ±0.5 °C as typical. In addition, data acquisition systems may include a microprocessor that converts compensated voltage readings directly into temperature output based on the standard thermocouple voltages. Combining a thermocouple, reference junction compensation, and a microprocessor results in a device termed a "smart sensor."

Thermocouple Standards

The National Institute of Standards and Technology (NIST) provides specifications for the materials and construction of standard thermocouple circuits for temperature measurement (9). Many material combinations exist for thermocouples; these material combinations are identified by a thermocouple type and denoted by a letter. Table 8.4 shows the letter designations and the polarity of common thermocouples, along with some basic application information for each type. The choice of a type of thermocouple depends on the temperature range to be measured, the particular application, and the desired uncertainty level.

Table 8.4 Thermocouple Designations

| Type | Material Combination | | Applications |
	Positive	Negative	
E	Chromel(+)	Constantan(−)	Highest sensitivity (<1,000 °C)
J	Iron(+)	Constantan(−)	Nonoxidizing environment (<760 °C)
K	Chromel(+)	Alumel(−)	High temperature (<1,372 °C)
S	Platinum/10% rhodium	Platinum(−)	Long-term stability high temperature (<1,768 °C)
T	Copper(+)	Constantan(−)	Reducing or vacuum environments (<400 °C)

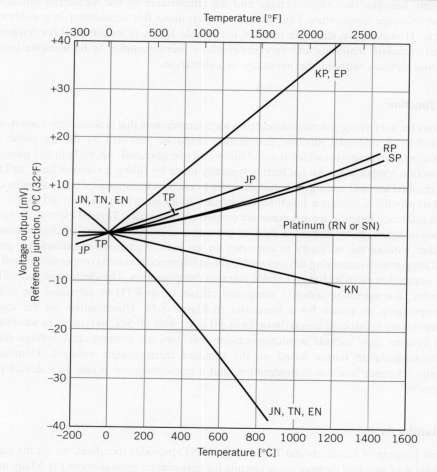

Figure 8.19 Thermal voltage of thermocouple materials relative to platinum-67. Note: For example, JP indicates the positive leg of a J thermocouple, or iron. JN indicates the negative leg of a J thermocouple, or constantan. All other notations are similar for each type of thermocouple. (Adapted from Benedict, R. P., *Fundamentals of Temperature, Pressure, and Flow Measurements*, 3rd ed. Copyright © 1984 by John Wiley and Sons, New York. Reprinted by permission.)

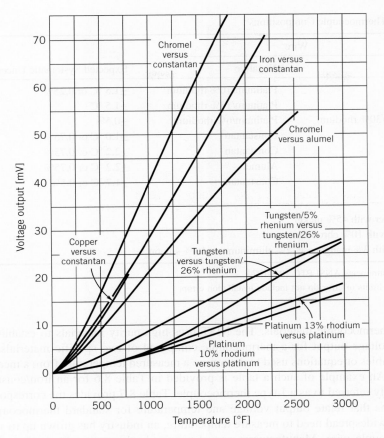

Figure 8.20 Thermocouple voltage output as a function of temperature for some common thermocouple materials. Reference junction is at 32 °F (0 °C). (Adapted from Benedict, R. P., *Fundamentals of Temperature, Pressure and Flow Measurements*, 3rd ed. Copyright © 1984 by John Wiley and Sons, New York. Reprinted by permission.)

To determine the voltage output of a particular material combination, a thermocouple is formed from a candidate material and a standard platinum alloy to form a thermocouple circuit having a 0 °C reference temperature. Figure 8.19 shows the output of various materials in combination with platinum-67. The notation indicates the thermocouple type. The law of intermediate temperatures then allows the voltage of any two materials, whose voltage relative to platinum is known, to be determined. Figure 8.20 shows a plot of the voltage as a function of temperature for some common thermocouple material combinations. The slope of the curves in this figure corresponds to the static sensitivity of the thermocouple measuring circuit.

Standard Thermocouple Voltage

Table 8.5 provides the standard composition of thermocouple materials, along with standard limits of error for the various material combinations. These limits specify the expected maximum errors

Table 8.5 Standard Thermocouple Compositions[a]

Type	Wire Positive	Wire Negative	Expected Systematic Uncertainty[b]
S	Platinum	Platinum/10% rhodium	±1.5 °C or 0.25%
R	Platinum	Platinum/13% rhodium	±1.5 °C
B	Platinum/30% rhodium	Platinum/6% rhodium	±0.5%
T	Copper	Constantan	±1.0 °C or 0.75%
J	Iron	Constantan	±2.2 °C or 0.75%
K	Chromel	Alumel	±2.2 °C or 0.75%
E	Chromel	Constantan	±1.7 °C or 0.5%

Alloy Designations

Constantan: 55% copper with 45% nickel

Chromel: 90% nickel with 10% chromium

Alumel: 94% nickel with 3% manganese, 2% aluminum, and 1% silicon

[a]From Temperature Measurements ANSI PTC 19.3-1974.
[b]Use greater value; these limits of error do not include installation errors.

resulting from the thermocouple materials. The NIST uses high-purity materials to establish the standard value of voltage output for a thermocouple composed of two specific materials. This results in standard tables or equations used to determine a measured temperature from a measured output voltage (9). An example of such a table is provided in Table 8.6 for an iron/constantan thermocouple, usually referred to as a J-type thermocouple. Table 8.7 provides the corresponding polynomial equations that relate output voltage and temperature for standard thermocouples.

Because of the widespread need to measure temperature, an industry has grown up to supply high-grade thermocouple wire. Manufacturers can also provide thermocouples having special tolerance limits relative to the NIST standard voltages with uncertainties in temperature ranging from ±1.0 °C to perhaps ±0.1 °C. Thermocouples constructed of standard thermocouple wire do not require calibration to provide measurement of temperature within the tolerance limits given in Table 8.5.

Thermocouple Voltage Measurement

The Seebeck voltage for a thermocouple circuit is measured with no current flow in the circuit. From our discussion of the Thomson and Peltier effects, it is clear that the voltage is different from the open-circuit value when there is a current flow in the thermocouple circuit. As such, the best method for the measurement of thermocouple voltages is a device that minimizes current flow. For many years, the potentiometer was the laboratory standard for voltage measurement in thermocouple circuits. A potentiometer, as described in Chapter 6, has nearly zero loading error at a balanced condition. However, modern voltage-measuring devices, such as digital voltmeters or data acquisition cards, have sufficiently large values of input impedance that they can be used with minimal loading error. These devices can also be used in either static or dynamic measuring situations where the loading error created by the measurement device is acceptable for the particular application.

Table 8.6   Thermocouple Reference Table for Type-J Thermocouple[a]

Temperature (°C)								Thermocouple voltage (mV)		
	0	−1	−2	−3	−4	−5	−6	−7	−8	−9
−210	−8.095									
−200	−7.890	−7.912	−7.934	−7.955	−7.976	−7.996	−8.017	−8.037	−8.057	−8.076
−190	−7.659	−7.683	−7.707	−7.731	−7.755	−7.778	−7.801	−7.824	−7.846	−7.868
−180	−7.403	−7.429	−7.456	−7.482	−7.508	−7.534	−7.559	−7.585	−7.610	−7.634
−170	−7.123	−7.152	−7.181	−7.209	−7.237	−7.265	−7.293	−7.321	−7.348	−7.376
−160	−6.821	−6.853	−6.883	−6.914	−6.944	−6.975	−7.005	−7.035	−7.064	−7.094
−150	−6.500	−6.533	−6.566	−6.598	−6.631	−6.663	−6.695	−6.727	−6.759	−6.790
−140	−6.159	−6.194	−6.229	−6.263	−6.298	−6.332	−6.366	−6.400	−6.433	−6.467
−130	−5.801	−5.838	−5.874	−5.910	−5.946	−5.982	−6.018	−6.054	−6.089	−6.124
−120	−5.426	−5.465	−5.503	−5.541	−5.578	−5.616	−5.653	−5.690	−5.727	−5.764
−110	−5.037	−5.076	−5.116	−5.155	−5.194	−5.233	−5.272	−5.311	−5.350	−5.388
−100	−4.633	−4.674	−4.714	−4.755	−4.796	−4.836	−4.877	−4.917	−4.957	−4.997
−90	−4.215	−4.257	−4.300	−4.342	−4.384	−4.425	−4.467	−4.509	−4.550	−4.591
−80	−3.786	−3.829	−3.872	−3.916	−3.959	−4.002	−4.045	−4.088	−4.130	−4.173
−70	−3.344	−3.389	−3.434	−3.478	−3.522	−3.566	−3.610	−3.654	−3.698	−3.742
−60	−2.893	−2.938	−2.984	−3.029	−3.075	−3.120	−3.165	−3.210	−3.255	−3.300
−50	−2.431	−2.478	−2.524	−2.571	−2.617	−2.663	−2.709	−2.755	−2.801	−2.847
−40	−1.961	−2.008	−2.055	−2.103	−2.150	−2.197	−2.244	−2.291	−2.338	−2.385
−30	−1.482	−1.530	−1.578	−1.626	−1.674	−1.722	−1.770	−1.818	−1.865	−1.913
−20	−0.995	−1.044	−1.093	−1.142	−1.190	−1.239	−1.288	−1.336	−1.385	−1.433
−10	−0.501	−0.550	−0.600	−0.650	−0.699	−0.749	−0.798	−0.847	−0.896	−0.946
0	0.000	−0.050	−0.101	−0.151	−0.201	−0.251	−0.301	−0.351	−0.401	−0.451
	0	+1	+2	+3	+4	+5	+6	+7	+8	+9
0	0.000	0.050	0.101	0.151	0.202	0.253	0.303	0.354	0.405	0.451
10	0.507	0.558	0.609	0.660	0.711	0.762	0.814	0.865	0.916	0.968
20	1.019	1.071	1.122	1.174	1.226	1.277	1.329	1.381	1.433	1.485
30	1.537	1.589	1.641	1.693	1.745	1.797	1.849	1.902	1.954	2.006
40	2.059	2.111	2.164	2.216	2.269	2.322	2.374	2.427	2.480	2.532
50	2.585	2.638	2.691	2.744	2.797	2.850	2.903	2.956	3.009	3.062
60	3.116	3.169	3.222	3.275	3.329	3.382	3.436	3.489	3.543	3.596
70	3.650	3.703	3.757	3.810	3.864	3.918	3.971	4.025	4.079	4.133
80	4.187	4.240	4.294	4.348	4.402	4.456	4.510	4.564	4.618	4.672
90	4.726	4.781	4.835	4.889	4.943	4.997	5.052	5.106	5.160	5.215
100	5.269	5.323	5.378	5.432	5.487	5.541	5.595	5.650	5.705	5.759
110	5.814	5.868	5.923	5.977	6.032	6.087	6.141	6.196	6.251	6.306
120	6.360	6.415	6.470	6.525	6.579	6.634	6.689	6.744	6.799	6.854
130	6.909	6.964	7.019	7.074	7.129	7.184	7.239	7.294	7.349	7.404
140	7.459	7.514	7.569	7.624	7.679	7.734	7.789	7.844	7.900	7.955
150	8.010	8.065	8.120	8.175	8.231	8.286	8.341	8.396	8.452	8.507
160	8.562	8.618	8.673	8.728	8.783	8.839	8.894	8.949	9.005	9.060
170	9.115	9.171	9.226	9.282	9.337	9.392	9.448	9.503	9.559	9.614

(*continued*)

Table 8.6 (*Continued*)

Temperature (°C)										Thermocouple voltage (mV)
	0	+1	+2	+3	+4	+5	+6	+7	+8	+9
180	9.669	9.725	9.780	9.836	9.891	9.947	10.002	10.057	10.113	10.168
190	10.224	10.279	10.335	10.390	10.446	10.501	10.557	10.612	10.668	10.723
200	10.779	10.834	10.890	10.945	11.001	11.056	11.112	11.167	11.223	11.278
210	11.334	11.389	11.445	11.501	11.556	11.612	11.667	11.723	11.778	11.8.34
220	11.889	11.945	12.000	12.056	12.111	12.167	12.222	12.278	12.334	12.389
230	12.445	12.500	12.556	12.611	12.667	12.722	12.778	12.833	12.889	12.944
240	13.000	13.056	13.111	13.167	13.222	13.278	13.333	13.389	13.444	13.500
250	13.555	13.611	13.666	13.722	13.777	13.833	13.888	13.944	13.999	14.055
260	14.110	14.166	14.221	14.277	14.332	14.388	14.443	14.499	14.554	14.609
270	14.665	14.720	14.776	14.831	14.887	14.942	14.998	15.053	15.109	15.164
280	15.219	15.275	15.330	15.386	15.441	15.496	15.552	15.607	15.663	15.718
290	15.773	15.829	15.884	15.940	15.995	16.050	16.106	16.161	16.216	16.272
300	16.327	16.383	16.438	16.493	16.549	16.604	16.659	16.715	16.770	16.825
310	16.881	16.936	16.991	17.046	17.102	17.157	17.212	17.268	17.323	17.378
320	17.434	17.489	17.544	17.599	17.655	17.710	17.765	17.820	17.876	17.931
330	17.986	18.041	18.097	18.152	18.207	18.262	18.318	18.373	18.428	18.483
340	18.538	18.594	18.649	18.704	18.759	18.814	18.870	18.925	18.980	19.035
350	19.090	19.146	19.201	19.256	19.311	19.366	19.422	19.477	19.532	19.587
360	19.642	19.697	19.753	19.808	19.863	19.918	19.973	20.028	20.083	20.139
370	20.194	20.249	20.304	20.359	20.414	20.469	20.525	20.580	20.635	20.690
380	20.745	20.800	20.855	20.911	20.966	21.021	21.076	21.131	21.186	21.241
390	21.297	21.352	21.407	21.462	21.517	21.572	21.627	21.683	21.738	21.793
400	21.848	21.903	21.958	22.014	22.069	22.124	22.179	22.234	22.289	22.345
410	22.400	22.455	22.510	22.565	22.620	22.676	22.731	22.786	22.841	22.896
420	22.952	23.007	23.062	23.117	23.172	23.228	23.283	23.338	23.393	23.449
430	23.504	23.559	23.614	23.670	23.725	23.780	23.835	23.891	23.946	24.001
440	24.057	24.112	24.167	24.223	24.278	24.333	24.389	24.444	24.499	24.555
450	24.610	24.665	24.721	24.776	24.832	24.887	24.943	24.998	25.053	25.109
460	25.164	25.220	25.275	25.331	25.386	25.442	25.497	25.553	25.608	25.664
470	25.720	25.775	25.831	25.886	25.942	25.998	26.053	26.109	26.165	26.220
480	26.276	26.332	26.387	26.443	26.499	26.555	26.610	26.666	26.722	26.778
490	26.834	26.889	26.945	27.001	27.057	27.113	27.169	27.225	27.281	27.337
500	27.393	27.449	27.505	27.561	27.617	27.673	27.729	27.785	27.841	27.897
510	27.953	28.010	28.066	28.122	28.178	28.234	28.291	28.347	28.403	28.460
520	28.516	28.572	28.629	28.685	28.741	28.798	28.854	28.911	28.967	29.024
530	29.080	29.137	29.194	29.250	29.307	29.363	29.420	29.477	29.534	29.590
540	29.647	29.704	29.761	29.818	29.874	29.931	29.988	30.045	30.102	30.159
550	30.216	30.273	30.330	30.387	30.444	30.502	30.559	30.616	30.673	30.730
560	30.788	30.845	30.902	30.960	31.017	31.074	31.132	31.189	31.247	31.304
570	31.362	31.419	31.477	31.535	31.592	31.650	31.708	31.766	31.823	31.881
580	31.939	31.997	32.055	32.113	32.171	32.229	32.287	32.345	32.403	32.461

Table 8.6   (*Continued*)

Temperature (°C)										Thermocouple voltage (mV)
	0	+1	+2	+3	+4	+5	+6	+7	+8	+9
590	32.519	32.577	32.636	32.694	32.752	32.810	32.869	32.927	32.985	33.044
600	33.102	33.161	33.219	33.278	33.337	33.395	33.454	33.513	33.571	33.630
610	33.689	33.748	33.807	33.866	33.925	33.984	34.043	34.102	34.161	34.220
620	34.279	34.338	34.397	34.457	34.516	34.575	34.635	34.694	34.754	34.813
630	34.873	34.932	34.992	35.051	35.111	35.171	35.2.30	35.290	35.350	35.410
640	35.470	35.530	35.590	35.650	35.710	35.770	35.830	35.890	35.950	36.010
650	36.071	36.131	36.191	36.191	36.252	36.373	36.433	36.494	36.554	36.615
660	36.675	36.736	36.797	36.858	36.918	36.979	37.040	37.101	37.162	37.223
670	37.284	37.345	37.406	37.467	37.528	37.590	37.651	37.712	37.773	37.835
680	37.896	37.958	38.019	38.081	38.142	38.204	38.265	38.327	38.389	38.450
690	38.512	38.574	38.636	38.698	38.760	38.822	38.884	38.946	39.008	39.070
700	39.132	39.194	39.256	39.318	39.381	39.443	39.505	39.568	39.630	39.693
710	39.755	39.818	39.880	39.943	40.005	40.068	40.131	40.193	40.256	40.319
720	40.382	40.445	40.508	40.570	40.633	40.696	40.759	40.822	40.886	40.949
730	41.012	41.075	41.138	41.201	41.265	41.328	41.391	41.455	41.518	41.581
740	41.645	41.708	41.772	41.835	41.899	41.962	42.026	42.090	42.153	42.217
750	42.281	42.344	42.408	42.472	42.536	42.599	42.663	42.727	42.791	42.855
760	42.919	42.983	43.047	43.110	43.174	43.238	43.303	43.367	43.431	43.495

[a]Reference junction at 0 °C.

Table 8.7   Reference Functions for Selected Letter Designated Thermocouples

The relationship between voltage and temperature is provided in the form of a polynomial in temperature [9]

$$E = \sum_{i=0}^{n} c_i T^i$$

where E is in μV and T is in °C. Constants are provided below.

Thermocouple Type	Temperature Range	Constants
J-type	−210 to 760 °C	$c_0 = 0.000\,000\,000\,0$ $c_1 = 5.038\,118\,7815 \times 10^{1}$ $c_2 = 3.047\,583\,693\,0 \times 10^{-2}$ $c_3 = -8.568\,106\,572\,0 \times 10^{-5}$ $c_4 = 1.322\,819\,529\,5 \times 10^{-7}$ $c_5 = -1.705\,295\,833\,7 \times 10^{-10}$ $c_6 = 2.094\,809\,069\,7 \times 10^{-13}$ $c_7 = -1.253\,839\,533\,6 \times 10^{-16}$ $c_8 = 1.563\,172\,569\,7 \times 10^{-20}$

(*continued*)

Table 8.7 (*Continued*)

The relationship between voltage and temperature is provided in the form of a polynomial in temperature [9]

$$E = \sum_{i=0}^{n} c_i T^i$$

where E is in μV and T is in °C. Constants are provided below.

Thermocouple Type	Temperature Range	Constants
T-type	−270 to 0 °C	$c_0 = 0.000\,000\,000\,0$
		$c_1 = 3.874\,810\,6364 \times 10^1$
		$c_2 = 4.419\,443\,434\,7 \times 10^{-2}$
		$c_3 = 1.184\,432\,310\,5 \times 10^{-4}$
		$c_4 = 2.003\,297\,355\,4 \times 10^{-5}$
		$c_5 = 9.013\,801\,955\,9 \times 10^{-7}$
		$c_6 = 2.265\,115\,659\,3 \times 10^{-8}$
		$c_7 = 3.607\,115\,420\,5 \times 10^{-10}$
		$c_8 = 3.849\,393\,988\,3 \times 10^{-12}$
		$c_9 = 2.821\,352\,192\,5 \times 10^{-14}$
		$c_{10} = 1.425\,159\,477\,9 \times 10^{-16}$
		$c_{11} = 4.876\,866\,228\,6 \times 10^{-19}$
		$c_{12} = 1.079\,553\,927\,0 \times 10^{-21}$
		$c_{13} = 1.394\,502\,706\,2 \times 10^{-24}$
		$c_{14} = 7.979\,515\,392\,7 \times 10^{-28}$
T-type	0 to 400 °C	$c_0 = 0.000\,000\,000\,0$
		$c_1 = 3.874\,810\,636\,4 \times 10^1$
		$c_2 = 3.329\,222\,788\,0 \times 10^{-2}$
		$c_3 = 2.061\,824\,340\,4 \times 10^{-4}$
		$c_4 = -2.188\,225\,684\,6 \times 10^{-6}$
		$c_5 = 1.099\,688\,092\,8 \times 10^{-8}$
		$c_6 = -3.081\,575\,877\,2 \times 10^{-11}$
		$c_7 = 4.547\,913\,529\,0 \times 10^{-14}$
		$c_8 = -2.751\,290\,167\,3 \times 10^{-17}$

For such needs, high-impedance voltmeters have been incorporated into commercially available temperature indicators, temperature controllers, and digital data acquisition systems (DAS).

Example 8.7

The thermocouple circuit shown in Figure 8.21 is used to measure the temperature T_1. The thermocouple reference junction labeled 2 is at a temperature of 0 °C. The voltage output is measured and found to be 9.669 mV. What is T_1?

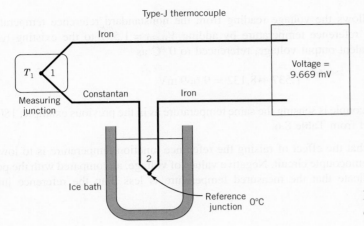

Figure 8.21 Thermocouple circuit for Example 8.7.

KNOWN A thermocouple circuit having one junction at 0 °C and a second junction at an unknown temperature. The circuit produces a voltage of 9.669 mV.

FIND The temperature T_1.

ASSUMPTION Thermocouple follows NIST standard voltage behavior.

SOLUTION Standard thermocouple tables such as Table 8.6 are referenced to 0 °C. The temperature of the reference junction for this case is 0 °C. Therefore, the temperature corresponding to the output voltage may simply be determined from Table 8.6, in this case as 180.0 °C.

COMMENT Because of the law of intermediate metals, the junctions formed at the voltage measuring device do not affect the voltage measured for the thermocouple circuit, and the voltage output reflects accurately the temperature difference between junctions 1 and 2.

Example 8.8

Suppose the thermocouple circuit in the previous example (Ex. 8.7) now has junction 2 maintained at a temperature of 30 °C, and produces an output voltage of 8.132 mV. What temperature is sensed by the measuring junction?

KNOWN The value of T_2 is 30 °C, and the output is 8.132 mV.

ASSUMPTION Thermocouple follows NIST standard voltage behavior.

FIND The temperature of the measuring junction

SOLUTION By the law of intermediate temperatures the output voltage for a thermocouple circuit having two junctions, one at 0 °C and the other at T_1, would be the sum of the voltages for a thermocouple circuit between 0 and 30 °C and between 30 °C and T_1. Thus,

$$V_{0-30} + V_{30-T_1} = V_{0-T_1}$$

This relationship allows the voltage reading from the nonstandard reference temperature to be converted to a 0 °C reference temperature by adding $V_{0-30} = 1.537$ to the existing reading. This results in an equivalent output voltage, referenced to 0 °C as

$$1.537 + 8.132 = 9.669 \text{ mV}$$

Clearly, this thermocouple is sensing the same temperature as in the previous example, 180.0 °C. This value is determined from Table 8.6.

COMMENT   Note that the effect of raising the reference junction temperature is to lower the output voltage of the thermocouple circuit. Negative values of voltage, as compared with the polarity listed in Table 8.4, indicate that the measured temperature is less than the reference junction temperature.

Example 8.9

A J-type thermocouple measures a temperature of 100 °C and is referenced to 0 °C. The thermocouple is AWG 30 (American wire gauge [AWG]; AWG 30 is 0.010-in. wire diameter) and is arranged in a circuit as shown in Figure 8.17a. The length of the thermocouple wire is 10 ft in order to run from the measurement point to the ice bath and to a voltmeter. The resolution of the voltmeter is 0.005 mV. If the thermocouple wire has a resistance per unit length, as specified by the manufacturer, of 5.6 Ω/ft, estimate the residual current in the thermocouple, assuming that the voltmeter has an infinite input impedance.

KNOWN   A voltmeter having a resolution of 0.005mV is used to measure the output voltage of a J-type thermocouple that is 10 ft long.

FIND   The residual current in the thermocouple circuit

SOLUTION   The total resistance of the thermocouple circuit is 56 Ω for 10 ft of thermocouple wire. The residual current is then found from Ohm's law as

$$I = \frac{E}{R} = \frac{0.005 \text{ mV}}{56 \, \Omega} = 8.9 \times 10^{-8} \text{ A}$$

COMMENT   The loading error due to this current flow is ~0.005 mV/54.3 μV/ °C ≈ 0.09 °C.

Example 8.10

Suppose that instead of assuming an infinite input impedance for the voltmeter, we assume that a high-impedance voltmeter is used. Determine the minimum input impedance required for the voltmeter that will limit the loading error to the same level as the infinite impedance case.

KNOWN   Loading error should be less than 8.9×10^{-8} A.

FIND   Input impedance for a voltmeter that would produce the same current flow or loading error as in Example 8.9

SOLUTION At 100 °C a J-type thermocouple referenced to 0 °C has a Seebeck voltage of $E_s = 5.269$ mV. At this temperature, the required voltmeter impedance to limit the current flow to 8.9×10^{-8} A is found from Ohm's law:

$$\frac{E_s}{I} = 5.269 \times 10^{-3} \text{ V}/8.9 \times 10^{-8} \text{ A} = 59.2 \text{ k}\Omega$$

COMMENT This input impedance is not at all high for a modern nanovoltmeter, and such a voltmeter would be a reasonable choice in this situation. As always, the allowable loading error should be determined based on the required uncertainty in the measured temperature.

Multiple-Junction Thermocouple Circuits

A thermocouple circuit composed of two junctions of dissimilar metals produces an open-circuit voltage that is related to the temperature difference between the two junctions. More than two junctions can be employed in a thermocouple circuit, and thermocouple circuits can be devised to measure temperature differences, or average temperature, or to amplify the output voltage of a thermocouple circuit.

Thermopiles

Thermopile is the term used to describe a multiple-junction thermocouple circuit that is designed to amplify the output of the circuit. Because thermocouple voltage outputs are typically in the millivolt range, increasing the voltage output may be a key element in reducing the uncertainty in the temperature measurement or may be necessary to allow transmission of the thermocouple signal to the recording device. Figure 8.22 shows a thermopile for providing an amplified output signal; in this case the output voltage would be N times the single thermocouple output, where N is the number of measuring junctions in the circuit. The average output voltage corresponds to the average temperature level sensed by the N junctions. Thus local point measurements entail consideration of the physical size of a thermopile, as compared to a single thermocouple. In transient measurements, a thermopile may have a more limited frequency range than a single thermocouple owing to its increased thermal capacitance. Thermopiles are particularly useful for reducing the uncertainty in measuring small

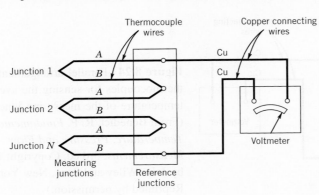

Figure 8.22 Thermopile arrangement. (From Benedict, R. P. *Fundamentals of Temperature, Pressure and Flow Measurements*, 3rd ed. Copyright © 1984 by John Wiley and Sons, New York. Reprinted by permission.)

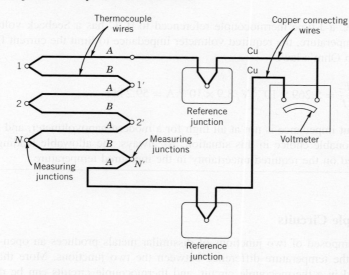

Figure 8.23 Thermocouples arranged to sense temperature differences. (From Benedict, R. P. *Fundamentals of Temperature, Pressure and Flow Measurements*, 3rd ed. Copyright © 1984 by John Wiley and Sons, New York. Reprinted by permission.)

temperature differences between the measuring and reference junctions. The principle has also been used to generate small amounts of power in spacecraft and to provide thermoelectric cooling using the Peltier effect.

Figure 8.23 shows a series arrangement of thermocouple junctions designed to measure the average temperature difference between junctions. This thermocouple circuit could also be used to measure deviations from a uniform temperature. In that case, any voltage output would indicate that a temperature difference existed between two of the thermocouple junctions. Alternatively, junctions $1, 2, \ldots, N$ could be located at one physical location, whereas junctions $1', 2', \ldots, N'$ could be located at another physical location. Such a circuit might be used to measure the heat flux through a solid.

Thermocouples in Parallel

When a spatially averaged temperature is desired, multiple thermocouple junctions can be arranged, as shown in Figure 8.24. In such an arrangement of N junctions, a mean voltage is

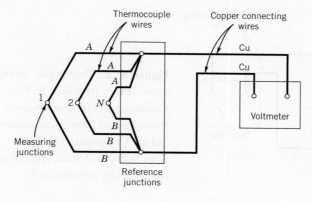

Figure 8.24 Parallel arrangement of thermocouples for sensing the average temperature of the measuring junctions. (From Benedict, R. P. *Fundamentals of Temperature, Pressure and Flow Measurements*, 3rd ed. Copyright © 1984 by John Wiley and Sons, New York. Reprinted by permission.)

produced, given by

$$\overline{V} = \frac{1}{N} \sum_{i-1}^{N} (V)_i \tag{8.17}$$

The mean voltage is indicative of a mean temperature,

$$\overline{T} = \frac{1}{N} \sum_{i-1}^{N} T_i \tag{8.18}$$

Applications for Thermoelectric Temperature Measurement: Heat Flux

Now that we understand measuring temperature using thermocouples and thermopiles, we will consider an important application for the use of these devices. Figure 8.25 illustrates conceptually how we might construct a *heat flux sensor*. We wish to measure the heat flux created by the temperature difference indicated in the figure, which flows from surface 1 to surface 2.

In this arrangement, thermopile junctions are placed on each side of a solid material, sometimes termed a barrier material. The thermal conductivity, k, of the barrier material is precisely known. We may compute the averaged heat flux through the barrier material as

$$|\dot{q}| = k \frac{|T_1 - T_2|}{L} \tag{8.19}$$

where

$|\dot{q}|$ - absolute value of heat flux(W/m^2)

k - thermal conductivity of the barrier material(W/m-K)

L - thickness of the barrier material(m)

$|T_1 - T_2|$ - absolute value of the temperature difference across the barrier material(K)

One embodiment of this concept is a thin-film heat flux sensor as illustrated in Figure 8.26. This sensor consists of thin-film thermocouples bonded to a flexible barrier material, such as Kapton.

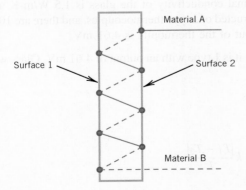

Figure 8.25 Construction of a heat flux sensor.

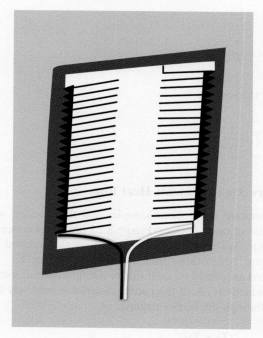

Figure 8.26   Thin-film heat flux sensor

When properly mounted on a surface, the heat flux through the sensor represents the heat flux through the surface. A typical sensitivity of a thin-film heat flux sensor is $3\,\mu V/(W/m^2)$.

Consider measuring the heat flux through the wall of a residence by mounting the heat flux sensor on the inside dry wall. Under these circumstances, the most accurate heat flux measurement would result if the sensor had the same surface characteristics as the dry wall. For example, so that the radiative component of heat transfer would match that of the dry wall, the sensor could be painted to match the wall.

Example 8.11

Consider using a thermopile arranged to measure temperature difference to determine the heat flux across a single pane glass window. The thermal conductivity of the glass is 1.5 W/m-K and the window is 5 mm thick. The thermopile is constructed of J-type thermocouples, and there are 10 active junctions. Determine the heat flux if the output of the thermopile is 4.61 mV.

KNOWN   For the thermopile, $N = 10$ and it is J-type with an output of 4.61 mV. Glass window has $L = 0.005$ m $k = 1.5$ W/m-K

FIND   The heat flux, $\dot{q}$ through the glass

SOLUTION   The heat flux is given by

$$|q| = k\frac{|T_1 - T_2|}{L} \tag{8.19}$$

The temperature difference can be determined from the thermopile output and Table 8.6. With 10 active junctions, the single junction equivalent output voltage would be 4.61 mV/10 = 0.461 mV At temperatures near room temperature the sensitivity of a J-type thermocouple is approximately 0.0512 mV/°C. Therefore the temperature difference may be determined as

$$T_2 - T_1 = 0.461 \text{ mV}/0.0512 \text{ mV}/°C = 9°C$$

And the resulting heat flux is

$$|q| = k\frac{|T_1 - T_2|}{L} = (1.5 \text{ W/m} - \text{K})\frac{9\,°C}{0.005 \text{ m}} = 2700 \text{ W/m}^2$$

COMMENT Note that the size of the degree, whether Kelvin or Celsius, is the same, and no units conversion is required.

Example 8.12: Heat Flux Case Study

Phase-change gel encapsulated within beads is being embedded in various bedding products to better regulate temperature. A person's body heat is absorbed by a mattress, warming the surface. However, a person can feel thermal discomfort if this heat is not conducted away sufficiently, in effect insulating the sleeper. The most common targets are mattresses made of foam known to cause thermal discomfort for many users. The idea is to absorb body energy within the gel more rapidly so as to maintain a cooler body feel longer. Supplemental air gaps in the foam help to move trapped energy away out of the gel and from the sleeper's body.

To quantify the effect of encapsulated gel on the cooling effect for a mattress, a major supplier of medical mattresses conducted a test using three different gel materials. A square sample of a gel mattress approximately 0.4 m square was used for testing. The top surface of the mattress was

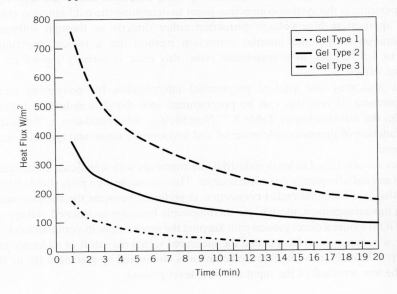

Figure 8.27 Heat flux as a function of time for three gel types employed in a foam mattress.

instrumented with a thin-film heat flux gauge, and a weighted constant-temperature heat source of the same size as the mattress sample was placed on top of the mattress sample. The conditions were designed to simulate a person's torso and the step-change in heat flux associated with lying down onto the mattress. The heat flux sensor output was appropriately amplified and recorded. Figure 8.27 shows the results of this test for three different gel types. The larger the heat flux, the greater the cooling effect. It is clear that the choice of particular gel material has a significant effect on the heat flux, both during early times and in steady state. Within 10 minutes the heat flux is approaching a steady-state value, which varies between samples. The experience of lying down on a mattress containing encapsulated phase-change gel creates a perception that the surface is initially quite cool, but within a short period the perceived level of cooling is significantly reduced. The test shows that some designs maintain the cooler perception longer, supposedly an aid to better sleeping. Tests such as this are intended to test the effectiveness of preliminary designs as an aide to making informed decisions to improve them.

Data Acquisition Considerations

Measuring temperatures by using thermocouples connected to data acquisition systems is common practice. However, the characteristics of the thermocouple, including the need for a reference or cold junction and the low signal voltages produced, complicate its use. Nevertheless, with a little attention and realistic expectation of achievable accuracy, the systems are quite acceptable for most monitoring and moderate accuracy measurements.

Once the appropriate thermocouple type is selected, attention must be given to the cold junction compensation method. The two connection points of the thermocouple to the data acquisition system (DAS) board form two new thermocouple junctions. Use of external cold junction methods between the thermocouple and the board eliminates this problem, but more frequently the thermocouple is connected directly to the board and uses built-in electronic cold junction compensation. This is usually accomplished by using a separate thermistor sensor, which measures the temperature at the system connection point to determine the cold junction error, and by providing an appropriate bias voltage correction either directly or through software. An important consideration is that the internal correction method has a typical uncertainty of the order of 0.5 to 1.5 °C and, as a systematic error, this error is directly passed on to the measurement as an offset.

These boards also may use internal polynomial interpolation for converting measured voltage into temperature. If not, this can be programmed into the data reduction scheme by using, for example, the information of Table 8.7. Nonetheless, this introduces a "linearization" error, which is a function of thermocouple material and temperature range and typically specified with the DAS board.

Thermocouples are often used in harsh industrial environments with significant electromagnetic interference (EMI) and radiofrequency (rf) noise sources. Thermocouple wire pairs should be twisted to reduce noise. Also, a differential-ended connection is preferred between the thermocouple and the DAS board. In this arrangement, though, the thermocouple becomes an isolated voltage source, meaning that there is no longer a direct ground path keeping the input within its common mode range. As a consequence, a common complaint is that the measured signal may drift or suddenly jump in the level of its output. Usually, this interference behavior is eliminated by placing a 10- to 100-kΩ resistor between the low terminal of the input and low-level ground.

Because most DAS boards use analog-to-digital (A/D) converters having a ± 5 V full scale, the signal must be conditioned using an amplifier. Usually a gain of 100 to 500 is sufficient. High-gain, very low noise amplifiers are important for accurate measurements. Consider a J-type thermocouple using a 12-bit A/D converter with a signal conditioning gain of 100. This allows for a full-scale input range of 100 mV, which is a suitable range for most measurements. Then, the A/D conversion resolution is

$$\frac{E_{FSR}}{(G)(2^M)} = \frac{10 \text{ V}}{(100)(1^{12})} = 24.4 \text{ } \mu\text{V/bit}$$

A J-type thermocouple has a sensitivity of ~ 55 μV/°C. Thus the measurement resolution becomes

$$(24.4 \text{ } \mu\text{V/bit})/(55 \text{ } \mu\text{V/°C}) = 0.44 \text{ °C/bit}.$$

Thermocouples tend to have long time constants relative to the typical sample rate capabilities of general-purpose DAS boards. If temperature measurements show greater than expected fluctuations, high-frequency sampling noise is a likely cause. Reducing the sample rate or using a smoothing filter are simple solutions. The period of averaging should be on the scale of the time constant of the thermocouple.

Example 8.13

It is desired to create an off-the-shelf temperature measuring system for a personal computer (PC)–based control application. The proposed temperature measuring system is illustrated schematically in Figure 8.28. The temperature measurement system consists of the following:

- A PC-based DAS composed of a data acquisition board, a computer, and appropriate software to allow measurement of analog input voltage signals.
- A J-type thermocouple and reference junction compensator. The reference junction compensator serves to provide an output voltage from the thermocouple equivalent to the value that would exist between the measuring junction and a reference temperature of 0 °C.
- A J-type thermocouple, which is uncalibrated but meets NIST standard limits of error.

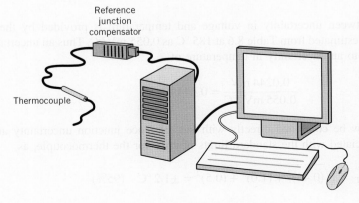

Figure 8.28 Personal computer (PC)-based temperature measurement system.

The system is designed to measure and control a process temperature that varies slowly in time compared to sampling rate of the DAS. The process nominally operates at 185 °C. The following specifications are applicable to the measurement system components:

Component	Characteristics	Accuracy Specifications
Data acquisition board	Analog voltage input range: 0 to 0.1 V	12-bit A/D converter accuracy: ±0.01% of reading
Reference junction compensator	J-type compensation range from 0 to 50 °C	±0.5 °C over the range 20 to 36 °C
Thermocouple (J-type)	Stainless steel sheathed ungrounded junction	Accuracy: ±1.0 °C based on NIST standard limits of error

The purpose of the measurement system requires that the temperature measurement have a total uncertainty of less than 1.5 °C. Based on a design stage uncertainty analysis, does this measurement system meet the overall accuracy requirement?

SOLUTION The design stage uncertainty for this measurement system is determined by expressing the uncertainty of each system component as an equivalent uncertainty in temperature, and then combining these design stage uncertainties.

The 12-bit A/D converter divides the full-scale voltage range into 2^{12}, or 4,096, equal-sized intervals. Thus, the resolution (quantization error) of the A/D in measuring voltage is

$$\frac{0.1 \text{ V}}{4096 \text{ intervals}} = 0.0244 \text{ mV}$$

The uncertainty of the DAS is specified as 0.01% of the reading. A nominal value for the thermocouple voltage must be known or established. In the present case, the nominal process temperature is 185 °C, which corresponds to a thermocouple voltage of approximately 10 mV. Thus the calibration uncertainty of the DAS is 0.001 mV.

The contribution to the total uncertainty of the temperature measurement system from the DAS can now be determined. First combine the resolution and calibration uncertainties as

$$u_{DAS} = \sqrt{(0.0244)^2 + (0.001)^2} = 0.0244 \text{ mV}$$

The relationship between uncertainty in voltage and temperature is provided by the static sensitivity, which can be estimated from Table 8.6 at 185 °C as 0.055 mV/°C. Thus an uncertainty of 0.0244 mV corresponds to an uncertainty in temperature of

$$\frac{0.0244 \text{ mV}}{0.055 \text{ mV/°C}} = 0.444 \text{ °C}$$

This uncertainty can now be combined directly with the reference junction uncertainty and the uncertainty interval associated with the standard limits of error for the thermocouple, as

$$u_T = \sqrt{(0.44)^2 + (1.0)^2 + (0.5)^2} = \pm 1.2 \text{ °C} \quad (95\%)$$

COMMENT   It might be appropriate in certain cases to calibrate the thermocouple against a laboratory standard, such as an RTD, which has a calibration traceable to NIST standards. With reasonable expense and care, an uncertainty level in the thermocouple of ±0.1 °C can be achieved, compared to the uncalibrated value of ±1 °C. If the thermocouple were so calibrated, the resulting uncertainty in the overall measurement of temperature is reduced to ±0.68 °C, so that by reducing the uncertainty contribution of the thermocouple by a factor of 10, the system uncertainty would be reduced by a factor of 2.

8.6   RADIATIVE TEMPERATURE MEASUREMENTS

There is a distinct advantage to measuring temperature by detecting thermal radiation. The sensor for thermal radiation need not be in contact with the surface to be measured, making this method attractive for a wide variety of applications. The basic operation of a radiation thermometer is predicated upon some knowledge of the radiation characteristics of the surface whose temperature is being measured, relative to the calibration of the thermometer. The spectral characteristics of radiative measurements of temperature are beyond the scope of the present discussion; an excellent source for further information is Dewitt and Nutter (10).

Radiation Fundamentals

Radiation refers to the emission of electromagnetic waves from the surface of an object. This radiation has characteristics of both waves and particles, which leads to a description of the radiation as being composed of photons. The photons generally travel in straight lines from points of emission to another surface, where they are absorbed, reflected, or transmitted. This electromagnetic radiation exists over a large range of wavelengths that includes x-rays, ultraviolet radiation, visible light, and infrared or thermal radiation, as shown in Figure 8.29. The thermal radiation emitted from an object

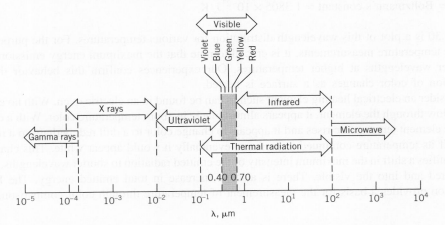

Figure 8.29   The electromagnetic spectrum. (From Incropera F. P., and D. P. DeWitt, *Fundamentals of Heat and Mass Transfer*, 2nd ed. Copyright © 1985 by John Wiley & Sons, New York. Reprinted by permission.)

is related to its temperature and has wavelengths ranging from approximately 10^{-7} to 10^{-3} m. It is necessary to understand two key aspects of radiative heat transfer in relation to temperature measurements. First, the radiation emitted by an object is proportional to the fourth power of its temperature. In the ideal case, this may be expressed as

$$E_b = \sigma T^4 \tag{8.20}$$

where E_b is the flux of energy radiating from an ideal surface, or the blackbody emissive power. The emissive power of a body is the energy emitted per unit area and per unit time. The term "blackbody" implies a surface that absorbs all incident radiation and as a result emits radiation in an "ideal" manner.

Second, the emissive power is a direct measure of the total radiation emitted by an object. However, energy is emitted by an ideal radiator over a range of wavelengths, and at any given temperature the distribution of the energy emitted as a function of wavelength is unique. Max Planck (1858–1947) developed the basis for the theory of quantum mechanics in 1900 as a result of examining the wavelength distribution of radiation. He proposed the following equation to describe the wavelength distribution of thermal radiation for an ideal or blackbody radiator:

$$E_{b\lambda} = \frac{2\pi h_p c^2}{\lambda^5 \left[\exp\left(h_p c / k_B \lambda T \right) - 1 \right]} \tag{8.21}$$

where

$E_{b\lambda}$ = total emissive power at the wavelength λ
λ = wavelength
c = speed of light in a vacuum = 2.998×10^8 m/s
h_p = Planck's constant = 6.6256×10^{-34} J-s
k_B = Boltzmann's constant = 1.3805×10^{-23} J/K

Figure 8.30 is a plot of this wavelength distribution for various temperatures. For the purposes of radiative temperature measurements, it is crucial to note that the maximum energy emission shifts to shorter wavelengths at higher temperatures. Our experiences confirm this behavior through observation of color changes as a surface is heated.

Consider an electrical heating element such as can be found in an electric oven. With no electric current flow through the element, it appears almost black, its room temperature color. With a current flow, the element temperature rises and it appears to change color to a dull red, perhaps to a reddish orange. If its temperature continued to increase, eventually it would appear white. This change in color signifies a shift in the maximum intensity of the emitted radiation to shorter wavelengths, out of the infrared and into the visible. There is also an increase in total emitted energy. The Planck distribution provides a basis for the measurement of temperature through color comparison.

Radiation Detectors

Radiative energy flux can be detected in a sensor by two basic techniques. The detector is subject to radiant energy from the source whose temperature is to be measured. The first technique involves a

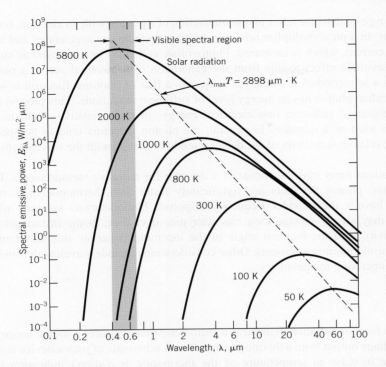

Figure 8.30 Planck distribution of blackbody emissive power as a function of wavelength. (From Incropera F. P., and D. P. DeWitt, *Fundamentals of Heat and Mass Transfer*, 2nd ed. Copyright © 1985 by John Wiley & Sons, New York. Reprinted by permission.)

thermal detector in which absorbed radiative energy elevates the detector temperature, as shown in Figure 8.31. These thermal detectors are certainly the oldest sensors for radiation, and the first such detector can probably be credited to Sir William Herschel, who verified the presence of infrared radiation using a thermometer and a prism. The equilibrium temperature of the detector is a direct measure of the amount of radiation absorbed. The resulting rise in temperature must then be measured. Thermopile detectors provide a thermoelectric power resulting from a change in temperature. A thermistor can also be used as the detector and results in a change in resistance with temperature.

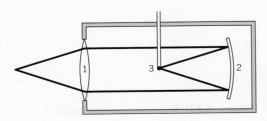

Figure 8.31 Schematic of a basic radiometer: (1) lens; (2) focusing mirror; (3) detector (thermopile or thermistor).

A second basic type of detector relies on the interaction of a photon with an electron, resulting in an electric current. In a photomultiplier tube, the emitted electrons are accelerated and used to create an amplified current, which is measured. Photovoltaic cells may be employed as radiation detectors. The photovoltaic effect results from the generation of a potential across a positive–negative junction in a semiconductor when it is subject to a flux of photons. Electron-hole pairs are formed if the incident photon has an energy level of sufficient magnitude. This process results in the direct conversion of radiation into electrical energy, in high sensitivity, and in a fast response time when used as a detector. In general, the photon detectors tend to be spectrally selective, so that the relative sensitivity of the detector tends to change with the wavelength of the measured radiation.

Many considerations enter into the choice of a detector for radiative measurements. If time response is important, photon detectors are significantly faster than thermopile or thermistor detectors and thus have a much wider frequency response. Photodetectors saturate, whereas thermopile sensors may slowly change their characteristics over time. Some instruments have variations of sensitivity with the incident angle of the incoming radiation; this factor may be important for solar insolation measurements. Other considerations include wavelength sensitivity, cost, and allowable operating temperatures.

Radiometer

Perhaps the simplest form of radiation detector is a radiometer that measures a source temperature by measuring the voltage output from a thermopile detector. A schematic of such a device is shown in Figure 8.31. The increase in temperature of the thermopile is a direct indication of the temperature of the radiation source. One application of this principle is in the measurement of total solar radiation incident upon a surface. Figure 8.32 shows a schematic of a pyranometer, used to measure global solar irradiance. It would have a hemispherical field of view and measures both the direct or beam radiation and diffuse radiation. The diffuse and beam components of radiation can be separated by shading the pyranometer from the direct solar radiation, thereby measuring the diffuse component.

Figure 8.33 shows infrared (IR) thermopile sensors that are manufactured using micromachining and advanced semiconductor processing methods. The hot junction of the thermopile is placed under the IR filter and the cold junctions under the IR mask. Manufacturing methods allow hundreds of junctions to be created in a micro-device. These sensors form the basis for many practical applications, ranging from timpanic thermometers to automatic climate control systems for automotive applications.

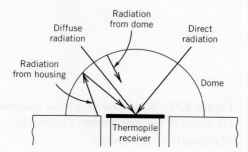

Figure 8.32 Pyranometer construction.

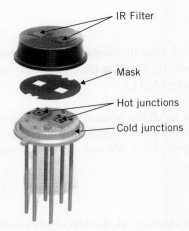

Figure 8.33   Industrial infrared (IR) thermopile sensor.

Pyrometry

Optical pyrometry identifies the temperature of a surface by its color, or more precisely the color of the radiation it emits. A schematic of an optical pyrometer is shown in Figure 8.34. A standard lamp is calibrated so that the current flow through its filament is controlled and calibrated in terms of the filament temperature. Comparison is made optically between the color of this filament and the surface of the object whose temperature is being measured. The comparator can be the human eye. Uncertainties in the measurement may be reduced by appropriately filtering the incoming light. Corrections must be applied for surface emissivity associated with the measured radiation; uncertainties vary with the skill of the user, and generally are on the order of 5 °C. Replacing the human eye with a different detector extends the range of useful temperature measurement and reduces the random uncertainty.

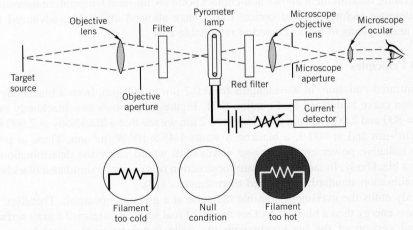

Figure 8.34   Schematic diagram of a disappearing filament optical pyrometer.

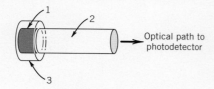

Optical path to photodetector

Figure 8.35 Optical fiber thermometer: (1) blackbody cavity (iridium film); (2) sapphire fiber (single crystal); (3) protective coating (Al_2O_3).

The major advantage of an optical pyrometer lies in its ability to measure high temperatures remotely. For example, it could be used to measure the temperature of a furnace without having any sensor in the furnace itself. For many applications this provides a safe and economical means of measuring high temperatures.

Optical Fiber Thermometers

The optical fiber thermometer is based on the creation of an ideal radiator that is optically coupled to a fiber-optic transmission system (11,12), as shown in Figure 8.35. The temperature sensor in this system is a thin, single-crystal aluminum oxide (sapphire) fiber; a metallic coating on the tip of the fiber forms a blackbody radiating cavity, which radiates directly along the sapphire crystal fiber. The single-crystal sapphire fiber is necessary because of the high-temperature operation of the thermometer. The operating range of this thermometer is 300 to 1,900 °C. Signal transmission is accomplished using standard, low-temperature fiber optics. A specific wavelength band of the transmitted radiation is detected and measured, and these raw data are reduced to yield the temperature of the blackbody sensor.

The absence of electrical signals associated with the sensor signal provides excellent immunity from electromagnetic and radiofrequency interference. The measurement system has superior frequency response and sensitivity. The system has been employed for measurement in combustion applications. Temperature resolution of 0.1 mK is possible.

Narrow Band Infrared Temperature Measurement

Infrared temperature measurement allows non-contact point or full-field temperature measurements. Advances in detector technology and optical filters have allowed affordable, advanced infrared thermometers and cameras to be developed at reasonable cost.

Fundamental Principles

Consider the infrared emission, in wavelengths from 0.7 μm to 100 μm, from a blackbody whose spectral emission curve is defined by Equation 8.21. Figure 8.36 shows two blackbody emission curves, for $T = 800$ and 2,000 K. At a wavelength of 2 μm we see that a blackbody at 2,000 K emits 3.29×10^5 W/m²-μm and at 800 K a blackbody emits 1.45×10^3 W/m²-μm. Thus, in principle, measuring the radiative power emitted at one wavelength would allow the determination of the temperature of a blackbody. In fact, appropriate construction of a cavity to simulate a blackbody can be used as a calibration standard for infrared thermometers (13).

A blackbody emits the maximum possible radiation at a given temperature. Therefore all real surfaces emit less energy than a blackbody. One model of real surfaces is termed a gray surface; it is simply a scaled version of the blackbody with the scale factor being the total hemispherical emissivity. Figure 8.36 shows a blackbody emission curve and a gray surface emission curve at 800 K; the emissivity of the gray surface, ε, is 0.6. Thus for real surfaces it is necessary to either know

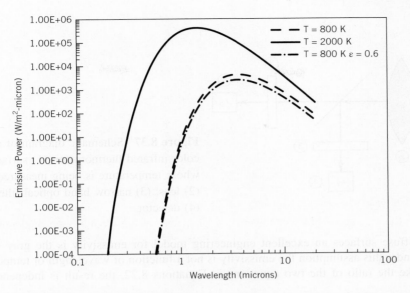

Figure 8.36 Blackbody emission curves for 800 K and 2,000 K and gray surface emission curve for 800 K and an emissivity of 0.6.

the emissivity or develop some means of eliminating the emissivity from the temperature measurement. Modern infrared thermometers have user input for surface emissivity and make corrections to the reported temperature based on this input.

The practical application of infrared thermometry requires an understanding of the conditions under which the measurement is being made, and the surface characteristics of the object for which the temperature is being measured. An infrared thermometer gathers all radiation from the surface being measured, both emitted and reflected. Thus for a surface having a significant reflectivity the temperature of the surroundings can affect the measured value. Absorption in the intervening medium can also affect the temperature measurement. For example, CO_2 and H_2O have several absorption bands between 1.5 and 3 μm; these particular wavelengths should not be selected for temperature measurement. The selection of a wavelength band depends on the expected temperature and the particular application.

Two-Color Thermometry

Two-color infrared thermometry refers to a technique that uses measurement of radiative emissive power at two different narrow wavelength bands. Equations 8.22 show the theoretical emissive power for two different wavelengths where we incorporated spectral and temperature dependent hemispherical emissivity.

$$E_{\lambda_1} = \varepsilon(\lambda_1, T) \frac{2\pi h_p c^2}{\lambda_1^5 \left[\exp\left(h_p c / k_B \lambda_1 T\right) - 1\right]}$$

$$E_{\lambda_2} = \varepsilon(\lambda_2, T) \frac{2\pi h_p c^2}{\lambda_2^5 \left[\exp\left(h_p c / k_B \lambda_2 T\right) - 1\right]}$$

$$(8.22)$$

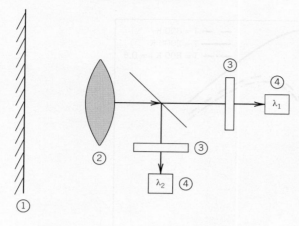

Figure 8.37 Schematic diagram of a two color infrared thermometer. (1) surface whose temperature is being measured, (2) lens, (3) narrow band optical filter, (4) detector.

For opaque diffuse surfaces an excellent engineering model for emissivity is the gray surface assumption. Under this assumption the emissivity is not a function of wavelength or temperature. Thus if we take the ratio of the two equations in Equations 8.22, the result is independent of the emissivity:

$$\frac{E_{\lambda_1}}{E_{\lambda_2}} = \frac{\lambda_2^5 \left[\exp\left(h_p c / k_B \lambda_2 T\right) - 1\right]}{\lambda_1^5 \left[\exp\left(h_p c / k_B \lambda_1 T\right) - 1\right]} \tag{8.23}$$

In practice, two independent measurements of emissive power at two different narrow wavelength bands are made, and the signal ratio is a direct indicator of the surface temperature without knowing the surface emissivity. Figure 8.37 shows a conceptual arrangement of the optics and detectors required to implement the two-color measurement method.

Full Field IR Imaging

Advances in CCD detector and microprocessor technology has allowed infrared cameras to gain wide usage in industry and for research purposes. Applications include assessing the effectiveness of insulation in buildings, finding victims in a fire, and mapping body surface temperature for a variety of medical purposes (14). Other applications include night vision, including both security and military uses, and maritime night navigation.

Example 8.14: Infrared Imaging of Environmental Water Systems to Visualize Fluid Motion Patterns

A variety of situations exist where it may be useful to monitor the discharge of warm water into an environmental water system, such as occurs at power plants that use lakes for a cooling water source. Figure 8.38 is an infrared image obtained using an IR camera sensitive to the 3-5 μm wavelength band (15). The image is essentially the temperature field of a warm water surface that is evaporating into still air. The evaporation of the water that occurs at the interface between the water and air cools the water at the interface and creates a buoyancy-driven flow in the water. In Figure 8.38, the darker colored pixels represent cooler temperatures. The complex pattern of cool

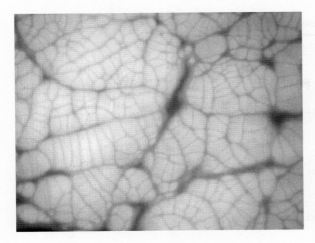

Figure 8.38 Infrared image of water surface under conditions where water is evaporating into still air. (Reprinted from Bower, S. M., and Saylor, J.,R., "The effects of surfactant monolayers on free surface natural convection," *International Journal of Heat and Mass Transfer*, 54, 5348-5358, (2011), with permission from Elsevier.)

and warm regions in this image reveal the underlying pattern of liquid motion created by the buoyancy forces resulting from the temperature difference between the fluid at the interface and the underlying, warmer fluid.

8.7   PHYSICAL ERRORS IN TEMPERATURE MEASUREMENT

In general, errors in temperature measurement derive from two fundamental sources. The first source of errors derives from uncertain information about the temperature of the sensor itself. Such uncertainties can result from random interpolation errors, calibration systematic errors, or a host of other error sources. Instrument and procedural uncertainty in the sensor temperature can be reduced by improved calibration or by changes in the measuring and recording instruments. However, errors in temperature measurement can still occur even if the temperature of the sensor was measured exactly. In such cases, the probe does not sense accurately the temperature it was intended to measure.

A list of typical errors associated with the use of temperature sensors is provided in Table 8.8. Random errors in temperature measurements are a result of resolution limits of measuring and recording equipment, time variations in extraneous variables, and other sources of variation. Thermocouples have some characteristics that can lead to both systematic and random errors, such as the effect of extension wires and connectors. Another major source of error for thermocouples involves the accuracy of the reference junction. Ground loops can lead to spurious readings, especially when thermocouple outputs are amplified for control or data acquisition purposes. As with any measurement system, calibration of the entire measurement system, in place if possible, is the best means of identifying error sources and reducing the resulting measurement uncertainty to acceptable limits.

Insertion Errors

This discussion focuses on ensuring that a sensor output accurately represents the temperature it is intended to measure. For example, suppose it is desired to measure the outdoor temperature. This measurement could employ a large dial thermometer, which might be placed on a football field or a

Table 8.8   Measuring Errors Associated with Temperature Sensors

Random Errors
1. Imprecision of readings
2. Time and spatial variations

Systematic Errors
1. Insertion errors, heating or cooling of junctions
   a. Conduction errors
   b. Radiation errors
   c. Recovery errors
2. Effects of plugs and extension wires
   a. Nonisothermal connections
   b. Loading errors
3. Ignorance of materials or material changes during measurements
   a. Aging following calibration
   b. Annealing effects
   c. Cold work hardening
4. Ground loops
5. Magnetic field effects
6. Galvanic error
7. Reference junction inaccuracies

tennis court in the direct sunlight, and assumed to represent "the temperature," perhaps as high as 50 °C (120 °F). But what temperature is being indicated by this thermometer? Certainly the thermometer is not measuring the air temperature, nor is it measuring the temperature of the field or the court. The thermometer is subject to the very sources of error we wish to describe and analyze. *The thermometer, very simply, indicates its own temperature!*

In our example, the temperature of the thermometer is the thermodynamic equilibrium temperature that results from the radiant energy gained from the sun, convective exchange with the air, and conduction heat transfer with the surface on which it is resting. Considering the fact that these thermometers typically have a glass cover that ensures a greenhouse effect, it is very likely that the thermometer temperature is significantly higher than the air temperature.

The physical mechanisms that may cause a temperature probe to indicate a temperature different from that intended include conduction, radiation, and velocity recovery errors. In any real measurement system, their effects could be coupled, and therefore should not be considered independently. However, for simplicity, we will consider each separately because our purpose is to provide only estimates of the errors, and not to provide predictive techniques for correcting measured temperatures. The goal of the measurement engineer should be to minimize these errors, as far as possible, through the careful installation and design of temperature probes.

Conduction Errors

Errors that result from conduction heat transfer between the measuring environment and the ambient are often called immersion errors. Consider the temperature probe shown in Figure 8.39. In many circumstances, a temperature probe extends from the measuring environment through a wall into the

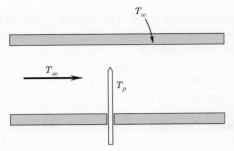

Figure 8.39　Temperature probe inserted into a measuring environment.

ambient environment, where indicating or recording systems are located. The probe and the electrical leads form a path for the conduction of energy from the measuring environment to the ambient. The fundamental nature of the error created by conduction in measured temperatures can be illustrated by the model of a temperature probe shown in Figure 8.40. The essential physics of immersion errors associated with conduction can be discerned by modeling the temperature probe as a fin. Suppose we assume that the measured temperature is higher than the ambient temperature. If we consider a differential element of the fin, as shown in Figure 8.40b, at steady state there is energy conducted along the fin and transferred by convection from its surface. The surface area for convection is Pdx, where P is the perimeter or circumference. Applying the first law of thermodynamics to this differential element yields

$$q_{x+dx} - q_x = hP\,dx[T(x) - T_\infty] \tag{8.24}$$

where h is the convection coefficient. If q is expanded in a Taylor series about the point x and the substitutions

$$\theta = T - T_\infty \qquad q = -kA\frac{dT}{dx} \qquad m = \sqrt{\frac{hP}{kA}} \tag{8.25}$$

are made, then the governing differential equation becomes

$$\frac{d^2\theta}{dx^2} - m^2\theta = 0 \tag{8.26}$$

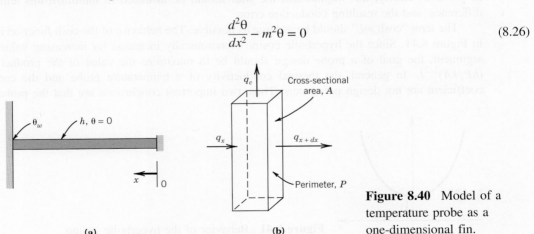

(a)　**(b)**

Figure 8.40　Model of a temperature probe as a one-dimensional fin.

Here k is the effective thermal conductivity of the temperature probe. The solution to this differential equation for the boundary conditions that the wall has a temperature T_w, or a normalized value $\theta_w = T_w - T_\infty$, and the end of the fin is small in surface area, is

$$\frac{\theta(x)}{\theta_w} = \frac{\cosh mx}{\cosh mL} \tag{8.27}$$

The point $x = 0$ is the location where the temperature is assumed to be measured, and so the solution is evaluated at $x = 0$ as

$$\frac{\theta(0)}{\theta_w} = \frac{T(0) - T_\infty}{T_w - T_\infty} = \frac{1}{\cosh mL} \tag{8.28}$$

From this analysis, the error arising from conduction, e_c, can be estimated. An ideal sensor would indicate the fluid temperature T_∞; thus, if the sensor temperature is $T_p = T(0)$, then the conduction error is

$$e_c = T_p - T_\infty = \frac{T_w - T_\infty}{\cosh mL} \tag{8.29}$$

Normally, the uncertainty due to conduction error is set at $u = e_c$. The uncertainty interval may not be symmetric.

Probe Design

The purpose of the preceding analysis is to gain some physical understanding of ways to minimize conduction errors (not to correct inaccurate measurements). The behavior of this solution is such that the ideal temperature probe would have $T_p = T_\infty$, or $\theta(0) = 0$, implying that $e_c = 0$. Equation 8.29 shows that a value of $\theta(0) \neq 0$ results from a nonzero value of θ_w, and a finite value of $\cosh mL$. The difference between the fluid temperature being measured and the wall temperature should be as small as possible; clearly, this implies that the wall should be insulated to minimize this temperature difference, and the resulting conduction error.

The term "$\cosh mL$" should be as large as possible. The behavior of the cosh function is shown in Figure 8.41. Since the hyperbolic cosine monotonically increases for increasing values of the argument, the goal of a probe design should be to maximize the value of the product mL, or $(hP/kA)^{1/2}L$. In general, the thermal conductivity of a temperature probe and the convection coefficient are not design parameters. Thus, two important conclusions are that the probe should

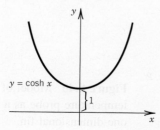

Figure 8.41 Behavior of the hyperbolic cosine.

be as small in diameter as possible, and should be inserted as far as possible into the measuring environment away from the bounding surface to make L large. A small diameter increases the ratio of the perimeter P to the cross-sectional area A. For a circular cross section, this ratio is $4/D$, where D is the diameter. A good rule of thumb based on Equation 8.29 is to have an $L/D > 50$ for negligible conduction error.

Although this analysis clearly indicates the fundamental aspects of conduction errors in temperature measurements, it does not provide the capability to correct measured temperatures for conduction errors. Conduction errors should be minimized through appropriate design and installation of temperature probes. Usually, the physical situation is sufficiently complex to preclude accurate mathematical description of the measurement errors. Additional information on modeling conduction errors may be found in Sparrow (16).

Radiation Errors

Consider a temperature probe used to measure a gas temperature. In the presence of significant radiation heat transfer, the equilibrium temperature of a temperature probe may be different from the fluid temperature being measured. Because radiation heat transfer is proportional to the fourth power of temperature, the importance of radiation effects increases as the absolute temperature of the measuring environment increases. The error due to radiation can be estimated by considering steady-state thermodynamic equilibrium conditions for a temperature sensor. Consider the case where energy is transferred to/from a sensor by convection from the environment and from/to the sensor by radiation to a body at a different temperature, such as a pipe or furnace wall. For this analysis conduction is neglected. A first law analysis of a system containing the probe, at steady-state conditions, yields

$$q_c + q_r = 0 \tag{8.30}$$

where

$$q_c = \text{convective heat transfer}$$
$$q_r = \text{radiative heat transfer}$$

The heat-transfer components can then be expressed in terms of the appropriate fundamental relations as

$$q_c = hA_s(T_\infty - T) \quad q_r = FA_s \varepsilon \sigma (T_w^4 - T^4) \tag{8.31}$$

Assuming that the surroundings may be treated as a blackbody, the first law for a system consisting of the temperature probe is

$$hA_s(T_\infty - T_p) = FA_s \varepsilon \sigma (T_p^4 - T_w^4) \tag{8.32}$$

A temperature probe is generally small compared to its surroundings, which justifies the assumption that the surroundings may be treated as black. The radiation error e_r is estimated by

$$e_r = (T_p - T_\infty) = \frac{F\varepsilon\sigma}{h}(T_w^4 - T_p^4) \tag{8.33}$$

where

σ = the Stefan-Boltzmann constant ($\sigma = 5.669 \times 10^{-8}$ W/m^{2}K^4)

ε = emissivity of the sensor

F = radiation view factor

T_p = probe temperature

T_w = temperature of the surrounding walls

T_∞ = fluid temperature being measured

If the sensor is small compared to the scale of the surroundings, the view factor from the sensor to the surroundings may be taken as 1. Normally, the uncertainty due to radiation error is set as $u = e_r$. The uncertainty interval is not symmetrical and might be modeled as a uniform (rectangular) distribution with bounds of 0 and e_r.

Example 8.15

A typical situation where radiation would be important occurs in measuring the temperature of a furnace. Figure 8.42 shows a small temperature probe for which conduction errors are negligible, which is placed in a high-temperature enclosure, where the fluid temperature is T_∞ and the walls of the enclosure are at T_w.

Convection and radiation are assumed to be the only contributing heat transfer modes at steady state. Develop an expression for the equilibrium temperature of the probe. Also determine the equilibrium temperature of the probe and the radiation error in the case where $T_\infty = 800\,°$C, $T_w = 500\,°$C, and the emissivity of the probe is 0.8. The convective heat transfer coefficient is 100 W/m^2 °C.

KNOWN   Temperatures $T_\infty = 800\,°$C and $T_w = 500\,°$C, with $h = 100$ W/m^2°C. The emissivity of the probe is 0.8.

FIND   An expression for the equilibrium temperature of the probe, and the resulting probe temperature for the stated conditions

ASSUMPTIONS   The surroundings may be treated as black, and conduction heat transfer is neglected

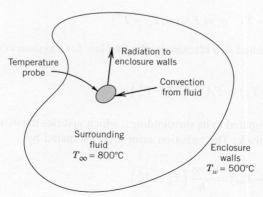

Figure 8.42   Analysis of a temperature probe in a radiative and convective environment.

SOLUTION   The probe is modeled as a small, spherical body within the enclosed furnace; the view factor from the probe to the furnace is 1. At steady state, an equilibrium temperature may be found from an energy balance. The first law for a system consisting of the temperature probe from Equation 8.32 is

$$h\left(T_\infty - T_p\right) + \sigma\varepsilon\left(T_w^4 - T_p^4\right) = 0 \tag{8.34}$$

The probe is attempting to measure T_∞. Equation 8.34 yields an equilibrium temperature for the probe of $T_p = 642.5\,°C$. The predicted radiation error in this case can be as large as (from Equation 8.33) $-157.5\,°C$. This result indicates that radiative heat transfer can create significant measurement errors at elevated temperatures.

COMMENT   Because the radiation error in this case can only lower the indicated temperature, it is an example of a nonsymmetrical error (see Section 5.10). We can use this predicted value as an estimate of the uncertainty resulting from this effect, or we can use the predicted value to correct the indicated value and assign a smaller uncertainty based on the uncertainty in the correction.

Radiation Shielding

Radiation shielding is a key concept in controlling radiative heat transfer; shielding for radiation is analogous to insulation to reduce conduction heat transfer. A radiation shield is an opaque surface interposed between a temperature sensor and its radiative surroundings so as to reduce electromagnetic wave interchange. In principle, the shield attains an equilibrium temperature closer to the fluid temperature than the surroundings. Because the probe can no longer "see" the surroundings, with the radiation shield in place, the probe temperature is closer to the fluid temperature. Additional information on radiation error in temperature measurements may be found in Sparrow (16). The following example serves to demonstrate radiation errors and the effect of shielding.

Example 8.16

Consider again Example 8.15 where the oven is maintained at a temperature of 800 °C. Because of energy losses, the walls of the oven are cooler, having a temperature of 500 °C. For the present case, consider the temperature probe as a small, spherical object located in the oven, having no thermal conduction path to the ambient. (All energy exchange is through convection and radiation.) Under these conditions, the probe temperature is 642.5 °C, as found in Example 8.15.

Suppose a radiation shield is placed between the temperature probe and the walls of the furnace, which blocks the path for radiative energy transfer. Examine the effect of adding a radiation shield on the probe temperature.

KNOWN   A radiation shield is added to a temperature probe in an environment with $T_\infty = 800\,°C$ and $T_w = 500\,°C$.

FIND   The radiation error in the presence of the shield.

ASSUMPTIONS   The radiation shield completely surrounds the probe, and the surroundings may be treated as a blackbody.

SOLUTION   The shield equilibrium temperature is higher than the wall temperature by virtue of convection with the fluid. As a result, the probe "sees" a higher temperature surface, and the probe temperature is closer to the fluid temperature, resulting in less measurement error.

For a single radiation shield placed so that it completely surrounds the probe, which is small compared to the size of the enclosure and has an emissivity of 1, the equilibrium temperature of the shield can be determined from Equation 8.32. The temperature of the shield is found to be 628 °C. Because the sensor now "sees" the shield rather than the wall, the temperature measured by the probe is 697 °C, which is also determined from Equation 8.32.

One shield with an emissivity of 1 provides for an improvement over the case of no shields, but a better choice of the surface characteristics of the shield material can result in much better performance. If the shield has an emissivity of 0.1, the shield temperature rises to 756 °C and the probe temperature to 771 °C.

COMMENT   Shielding provides improved temperature measurements by reducing radiative heat transfer. Another area for improvement in this temperature measurement could be the elevation of the wall temperature through insulation. This discussion of radiation shielding serves to demonstrate the usefulness of shielding as a means of improving temperature measurements in radiative environments. As with conduction errors, the development should be used to guide the design and installation of temperature sensors rather than to correct measured temperatures.

Recovery Errors in Temperature Measurement

The kinetic energy of a gas moving at high velocity can be converted to sensible energy by reversibly and adiabatically bringing the flow to rest at a point. The temperature resulting from this process is called the stagnation or total temperature T_t. On the other hand, the static temperature of the gas, T_∞, is the temperature that would be measured by an instrument moving at the local fluid velocity. From a molecular point of view, the static temperature measures the magnitude of the random kinetic energy of the molecules that comprise the gas, while the stagnation temperature includes both the directed and random components of kinetic energy. Generally, the engineer would be content with knowledge of either temperature, but in high-speed gas flows the sensor indicates neither temperature.

For negligible changes in potential energy, and in the absence of heat transfer or work, the energy equation for a flow may be written in terms of enthalpy and kinetic energy as

$$h_1 + \frac{U^2}{2} = h_2 \tag{8.35}$$

where state 2 refers to the stagnation condition and state 1 to a condition whereby the gas is flowing with the velocity U. Assuming ideal gas behavior, the enthalpy difference $h_2 - h_1$ may be expressed as $c_p(T_2 - T_1)$, or in terms of static and stagnation temperatures

$$\frac{U^2}{2c_p} = T_t - T_\infty \tag{8.36}$$

The term $U^2/2c_p$ is called the *dynamic temperature*.

What implication does this have for the measurement of temperature in a flowing gas stream? The physical nature of gases at normal pressures and temperatures is such that the velocity of the gas on a solid surface is zero, because of the effects of viscosity. Thus, when a temperature probe is placed in a moving fluid, the fluid is brought to rest on the surface of the probe. Deceleration of the flow by the probe converts some portion of the directed kinetic energy of the flow to thermal energy and elevates the temperature of the probe above the static temperature of the gas. The fraction of the kinetic energy recovered as thermal energy is called the recovery factor, r, defined as

$$r \equiv \frac{T_p - T_\infty}{U^2/2c_p} \tag{8.37}$$

where T_p represents the equilibrium temperature of the stationary (with respect to the flow) real temperature probe. In general, r may be a function of the velocity of the flow, or, more precisely, the Mach number and Reynolds number of the flow, and the shape and orientation of the temperature probe. For thermocouple junctions of round wire, Moffat (17) reports values of

$$r = 0.68 \pm 0.07 \quad (95\%) \quad \text{for wires normal to the flow}$$
$$r = 0.86 \pm 0.09 \quad (95\%) \quad \text{for wires parallel to the flow}$$

These recovery factor values tend to be constant at velocities for which temperature errors are significant, usually flows where the Mach number is greater than 0.1. For thermocouples having a welded junction, a spherical weld bead significantly larger than the wire diameter tends to a value of the recovery factor of 0.75, for the wires parallel or normal to the flow. The relationships between temperature and velocity for temperature probes with known recovery factors are

$$T_p = T_\infty + \frac{rU^2}{2c_p} \tag{8.38}$$

or in terms of the recovery error, e_U,

$$e_U = T_p - T_\infty = \frac{rU^2}{2c_p} \tag{8.39}$$

The probe temperature is related to the stagnation temperature by

$$T_p = T_t - \frac{(1-r)U^2}{2c_p} \tag{8.40}$$

Fundamentally, in liquids the stagnation and static temperatures are essentially equal, and the recovery error may generally be taken as zero for liquid flows. In any case, high-velocity flows are rarely encountered in liquids.

Normally, the uncertainty assigned to the recovery error is set at $u = e_U$. The uncertainty interval is often not symmetrical.

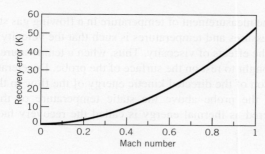

Figure 8.43 Behavior of recovery error as a function of Mach number.

Example 8.17

A temperature probe having a recovery factor of 0.86 is to be used to measure a flow of air at velocities up to the sonic velocity, at a pressure of 1 atm and a static temperature of 30 °C. Calculate the value of the recovery error in the temperature measurement as the velocity of the air flow increases, from 0 to the speed of sound, using Equation 8.39.

KNOWN $r = 0.86$ $p_\infty = 1$ atm abs $= 101$ kPa abs

$M \leq 1$ $T_\infty = 30\,°C = 303$ K

FIND   The recovery error as a function of air velocity

ASSUMPTION   Air behaves as an ideal gas

SOLUTION   Assuming that air behaves as an ideal gas, the speed of sound is expressed as

$$a = \sqrt{kRT} \tag{8.41}$$

For air at 101 kPa and 303 K, with $R = 0.287$ kJ/kg K, the speed of sound is approximately 349 m/s.

Figure 8.43 shows the error in temperature measurement as a function of Mach number for this temperature probe. The *Mach number* is defined as the ratio of the flow velocity to the speed of sound.

COMMENT   Typically, the static and total temperatures of the flowing fluid stream are to be determined from the measured probe temperature. In this case a second independent measurement of the velocity is necessary.

8.8   SUMMARY

Temperature is a fundamentally important quantity in science and engineering, both in concept and practice. Thus temperature is one of the most widely measured engineering variables, providing the basis for a variety of control and safety systems. This chapter provides the basis for the selection and installation of temperature sensors.

Temperature is defined for practical purposes through the establishment of a temperature scale, such as the Kelvin scale, that encompasses fixed reference points and interpolation standards. The International Temperature Scale 1990 is the accepted standard for temperature measurement.

The two most common methods of temperature measurement employ thermocouples and resistance temperature detectors. Standards for the construction and use of these temperature measuring devices have been established and provide the basis for selection and installation of commercially available sensors and measuring systems.

Installation effects on the accuracy of temperature measurements are a direct result of the influence of radiation, conduction, and convection heat transfer on the equilibrium temperature of a temperature sensor. The installation of a temperature probe into a measuring environment can be accomplished in such a way as to minimize the uncertainty in the resulting temperature measurement.

REFERENCES

1. Patterson, E. C., Eponyms: Why Celsius? *American Scientist* 77(4): 413, 1989.
2. Klein, H. A., *The Science of Measurement: A Historical Survey*, Dover, Mineola, NY, 1988.
3. Committee Report, the International Temperature Scale of 1990,[4] *Metrologia* 27(1): 3–10, 1990. (An Erratum appears in *Metrologia* 27(2): 107, 1990).
4. *Temperature Measurement*, Supplement to American Society of Mechanical Engineers PTC 19.3, 1974.
5. Diehl, W., Thin-film PRTD, *Measurements and Control*, 155–159, December 1982.
6. Thermistor Definitions and Test Methods, Electronic Industries Association Standard RS-275-A (ANSI Standard C83. 68–1972), June 1971.
7. Thermometrics, Inc., Vol. 1. NTC Thermistors, 1993.
8. Brown, D. R., N. Fernandez, J. A. Dirks, and T. B. Stout (March 2010). "The Prospects of Alternatives to Vapor Compression Technology for Space Cooling and Food Refrigeration Applications." Pacific Northwest National Laboratory (PNL). U.S. Department of Energy.
9. Burns, G. W., M. G. Scroger, and G. F. Strouse, Temperature-Electromotive Force Reference Functions and Tables for the Letter-Designated Thermocouple Types Based on the ITS-90, *NIST Monograph* 175, April 1993 (supersedes NBS Monograph 125).
10. Dewitt, D. P., and G. D. Nutter, *Theory and Practice of Radiation Thermometry*, Wiley-Interscience, New York, 1988.
11. Dils, R. R., High-temperature optical fiber thermometer, *Journal of Applied Physics*, 54(3): 1198–1201 1983.
12. Optical fiber thermometer, *Measurements and Control*, April 1987.
13. Dalta, R. U., M. C. Croarkin, and A. C. Parr, Cryogenic blackbody Calibrations at the NIST Low Background Infrared Calibration Facility, *Journal of Research of NIST*, 99(1), 1994.
14. Lahiri, B. B., S. Bagavathiappan, T. Jayakumar, and John Philip, Medical applications of infrared thermometry: A review, *Infrared Physics and Technology*, 55 (2012): 221–235.

---

[4] The text of the superseded International Practical Temperature Scale of 1968 appears as an appendix in the National Bureau of Standards monograph 124. An amended version was adopted in 1975, and the English text published: *Metrologia* 12: 7–17, 1976.

15. Bower, S. M., and J. R. Saylor, The effects of surfactant monolayers on free surface natural convection, *International Journal of Heat and Mass Transfer*, 54 (2011), 5348–5358.

16. Sparrow, E. M., Error estimates in temperature measurement. In E. R. G. Eckert and R. J. Goldstein (Eds.), *Measurements in Heat Transfer*, 2nd ed., Hemisphere, Washington, DC, 1976.

17. Moffat, R. J., Gas temperature measurements. In *Temperature—Its Measurement and Control in Science and Industry*, Vol. 3, Part 2, Reinhold, New York, 1962.

SUGGESTED READING

Benedict, R. P., *Fundamentals of Temperature, Pressure, and Flow Measurements*, 3rd ed., Wiley, New York, 1984.

NOMENCLATURE

a speed of sound (lt^{-1})

$b_{\bar{x}}$ systematic standard uncertainty in x

c speed of light in a vacuum (lt^{-1})

c_p specific heat ($l^2 t^{-2}/°$)

d thickness; diameter (l)

e_c conduction temperature error (°)

e_U recovery temperature error (°)

e_r radiation temperature error (°)

h convective heat transfer coefficient ($mt^{-3}/°$); enthalpy ($l^2 t^{-2}$)

k thermal conductivity ($mlt^{-3}/°$)

k ratio of specific heats

l length (l)

m $\sqrt{hP/kA}$ (eq. 8.25) (l^{-2})

q heat flux (mt^3) or heat transfer (ml^2/t^2)

r recovery factor

r_c radius of curvature (l)

$s_{\bar{x}}$ random standard uncertainty in x

u uncertainty

u_d design-stage uncertainty

A_c cross-sectional area (l^2)

A_s surface area (l^2)

B systematic uncertainty

C_α coefficient of thermal expansion ($ll^{-1}/°$)

D diameter (l)

E_b blackbody emissive power (mt^{-3})

$E_{b\lambda}$ spectral emissive power (t^{-3})

E_i input voltage (V)

E_1 voltage drop across R_1 (V)

E_o output voltage (V)

F radiation view factor

I current (A)

L length or thickness (l)

P perimeter (l)

R resistance (Ω); gas constant ($l^2 t^{-2}/°$)

R_0 reference resistance (Ω)

R_T thermistor resistance (Ω)

T temperature (°)

T_0 reference temperature (°)

T_p probe temperature (°)

T_w wall or boundary temperature (°)

T_∞ fluid temperature (°)

U fluid velocity (lt^{-1})

V voltage (V)

α temperature coefficient of resistivity ($\Omega/°$)

α_{AB} Seebeck coefficient

β material constant for thermistor resistance (°); constant in polynomial expansion

π_{AB} Peltier coefficient ($m\, l^2 t^{-3}\, \text{A}^{-1}$)

σ Thomson coefficient ($ml^2 t^{-3}\, \text{A}^{-1}$); Stefan-Boltzmann constant for radiation ($mt^{-3}(°)^4$)

θ nondimensional temperature

θ_x sensitivity index for variable x (uncertainty analysis)

ρ_e resistivity (Ω, l)

δ thermistor dissipation constant ($ml^2 t^{-3}/°$)

ε emissivity

PROBLEMS

8.1 Define and discuss the significance of the following terms as they apply to temperature and temperature measurements:

    a. Temperature scale

    b. Temperature standards

    c. Fixed points

    d. Interpolation

8.2 Fixed temperature points in the International Temperature Scale are phase equilibrium states for a variety of pure substances. Discuss the conditions necessary within an experimental apparatus to accurately reproduce these fixed temperature points. How would elevation, weather, and material purity affect the uncertainty in these fixed points?

8.3 Answers to the following questions may be found in the International Temperature Scale of 1990[5]:

    a. What are the minimum and maximum temperatures defined in ITS-90?

    b. How are temperatures between 0.65 K and 5.0 K defined?

    c. What fixed points define 24.5561 K, 505.078 K, and 1,234.93 K?

    d. What is the interpolation instrument in the temperature range from 13.8033K to 961.78K? What is the output value of this instrument at 400 K?

    e. Above 961.78 °C, what is the standard for interpolation?

8.4 Calculate the resistance of a platinum wire that is 2 m in length and has a diameter of 0.1 cm. The resistivity of platinum at 25 °C is 9.83×10^{-6} Ω-cm. What implications does this result have for the construction of a resistance thermometer using platinum?

8.5 Plot the resistance of a platinum wire that is 5 m long as a function of wire diameter ranging from 0.1 mm to 2 mm. Estimate the allowable tensile load for the 0.1 mm wire. (Hint: The tensile strength of platinum is 120 MPa.)

8.6 An RTD forms one arm of a Wheatstone bridge, as shown in Figure 8.44. The RTD is used to measure a constant temperature with the bridge operated in a balanced mode. The RTD has a resistance of 100 Ω at a temperature of 0 °C, and a thermal coefficient of resistance, $\alpha = 0.003925\,°C^{-1}$. The value of the variable resistance R_1 must be set to 171 Ω to balance the bridge circuit.

    a. Determine the temperature of the RTD.

    b. Compare this circuit to the equal-arm bridge in Example 8.2. Which circuit provides the greater static sensitivity?

8.7 An RTD forms one arm (R_4) of a Wheatstone bridge, as shown in Figure 8.8. The RTD is used to measure a constant temperature, with the bridge operated in deflection mode. The RTD has a resistance of 25 Ω at a temperature of 0 °C, and a thermal coefficient of resistance, $\alpha = 0.003925\,°C^{-1}$. If the RTD is subjected to a temperature of 100 °C

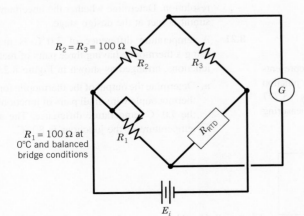

$R_2 = R_3 = 100\ \Omega$

R_2

R_3

G

$R_1 = 100\ \Omega$ at 0°C and balanced bridge conditions

R_1

R_{RTD}

E_i

Figure 8.44 Wheatstone bridge circuit for Problem 8.6.

[5] The ITS-90 document is readily accessible through an Internet search.

and the input voltage to the bridge is 5V, what is the output voltage?

8.8   Research and describe current state-of-the-art electronic modules to measure temperature using an RTD sensor. List typical specifications that could be used in a design stage uncertainty analysis.

8.9   Research and describe current state-of-the art electronic instruments to measure the resistance of an RTD using both the 2-wire and 4-wire methods. The 2-wire method simply connects the ohmmeter directly across the RTD using lead wires. What are the advantages and disadvantages of the 2-wire method?

8.10   Estimate the required level of uncertainty in the measurement of resistance for a platinum RTD if it is to serve as a local standard for the calibration of a temperature measurement system for an uncertainty of $\pm 0.005\,°C$. Assume $R(0\,°C) = 100\,\Omega$.

8.11   A thermistor is placed in a $100\,°C$ environment and its resistance measured as $20,000\,\Omega$. The material constant, β, for this thermistor is $3,650\,°C$. If the thermistor is then used to measure a particular temperature, and its resistance is measured as $500\,\Omega$, determine the thermistor temperature.

8.12   Using a spreadsheet or similar software, reproduce Figure 8.11.

8.13   Define and discuss the following terms related to thermocouple circuits:

a. Thermocouple junction

b. Thermocouple laws

c. Reference junction

d. Peltier effect

e. Seebeck coefficient

8.14   The thermocouple circuit in Figure 8.45 represents a J-type thermocouple with the reference junction having $T_2 = 0\,°C$. The output voltage is 13.777 mV. What is the temperature of the measuring junction, T_1?

8.15   The thermocouple circuit in Figure 8.45 represents a J-type thermocouple. The circuit produces an output voltage of 14.23 mV for $T_1 = 600\,°C$. What is T_2?

8.16   The thermocouple circuit in Figure 8.45 is composed of copper and constantan and has an output voltage of 6 mV for $T_1 = 200\,°C$. What is T_2?

8.17   **a.** The thermocouple shown in Figure 8.46a yields an output voltage of 7.947 mV. What is the temperature of the measuring junction?

b. Suppose the reference junction temperature changes to 25 °C. If the measuring junction of (a) remains at the same temperature, what voltage would be measured by the voltmeter?

c. Copper extension leads are installed as shown in Figure 8.46b. For an output voltage of 7.947 mV and a reference junction temperature of 0 °C, what is the temperature of the measuring junction?

8.18   A J-type thermocouple referenced to 70 °F has a measured output voltage of 2.878 mV. What is the temperature of the measuring junction?

8.19   A J-type thermocouple referenced to 0 °C indicates 4.115 mV. What is the temperature of the measuring junction?

8.20   A temperature measurement requires an uncertainty of $\pm 2\,°C$ (5%) at a temperature of $200\,°C$. A standard T-type thermocouple is to be used with a readout device that provides electronic ice-point reference junction compensation, and has a stated instrument uncertainty of 0.5 °C (95%) with 0.1 °C resolution. Determine whether the uncertainty constraint is met at the design stage.

8.21   A temperature difference of $3.0\,°C$ is measured using a thermopile having three pairs of measuring junctions, arranged as shown in Figure 8.23.

a. Determine the output of the thermopile for J-type thermocouple wire, if all pairs of junctions sense the 3.0 °C temperature difference. The average temperature of the junctions is 80 °C.

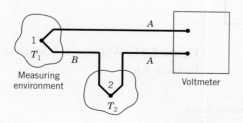

Figure 8.45   Thermocouple circuit for Problems 8.14 through 8.16; (1) measuring junction; (2) reference junction.

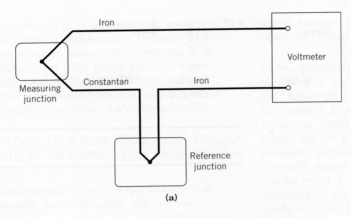

(a)

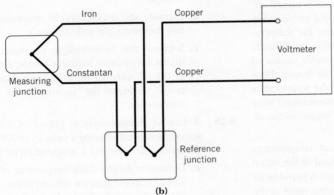

(b)

Figure 8.46 Schematic diagram for Problem 8.17.

b. If the thermopile is constructed of wire that has a maximum emf variation from the NIST standard values of $\pm 0.8\%$ (95%), and the voltage measuring capabilities in the system are such that the uncertainty is ± 0.0005 V (95%), perform an uncertainty analysis to estimate the uncertainty in the measured temperature difference at the design stage.

8.22 Complete the following table for a J-type thermocouple:

Temperature (°C)		
Measured	Reference	emf (mV)
100	0	
	0	−0.5
100	50	
	50	2.5

8.23 Complete the following table for a T-type thermocouple using appropriate computer software or programmable calculator:

Temperature (°C)		
Measured	Reference	emf (mV)
100	0	
	0	0.6
100	50	
	50	3

8.24 A J-type thermopile is constructed as shown in Figure 8.22, to measure a single temperature. For a four-junction thermopile referenced to 0 °C, what would be the voltage produced at a temperature of 125 °C? If a voltage measuring device was available that had a total uncertainty of ± 0.0001 V (95%), how many junctions would be required in the thermopile to reduce the uncertainty in the measured temperature to $\pm 0.1\,^{\circ}$C (95%)?

8.25 You are employed as a heating, ventilating, and air conditioning engineer. Your task is to decide where in a residence to place a thermostat, and how it is to be mounted on the wall. A thermostat contains a bimetallic temperature measuring device that serves as the sensor for the control logic of the heating and air conditioning system for the house. Consider the heating season. When the temperature of the sensor falls 1 °C below the set-point temperature of the thermostat, the furnace is activated; when the temperature rises 1 °C above the set-point temperature, the furnace is turned off. Discuss where the thermostat should be placed in the house, what factors could cause the temperature of the sensor in the thermostat to be different from the air temperature, and possible causes of discomfort for the occupants of the house. How does the thermal capacitance of the temperature sensor affect the operation of the thermostat? Why are thermostats typically set 5 °C higher in the air conditioning season?

8.26 A J-type thermocouple for use at temperatures between 0 and 100 °C was calibrated at the steam point in a device called a hypsometer. A hypsometer creates a constant temperature environment at the saturation temperature of water, at the local barometric pressure. The steam-point temperature is strongly affected by barometric pressure variations. Atmospheric pressure on the day of this calibration was 30.1 in. Hg. The steam-point temperature as a function of barometric pressure may be expressed as

$$T_{st} = 212 + 50.422\left(\frac{p}{p_0} - 1\right) - 20.95\left(\frac{p}{p_0} - 1\right)^2 \ [^{\circ}F]$$

where $p_0 = 29.921$ in. Hg. At the steam point, the output voltage produced by the thermocouple, referenced to 0 °C, is measured as 5.310 mV. Construct a calibration curve for this thermocouple by plotting the difference between the thermocouple reference table value and the measured value $(V_{ref} - V_{meas})$ versus temperature. What is this difference at 0 °C? Suggest a means for measuring temperatures between 0 and 100 °C using this calibration, and estimate the contribution to the total uncertainty.

8.27 A J-type thermocouple is calibrated against an RTD standard within $\pm 0.01\,^{\circ}$C (95%) between 0 °C and 200 °C. The output voltage is measured with a voltmeter having 0.001 mV resolution and less than 0.015 mV (95%) systematic uncertainty. The reference junction temperature is 0 °C. The calibration procedure yields the following results:

T_{RTD} (°C)	0.00	20.50	40.00	60.43	80.25	100.65
emf (mV)	0.010	1.038	2.096	3.207	4.231	5.336

a. Determine a polynomial to describe the relation between the temperature and thermocouple voltage.

b. Estimate the uncertainty in temperature using this thermocouple and voltmeter.

c. Suppose the thermocouple is connected to a digital temperature indicator having a resolution of 0.1 °C with 0.3 °C (95%) systematic uncertainty. Estimate the uncertainty in indicated temperature.

8.28 A beaded thermocouple is placed in a duct in a moving gas stream having a velocity of 200 ft/s. The thermocouple indicates a temperature of 1,400 R.

a. Determine the true static temperature of the fluid based on an estimate for velocity errors. Take the specific heat of the fluid to be 0.6 Btu/lbm°R and the recovery factor to be 0.22.

b. Estimate the possible error (uncertainty) in the thermocouple reading due to radiation if the walls of the duct are at 1,200 °R. The view factor from the probe to the duct walls is 1, the convective heat-transfer coefficient, h, is 30 Btu/hr-ft^2-°R, and the emissivity of the temperature probe is 1.

8.29 Consider a welded thermocouple bead that experiences an air flow at 90 m/s. The thermocouple indicates a temperature of 600 °C. Determine the true static temperature of the air by correcting for recovery errors assuming a recovery factor of 0.25.

8.30 It is desired to measure the static temperature of the air outside of an aircraft flying at 20,000 ft with a speed of 300 miles per hour, or 438.3 ft/s. A temperature probe is used that has a recovery factor, r, of 0.75. If the static temperature of the air is 413 R and the specific heat is 0.24 Btu/lbm R, what is the

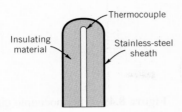

Figure 8.47 Typical construction of a sheathed thermocouple.

temperature indicated by the probe? Local atmospheric pressure at 20,000 ft is approximately 970 lb/ft², which results in an air density of 0.0442 lbm/ft³. Discuss additional factors that might affect the accuracy of the static temperature reading.

8.31 The static temperature of the air outside of an aircraft flying at 7,000 m with a speed of 150 m/s is to be measured. A temperature probe is used that has a recovery factor, r, of 0.75. If the static temperature of the air is $-35\,°C$ and the specific heat is 1,000 J/kg-K, what is the temperature indicated by the probe? Local atmospheric pressure at 7,000 m is approximately 45 kPa, which results in an air density of 0.66 kg/m³. Discuss additional factors that might affect the accuracy of the static temperature reading.

8.32 Consider the typical construction of a sheathed thermocouple, as shown in Figure 8.47. Analysis of this geometry to determine conduction errors in temperature measurement is difficult. Suggest a method for placing a realistic upper limit on the conduction error for such a probe, for a specified immersion depth into a convective environment.

8.33 An iron-constantan thermocouple is placed in a moving air stream in a duct, as shown in Figure 8.48. The thermocouple reference junction is maintained at 212 °F. The wall temperature, T_w, is 500 °F, and the air velocity is 200 ft/s. The output voltage from the thermocouple is 14.143 mV.

a. Determine the thermocouple junction temperature.

b. By considering recovery and radiation errors, estimate the possible value for total error in the indicated temperature. Discuss whether this estimate of the measurement error is conservative and why, or why not. The heat-transfer coefficient may be taken as 70 Btu/hr-ft²-°F.

Air Properties	Thermocouple Properties
$c_p = 0.24$ Btu/lbm °F	$r = 0.7$
$U = 200$ ft/s	$\varepsilon = 0.25$

8.34 An iron-constantan thermocouple is placed in a moving air stream in a duct, as shown in Figure 8.48. The air flows at 70 m/s. The emissivity of the thermocouple is 0.5 and the recovery factor is 0.6. The wall temperature, T_w is 300 °C. The thermocouple reference junction is maintained at 100 °C. The emf output from the thermocouple is 16 mV.

a. Determine the thermocouple junction temperature.

b. By considering recovery and radiation errors, estimate the possible value for total error in the indicated temperature. Discuss whether this estimate of the measurement error is conservative and why, or why not. The heat-transfer coefficient may be taken as 100 W/m²K.

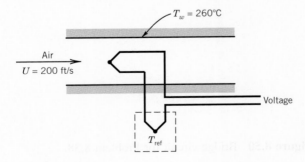

Figure 8.48 Schematic diagram for Problems 8.33, 34, and 35.

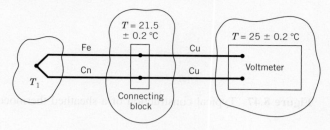

Figure 8.49   Thermocouple circuit for Problem 8.37.

8.35   An iron-constantan thermocouple is placed in a moving air stream in a duct, as shown in Figure 8.48. The air flows at 180 ft/sec. The emissivity of the thermocouple is 0.7 and the recovery factor is 0.7. The wall temperature, T_w is 500 °F. The thermocouple reference junction is maintained at 32 °F. The emf output from the thermocouple is 20 mV.

   a. Determine the thermocouple junction temperature.

   b. By considering recovery and radiation errors, estimate the possible value for total error in the indicated temperature. Discuss whether this estimate of the measurement error is conservative and why, or why not. The heat-transfer coefficient may be taken as 40 Btu/hr-ft² °F, and the specific heat as $c_p = 0.24$ Btu/lbm °F.

8.36   In Example 8.5, an uncertainty value for R_T was determined at 125 °C as $B_{R_T} = 247\ \Omega$ (95%). Show that this value is correct by performing an uncertainty analysis on R_T. In addition, determine the value of B_{R_T} at the temperatures 150 °C and 100 °C. What error is introduced into the uncertainty analysis for β by using the value of B_{R_T} at 125 °C?

8.37   The thermocouple circuit shown in Figure 8.49 measures the temperature T_1. The voltmeter limits of error are given as

Limits of error: ±0.05% (95%) of reading +15 µV at 25°C

Resolution: 5 µV

   Give a best estimate for the temperature T_1, if the output voltage is 9 mV.

8.38   A platinum RTD ($\alpha = 0.00392$ °C^{-1}) is to be calibrated in a fixed-point environment. The probe is used in a balanced mode with a Wheatstone bridge, as shown in Figure 8.50. The bridge resistances are known to an uncertainty of ±0.001 Ω (95%). At 0 °C the bridge balances when $R_c = 100.000\ \Omega$. At 100 °C the bridge balances when $R_c = 139.200\ \Omega$.

   a. Find the RTD resistance corresponding to 0 °C and 100 °C, as well as the uncertainty in each value.

   b. Calculate the uncertainty in determining a temperature using this RTD–bridge system for a measured temperature that results in $R_c = 300\ \Omega$. Assume $u_\alpha = \pm 1 \times 10^{-5}$/°C (95%).

8.39   A thin-film heat flux sensor employs a K-type thermopile to measure temperature difference. This thermopile yields a sensitivity of $5\ \dfrac{\mu V}{W/m^2}$. Consider using this heat flux sensor to measure the heat flux for the following applications, and

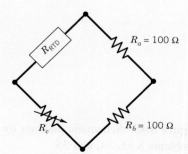

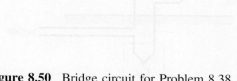

Figure 8.50   Bridge circuit for Problem 8.38.

determine the expected output voltage and comment on the significance of the result for:

a. A single-pane glass window that is 6 mm thick has a thermal conductivity $k = 1.5$ W/m-K and experiences a temperature difference of 20 °C.

b. A brick wall that is 6 cm thick has a thermal conductivity $k = 0.7$ W/m-K and experiences a temperature difference of 15 °C.

c. The wall of a polystyrene cooler that is 3.8 cm thick has a thermal conductivity $k = 0.05$ W/m-K and experiences a temperature difference of 15 °C.

8.40 A thin-film heat flux sensor has a sensitivity $\theta_q = 5\, \dfrac{\mu V}{W/m^2}$. Consider using this heat flux sensor to measure heat flux such that the nominal voltage output is 40 μV The voltage output is measured by a nano-voltmeter having an uncertainty of 0.003% of the reading. The uncertainty in the sensitivity of the heat flux sensor is ±10% (this assumes the sensor is employed without local calibration). Determine the uncertainty in a measured heat flux having the parameters listed above.

8.41 A T-type thermopile is used to measure temperature difference across insulation in the ceiling of a residence in an energy monitoring program. The temperature difference across the insulation is used to calculate energy loss through the ceiling from the relationship

$$Q = kA_c(\Delta T/L)$$

where

A_c = ceiling area = 15 m²
k = insulation thermal conductivity
  = 0.4W/m-°C
L = insulation thickness = 0.25 m
ΔT = temperature difference = 5 °C
Q = heat loss(W)

The value of the temperature difference is expected to be 5 °C, and the thermocouple emf is measured with an uncertainty of ±0.04 mV.

Determine the required number of thermopile junctions to yield an uncertainty in Q of ±5% (95%), assuming the uncertainty in all variables other than ΔT may be neglected.

8.42 A T-type thermocouple referenced to 0 °C is used to measure the temperature of boiling water. What is the output voltage of this circuit at 100 °C?

8.43 A T-type thermocouple referenced to 0 °C develops an output voltage of 1.2 mV. What is the temperature sensed by the thermocouple?

8.44 A temperature measurement system consists of a digital voltmeter and a T-type thermocouple. The thermocouple leads are connected directly to the voltmeter, which is placed in an air-conditioned space at 25 °C. The output voltage from the thermocouple is 10 mV. What is the measuring junction temperature?

8.45 The result of more than 60 temperature measurements over time during fixed conditions in a furnace determines $\overline{T} = 624.7\,°C$ with $s_{\overline{T}} = 2.4\,°C$. The engineer suspects a radiation systematic error. The walls of the furnace are at 400 °C, so that the actual temperature in the furnace may be larger than 624.7, but not less. As such, the lower bound of error is 0 °C relative to the measured mean value. Estimate the radiation error in the measurement; assume reasonable values for the heat transfer parameters. Develop a statement for the true mean temperature with its confidence interval assuming the systematic errors follow a rectangular distribution with upper bound from your estimate and lower bound 0 °C relative to the measured mean value.

8.46 Flexivity, k, is a property of bimetallic strips that if known allows the radius of curvature to be calculated directly as

$$r_c = \frac{d}{2k(T_2 - T_1)}$$

where T_2 is the temperature of interest, T_1 is the assembly temperature of the bimetallic strip, and r_c is infinite. If a 3-cm strip is assembled at 25 °C, plot the shape of the strip at temperatures of 50 °C, 75 °C, and 100 °C for a k value of 7.3×10^{-5} K^{-1}.

Chapter 9

Pressure and Velocity Measurements

9.1 INTRODUCTION

This chapter introduces methods for measuring pressure and the velocity within fluids. Pressure is measured in static and dynamic systems, in acoustics, and in fluid systems. Instruments and procedures for establishing known values of pressure for calibration purposes, as well as various types of transducers for pressure measurement, are discussed. We also discuss well-established methods measuring the local and full-field velocity within a moving fluid. Finally, we present practical considerations, including common error sources, for pressure and velocity measurements. Although there are various practical test standards for pressure, many of which are applied to a specific application or measuring device, the American Society of Mechanical Engineers' Performance Test Code (ASME PTC) 19.2 provides an overview of basic pressure concepts and measuring instruments and is an accepted test standard (1).

Upon completion of this chapter, the reader will be able to

- Explain absolute and gauge pressure concepts and describe the working standards that measure pressure directly,

- Explain the physical principles underlying mechanical pressure measurements and the various types of transducers used to measure pressure

- Explain pressure concepts related to static systems and with moving fluids

- Analyze the dynamic behavior of pressure system response due to transmission line effects

- Explain the physical principles underlying various velocity measurement methods and their practical use

9.2 PRESSURE CONCEPTS

Pressure represents a contact force per unit area. It acts inwardly (compressive) and normally to a surface. A pure vacuum, which contains no molecules, provides the limit for a primary standard for absolute zero pressure. As shown in Figure 9.1, the absolute pressure scale is quantified relative to this absolute zero pressure. The pressure under standard atmospheric conditions is defined as 1.01320×10^5 Pa absolute (where $1\,\text{Pa} = 1\,\text{N/m}^2$) (2). This is equivalent to

101.32 kPa absolute

1 atm absolute

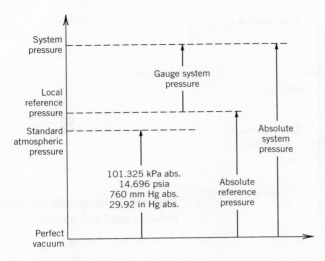

Figure 9.1 Relative pressure scales.

14.696 lb/in.^{2} absolute (written as psia)

1.013 bar absolute   (where 1 bar = 100 kPa)

The term "absolute" might be abbreviated as "a" or "abs."

Also indicated in Figure 9.1 is a gauge pressure scale. The gauge pressure scale is measured relative to some absolute reference pressure, which is defined in a manner convenient to the measurement. The relation between an absolute pressure, p_{abs}, and its corresponding gauge pressure, p_{gauge}, is given by

$$p_{gauge} = p_{abs} - p_0 \tag{9.1}$$

where p_0 is a reference pressure. A commonly used reference pressure is the local absolute atmospheric pressure. Absolute pressure is a positive number. Gauge pressure can be positive or negative depending on the value of measured pressure relative to the reference pressure. A differential pressure, such as $p_1 - p_2$, is a relative measure of pressure.

Pressure can also be described in terms of the pressure exerted on a surface that is submerged in a column of fluid at depth h, as depicted in Figure 9.2. From hydrostatics, the pressure at any depth within a fluid of specific weight γ can be written as

$$p_{abs}(h) = p(h_0) + \gamma h = p_0 + \gamma h \tag{9.2}$$

where p_0 is the pressure at an arbitrary datum line at h_0, and h is measured relative to h_0. The fluid specific weight is given by $\gamma = \rho g$ where ρ is the density. When Equation 9.2 is rearranged, the equivalent head of fluid at depth h becomes

$$h = \left[p_{abs}(h) - p(h) \right] / \gamma = (p_{abs} - p_0)/\gamma \tag{9.3}$$

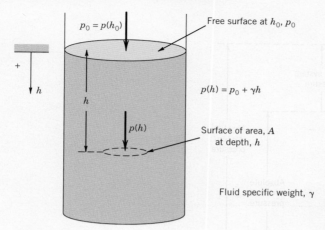

Figure 9.2   Hydrostatic-equivalent pressure head and pressure.

The equivalent pressure head at one standard atmosphere ($p_0 = 0$ absolute) is

$$760 \text{ mm Hg abs} = 760 \text{ torr abs} = 1 \text{ atm abs}$$
$$= 10{,}350.8 \text{ mm H}_2\text{O abs} = 29.92 \text{ in Hg abs}$$
$$= 407.513 \text{ in H}_2\text{O abs}$$

The standard is based on mercury (Hg) with a density of $0.0135951 \text{ kg/cm}^3$ at 0 °C and water at $0.000998207 \text{ kg/cm}^3$ at 20 °C (2).

Example 9.1

Determine the absolute and gauge pressures and the equivalent pressure head at a depth of 10 m below the free surface of a pool of water at 20 °C.

    KNOWN $h = 10$ m; where $h_0 = 0$ is the free surface

               $T = 20$ °C

               $\rho_{\text{H}_2\text{O}} = 998.207 \text{ kg/m}^3$

               Specific gravity of mercury, $S_{Hg} = 13.57$

    ASSUMPTION $p(h_0) = 1.0132 \times 10^5$ Pa abs

    FIND p_{abs}, p_{gauge}, and h

    SOLUTION   The absolute pressure is found directly from Equation 9.2. Using the pressure at the free surface as the reference pressure and the datum line for h_0, the absolute pressure must be

$$p_{\text{abs}}(h) = 1.0132 \times 10^5 \text{ N/m}^2 + \frac{(997.4 \text{ kg/m}^3)\,(9.8 \text{ m/s}^2)(10 \text{ m})}{1 \text{ kg-m/N-s}^2}$$
$$= 1.991 \times 10^5 \text{ N/m}^2 \text{ abs}$$

This is equivalent to 199.1 kPa abs or 1.96 atm abs or 28.80 lb/in.^{2} abs or 1.99 bar abs.

The pressure can be described as a gauge pressure by referencing it to atmospheric pressure. From Equation 9.1,

$$p(h) = p_{\text{abs}} - p_0 = \gamma h$$
$$= 9.775 \times 10^4 \text{ N/m}^2$$

which is also equivalent to 97.7 kPa or 0.96 atm or 14.1 lb/in.[2] or 0.98 bar.

We can express this pressure as an equivalent column of liquid,

$$h = \frac{p_{\text{abs}} - p_0}{\rho g} = \frac{\left(1.991 \times 10^5\right) - \left(1.0132 \times 10^5\right) \text{ N/m}^2}{(998.2 \text{ kg/m}^3)(9.8 \text{ m/s}^2)(1 \text{ N-s}^2/\text{kg-m})}$$
$$= 10 \text{ m H}_2\text{O}$$

9.3   PRESSURE REFERENCE INSTRUMENTS

The units of pressure can be defined through the standards of the fundamental dimensions of mass, length, and time. In practice, pressure transducers are calibrated by comparison against certain reference instruments, which also serve as pressure measuring instruments. This section discusses several basic pressure reference instruments that can serve either as working standards or as laboratory instruments.

McLeod Gauge

The McLeod gauge, originally devised by Herbert McLeod in 1874 (3), is a pressure-measuring instrument and laboratory reference standard used to establish gas pressures in the subatmospheric range of 1 mm Hg abs down to 0.1 mm Hg abs. A pressure that is below atmospheric pressure is also called a vacuum pressure. One variation of this instrument is sketched in Figure 9.3a, in which the gauge is connected directly to the low-pressure source. The glass tubing is arranged so that a sample of the gas at an unknown low pressure can be trapped by inverting the gauge from the sensing position, depicted as Figure 9.3a, to that of the measuring position, depicted as Figure 9.3b. In this way, the gas trapped within the capillary is isothermally compressed by a rising column of mercury. Boyle's law is then used to relate the two pressures on either side of the mercury to the distance of travel of the mercury within the capillary. Mercury is the preferred working fluid because of its high density and very low vapor pressure.

At the equilibrium and measuring position, the capillary pressure, p_2, is related to the unknown gas pressure to be determined, p_1, by $p_2 = p_1(\forall_1/\forall_2)$, where $\forall_1$ is the gas volume of the gauge in Figure 9.3a (a constant for a gauge at any pressure) and $\forall_2$ is the capillary volume in Figure 9.3b. But $\forall_2 = Ay$, where A is the known cross-sectional area of the capillary and y is the vertical length of the capillary occupied by the gas. With γ as the specific weight of the mercury, the difference in pressures is related by $p_2 - p_1 = \gamma y$ such that the unknown gas pressure is just a function of y:

$$p_1 = \gamma A y^2 / (\forall_1 - Ay) \tag{9.4}$$

In practice, a commercial McLeod gauge has the capillary etched and calibrated to indicate either pressure, p_1, or its equivalent head, p_1/γ, directly. The McLeod gauge generally does not require

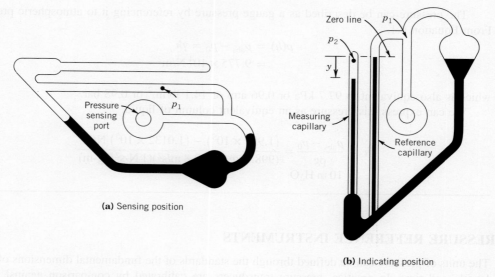

(a) Sensing position

(b) Indicating position

Figure 9.3 McLeod gauge.

correction. The reference stem offsets capillary forces acting in the measuring capillary. Instrument systematic uncertainty is on the order of 0.5% (95%) at 1 mm Hg abs and increases to 3% (95%) at 0.1 mm Hg abs.

Barometer

A barometer consists of an inverted tube containing a fluid and is used to measure atmospheric pressure. To create the barometer, the tube, which is sealed at only one end, is evacuated to zero absolute pressure. The tube is immersed with the open end down within a liquid-filled reservoir as shown in the illustration of the Fortin barometer in Figure 9.4. The reservoir is open to atmospheric pressure, which forces the liquid to rise up the tube.

From Equations 9.2 and 9.3, the resulting height of the liquid column above the reservoir free surface is a measure of the absolute atmospheric pressure in the equivalent head (Eq. 9.3). Evangelista Torricelli (1608–1647), a colleague of Galileo, can be credited with developing and interpreting the working principles of the barometer in 1644.

As Figure 9.4 shows, the closed end of the tube is at the vapor pressure of the barometric liquid at room temperature. So the indicated pressure is the atmospheric pressure minus the liquid vapor pressure. Mercury is the most common liquid used, because it has a very low vapor pressure and so, for practical use, the indicated pressure can be taken as the local absolute barometric pressure. However, for very accurate work, the barometer needs to be corrected for temperature effects on the vapor pressure, for temperature and altitude effects on the weight of mercury, and for deviations from standard gravity (9.80665 m/s^2 or 32.17405 ft/s^2). Correction curves are provided by instrument manufacturers.

Barometers are used as local standards for the measurement of absolute atmospheric pressure. Under standard conditions for pressure temperature and gravity, the mercury rises 760 mm (29.92 in.)

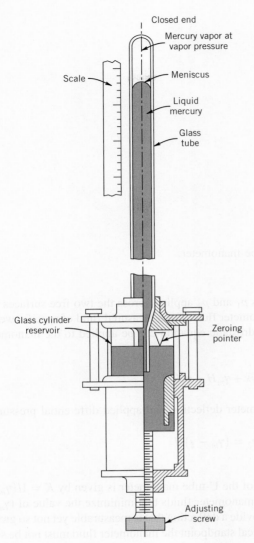

Figure 9.4 Fortin barometer.

above the reservoir surface. The U.S. National Weather Service always reports a barometric pressure that has been corrected to sea level elevation.

Manometer

A manometer is an instrument used to measure differential pressure based on the relationship between pressure and the hydrostatic equivalent head of fluid. Several design variations are available, allowing measurements ranging from the order of 0.001 mm of manometer fluid to several meters.

The U-tube manometer in Figure 9.5 consists of a transparent tube filled with an indicating manometer fluid, which is a liquid of specific weight γ_m. This forms two free surfaces of the

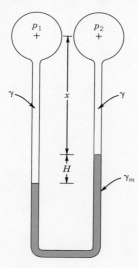

Figure 9.5   U-tube manometer.

manometer fluid. The difference in pressures p_1 and p_2 applied across the two free surfaces brings about a deflection, H, in the level of the manometer fluid. For a measured fluid of specific weight γ, which can be either a gas or a liquid, the hydrostatic equation can be applied to the manometer of Figure 9.5 as

$$p_1 = p_2 + \gamma x + \gamma_m H - \gamma(H + x)$$

which yields the relation between the manometer deflection and applied differential pressure,

$$p_1 - p_2 = (\gamma_m - \gamma)H \tag{9.5}$$

From Equation 9.5, the static sensitivity of the U-tube manometer is given by $K = 1/(\gamma_m - \gamma)$. To maximize manometer sensitivity, choose manometer fluids that minimize the value of $(\gamma_m - \gamma)$. The manometer fluid should be selected to provide a deflection that is measurable yet not so great that it becomes awkward to observe. From a practical standpoint the manometer fluid must not be soluble with the measured fluid.

A variation in the U-tube manometer is the micromanometer shown in Figure 9.6. These special-purpose instruments are used to measure very small differential pressures, down to 0.005 mm H_2O (0.0002 in. H_2O). To null the manometer, the reservoir is moved up or down until the level of the manometer fluid within the reservoir is at the same level as a set mark within a magnifying sight glass. Changes in pressure displace the fluid so that the reservoir must be moved up or down to bring the meniscus back to the set mark. The height of repositioning is equal to the change in the equivalent pressure head. The position of the reservoir is controlled by a micrometer or other calibrated displacement measuring device.

The inclined tube manometer is also used to measure small changes in pressure. It is essentially a U-tube manometer with one leg inclined at an angle θ, typically from 10 to 30 degrees relative to the horizontal. As indicated in Figure 9.7, a change in pressure equivalent to a deflection of height

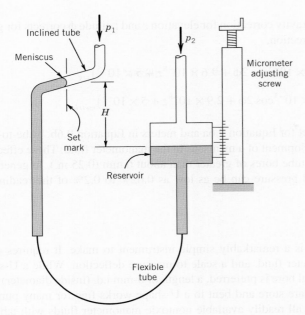

Inclined tube

Meniscus

p_1

p_2

Micrometer
adjusting
screw

H

Set
mark

Reservoir

Flexible
tube

Figure 9.6   Micromanometer.

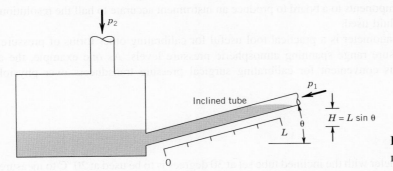

p_2

p_1

Inclined tube

$H = L \sin \theta$

θ

L

0

Figure 9.7   Inclined tube
manometer.

H in a U-tube manometer would bring about a change in position of the meniscus in the inclined leg
of $L = H/\sin \theta$. This provides an increased sensitivity over the conventional U-tube by the factor
of $1/\sin \theta$.

A number of elemental errors affect the instrument uncertainty of all types of manometers.
These include scale and alignment errors, zero (null) error, temperature error, gravity error, and
capillary and meniscus errors. The specific weight of the manometer fluid varies with temperature
but can be corrected. For example, the temperature dependence of mercury specific weight is
approximated by

$$\gamma_{Hg} = \frac{133.084}{1 + 0.00006T} \left[N/m^3 \right] = \frac{848.707}{1 + 0.000101(T - 32)} \left[lb/ft^3 \right]$$

with T in °C or °F, respectively. A gravity correction for elevation z and latitude ϕ corrects for gravity errors using the dimensionless correction,

$$e_1 = -(2.637 \times 10^{-3}\cos 2\phi + 9.6 \times 10^{-8}z + 5 \times 10^{-5})_{US} \tag{9.6a}$$

$$= -(2.637 \times 10^{-3}\cos 2\phi + 2.9 \times 10^{-8}z + 5 \times 10^{-5})_{metric} \tag{9.6b}$$

where ϕ is in degrees and z is in feet for Equation 9.6a and meters in Equation 9.6b. Tube-to-liquid capillary forces can lead to the development of a meniscus in the manometer fluid. These effects can be minimized by using manometer tube bores of greater than about 6 mm (0.25 in.). In general, the instrument uncertainty in measured pressure can be as low as 0.02% to 0.2% of the reading.

Example 9.2: U-tube Manometer

A high-quality U-tube manometer is a remarkably simple instrument to make. It requires only a transparent U-shaped tube, manometer fluid, and a scale to measure deflection. While a U-shaped glass tube of 6 mm or greater internal bore is preferred, a length of 6-mm i.d. (inside diameter) thick-walled clear tubing from the hardware store and bent to a U-shape works fine for many purposes. Water, alcohol, and mineral oil are all readily available nontoxic manometer fluids with tabulated properties. A sheet of graph paper or a ruler serves to measure meniscus deflection. There is a limit to the magnitude of pressure differential that can be measured, although use of a step stool extends this range. Tack the components to a board to produce an instrument accurate to half the resolution of the scale in terms of fluid used.

The U-tube manometer is a practical tool useful for calibrating other forms of pressure transducers in the pressure range spanning atmospheric pressure levels. As one example, the ad hoc U-tube described is convenient for calibrating surgical pressure transducers over physiological pressure ranges.

Example 9.3

An inclined manometer with the inclined tube set at 30 degrees is to be used at 20 °C to measure an air pressure of nominal magnitude of 100 N/m² relative to ambient. Manometer "unity" oil ($S = 1$) is to be used. The specific weight of the oil is 9,770 ± 0.5% N/m² (95%) at 20 °C, the angle of inclination can be set to within 1 degree using a bubble level, and the manometer resolution is 1 mm with a manometer zero error equal to its interpolation error. Estimate the uncertainty in indicated differential pressure at the design stage.

> **KNOWN** $p = 100$ N/m² (nominal)
>
> Manometer
>
> Resolution: 1 mm
>
> Zero error: 0.5 mm (1/2 of resolution, 95%)
>
> $\theta = 30 \pm 1°$ (95% assumed)
>
> $\gamma_m = 9,770 \pm 0.5\%$ N/m³ (95%)

ASSUMPTIONS Temperature and capillary effects in the manometer are negligible. The degrees of freedom in the stated uncertainties are large (see Chapter 5).

FIND u_d

SOLUTION The relation between pressure and manometer deflection is given by Equation 9.5 with $H = L \sin \theta$:

$$\Delta p = p_1 - p_2 = L(\gamma_m - \gamma) \sin \theta$$

where p_2 is the ambient pressure so that Δp is the nominal pressure relative to the ambient. For a nominal $\Delta p = 100 \text{ N/m}^2$, the nominal manometer rise L would be

$$L = \frac{\Delta p}{(\gamma_m - \gamma)\sin \theta} \approx \frac{\Delta p}{\gamma_m \sin \theta} = 21 \text{ mm}$$

where $\gamma_m \gg \gamma$ and the value for γ and its uncertainty are neglected. For the design stage analysis, $p = f(\gamma_m, L, \theta)$, so that the uncertainty in pressure, Δp, is estimated by

$$(u_d)_p = \sqrt{\left[\frac{\partial \Delta p}{\partial \gamma_m}(u_d)_{\gamma_m}\right]^2 + \left[\frac{\partial \Delta p}{\partial L}(u_d)_L\right]^2 + \left[\frac{\partial \Delta p}{\partial \theta}(u_d)_\theta\right]^2}$$

At assumed 95% confidence levels, the manometer specific weight uncertainty and angle uncertainty are estimated from the problem statement as

$$(u_d)_{\gamma_m} = (9770 \text{ N/m}^3)(0.005) \approx 49 \text{ N/m}^3$$
$$(u_d)_\theta = 1 \text{ degree} = 0.0175 \text{ rad}$$

The uncertainty in estimating the pressure from the indicated deflection arises from both the manometer resolution, u_o, and the zero point offset error, which we take as its instrument error, u_c. Using the uncertainties associated with these errors,

$$(u_d)_L = \sqrt{u_o^2 + u_c^2} = \sqrt{(0.5 \text{ mm})^2 + (0.5 \text{ mm})^2} = 0.7 \text{ mm}$$

Evaluating the derivatives and substituting values gives a design-stage uncertainty interval of a measured Δp of

$$(u_d)_{\Delta p} = \sqrt{(0.26)^2 + (3.42)^2 + (3.10)^2} = \pm 4.6 \text{ N/m}^2 \quad (95\%)$$

COMMENT At a 30-degree inclination and for this pressure, the uncertainty in pressure is affected almost equally by the instrument inclination and the deflection uncertainties. As the manometer inclination is increased to a more vertical orientation—that is, toward the U-tube manometer—inclination uncertainty becomes less important and is negligible near 90 degrees. However, for a U-tube manometer, the deflection is reduced to less than 11 mm, a 50% reduction in manometer sensitivity, with an associated design-stage uncertainty of 6.8 N/m^2 (95%).

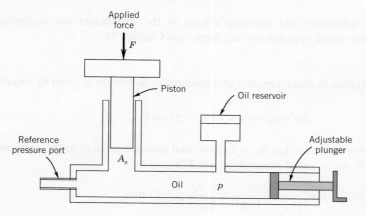

Figure 9.8   Deadweight tester.

Deadweight Testers

The deadweight tester makes direct use of the fundamental definition of pressure as a force per unit area to create and to determine the pressure within a sealed chamber. These devices are a common laboratory standard for the calibration of pressure-measuring devices over the pressure range from 70 to 7×10^7 N/m^2 (0.01 to 10,000 psi). A deadweight tester, such as that shown in Figure 9.8, consists of an internal chamber filled with oil, a close-fitting piston and cylinder, and a set of calibrated masses. Chamber pressure acts on the end of the carefully machined piston. A static equilibrium exists when the external pressure, owing to the force exerted on the fluid by the piston acting over the equivalent area A_e, balances with the chamber pressure. The known weight of the piston plus the additional weight of added calibrated masses provide this external force F. At static equilibrium the piston floats and the chamber pressure can be deduced as

$$p = \frac{F}{A_e} + \sum \text{error corrections} \qquad (9.7)$$

A pressure transducer connected to a reference port is calibrated by comparison to the chamber pressure. For most calibrations, the error corrections can be ignored.

When error corrections are applied, the instrument uncertainty in the chamber pressure using a deadweight tester can be as low as 0.005% to 0.01% of the reading. A number of elemental errors contribute to Equation 9.7, including air buoyancy effects, variations in local gravity, uncertainty in the known mass of the piston and added masses, shear effects, thermal expansion of the piston area, and elastic deformation of the piston (1).

An indicated pressure, p_i, can be corrected for gravity effects, e_1, from Equation 9.6a or 9.6b, and for air buoyancy effects, e_2, by

$$p = p_i(1 + e_1 + e_2) \qquad (9.8)$$

where

$$e_2 = -\gamma_{\text{air}}/\gamma_{\text{masses}} \qquad (9.9)$$

The tester fluid lubricates the piston so that the piston is partially supported by the shear forces in the oil in the gap separating the piston and the cylinder. This error varies inversely with the tester fluid viscosity, so high-viscosity fluids are preferred. In a typical tester, the uncertainty due to this error is less than 0.01% of the reading. At high pressures, elastic deformation of the piston affects the actual piston area. For this reason, the effective area is based on the average of the piston and cylinder diameters.

Example 9.4

A deadweight tester indicates 100.00 lb/in.^{2} (i.e., 100.00 psi), at 70 °F in Clemson, SC ($\phi = 34°$, $z = 841$ ft). Manufacturer specifications for the effective piston area were stated at 72 °F so that thermal expansion effects remain negligible. Take $\gamma_{air} = 0.076$ lb/ft^3 and $\gamma_{mass} = 496$ lb/ft^3. Correct the indicated reading for known errors.

KNOWN $p_i = 100.00$ psi; $z = 841$ ft; $\phi = 34°$

ASSUMPTION   Systematic error corrections for altitude and latitude apply

FIND p

SOLUTION   The corrected pressure is found using Equation 9.8. From Equation 9.9, the correction for buoyancy effects is

$$e_2 = -\gamma_{air}/\gamma_{masses} = -0.076/496 = -0.000154$$

The correction for gravity effects is from Equation 9.6a:

$$e_1 = -(2.637 \times 10^{-3} \cos 2\phi + 9.6 \times 10^{-8} z + 5 \times 10^{-5})$$
$$= -(0.0010 + 8 \times 10^{-5} + 5 \times 10^{-5}) = -0.001119$$

From Equation 9.8, the corrected pressure becomes

$$p = 100.00 \times (1 - 0.000154 - 0.001119) \text{ lb/in.}^2 = 99.87 \text{ lb/in.}^2$$

COMMENT   This amounts to correcting an indicated signal for known systematic errors. Here that correction is $\approx 0.13\%$.

9.4   PRESSURE TRANSDUCERS

A pressure transducer is a device that converts a measured pressure into a mechanical or electrical signal. The primary sensor is usually an elastic element that deforms or deflects under the measured pressure relative to a reference pressure. Several common elastic elements used, as shown in Figure 9.9, include the Bourdon tube, bellows, and diaphragm. A transducer element converts the elastic element deflection into a readily measurable signal such as an electrical voltage or mechanical rotation of a pointer. There are many methods available to perform this transducer function, and we examine a few common ones.

General categories for pressure transducers are absolute, gauge, vacuum, and differential. These categories reflect the application and reference pressure used. Absolute transducers have a sealed

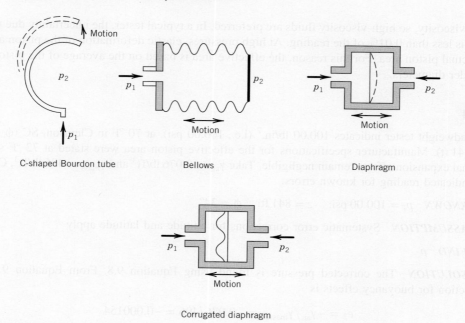

Figure 9.9 Elastic elements used as pressure sensors.

reference cavity held at a pressure of absolute zero, enabling absolute pressure measurements. Gauge transducers have the reference cavity open to atmospheric pressure and measure pressure relative to this. Differential transducers measure the difference between two applied pressures. Vacuum transducers are a special form of absolute transducer for low-pressure measurements.

Pressure transducers are subject to some or all of the following elemental errors: resolution, zero error, linearity error, sensitivity error, hysteresis, noise, and drift resulting from environmental temperature changes. Electrical transducers are also subject to loading error between the transducer output and its indicating device (see Chapter 6). When this is a consideration, a voltage follower (see Chapter 6) can be inserted at the output of the transducer to isolate transducer load.

Bourdon Tube

The Bourdon tube is a curved metal tube having an elliptical cross section that mechanically deforms under pressure. In practice, one end of the tube is held fixed and the input pressure applied internally. A pressure difference between the outside of the tube and the inside of the tube brings about tube deformation and a deflection of the tube free end. This action of the tube under pressure can be likened to the action of a deflated balloon that is subsequently inflated. The magnitude of the deflection of the tube end is proportional to the magnitude of the pressure difference. Several variations exist, such as the C shape (Fig. 9.9), the spiral, and the twisted tube. The exterior of the tube is usually open to atmosphere (hence, the origin of the term "gauge" pressure referring to pressure referenced to atmospheric pressure), but in some variations the tube may be placed within a sealed housing and the tube exterior exposed to some other reference pressure, allowing for absolute and for differential designs.

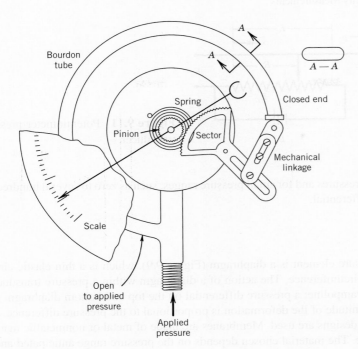

Figure 9.10 Bourdon tube pressure gauge.

The Bourdon tube mechanical dial gauge is a commonly used pressure transducer. A typical design is shown in Figure 9.10, in which the secondary element is a mechanical linkage that converts the tube displacement into a rotation of a pointer. Designs exist for low or high pressures, including vacuum pressures, and selections span a wide choice in range. The best Bourdon tube gauges have instrument uncertainties as low as 0.1% of the full-scale deflection of the gauge, but values of 0.5% to 2% are more common. However, the attractiveness of this device is that it is simple, portable, and robust, lasting for years of use.

Bellows and Capsule Elements

A bellows sensing element is a thin-walled, flexible metal tube formed into deep convolutions and sealed at one end (Fig. 9.9). One end is held fixed while pressure is applied internally. A difference between the internal and external pressures causes the bellows to change in length. The bellows is housed within a chamber that can be sealed and evacuated for absolute measurements, vented through a reference pressure port for differential measurements, or opened to atmosphere for gauge pressure measurements. A similar design, the capsule sensing element, is also a thin-walled, flexible metal tube whose length changes with pressure, but its shape tends to be more narrow in diameter as well as longer.

A mechanical linkage is used to convert the translational displacement of the bellows or capsule sensors into a measurable form. A common transducer is the sliding arm potentiometer (voltage-divider, Chapter 6) found in the potentiometric pressure transducer shown in Figure 9.11. Another type uses a linear variable displacement transducer (LVDT; see Chapter 12) to measure the bellows or capsule displacement. The LVDT design has a high sensitivity and is commonly found in pressure

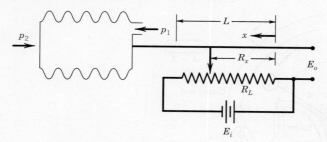

Figure 9.11   Potentiometer pressure transducer.

transducers rated for low pressures and for small pressure ranges, such as zero to several hundred mm Hg absolute, gauge, or differential.

Diaphragms

An effective primary pressure element is a diaphragm (Figure 9.9), which is a thin elastic circular plate supported about its circumference. The action of a diaphragm within a pressure transducer is similar to the action of a trampoline; a pressure differential on the top and bottom diaphragm faces acts to deform it. The magnitude of the deformation is proportional to the pressure difference. Both membrane and corrugated designs are used. Membranes are made of metal or nonmetallic material, such as plastic or neoprene. The material chosen depends on the pressure range anticipated and the fluid in contact with it. Corrugated diaphragms contain a number of corrugations that serve to increase diaphragm stiffness and to increase the diaphragm effective surface area.

Diaphragm-based pressure transducers have good linearity and resolution over their useful range. An advantage of the diaphragm sensor is that the very low mass and relative stiffness of the thin diaphragm give the sensor a very high natural frequency with a small damping ratio for accurate transient or dynamic measurements. Hence, these transducers can have a very wide frequency response and very short 90% rise and settling times. The natural frequency of a circular diaphragm can be estimated by (4)

$$\omega_n(\text{rad/s}) = 10.21 \sqrt{\frac{E_m t^2}{12\left(1 - \upsilon_p^2\right)\rho r^4}} \tag{9.10}$$

where E_m is the material bulk modulus (N/m^2), t the thickness (m), r the radius (m), ρ the material density (kg/m^3), and υ_p the Poisson's ratio for the diaphragm material. The maximum elastic deflection of a uniformly loaded, circular diaphragm supported about its circumference occurs at its center and can be estimated by

$$y_{\text{max}} = \frac{3(p_1 - p_2)\left(1 - \upsilon_p^2\right)r^4}{16E_m t^3} \tag{9.11}$$

provided that the deflection does not exceed one-third the diaphragm thickness. Various secondary elements are available to translate this displacement of the diaphragm into a measurable signal. Several methods are discussed below.

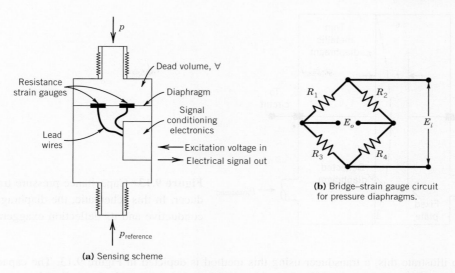

Figure 9.12 Diaphragm pressure transducer using four active resistance strain gauges.

Strain Gauge Elements

A common method for converting diaphragm displacement into a measurable signal is to sense the strain induced on the diaphragm surface as it is displaced. Strain gauges, devices whose measurable resistance is proportional to their sensed strain (see Chapter 11), can be bonded directly onto the diaphragm, integrated within the diaphragm material or onto a deforming element (such as a thin beam) attached to the diaphragm so as to deform with the diaphragm and to sense strain. Metal strain gauges can be used with liquids. Strain gauge resistance is reasonably linear over a wide range of strain and can be directly related to the sensed pressure (5). A diaphragm transducer using strain gauge detection is depicted in Figure 9.12.

By using semiconductor technology in pressure transducer construction, we have a variety of very fast, very small, highly sensitive strain gauge diaphragm transducers for measuring both static and dynamic pressures. Silicone piezoresistive strain gauges can be diffused into a single crystal of silicone wafer, which forms the diaphragm. Semiconductor strain gauges have a static sensitivity that is 50 times greater than conventional metallic strain gauges. Because the piezoresistive gauges are integral to the diaphragm, they are relatively immune to the thermoelastic strains prevalent in conventional metallic strain gauge–diaphragm constructions. Furthermore, a silicone diaphragm does not creep with age (as does a metallic gauge), thus minimizing calibration drift over time. However, uncoated silicone does not tolerate liquids.

Capacitance Elements

Another common method to convert diaphragm displacement to a measurable signal is a capacitance sensor. One version uses a thin metallic diaphragm as one plate of a capacitor paired with a fixed plate to complete the capacitor. The diaphragm is exposed to the process pressure on one side and to a reference pressure on the other or to a differential pressure. When pressure changes, this deflects the diaphragm and the gap between the plates changes, causing a change in capacitance.

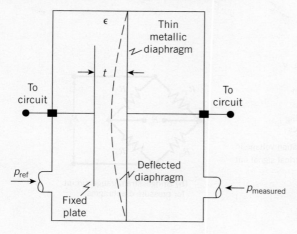

Figure 9.13 Capacitance pressure transducer. In this schematic, the diaphragm is conductive and its deflection exaggerated.

To illustrate this, a transducer using this method is depicted in Figure 9.13. The capacitance C developed between two parallel plates separated by average gap t is determined by

$$C = c\varepsilon A/t \qquad (9.12)$$

where the product $c\varepsilon$ is the permittivity of the material between the plates relative to a vacuum ($\varepsilon = 8.85 \times 10^{-12}$ F/m; c = dielectric constant), and A is the overlapping area of the two plates. The dielectric constant depends on the material in the gap, which for air is $c = 1$ but for water is $c = 80$. The capacitance responds to an instantaneous change in the area-averaged plate gap separation from which the time-dependent pressure is determined. However, the capacitance change is small relative to the absolute capacitance. Oscillator and bridge circuits are commonly used to operate the circuit and to measure the small capacitance change providing an output voltage E_0.

Example 9.5

Estimate the theoretical capacitance of a sensor similar to that of Figure 9.13 if the plate area is 1 mm^2 and the instantaneous plate separation is 1 mm. What is the sensor sensitivity? Assume the plates are separated by air.

SOLUTION With $A = 1$ mm^2, $t = 1$ mm, $\varepsilon = 8.85 \times 10^{-15}$ F/mm, $c = 1$:

$$C = c\varepsilon A/t = (1)\left(8.85 \times 10^{-15}\ \text{F/mm}\right)\left(1\ \text{mm}^2\right)/(1\ \text{mm}) = 8.85 \times 10^{-15}\ \text{F}$$

or $C = 8.85 \times 10^{-3}$ pF (picofarads). For small pressure changes, the diaphragm curvature has a negligible effect on the area-averaged plate gap so that pressure can be treated as linear with the area-average gap value.

The sensor sensitivity is $K_s = \partial C/\partial t = -8.85 \times 10^{-3} \times \left(cA/t^2\right)$ pF/mm.

The capacitance pressure transducer has the attractive features of other diaphragm transducers, including small size and a wide operating range, and is suitable for both static and dynamic pressure measurements. When the technique is used to measure acoustical sound pressure, the device is known as a condenser microphone. Many inexpensive pressure transducers use this measuring principle and are suitable for most engineering measurement demands, including a niche as the on-board car tire pressure sensor. However, the principle is sensitive to temperature changes and has a relatively high impedance output. If attention is paid to these shortcomings, an accurate and stable transducer can be made.

Piezoelectric Crystal Elements

Piezoelectric crystals form effective secondary elements for dynamic pressure measurements, including measuring acoustical sound pressure and dynamic mechanical surface-to-surface contact pressure. They are not intended for static measurements.

Under the action of compression, tension, or shear, a piezoelectric crystal deforms and develops a surface charge q, which is proportional to the force acting to bring about the deformation. In a piezoelectric pressure transducer, a preloaded crystal or stack of crystals is mounted to the diaphragm sensor as indicated in Figure 9.14. Pressure acts normal to the crystal axis and changes the crystal thickness t by a small amount Δt. This sets up a charge, $q = K_q pA$, where p is the pressure acting over the electrode area A and K_q is the crystal charge sensitivity, a material property. A charge amplifier (see Chapter 6) is used to convert charge to voltage so that the voltage developed across the electrodes is

$$E_o = q/C \tag{9.13}$$

where C is the capacitance of the crystal-electrode combination, again given by Equation 9.12. The operating equation becomes

$$E_o = K_q tp/c\varepsilon = Kp \tag{9.14}$$

where K is the overall transducer gain. The crystal sensitivity for quartz, the most common material used, is $K_q = 2.2 \times 10^{-9}$ coulombs/N.

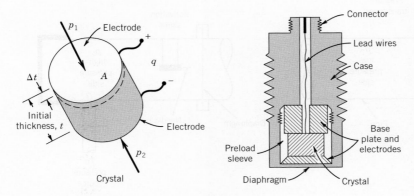

Figure 9.14 Piezoelectric pressure transducer.

9.5 PRESSURE TRANSDUCER CALIBRATION

Static Calibration

A static calibration of a pressure transducer is usually accomplished either by direct comparison against any of the pressure reference instruments discussed (Section 9.3) or against the output from a certified laboratory standard transducer. For the low-pressure range, the McLeod gauge or the appropriate manometric instruments along with the laboratory barometer serve as convenient working standards. The approach is to pressurize (or evacuate) a chamber and expose both the reference instrument, which serves as the standard, and the candidate transducer to the same pressure for a side-by-side measurement. For the high-pressure range, the deadweight tester is a commonly used pressure reference standard.

Dynamic Calibration

The rise time and frequency response of a pressure transducer are found by dynamic calibration, and there are a number of clever ways to accomplish this (6). As discussed in Chapter 3, the rise time of an instrument is found through a step change in input. For underdamped systems, natural frequency and damping ratio can be determined based on the ringing behavior (see Chapter 3) from either a step or an impulse test. The frequency response can also be found directly by applying a constant amplitude periodic input signal and varying its frequency. Although a flush-mounted transducer places the sensor in direct contact with the measurement site, transducers could also be attached by means of a pressure tap or by connecting tubing between the transducer and the tap. This length (called the transmission line) affects the overall response and should be included as part of the dynamic calibration.

An electrical switching valve or flow control valve can create a step change in pressure. But the mechanical lag of the valve limits its use to transducers having an expected rise time of 50 ms or more. Faster applications might use a shock tube calibration or some equivalent diaphragm burst test.

As shown in Figure 9.15, the shock tube consists of a long pipe separated into two chambers by a thin diaphragm. The transducer is mounted into the pipe wall of one chamber, the expansion section,

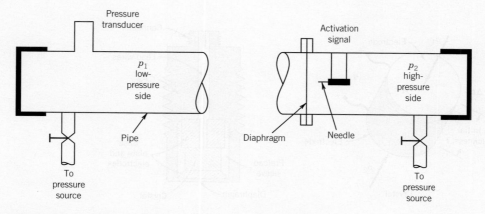

Figure 9.15 Schematic of a shock tube facility.

at pressure p_1. The pressure in the other chamber, the driver section, is raised from p_1 to p_2. Some mechanism, such as a mechanically controlled needle, is used to burst the diaphragm on command. Upon bursting, the pressure differential causes a pressure shock wave to move down the low-pressure chamber. A shock wave has a thickness on the order of 1 µm and moves at the speed of sound, a. So as the shock passes the transducer, the transducer experiences a sudden rise in pressure from p_1 to p_3 over a time $t = d/a$, where d is the diameter of the transducer pressure port, and pressure p_3 is

$$p_3 = p_1\left[1 + (2k/k + 1)(M_1^2 - 1)\right] \tag{9.15}$$

where k is the gas-specific heat ratio and M_1 is the Mach number calculated using normal shock wave tables and absolute pressure p_1. The velocity of the shock wave can also be deduced from the output of fast-acting standard pressure sensors mounted in the shock tube wall. Typical values of t are on the order of 1 to 10 µs, so this method is at least four orders of magnitude faster than a switching valve. The transducer rise time is calculated from the output record.

A common verification check for system response is the "pop test," which is well suited for liquids or gases and systems needing just moderate response times. In this situation, the transducer and connecting tubing are pressurized to a steady value, perhaps by using a syringe or hand pump. The system is suddenly vented to atmosphere. The recorded transducer response is indicative of the system rise time and ringing behavior. One variation of this approach attaches a balloon or similar flexible material to one end of the connecting tubing/transducer system. After pressurizing, the balloon is popped to suddenly vent the system.

Other techniques include an encased loudspeaker or an acoustically resonant enclosure, which serves as a frequency driver, a reciprocating piston and cylinder to develop a sinusoidal pressure variation, or using an oscillating flow control valve to vary system pressure with time (6).

Example 9.6: Pop Test

A surgical pressure transducer is attached to a stiff-walled catheter and a small balloon is attached at its other end. The catheter is filled with saline using a syringe to pressurize the system to 80 mm Hg. At $t = 0$ s, the balloon is popped, forcing a step change in pressure from 80 to 0 mm Hg. The time-based signal is recorded (Fig. 9.16). From the data, the ringing period is measured to be 45.5 ms at second peak amplitude, $y(0.0455) = 5.152$ mV. Would this system be suitable for measuring physiological pressures having frequency content up to 5 Hz? Static sensitivity $K = 1$ mV/mm Hg.

> **KNOWN** $A = 80$ mm Hg, $K = 1$ mV/mm Hg,
>
> $\qquad y(0) = KA = 80$ mV; $y(0.0455) = 5.152$ mV
>
> $\qquad T_d = 0.0455$ s
>
> **FIND** $M(f = 5 \text{ Hz})$
>
> **SOLUTION**   We use methods developed in Chapter 3. The step-function response has the form
>
> $$y(t) = Ce^{-\omega_n \zeta t}\cos\left(\omega_n\sqrt{1 - \zeta^2} + \phi\right)$$

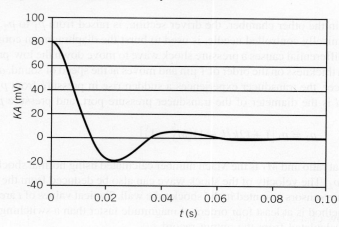

Figure 9.16 Recorded output signal from the pop test of Example 9.6.

where for $y(0) = 80$ mV, $C = 80$ mV, which is consistent with the recorded signal. The steady-state value is $y(\infty) = 0$. The first peak amplitude is at $t = 0$, $y_1 = y(0) = 80$ mV. The second peak amplitude is $y_2 = y(0.0455) = 5.152$ mV. Using logarithmic decrement,

$$\zeta = \frac{1}{\sqrt{1 + \left(2\pi/\ln\left(y_1/y_2\right)\right)^2}} = \frac{1}{\sqrt{1 + (2\pi/\ln(80/5.152))^2}} = 0.40$$

$$f_n = 1/T_d\sqrt{1 - \zeta^2} = 23.98 \text{ Hz} \quad (\text{i.e., } \omega_n = 150 \text{ rad/s})$$

Then,

$$M(f = 5) = \frac{1}{\sqrt{\left[1 - (f/f_n)^2\right]^2 + (2\zeta f/f_n)^2}} = 1.03$$

With a dynamic error $(\delta(f) = M(f) - 1)$ of only about 3%, this system should be suitable for the intended measurements.

9.6   PRESSURE MEASUREMENTS IN MOVING FLUIDS

Pressure measurements in moving fluids warrant special consideration. Consider the flow over the bluff body shown in Figure 9.17. Assume that the upstream flow is uniform and steady with negligible losses. Along streamline A, the upstream flow moves with a velocity U_1, such as at point 1. As the flow approaches point 2, it must slow down and finally stop at the front end of the body. Above streamline A, flow moves over the top of the bluff body, and below streamline A, flow moves under the body. Point 2 is known as the stagnation point and streamline A the stagnation streamline for this flow. Along streamline B, the velocity at point 3 is U_3 and because the upstream flow is considered to be uniform it follows that $U_1 = U_3$. As the flow along B approaches the body, it is deflected around the body. From the conservation of mass principles, $U_4 > U_3$. Application of conservation of energy

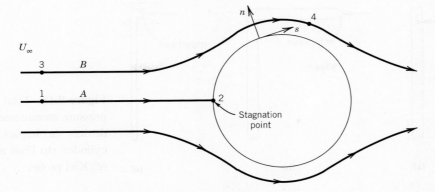

Figure 9.17   Streamline flow over a bluff body.

between points 1 and 2 and between 3 and 4 yields

$$p_1 + \rho U_1^2/2 = p_2 + \rho U_2^2$$
$$p_3 + \rho U_3^2/2 = p_4 + \rho U_4^2$$
(9.16)

However, because point 2 is the stagnation point, $U_2 = 0$, and

$$p_2 = p_t = p_1 + \rho U_1^2/2$$
(9.17)

it follows that $p_2 > p_1$ by an amount equal to $p_v = \rho U_1^2/2$, called the *dynamic pressure*, an amount equivalent to the kinetic energy per unit mass of the flow as it moves along the streamline. If no energy is lost through irreversible processes, such as through a transfer of heat,[1] this translational kinetic energy is transferred completely into p_2. The value of p_2 is known as the *stagnation* or the *total pressure* and is noted as p_t. The total pressure can be measured by bringing the flow to rest at a point in an isentropic manner.

The pressures at 1, 3, and 4 are known as static pressures[2], of the flow. The *static pressure* is that pressure sensed by a fluid particle as it moves with the local flow. The static pressure and velocity at points 1 and 3 are given the special names of the "freestream static pressure" and "freestream velocity." Because $U_4 > U_3$, Equation 9.16 shows that $p_4 < p_3$. It follows from Equation 9.17 that the total pressure is the sum of the static and dynamic pressures anywhere in the flow.

Total Pressure Measurement

In practice, the total pressure is measured using an impact probe such as those depicted in Figure 9.18. A small hole in the impact probe is aligned with the flow so as to cause the flow to come to rest at the hole. The sensed pressure is transferred through the impact probe to a pressure transducer or other

---

[1] This is a realistic assumption for subsonic flows. In supersonic flows, the assumption is not valid across a shock wave.
[2] The term "static pressure" conforms to common expression; the term "stream pressure" is also used.

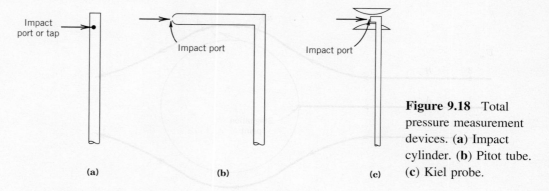

Figure 9.18 Total pressure measurement devices. (**a**) Impact cylinder. (**b**) Pitot tube. (**c**) Kiel probe.

pressure-sensing device such as a manometer. Alignment with the flow is somewhat critical, although the probes in Figure 9.18a, b are relatively insensitive (within ~1% error in indicated reading) to misalignment within a ±7 degree angle (1). A special type of impact probe shown in Figure 9.18c, known as a Kiel probe, uses a shroud around the impact port. The effect of the shroud is to force the local flow to align itself with the shroud axis so as to impact directly onto the impact port. This effectively eliminates total pressure sensitivity to misalignment up to ±40 degree (1). A thermo- couple temperature sensor is often built into the impact site so that the flow stagnation temperature can be measured simultaneously.

Static Pressure Measurement

The local value of static pressure in a moving stream is measured by sensing the pressure in the direction that is normal to the flow streamline. Within ducted flows, static pressure is sensed by wall taps, which are small, burr-free, holes drilled into the duct wall perpendicular to the flow direction at the measurement point. The tap is fitted with a hose or tube, which is connected to a pressure gauge or transducer. A recommended design for a wall tap is shown in Figure 9.19. The tap hole diameter d is typically between 1% and 10% of the pipe diameter, with the smaller size preferred (7). The tap must be perpendicular to the local tangent to the wall and with no drilling burrs (8).

Alternatively, a static pressure probe can be inserted into the flow to measure local stream pressure. It should be a streamlined design to minimize the disturbance of the flow. It should be

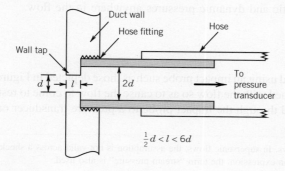

Figure 9.19 Anatomy of a static pressure wall tap.

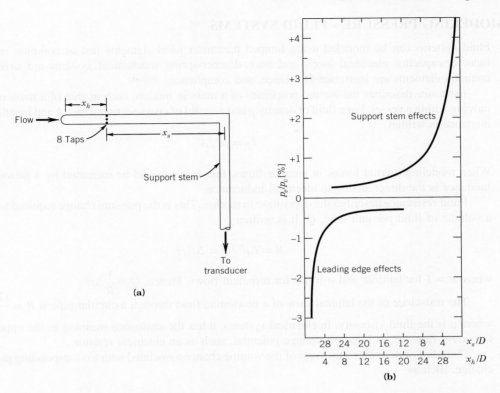

Figure 9.20 Improved Prandtl tube for static pressure. (**a**) Design. (**b**) Relative static error along tube length. D is the probe's tube diameter.

physically small so as not to cause more than a negligible increase in velocity in the vicinity of measurement. As a rule, the frontal area of the probe should not exceed 5% of the pipe flow area. The static pressure sensing port should be located well downstream of the leading edge of the probe so as to allow the streamlines to realign themselves parallel with the probe. Such a concept is built into the improved Prandtl tube design shown in Figure 9.20a.

A Prandtl tube probe consists of eight holes arranged about the probe circumference and positioned 8 to 16 probe tube diameters, D, downstream of the probe leading edge and 16 probe tube diameters upstream of its support stem. A pressure transducer or manometer is connected to the probe stem to measure the sensed pressure. The hole positions are chosen to minimize static pressure error caused by the disturbance to the flow streamlines due to the probe's leading edge and stem. This is illustrated in Figure 9.20b, where the relative static error, $p_e/p_v = (p_i - p)/(\frac{1}{2}\rho U^2)$, is plotted as a function of tap location along the probe body. Real viscous effects around the static probe cause a slight discrepancy between the actual static pressure and the indicated static pressure. To account for this, a correction factor, C_0, is used with $p = C_0 p_i$ where $0.99 < C_0 < 0.995$ and p_i is the indicated (measured) pressure.

9.7   MODELING PRESSURE - FLUID SYSTEMS

Fluid systems can be modeled using lumped parameter ideal elements just as common resistor-inductor-capacitor electrical loops and mass-damper-spring mechanical systems are used. The common elements are inertance, resistance, and compliance.

Inertance describes the inertial properties of a mass in motion, such as that of a mass of fluid moving within a vessel. For a fluid of density ρ and a vessel of cross-sectional area A and length ℓ, the inertance is written

$$L_f = \rho\ell/A \qquad (9.18)$$

When modeling inertial forces in laminar flows, this value should be increased by a factor of $\frac{4}{3}$. Inertance is the direct analog to electrical inductance.

Fluid *resistance* describes the opposition to motion. This is the pressure change required to move a volume of fluid per unit time, Q. It is written

$$R = \Delta p^n/Q = \Delta E/I \qquad (9.19)$$

where $n = 1$ for laminar and $n = 0.5$ for turbulent flows. Hence, $Q = \frac{1}{R}\Delta p^n$.

The resistance of the laminar flow of a newtonian fluid through a circular pipe is $R = \dfrac{128\,\mu\ell}{\pi d^4}$, where μ is the fluid viscosity. In electrical systems, it has the analogous meaning as the opposition to current flow for an imposed voltage potential, such as an electrical resistor.

Compliance describes a measure of the volume change associated with a corresponding pressure change, such as

$$C_{vp} = \Delta\forall/\Delta p \qquad (9.20)$$

It is a measure of flexibility of a structure, component, or substance, and so it is the inverse of the system *stiffness*. Compliance is the direct analog to electrical capacitance.

9.8   DESIGN AND INSTALLATION: TRANSMISSION EFFECTS

Consider the configuration depicted in Figure 9.21 in which a tube of volume $\forall_t$ with length ℓ and diameter d is used to connect a pressure tap to a pressure transducer of internal dead volume $\forall$ (e.g., Figure 9.12). Under static conditions, the pressure transducer indicates the static pressure at the tap. But if the pressure at the tap is a time-dependent pressure, $p_a(t)$, the response behavior of the tubing influences the time-indicated output from the transducer, $p(t)$.

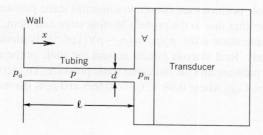

Figure 9.21 Wall tap to pressure transducer connection: the transmission line.

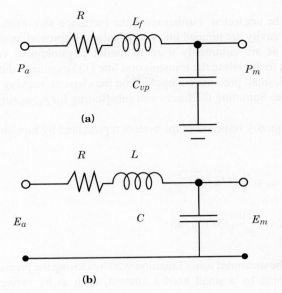

Figure 9.22 An equivalent lumped parameter network of the pressure transmission line model of Figure 9.21 using an electrical analogy.

By considering the one-dimensional pressure forces acting on a lumped mass of fluid within the connecting tube, balancing inertance, compliance, and resistance against forcing function, we can develop a model for the pressure system response. A network model is shown in Figure 9.22a in which the circuit is driven between two pressures, the applied pressure $p_a(t)$ at the tap and the measured pressure $p_m(t)$ at the transducer sensor. Using the electrical analog, inertance is modeled by the inductor, fluid resistance by a resistor, and compliance by a capacitor (Fig. 9.22b). The circuit analysis of the two loops gives

$$L\ddot{I} + RI + \frac{1}{C}\int I dt = E_a \quad \text{and} \quad \frac{1}{C}\int I dt = E_m \tag{9.21}$$

Taking the derivative of the second loop to get $\dot{E}_m$ and $\ddot{E}_m$ in terms of $\dot{I}$ and $\ddot{I}$, then substituting these back into Equation 9.21 with $E_a = p_a$, $E_m = p_m$, $L = L_f$, and $C = C_{vp}$ gives

$$L_f C_{vp} \ddot{p}_m + R C_{vp} \dot{p}_m + p_m = p_a(t) \tag{9.22}$$

Substituting Equations 9.18 to 9.20 into Equation 9.22 gives the working system response equation for the applied and measured pressures,

$$\frac{16\ell\rho C_{vp}}{3\pi d^2}\ddot{p}_m + \frac{128\,\mu\ell C_{vp}}{\pi d^4}\dot{p}_m + p_m = p_a(t) \tag{9.23}$$

in which we have augmented the fluid inertial force by $\frac{4}{3}$ (9).

In this simple model, the system compliance lumps the compliance of the fluid, tube walls, and transducer into a single value, C_{vp}. Because these individual compliances could be modeled separately by using capacitors in parallel, the total capacitance is simply the sum of each. If one compliance dominates, the others can be neglected. Furthermore, the inertance and resistance of the connecting tube and the transducer cavity are lumped into single values. Improved models use distributed, lumped parameters such as are commonly used to model physiological vascular systems (10). A longstanding approach to modeling the transmission line (11) examines the forces acting on a fluid element. In this model, small pressure changes act on the element, moving it back and forth by a distance x within the tube. Summing the forces and substituting for p_m again results in Equation 9.23 (12).[3]

We can study the transient and frequency response of the system represented by Equation 9.23 by extracting values for ω_n and ζ,

$$\omega_n = \frac{d}{4}\sqrt{3\pi/\rho\ell C_{vp}} \tag{9.24}$$

$$\zeta = \frac{16\,\mu}{d^3}\sqrt{3\ell C_{vp}/\pi\rho} \tag{9.25}$$

The total system compliance could be measured using Equation 9.20 by closing the pressure tap, increasing the fluid volume in the tubing by a small known amount, such as by syringe, and measuring the corresponding pressure change.

Liquids

Liquids are relatively incompressible, so the compression-restoring force in the transmission line results primarily from the compliance in the transducer, which is a stated transducer specification, for use in Equations 9.24 and 9.25. Connecting tubing can usually be considered rigid for the underlying assumptions of the above lumped parameter analysis. Thick-walled, flexible tubing is often used, but this is fairly rigid and its compliance can be ignored. If need be, the compliance can be measured.

Gases

For gases, we simplify by assuming that the system is rigid relative to the compressibility of the gas. Compliance is then modeled in terms of the fluid's adiabatic bulk modulus of elasticity, $E_m = \forall/C_{vp}$. This gives

$$\omega_n = \frac{d}{4}\sqrt{3\pi E_m/\rho\ell\forall} \tag{9.26}$$

$$\zeta = \frac{16\,\mu}{d^3}\sqrt{3\ell\forall/\pi\rho E_m} \tag{9.27}$$

Equations 9.26 and 9.27 can also be written in terms of the speed of sound for the gas, a, which is related to its compressibility by $a = \sqrt{E_m/\rho}$ and for a perfect gas by $a = \sqrt{kRT}$, where T is the gas

---

[3] Be aware that models are based on simplifying assumptions and should be used only as a guide in the design of a system, not as a replacement for in situ calibration.

absolute temperature, giving

$$\omega_n = \frac{ad}{4}\sqrt{3\pi/\ell\forall} \qquad (9.28)$$

$$\zeta = \frac{16\mu}{a\rho d^3}\sqrt{3\ell\forall/\pi} \qquad (9.29)$$

When the tube volume, $\forall_t \gg \forall$, then a series of standing pressure waves develop and we can expect $\omega \sim O(a/\ell)$. Hougen et al. (13) discuss an improved prediction as

$$\omega_n = \frac{a}{\ell\sqrt{0.5 + \forall/\forall_t}} \qquad (9.30)$$

$$\zeta = \frac{16\,\mu\ell}{\rho ad^2}\sqrt{0.5 + \forall/\forall_t} \qquad (9.31)$$

Note that in all cases, larger diameter and shorter length tubes improve pressure system response.

Example 9.7

The pressure in a water-filled pipe varies with time. A pressure transducer is connected to a wall tap using a length of non-rigid plastic tubing to measure this pressure. To estimate the pressure system compliance, the tubing is removed at the wall tap, filled with water, and purged of any residual air. The transducer output is noted. Using a syringe, 1 mL of water is then added to the system and the corresponding measured pressure increases by 100 mm Hg. Find the compliance.

SOLUTION   Using Equation 9.20, the compliance of the transducer-tubing system is

$$C_{vp} = \Delta\forall/\Delta p = 1\ \text{mL}/100\ \text{mm Hg} = 0.01\ \text{mL/mm Hg}$$

Example 9.8

A pressure transducer with a natural frequency of 100 kHz is connected to a 0.10-in. static wall pressure tap using a 0.10-in. i.d. rigid tube that is 5 in. long. The transducer dead volume is 1 in.^{3} Determine the system frequency response to fluctuating pressures of air at 72 °F if the pressure fluctuates about an average value of 1 atm abs. Express the frequency response in terms of $M(\omega)$. $R_{air} = 53.3$ ft-lb/lb$_m$-°R; $\mu = 4 \times 10^{-7}$ lb-s/ft^2; $E_m = 20.5$ lb/in.^{2}

KNOWN $\ell = 5$ in. $k = 1.4$

$\quad\quad\quad\quad\quad d = 0.1$ in. $T = 72°F = 532°R$

$\quad\quad\quad\quad\quad \forall = 1$ in.^{3} $\rho = p/RT = 0.075$ lb$_m$/ft^3

ASSUMPTION   Air behaves as a perfect gas.

FIND $M(\omega)$

SOLUTION The frequency-dependent magnitude ratio is given by Equation 3.22

$$M(\omega) = \frac{1}{\sqrt{\left[1 - (\omega/\omega_n)^2\right]^2 + \left[2\zeta(\omega/\omega_n)\right]^2}}$$

We need ω_n and ζ to solve for $M(\omega)$ for various frequencies.

With $\forall_t < \forall$, we use Equations 9.26 and 9.27, or alternately using Equations 9.28 and 9.29 with $a = 1,130$ ft/s, to find

$$\omega_n = \frac{d}{4}\sqrt{3\pi E_m/\rho\ell\forall} = 470 \text{ rad/s}$$

$$\zeta = \frac{16\,\mu}{d^3}\sqrt{3\ell\forall/\pi\rho E_m} = 0.06$$

This system is lightly damped. The magnitude frequency response, $M(\omega)$, is shown in Figure 9.23. The transmission line effects control the transducer response. In effect, the transmission line acts as a mechanical low-pass filter for the transducer.

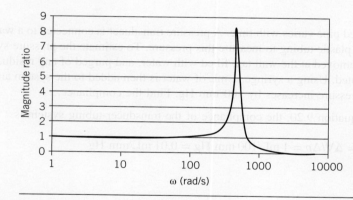

Figure 9.23 Frequency response of the transmission line for Example 9.10.

Heavily Damped Systems

In systems in which we estimate a damping ratio greater than 1.5, the frequency response model can be simplified further. The behavior of the pressure-measuring system closely follows that of a first-order system. Again, the typical pressure transducer–tubing system has a compliance C_{vp}, which is a measure of the transducer and tubing volume change relative to an applied pressure change. The response of the first-order system is indicated through its time constant, which is estimated by neglecting the second-order term in Equation 9.23 so that (12)

$$\tau = \frac{128\,\mu\ell C_{vp}}{\pi d^4} \tag{9.32}$$

Equation 9.32 shows that the time constant is proportional to ℓ/d^4. Long- and small-diameter connecting tubes promote a more sluggish system response to changes in pressure.

9.9   ACOUSTICAL MEASUREMENTS

Acoustical measurements involve the measurement of sound pressure. There are many applications that require measuring sound pressure. For example, sound measurements are used to amplify or record speech and music and to tune large rooms and theatre acoustics to achieve appropriate sound level reproduction (28). A common application is to monitor for increased sound levels or changes in frequency in rotating equipment, which are indicators of failing bearings, rotors, or gears (29). A significant engineering application involves measuring sound originating from appliances, machinery, or office equipment and within vehicles in an effort to reduce noise levels that affect the comfort and safety of occupants and workers, as well as to ensure compliance with national noise standards.

Sound is quantified by sound pressure (1 Pa = 1 µbar), a dynamic variation of pressure from the prevailing ambient pressure. It is described by a spectrum of signal amplitudes at corresponding frequencies (e.g., sound frequency spectra are shown in Figures 2.20 and 2.21). Sound pressure is usually time-integrated, with the resulting value referenced against the value taken as the threshold of human hearing at 1,000 Hz. This value is $p_0 = 2 \times 10^{-5}$ Pa = 0.0002 µbar. Sound power is the energy per unit time from a sound source. Sound intensity is power per unit area with a reference value of $I_0 = 10^{-16}$ W/cm^2. Sound pressure and intensity levels are usually expressed in decibels:

$$\text{Pressure level (dB)} = 20 \log (p/p_0)$$
$$\text{Intensity level (dB)} = 10 \log (I/I_0)$$

(9.33)

Hence, doubling of the sound pressure and intensity equates to changes of 6dB and 3 dB, respectively. Some typical sound pressure levels as detected by the human ear are listed in Table 9.1, a range that spans a factor of 1:10,000,000.

Signal Weighting

There are applications that require reporting the sound level exactly as it is measured across the different frequencies, such as in reporting the frequency response of a transducer or for measuring the force of sound pressure acting against a wall or boundary. The frequency bandwidth of transducers varies with individual design; some extend up to 1 GHz. There are other applications that require reporting the sound levels in ways that better describe how the human ear would perceive them, such as in noise isolation, noise abatement, or audio playback. Normal human hearing is sensitive to frequencies ranging from 20 Hz to 20,000 Hz, but the human ear does not respond to sound equally over this frequency range. The ear is most sensitive to sound at

Table 9.1   Sound Pressure Level References

0 dB = 0.00002 Pa	Threshold of Hearing	100 dB = 2 Pa	Jackhammer
60 dB = 0.02 Pa	Conversation	120 dB = 20 Pa	Airplane takeoff
94 dB = 1 Pa	Large truck passing	140 dB = 200 Pa	Threshold of pain

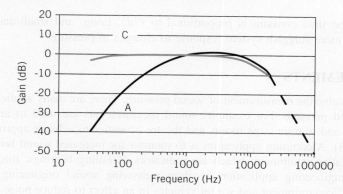

Figure 9.24 Frequency weighting: A- and C- scales.

frequencies between 1 kHz = 1000 Hz and 5 kHz and attenuates sound at all frequencies above and below these frequencies. Signal weighting designations were developed to simulate this frequency response. The A-weighting scale compensates sound measurements in a manner that mimics the human ear frequency response at common sound pressure levels. A-weighting is used in the reporting of industrial and environmental noise measurements, as well as in reporting audio equipment response and music reproduction. At higher sound levels (> 100 dB), the human ear tends to have a flatter frequency response and the C-weighting scale is used.

Both scales are shown in the spectral plot of Figure 9.24 with the corresponding octave-band center frequencies listed in Table 9.2. An octave is just the interval between two points where the frequency at the second point is twice the frequency of the first. The A- and C-weighting scales are both equal to 0dB at 1 kHz. Measurements using A-weighting or C-weighting are designated dB(A) or dB(C), respectively. Exact measurements are scaled according to these weightings. For example, to apply A-weighting to a measured sound pressure level at a particular frequency or octave center-band frequency, apply the correction value:

$$reported\ value\ in\ dB(A) = measured\ value\ in\ dB\ +\ A\text{-}weighting\ in\ dB \tag{9.34}$$

In this way, the weighting adjusts the measured value to the level that the human ear would hear it. When sound pressure is measured at several discrete octaves, an overall sound pressure can be calculated by logarithmic addition. The overall sound pressure level, SPL (dB), is formed from the corresponding sound pressures measured over each octave:

$$\overline{SPL} = 10\log_{10}\left(\sum_{i} 10^{(SPL_i/10)} \right) \tag{9.35}$$

Table 9.2 A-weighting and C-weighting Scales

Relative Response (dB)	Octave-Band Center Frequency f in Hz									
	31.25	62.5	125	250	500	1,000	2,000	4,000	8,000	16,000
dB(A)	−39.4	−26.2	−16.1	−8.6	−3.2	0	+1.2	+1.0	−1.1	−6.6
dB(C)	−3.0	−0.8	−0.2	0	0	0	−0.2	−0.8	−3.0	−8.5

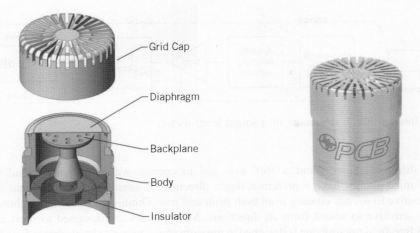

- Grid Cap
- Diaphragm
- Backplane
- Body
- Insulator

Figure 9.25 Condenser microphone schematic design and a free-field condenser microphone. (Courtesy of PCB Piezotronics, Inc., Depew, New York. Used with permission.)

Microphones

Acoustical sound pressure is measured using a microphone that measures pressure amplitude and frequency. There are different microphone designs available: dynamic, condenser, and piezoelectric.

Dynamic microphones are passive devices that operate on electromagnetic principles (Chapter 6) similar to those of a loudspeaker. A diaphragm attached to a voice coil acts as the sensor. The wire-wound voice coil is wrapped about a permanent magnet. Dynamic sound pressure contacting the diaphragm moves the coil generating an AC voltage signal. The output signal follows the frequency and amplitude of the measured sound pressure.

Condenser microphones (Figure 9.25) contain a metal diaphragm sensor and a metal backplane with an air gap between them to form a capacitor (also known as a condenser), a design that mirrors a capacitance pressure transducer (Section 9.4). Sound pressure waves move the diaphragm so as to change this air gap width. This alters the capacitance value, which in turn affects the value of the output signal. The output signal follows the frequency and amplitude of the sound pressure. A DC excitation voltage powers the backplane of the capacitor. An electret condenser microphone places a permanently charged electret material either in the diaphragm, in the backplane, or in the interval between them to provide a polarizing voltage caused by diaphragm motion, eliminating the need for the excitation voltage. The low-level signal output needs amplification.

Piezoelectric microphones contain a quartz or ceramic crystal coupled to a diaphragm sensor. As described with piezoelectric pressure transducers (Figure 9.14), the pressure force deforms the preloaded crystal, which develops a surface charge q, proportional to the sound pressure. A charge amplifier is used to convert charge to voltage. Despite low sensitivities, these sensors are quite durable, have a wide frequency bandwidth (up to 1 GHz), and are able to measure very high sound levels. These are the microphone of choice in acoustic emissions testing, such as for detecting mechanical bearing failure or structural cracks.

Microphones are calibrated using a reference noise source, usually a constant pressure (94 dB) signal at a fixed frequency. Microphones can be made to have different sensitivities to sound levels

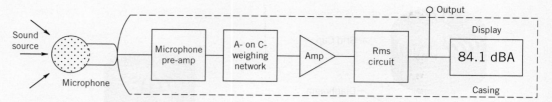

Figure 9.26 Internal block schematic of a sound level meter.

coming from different angles about a 360° axis and to operate within different sound fields. Unidirectional microphones have a preferred single direction of sensitivity. Bidirectional microphones are sensitive to sounds coming from both front and rear. Omni-directional microphones are nearly equally sensitive to sound from all directions. Microphones are designed to meet certain applications. A free-field microphone is designed to measure sound pressure levels that radiate from a single source or direction. These might be used to measure the sound coming from an appliance or machine. These microphones work best in open areas or within anechoic chambers. A pressure field microphone is designed to measure the sound pressure at a position or wall boundary. These are often mounted with only the diaphragm exposed and flush with the wall. These work well to measure the pressure exerted on a wall or structure like an airplane wing or within housings and cavities. A random incidence type of microphone is omnidirectional so as to measure sound pressure coming from multiple directions, multiple sources, and multiple reflections. Examples of their use include sound measurements within a car or a music hall.

Sound level meters (SLM) are self-contained, hand-held devices used in the sound measurement industry to measure sound pressure. An SLM consists of a microphone, signal conditioning including weighting scale algorithms and root-mean-square circuits, and a readout device (Figure 9.26). The standard IEC 61672-1 specifies sound level meter performance (30).

<table>
<tr><td colspan="10"> </td></tr>
</table>

Example 9.9

Apply A-weighting to these octave-band measurements. Find the overall average sound pressure level.

	Octave-Band Center Frequency f in Hz								
	31.25	62.5	125	250	500	1,000	2,000	4,000	8,000
Relative Response (dB)	94	95	92	95	97	97	102	97	92
Measured SPL (dB)									
A-weighting (dB)	−39.4	−26.2	−16.1	−8.6	−3.2	0	+1.2	+1.0	−1.1
Corrected Values dB(A)	55	69	76	86	94	97	103	98	91

The overall average sound pressure level is

$$\overline{SPL} = 10\log_{10}(10^{SPL_i/10})$$

$$= 10\log\left(10^{5.5} + 10^{6.9} + 10^{7.6} + 10^{8.6} + 10^{9.4} + 10^{9.7} + 10^{10.3} + 10^{9.8} + 10^{9.1}\right) = 105 \text{ dB(A)}$$

Example 9.10: Case Study: Active Noise Reduction Headsets

Traditional passive headsets are very good at attenuating external ambient high-frequency sounds (> 1 kHz), effectively reducing those background noise levels as heard by the user. They are less effective at attenuating low-frequency background sounds. Active noise reduction (ANR) is a method that improves a headset's ability to attenuate the low-frequency noise signals, including the subaural sounds that can fatigue a listener. They are particularly effective at reducing constant level background ambient noise, such as ventilation system or machinery noise. Developed in the 1950s and becoming popular with general aviation pilots in the 1990s to attenuate the constant drone of aircraft propeller noise, these types of headsets are now popular with the general public to filter out unwanted continuous sounds.

In practical design, a microphone measures the ambient low-frequency background noise present within or just outside the earpiece of the listener. The measured noise is passed through signal conditioning circuitry and broadcast through a speaker mounted within the earpiece at exactly 180° out of phase with itself. These measured and broadcast noise signals sum to zero, canceling each other out. Non-repetitive sounds and sounds at other frequencies are not canceled; these are transmitted through the earpiece and are unaffected by the out-of-phase signal. The net effect is that background noise levels that occur over a certain range of frequencies are reduced. The example spectrum in Figure 9.27 compares the sound reduction as heard using either a passive or ANR headset. The typical overall sound level reduction (using Equation 9.35) is −10 dB(A) beyond the reduction of a quality passive headset. Desirable sounds are clearer with ANR owing to the higher signal-to-noise ratio. A headset provides a controlled small-volume listening environment requiring very little power for implementing ANR methods. The future challenge is taking ANR to larger-volume environments, such as car cabins, appliance cabinets, and high noise devices in the work environment.

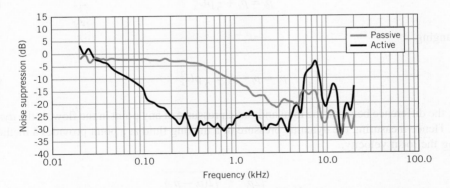

Figure 9.27 Typical ambient noise reduction as experienced by a listener using a headset of either passive only or active noise reduction design.

9.10   FLUID VELOCITY MEASURING SYSTEMS

Velocity measuring systems are used to measure the local velocity in a moving fluid. Desirable information can consist of the mean velocity as well as any of the dynamic components of the velocity. Dynamic components are found in pulsating, phasic, or oscillating flows or in turbulent flows. For most general engineering applications, information about the mean flow velocity is usually sufficient. The dynamic velocity information is often sought during applied and basic fluid mechanics research and development, such as in attempting to study airplane wing response to air turbulence, a complex periodic waveform as the wing sees it. In general, the instantaneous velocity can be written as

$$U(t) = \overline{U} + u \tag{9.36}$$

where $\overline{U}$ is the mean velocity and u is the time-dependent dynamic (fluctuating) component of the velocity. The instantaneous velocity can also be expressed in terms of a Fourier series:

$$U(t) = \overline{U} + \sum_i C_i \sin(\omega_i t + \phi_i') \tag{9.37}$$

so that the mean velocity and the amplitude and frequency information concerning the dynamic velocity component can be resolved with a Fourier analysis of the time-dependent velocity signal.

Pitot-Static Pressure Probe

For a steady, incompressible, isentropic flow, Equation 9.16 can be written at any arbitrary point x in the flow field as

$$p_t = p_x + \frac{1}{2}\rho U_x^2 \tag{9.38}$$

or, rearranging,

$$p_v = p_t - p_x = \frac{1}{2}\rho U_x^2 \tag{9.39}$$

Again p_v, the difference between the total and static pressures at any point in the flow, is the *dynamic pressure*. Hence measuring the dynamic pressure of a moving fluid at a point provides a method for estimating the local velocity,

$$U_x = \sqrt{\frac{2p_v}{\rho}} = \sqrt{\frac{2(p_t - p_x)}{\rho}} \tag{9.40}$$

In practice, Equation 9.40 is used through a device known as a *pitot-static pressure probe*. Such an instrument has an outward appearance similar to that of an improved Prandtl static pressure probe (Fig. 9.20a), except that the pitot–static probe contains an interior pressure tube attached to an impact port at the leading edge of the probe, as shown in Figure 9.28. This creates two coaxial internal cavities within the probe, one exposed to the total pressure and the second exposed to the static pressure. The two pressures are typically measured using a differential pressure transducer so as to indicate p_v directly.

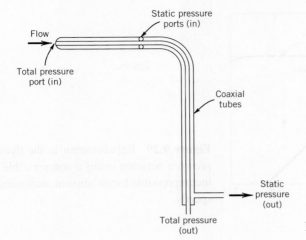

Figure 9.28   Pitot–static pressure probe.

The pitot–static pressure probe is relatively insensitive to misalignment over the yaw angle range of ±15 degrees (1). When possible, the probe can be rotated until a maximum signal is measured, a condition that indicates that it is aligned with the mean flow direction. However, the probes have a lower velocity limit of use that is brought about by strong viscous effects in the entry regions of the pressure ports. In general, viscous effects should not be a concern, provided that the Reynolds number based on the probe radius, $\text{Re}_r = \overline{U}r/v$, is greater than 500, where v is the kinematic viscosity of the fluid. For $10 < \text{Re}_r < 500$, a correction to the dynamic pressure should be applied, $p_v = C_v p_i$, where

$$C_v = 1 + (4/\text{Re}_r) \tag{9.41}$$

and p_i is the indicated dynamic pressure from the probe. However, even with this correction, the measured dynamic pressure has a systematic uncertainty on the order of 40% at $\text{Re}_r \approx 10$ but decreases to 1% for $\text{Re}_r \geq 500$.

In high-speed gas flows, compressibility effects near the probe leading edge require a closer inspection of the governing equation for a pitot-static pressure probe. An energy balance along a streamline for a perfect gas between any point x and the stagnation point can be written as

$$\frac{U^2}{2} = c_p(T_t - T_\infty) \tag{8.34}$$

For an isentropic process, the relationship between temperature, pressure, and density can be stated as

$$\frac{T_t}{T_x} = \left(\frac{p_t}{p_x}\right)^{\frac{k-1}{k}} = \left(\frac{\rho_t}{\rho_x}\right)^{k-1} = \left(1 + \frac{k-1}{2}M^2\right) \tag{9.42}$$

where k is the ratio of specific heats for the gas, $k = c_p/c_v$. The Mach number of a moving fluid relates its local velocity to the local speed of sound,

$$M = U/a \tag{9.43}$$

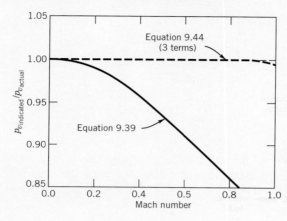

Figure 9.29 Relative error in the dynamic pressure between using a compressible and incompressible formulation at increasing flow speeds.

where the speed of sound, also called the acoustic wave speed, for a perfect gas is $a = \sqrt{kRT_x}$ where T_x is the absolute temperature of the gas at the point of interest. Combining these equations and using a binomial expansion gives a relationship between total pressure and static pressure at any point x in a moving compressible flow,

$$p_v = p_t - p_x = \frac{1}{2}\rho U_x^2 \left[1 + M^2/4 + (2 - k)M^4/24 + \cdots\right] \tag{9.44}$$

Equation 9.44 reduces to Equation 9.39 when $M \ll 1$. The error in between the two equations relative to the true dynamic pressure becomes significant for $M > 0.3$, as shown in Figure 9.29. Thus $M \sim 0.3$ is used as the incompressible limit for perfect gas flows.

For $M > 1$, the local velocity is found through iteration using the Rayleigh relation:

$$p_t/p = \left(\frac{(k+1)^2 M^2}{4kM^2 - 2(k-1)}\right)^{k/k-1} \left(\frac{1 - k + 2kM^2}{k + 1}\right) \tag{9.45}$$

where both p and p_t are the measured values.

Thermal Anemometry

The rate at which energy, $\dot{Q}$, is transferred between a warm body at T_s and a cooler moving fluid at T_f is proportional both to the temperature difference between them and to the thermal conductance of the heat transfer path, hA. This thermal conductance increases with fluid velocity, thereby increasing the rate of heat transfer at any given temperature difference. Hence a relationship between the rate of heat transfer and velocity exists, forming the working basis of a thermal anemometer.

A thermal anemometer uses a sensor, a metallic resistance temperature detector (RTD) element, that makes up one active leg of a Wheatstone bridge circuit, as indicated in Figure 9.30. The resistance–temperature relation for such a sensor was shown in Chapter 8 to be well represented by

$$R_s = R_0[1 + \alpha(T_s - T_0)] \tag{9.46}$$

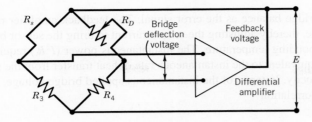

Figure 9.30   Thermal anemometer circuit, shown in constant resistance mode.

so that sensor temperature T_s can be inferred through a resistance measurement. A current is passed through the sensor to heat it to some desired temperature above that of the host fluid. The relationship between the rate of heat transfer from the sensor and the cooling fluid velocity is given by King's law (14) as

$$\dot{Q} = I^2 R = A + BU^n \tag{9.47}$$

where A and B are constants that depend on the fluid and sensor physical properties and operating temperatures and n is a constant that depends on sensor dimensions (15). Typically, $0.45 \leq n \leq 0.52$ (16). A, B, and n are found through calibration.

Two types of sensors are common: the hot wire and the hot film. As shown in Figure 9.31, the hot-wire sensor consists of a tungsten or platinum wire typically ranging 1–4 mm in length and 1.5–15 μm in diameter. The wire is supported between two rigid needles that protrude from a ceramic tube that houses the lead wires. A hot-film sensor usually consists of a thin (2 μm) platinum or gold film deposited onto a glass substrate and covered with a high thermal conductivity coating. The coating acts to electrically insulate the film and offers some mechanical protection. Hot wires are used in electrically nonconducting fluids, whereas hot films can be used in conducting fluids or wherever a more rugged sensor is needed.

Two anemometer bridge operating modes are possible: (1) constant current and (2) constant resistance. In constant-current operation, a fixed current is passed through the sensor to heat it. The sensor resistance, and thus its temperature, based on Equation 9.46, are permitted to vary with the rate of heat transfer between the sensor and its environment. Bridge-deflection voltage provides a measure of the cooling velocity. The more common mode of operation is constant resistance. In constant-resistance operation, the sensor resistance is maintained constant in measurement by using a differential feedback amplifier to sense small changes in bridge balance; that is, the circuit acts as a

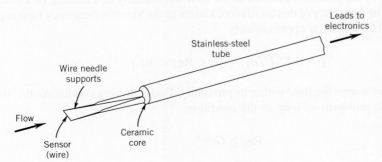

Figure 9.31   Schematic of a hot-wire probe.

closed loop controller using the bridge balance as the error signal. The feedback amplifier rapidly readjusts the bridge applied voltage, thereby adjusting the sensor current to bring the sensor back to its set-point resistance and corresponding temperature. The instantaneous power (I^2R_s) required to maintain constant temperature is equivalent to the instantaneous rate of heat transfer from the sensor ($\dot{Q}$), which varies with cooling velocity. In terms of the instantaneous applied bridge voltage, E, the velocity is found by the general correlation

$$E^2 = C + DU^n \tag{9.48}$$

where the values of C, D, and n are found by calibration under a fixed sensor and fluid temperature condition. An electronic or digital linearizing scheme is usually employed to condition the signal by performing the transformation

$$E_1 = K\left(\frac{E^2 - C}{D}\right)^{1/n} \tag{9.49}$$

such that the measured output from the linearizer, E_1, is

$$E_1 = KU \tag{9.50}$$

where K is found through a static calibration.

For mean velocity measurements, the thermal anemometer is a straightforward device to use. Multiple velocity components can be measured by using multiple sensors, each sensor aligned differently to the mean flow direction and operated by independent anemometer circuits (15, 17). Because it has a high-frequency response, fluctuating (dynamic) velocities can be measured. In highly turbulent flows with root-mean-square (rms) fluctuations of $\sqrt{u^2} \geq 0.1\overline{U}$, signal interpretation can become complicated, but it has been well investigated (17). Low-frequency fluid temperature fluctuations can be compensated for by placing resistor R_3 directly adjacent to the sensor and exposed to the flow. An extensive bibliography of thermal anemometry theory and signal interpretation can be found elsewhere (18).

In constant resistance, a hot-wire system can attain a frequency response that is flat up to 100,000 Hz, which makes it particularly useful in fluid mechanics turbulence research. However, less expensive and more rugged systems are commonly used for industrial flow monitoring, where a fast dynamic response is desirable. There is an upper frequency limit on a cylindrical sensor of diameter d that is brought about by the natural oscillation in the flow immediately downstream of a body. This vibrates the sensor. The frequency of this oscillation, known as the Strouhal frequency (and explained further in Section 10.6), occurs at approximately

$$f \approx 0.22[\overline{U}/d] \quad (10^2 < \mathrm{Re}_d < 10^7) \tag{9.51}$$

The heated sensor warms the fluid within its proximity. Under flowing conditions, this does not cause any measurable problems so long as the condition

$$\mathrm{Re}_d \geq Gr^{1/3} \tag{9.52}$$

is met where $\mathrm{Re}_d = \overline{U}d/v$, $Gr = d^3 g\beta(T_s - T_{\text{fluid}})/v^2$, and β is the coefficient of thermal expansion of the fluid. Equation 9.52 ensures that the inertial forces of the moving fluid dominate over the buoyant forces brought on by the heated sensor. This forms a lower velocity limit on the order of 0.6 m/s for using hot-wire sensors in air.

Doppler Anemometry

The Doppler effect describes the phenomenon experienced by an observer whereby the frequency of light or sound waves emitted from a source that is traveling away from or toward the observer is shifted from its original value and by an amount proportional to its speed. Most readers are familiar with the change in pitch of a train as heard by an observer as the train changes from approaching to receding. Any radiant energy source, such as a sound or light wave, demonstrates a Doppler effect, first recognized and modeled by Christian Johann Doppler (1803–1853). The observed shift in frequency, called the Doppler shift, is directly related to the speed of the emitter relative to the observer.

Doppler anemometry refers to a class of techniques that use the Doppler effect to measure the local velocity at a point in a moving fluid. The emission source is a stationary coherent narrow incident wave. Either acoustic waves or light waves are used. Small particles suspended in and moving with the fluid are used to generate the Doppler effect by scattering the incident waves. Here we discuss point and full-field velocity measurement methods. In Chapter 10, we discuss a flow rate measurement method using Doppler (ultrasonic) principles.

A laser Doppler anemometer (LDA) uses a laser beam as the emission source (19–22) to measure the time-dependent velocity at a point in the flow. As a moving particle suspended in the fluid passes through a laser beam, it scatters light in all directions. An observer viewing this encounter perceives the scattered light at a frequency, f_s:

$$f_s = f_i \pm f_D \qquad (9.53)$$

where f_i is the frequency of the incident laser beam and f_D is the Doppler shift frequency. Using visible light, an incident laser beam frequency is on the order of 10^{14} Hz. For most engineering applications, the velocities are such that the Doppler shift frequency, f_D, is on the order of 10^3 to 10^7 Hz. An operating mode that distinctly isolates the Doppler frequency and focuses on a small volume of flow is shown in Figure 9.32. In this dual-beam mode, a single laser beam is divided into two coherent beams using an optical beam splitter, which are then focused through a lens at a point in the flow. This focal point forms the measuring volume (sensor) of the instrument. Particles suspended

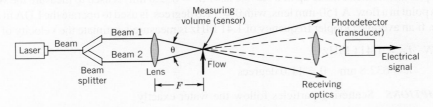

Figure 9.32 Laser Doppler anemometer, shown here in the dual-beam mode of operation.

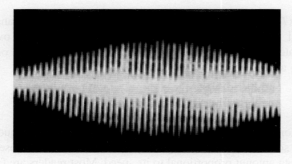

Figure 9.33 Oscilloscope trace of a photodiode output showing the Doppler frequency from a single particle moving through the measuring volume. (Photo courtesy of Richard S. Figliola.)

in and moving with the fluid scatter light as they pass through the beams. Within the measuring volume incident information from the two incident beams mix, a process known as optical heterodyne. The outcome of this mixing separates the incident frequency from the Doppler frequency into two distinct signal components.

The related technique of phase Doppler anemometry (24) is used to simultaneously measure particle sizes. It shares a similar optical setup but uses two or more photodetectors. Phase differences between the detected signals are proportional to the light-scattering particle size. Engineering applicationss include emissions monitoring, combustion, and pneumatic transport.

For the setup shown in Figure 9.32, the velocity is related directly to the Doppler shift by

$$U = \frac{\lambda}{2\sin\theta/2} f_D = d_f f_D \qquad (9.54)$$

The component of the velocity measured is that which is in the plane of and bisector to the crossing beams (21). By using beams of different color or polarization, different velocity components can be measured simultaneously. The output from the photodiode transducer is a current of a magnitude proportional to the square of the amplitude of the scattered light seen and of a frequency equal to f_D. This effect is seen as a Doppler "burst" shown in the trace of Figure 9.33. In a time-varying flow field, the Doppler shift from successive scatters will vary with time. The most common way to measure this time-dependent frequency variation is through a burst analyzer using either fast Fourier transform or autocorrelation methods. At very low light levels, photon correlation techniques are successful (20, 21). As a point measuring technique, the measuring volume needs to be translated about to map out the flow field.

Example 9.11

A laser Doppler anemometer made up of a He–Ne laser ($\lambda = 632.8$ nm) is used to measure the velocity of water at a point in a flow. A 150-mm lens, with $\theta = 11.0$ degrees, is used to operate the LDA in a dual-beam mode. If an average Doppler frequency of 1.41 MHz is measured, estimate the velocity of water.

KNOWN $\bar{f}_D = 1.411$ MHz

$\lambda = 632.8$ nm $\theta = 11.0$ degrees

ASSUMPTIONS Scattering particles follow the water exactly

FIND $\overline{U}$

SOLUTION   Applying Equation 9.54,

$$\overline{U} = \frac{1}{2\sin\theta/2}\overline{f}_D = \left(\frac{632.8 \times 10^{-9}\ \text{m}}{2\sin(11°/2)}\right)\left(1.411 \times 10^6\ \text{Hz}\right) = 4.655\ \text{m/s}$$

Particle Image Velocimetry

Particle image velocimetry (PIV) measures the full-field instantaneous velocities in a planar cross section of a flow. The technique tracks the time displacement of particles, which are assumed to follow the flow. Principle components for the technique are a coherent light source (laser beam), optics, a CCD camera, and dedicated signal interrogation software.

In a simple overview, the image of particles suspended in the flow are illuminated and recorded repetitively at very short intervals in time. Sequential image pairs are compared. The distance traveled by a particle during the period between images is a measure of its velocity. By repeatedly recording the particle field, the particle positions can be tracked and velocity as a function of space and time obtained.

In a typical layout, such as shown in Figure 9.34, a laser beam is passed through a cylindrical lens, which converts the beam into a two-dimensional (2D) sheet of light. This laser sheet is mechanically situated to illuminate an appropriate cross section of the flow field. The camera is positioned and focused to record the view of the illuminated field. The time interval between image pairs can be controlled either by using a pulsed-light laser synchronized with the camera shutter or by using a continuous light laser and a high speed motion camera. The acquired digital images are stored and processed by interrogation software, resulting in a full-field instantaneous velocity mapping of the flow.

The operating principle is based on particle displacement with time

$$\overrightarrow{U} = \Delta\overrightarrow{x}/\Delta t \tag{9.55}$$

where $\overrightarrow{U}$ is the instantaneous particle velocity vector based on its spatial position $\overrightarrow{x}(x, y, z, t)$. The camera records particle position into separate image frames. To obtain velocity data in a rapid manner, each image is divided into small areas, called interrogation areas. The corresponding interrogation areas between two images, I_1 and I_2, are cross-correlated with each other, on a

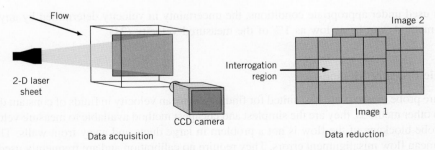

Figure 9.34 Basic layout of a digital particle image velocimeter.

pixel-by-pixel basis. A particular particle movement from position $\overrightarrow{x}_1$ to $\overrightarrow{x}_2$ shows up as a signal peak in the correlation $R_{12}(\Delta \overrightarrow{x})$, where

$$R_{12}(\Delta \overrightarrow{x}) = \int\!\!\int_A I_1(\Delta \overrightarrow{x}) I_2(\overrightarrow{x} + \Delta \overrightarrow{x}) d\overrightarrow{x} \tag{9.56}$$

identifying the common particle and allowing the estimate of the particle displacement $\Delta \overrightarrow{x}$. By repeating the cross-correlation between images for each interrogation area, a velocity vector map of the full area results.

A number of variations of this process have been developed but the concept remains the same. The technique works in gases or liquids. Three-dimensional information can be obtained by using two cameras. Particle size and properties should be chosen relative to the fluid and flow velocities expected so that the particles move with the fluid, but the particles must be large enough relative to camera pixel size to avoid peak-locking errors (23). The maximum flow speed measurable is limited by the interrogation area size. The resolution depends on laser flash width and separation time, flow velocity, camera recording time, and image magnification. By using a high magnification lens, called microscopic PIV, measurements with a very small length scale are possible. Raffel et al. (23) discuss the technique, its variations, and error estimates.

Selection of Velocity Measuring Methods

Selecting the best velocity measuring system for a particular application involves a number of factors that an engineer needs to weight accordingly:

1. Required spatial resolution
2. Required velocity range
3. Sensitivity to velocity changes only
4. Required need to quantify dynamic velocity
5. Acceptable probe blockage of flow
6. Ability to be used in hostile environments
7. Calibration requirements
8. Low cost and ease of use

When used under appropriate conditions, the uncertainty in velocity determined by any of the discussed methods can be as low as 1% of the measured velocity (24).

Pitot–Static Pressure Methods

The pressure probe methods are best suited for finding the mean velocity in fluids of constant density. Relative to other methods, they are the simplest and cheapest method available to measure velocity at a point. Probe blockage of the flow is not a problem in large ducts and away from walls. They are subject to mean flow misalignment errors. They require no calibration and are frequently used in the field and laboratory alike.

Thermal Anemometer

Thermal anemometers are best suited for use in very clean fluids of constant temperature and density. They are well suited for measuring dynamic velocities with very high resolution. However, signal interpretation in strongly dynamic flows can be complicated (17, 25). Hot-film sensors are less fragile and less susceptible to contamination than hot-wire sensors. Probe blockage is not significant in large ducts and away from walls. The cylindrical sensors are 180 degrees directionally ambiguous (i.e., flows from the left or right directions give the same output signal), an important factor in flows that may contain flow reversal regions. An industrial-grade system can be built rather inexpensively. The thermal anemometer is usually calibrated against pressure probes.

Doppler Anemometry

The laser Doppler anemometer (LDA) is a relatively expensive point velocity measuring technique for transparent fluids that is well suited to hostile, combusting, or dynamic (unsteady, pulsatile, or highly turbulent) flow environments. The method offers good frequency response, small spatial and temporal resolutions, no probe blockage, and simple signal interpretation, but requires optical access and the presence of scattering particles (26, 27).

Particle Image Velocimetry

Particle image velocimetry (PIV) is a relatively expensive and technically advanced full-field velocity measuring technique that can be used for most types of flows, including hostile and combusting flows. There is no probe blockage of the flow, but it requires optical access and the presence of scattering particles. The method provides an instantaneous snapshot of the flow, providing excellent views of flow structures. Time-dependent quantification of such dynamic flows is possible, but frequency bandwidth is limited by camera frame rate and spatial resolution. Careful planning is required in particle selection to ensure that the particle velocities represent the fluid velocity exactly and that particle size is properly matched to pixel resolution.

9.11   SUMMARY

Several reference pressure instruments have been presented that form the working standards for pressure transducer calibration. Pressure transducers convert sensed pressure into an output form that is readily quantifiable. These transducers come in many forms but tend to operate on hydrostatic principles, expansion techniques, or force-displacement methods.

In moving fluids, special care must be taken when measuring pressure to delineate between static and total pressure. Methods for the separate measurement of static and total pressure or for the measurement of the dynamic pressure are readily available and well documented. But improper measuring technique causes errors, lowering the total pressure or increasing the static pressure.

Measuring the local velocity within a moving fluid can be accomplished in a number of ways. Selecting the proper tool requires assessment of the need: mean or fluctuating velocity, point or full field measurement, optical access or opaque boundaries or fluid. Specifically, dynamic pressure, thermal anemometry, Doppler anemometry, and particle velocimetry methods have been presented. As discussed, each method offers advantages over the others, and the best technique must be carefully weighed against the needs and constraints of a particular application.

REFERENCES

1. American Society of Mechanical Engineers (ASME), *PTC 19.2–2010: Pressure Measurement*, ASME International, New York, 2010.

2. Brombacher, W. G., D. P. Johnson, and J. L. Cross, Mercury Barometers and Manometers, *National Bureau Standards Monograph 8*, 1960.

3. McLeod, H. G., Vacuum Gauge, *Philosophical Magazine*, *48*, 110, 1874. Reprinted in *History of Vacuum Science and Technology*, eds. T. E. Madey and W. C. Brown, American Vacuum Society, New York, 1984, pp. 102–105.

4. M. Hetenyi, ed., *Handbook of Experimental Stress Analysis*, Wiley, New York, 1950.

5. Way, S., Bending of circular plates with large deflection, *Transactions of the ASME* 56, 1934.

6. Instrumen Society of America (ISA), *A Guide for the Dynamic Calibration of Pressure Transducers*, ISA-37.16.01-2002, Instrument Society of America, 2002.

7. Franklin, R. E., and J. M. Wallace, Absolute measurements of static-hole error using flush transducers, *Journal of Fluid Mechanics* 42, 1970.

8. Rayle, R. E., Influence of orifice geometry on static pressure measurements, ASME Paper No. 59-A-234, 1959.

9. Munson, B., D. Young, T. Okishi, *Fundamentals of Fluid Mechanics*, 6th ed., Wiley, New York, 2009.

10. Migliavacca, F., et al., Multiscale modelling in biofluidynamics: Application to reconstructive paediatric cardiac surgery, *Journal of Biomechanics*, 39, 6, 2006.

11. Delio, G., G. Schwent, and R. Cesaro, Transient behavior of lumped-constant systems for sensing gas pressures, National Advisory Council on Aeronautics (NACA) TN-1988, 1949.

12. Doebelin, E. O., *Measurement Systems: Application and Design*, 5th ed., McGraw-Hill Science/Engineering/Math, New York, 2003.

13. Hougen, J., O. Martin, and R. Walsh, Dynamics of pneumatic transmission lines, *Control Engineering*, September 1963.

14. King, L. V., On the convection from small cylinders in a stream of fluid: Determination of the convection constants of small platinum wires with application to hot-wire anemometry, *Proceedings of the Royal Society, London* 90, 1914.

15. Hinze, J. O., *Turbulence*, McGraw-Hill, New York, 1959.

16. Collis, D. C., and M. J. Williams, Two-dimensional convection from heated wires at low Reynolds numbers, *Journal of Fluid Mechanics* 6, 1959.

17. Rodi, W., A new method for analyzing hot-wire signals in a highly turbulent flow and its evaluation in a round jet, *DISA Information*, Dantek Electronics, Denmark, 1975. See also Bruun, H. H., Interpretation of X-wire signals, *DISA Information*, Dantek Electronics, Denmark, 1975.

18. Freymuth, P., A bibliography of thermal anemometry, *TSI Quarterly* 4, 1978.

19. Yeh, Y., and H. Cummins, Localized fluid flow measurement with a He-Ne laser spectrometer, *Applied Physics Letters* 4, 1964.

20. Photon correlation and light beating spectroscopy, H. Z. Cummins and E. R. Pike, eds., *Proceedings of the NATO ASI*, Plenum, New York, 1973.

21. Durst, F., A. Melling, and J. H. Whitelaw, *Principles and Practice of Laser Doppler Anemometry*, Academic Press, New York, 1976.

22. R. J. Goldstein, ed., *Fluid Mechanics Measurements*, 2nd ed., CRC Press, New York, 1996.

23. Raffel, M., Willert, C. E., Wereiey, S. T., Kompenhans, J., *Particle Image Velocimetry*, 2nd ed, Springer, Heidelberg, 2007.

24. Goldstein, R. J., and D. K., Kried, Measurement of laminar flow development in a square duct using a laser doppler flowmeter, *Journal of Applied Mechanics* 34, 1967.

25. Yavuzkurt, S., A guide to uncertainty analysis of hot-wire data, *Transactions of the ASME, Journal of Fluids Engineering* 106, 1984.

26. Maxwell, B. R., and R. G. Seaholtz, Velocity lag of solid particles in oscillating gases and in gases passing through normal shock waves, National Aeronautics and Space Administration TN-D-7490.

27. Dring, R. P., and Suo, M. Particle trajectories in swirling flows, *Transactions of the ASME, Journal of Fluids Engineering* 104, 1982.

28. Acoustics—measurement of room acoustic parameters, *ISO 3382*, International Standards Organization, Geneva, Switzerland, 2012.

29. Mechanical vibration—Evaluation of machine vibration by measurements on non-rotating parts, ISO 10816, International Organization for Standardization, Geneva.

30. Electroacoustics—Sound level meters—Part 1: Specifications, IEC 61672, International Electrotechnical Commission (IEC). Geneva, Switzerland, September 2013.

NOMENCLATURE

d	diameter (l)		z	altitude (l)
e	elemental errors		C	capacitance (F); compliance ($l^4 t^2 m^{-1}$)
d_f	fringe spacing (l)		E	voltage (V)
f_D	doppler frequency (Hz)		E_m	bulk modulus of elasticity ($m^{-1} l t^{-2}$)
h	depth (l)		Gr	Grashof number
h_0	reference depth (l)		H	manometer deflection height (l)
k	ratio of specific heats		K	static sensitivity
p	pressure ($m^{-1} l t^{-2}$)		K_q	charge sensitivity ($m^{-1} l t^{-2}$)
p_a	applied pressure ($m^{-1} l t^{-2}$)		K_E	voltage sensitivity ($\mathrm{V} m^{-1} t^{-2}$)
p_{abs}	absolute pressure ($m^{-1} l t^{-2}$)		L	inductance (H); inertance (ml^{-4})
p_e	relative static pressure error ($m^{-1} l t^{-2}$)		M	Mach number
p_i	indicated pressure ($m^{-1} l t^{-2}$)		R	resistance (Ω)
p_m	measured pressure ($m^{-1} l t^{-2}$)		R	fluid resistance[4]
p_t	total or stagnation pressure ($m^{-1} l t^{-2}$)		Re$_d$	Reynolds number, Re = Vd/v
p_v	dynamic pressure ($m^{-1} l t^{-2}$)		S	specific gravity
q	charge (C)		U	velocity ($l t^{-1}$)
r	radius (l)		$\forall$	volume (l^3)
t	thickness (l)		γ	specific weight ($m l^{-2} t^{-2}$)
y	displacement (l)		ζ	damping ratio
ρ	density ($m l^{-3}$)		ε	dielectric constant
ι	time constant (t)		λ	wavelength (l)
ϕ	latitude		μ	absolute viscosity ($m t^{-1} l^{-1}$)
ω	frequency (t^{-1})		v	kinematic viscosity (l^2/t)
ω_n	natural frequency (t^{-1})		υ_p	Poisson ratio

---

[4] See Section 9.7 for definition and dimensions

PROBLEMS

9.1 State the following pressures to units of N/m^2:

   a. 500 mm Hg abs
   b. 720 torr
   c. 14.696 lb/in^2 abs
   d. 1 bar

9.2 State the following pressures as absolute pressure in pascals relative to one standard atmosphere:

   a. −1.0132 kPa
   b. 760 mm Hg
   c. 29.92 in Hg absolute
   d. 5 lb/in^2

9.3 A water-filled manometer is used to measure the pressure in an air-filled tank. One leg of the manometer is open to atmosphere. For a measured manometer deflection of 250 cm water, determine the tank static pressure. Barometric pressure is 101.3 kPa abs.

9.4 A deadweight tester is used to provide a standard reference pressure for the calibration of a pressure transducer. A combination of 25.3 kg_f of 7.62-cm-diameter stainless steel disks is needed to balance the tester piston against its internal pressure. For an effective piston area of 5.065 cm^2 and a piston weight of 5.35 kg_f, determine the standard reference pressure in bars, N/m^2, and Pa abs. Barometric pressure is 770 mm Hg abs, elevation is 20 m, and latitude is 42 degrees.

9.5 The pressure differential across an orifice plate meter is measured using an inclined tube manometer with one pressure attached to each end of the manometer. Under no flow conditions, the manometer deflection is zero. For a set flow rate, the manometer deflects 10 cm H_2O. With a manometer inclination of 30 degrees relative to horizontal, determine the pressure differential across the orifice meter.

9.6 Show that the static sensitivity of an inclined tube manometer is a factor of $1/\sin\theta$ higher than for a U-tube manometer.

9.7 Determine the static sensitivity of an inclined tube manometer set at an angle of 30 degrees. The manometer tube measures the pressure difference of air using mercury as the manometer fluid.

9.8 Show that the instrument (systematic) uncertainty in an inclined tube manometer in Example 9.3 increases to 6.8 N/m^2 as θ goes to 90 degrees.

9.9 A strain gauge, diaphragm pressure transducer (accuracy: <0.1% reading) is subjected to a pressure differential of 10 kPa. If the output is measured using a voltmeter having a resolution of 10 mV and accuracy of better than 0.1% of the reading, estimate the uncertainty in pressure at the design stage. How does this change at 100 and 1,000 kPa?

9.10 Select a practical fluid to use in a manometer to measure pressures up to 69 kPa of an inert gas ($\gamma = 10.4\,N/m^3$), if water ($\gamma = 9,800\,N/m^3$), oil ($S = 0.82$), and mercury ($S = 13.57$) are available. Discuss the rationale for your choice(s).

9.11 An air pressure over the 200- to 400-N/m^2 range is to be measured relative to atmosphere using a U-tube manometer with mercury ($S = 13.57$). Manometer resolution is 1 mm with a zero error uncertainty of 0.5 mm. Estimate the design-stage uncertainty in gauge pressure based on the manometer indication at 20 °C. Would an inclined manometer ($\theta = 30$ degrees) be a better choice if the inclination could be set to within 0.5 degree?

9.12 Calculate the design-stage uncertainty in estimating a nominal pressure of 10,000 N/m^2 using an inclined manometer (resolution: 1 mm; zero error uncertainty: 0.5 mm) with water at 20 °C for inclination angles of 10 to 90 degrees (using 10-degree increments). The inclination angle can be set to within 1 degree.

9.13 The pressure drop across a valve through which air flows is expected to be 10 kPa. If this differential were applied to the two legs of a U-tube manometer filled with mercury, estimate the manometer deflection. What is the deflection if a 30-degree inclined tube manometer were used? $S_{Hg} = 13.6$.

9.14 Estimate the sensitivity (pF/mm) of a capacitance transducer, such as in Figure 9.13, using water as the dielectric medium. The transducer plates have an overlap area of $8.1 \pm 0.01\,mm^2$ with an average gap separation of 1 mm.

9.15 A diaphragm pressure transducer is calibrated against a pressure standard that has been certified by the National Institute of Standards and Technology (NIST) (uncertainty: $b = 0.25$ psi). Both the standard and pressure transducer output a voltage signal, which is to be measured by a voltmeter ($b = 5\,\mu V$; resolution: 1 μV). A calibration curve fit yields: $p = 0.564 + 24.0E \pm 0.5$ psi (68%) based on 35 points over the 0- to 100-psi range. When the transducer is installed for its intended purpose, installation effects are estimated to affect its reading by 0.25 psi (68%). Estimate the uncertainty associated with a pressure measurement using the installed transducer-voltmeter system.

9.16 A diaphragm pressure transducer is coupled with a water-cooled sensor for high-temperature environments. Its manufacturer claims that it has a rise time of 10 ms, a ringing frequency of 200 Hz, and damping ratio of 0.8. Describe a test plan to verify the manufacturer's specifications. Would this transducer have a suitable frequency response to measure the pressure variations in a typical four-cylinder engine? Show your reasoning.

9.17 Find the natural frequency of a 1-mm-thick, 6-mm-diameter steel diaphragm to be used for high-frequency pressure measurements near atmospheric pressure. What would be the maximum operating pressure difference that could be applied? What is the effect of a larger diameter for this application?

9.18 A 2.0 mm thick circular steel diaphragm ($E_m = 200$ GPa, $v_p = 0.32$, $\rho = 4.49$ kg/m^3) within a high-pressure transducer is 20.0 cm in diameter. Determine the maximum deflection limit. Find its natural frequency. Determine its maximum differential pressure limit.

9.19 Estimate the differential pressure limit for a 0.5-mm-thick, 25-mm-diameter steel diaphragm pressure transducer. $v_p = 0.32$, $E_m = 200$ GPa.

9.20 The pressure fluctuations in a pipe filled with air at 20 °C at about 1 atm is to be measured using a static wall tap, rigid connecting tubing, and a diaphragm pressure transducer. The transducer has a natural frequency of 100,000 Hz. For a tap and tubing diameter of 3.5 mm, a tube length of 0.25 m, and a transducer dead volume of 1,600 mm^3, estimate the resonance frequency of the system. What is the maximum frequency that this system can measure with no more than a 10% dynamic error? Plot the frequency response of the system.

9.21 A pressure transducer has a range of ± 100 mm Hg, a sensitivity of 50 μV/mm Hg/V (excitation) and an allowable excitation voltage from 5 to 10 VDC. Estimate the required gain needed to maximize the values of the recorded signal using a ± 5 V, 16-bit data acquisition system.

9.22 What is the sensitivity of a pitot-static tube pressure relative to velocity?

9.23 A pitot-static pressure probe inserted within a large duct indicates a differential pressure of 20.3 cm H$_2$O. What is the velocity measured?

9.24 A tall pitot-static tube is mounted through and 1.5-m above the roof of a performance car such that it senses the freestream flow. Estimate the static, stagnation, and dynamic pressure sensed at 325 kph if (a) the car is moving on a long, straight section of road and (b) the car is stationary within an open-circuit wind tunnel in which the flow is blown over the car.

9.25 The pressure transmission line response equation of Equation 9.23 can also be derived by considering the forces acting on a fluid element within the connecting tubing (12). Develop a model for the system measured pressure response based on an applied pressure force, shear resistance force, and restoring compliance related force through Newton's momentum (second law) principles. Refer to Figure 9.35.

9.26 A Kiel probe and a static pressure tube are used to measure air pressures separately at a position within a duct. The measured stagnation pressure is 1.08 bar, and the static pressure is 1.02 bar. Air stream stagnation temperature is measured to be 45 °C. Find the measured duct velocity.

9.27 A pitot-static tube is used to measure the velocity in an air stream. Air speed is expected to vary from 0 to 90 m/s. The pitot-static tube is connected to a differential piezoelectric transducer that has a built-in sensitivity of 20 mV/mm Hg. If the signal is to be recorded on a 0–10 V, 16-bit data acquisition system, suggest a signal conditioning gain that makes best use of the data acquisition range capability.

9.28 A 1.5-mm i.d., 1-m-long catheter filled with saline is attached to a diaphragm pressure transducer. The system has a compliance of 2×10^{-4} mL/mm Hg. Estimate the natural frequency and damping ratio of the system. Use water for your calculations.

9.29 Compare the inertance of water in a 0.2-m-long tube to that of a 1-m-long tube each of 25-mm and then of 12.5-mm diameter.

9.30 The output from a resting healthy human adult heart is about 5 L/min. We can estimate that the mean systemic pressure is 95 mm Hg with a mean atrial pressure of 4 mm Hg. The mean pulmonary

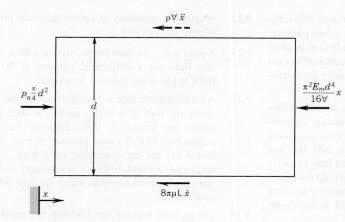

Figure 9.35 Freebody diagram on a fluid element (x-direction) for Problem 9.25.

pressure is 15 mm Hg with an atrial pressure of 2 mm Hg. Compare the vascular resistance of the left (systemic) circulation to that of the right (pulmonary) circulation. Assume that the flow remains laminar.

9.31 The left ventricle of a healthy man ejects 80 mL of blood into the aorta with each heartbeat. The pressure of the corresponding circulation varies between 120 and 80 mm Hg during each beat, known as the blood pressure, for a change of 40 mm Hg. Estimate the average compliance of the left circulation.

9.32 A pressure drop of 213 Pa is measured between two points along a vascular tube by intravenous catheterization. Flow rate is measured to be 16.7 cm³/s. Using lumped parameter methods, find the resistance in Woods units (1 WU = 1 mm Hg/Lpm). Woods units are the common unit used in cardiovascular measurements.

9.33 Wall pressure taps (e.g., Figs. 9.19 and 9.21) are often used to sense surface pressure and are connected to transducers by connecting tubing. Two race engineers discuss the preferred diameter of the tubing to measure pressure changes on the surface of a car as it moves along a track. The tubing length may be up to 2 m. Engineer A suggests very small 2-mm-diameter tubing to reduce air volume so as to increase response time. Engineer B disagrees and suggests 5-mm tubing to balance air friction with air volume to increase response time. Offer your opinion and its basis. (Hint: Look at length-to-diameter effects.)

9.34 Show that Equation 9.23 can be reduced to the fundamental form of Equation 9.22 in terms of system inertance, resistance, and compliance.

9.35 Apply a circuit analysis to an RLC analog of Figure 9.22 showing the steps to achieve Equations 9.21 through 9.23. Use the direct analogy equating E_a, E_m with p_a, p_m, respectively.

9.36 Determine the resolution of a manometer required to measure the velocity of air from 5 to 50 m/s using a pitot-static tube and a manometer fluid of mercury ($S = 13.57$) to achieve a zero-order uncertainty of 5% and 1%.

9.37 A long cylinder is placed into a wind tunnel and aligned perpendicular to an oncoming freestream. Static wall pressure taps are located circumferentially about the centerline of the cylinder at 45-degree increments with 0 degrees at the impact (stagnation) position. Each tap is connected to a separate manometer referenced to atmosphere. A pitot-static tube indicates an upstream dynamic pressure of 20.3 cm H_2O, which is used to determine the freestream velocity. The following static pressures are measured:

Tap (degrees)	p (cm H_2O)	Tap (degrees)	p (cm H_2O)
0	0.0	135	5.0
30	20.3	180	5.2
90	81.3		

Compute the local velocities around the cylinder if the total pressure in the flow remains constant. $p_{atm} = 101.3$ kPa abs, $T_{atm} = 16\,°C$.

9.38 The following octave band center frequency sound level measurements were taken during a song played at an indoors heavy metal rock concert. Estimate the average sound level after applying C-weighting.

Octave (Hz)	31.25	62.5	125	250	500	1,000	2,000	4,000	8,000	16,000
SPL (dB)	102	103	100	103	105	105	110	105	100	92

9.39 What is the sound pressure in pascals if the average sound level is reported as 70 dB when referenced to the threshold of hearing?

9.40 What is the sound pressure in pascals if the measured sound level is reported as 60.0 dB(A) at the octave band of 62.5 Hz?

9.41 Microphone sensitivity is rated at 1 kHz in terms of volts/pascal or dB = $20 \log(V/Pa)$ so that 1 V/Pa = -0 dB. Determine the output level from a microphone having a -34 dB sensitivity measuring a 1 kHz tone at 120 dB.

9.42 A 6-mm-diameter pitot-static tube is used as a working standard to calibrate a hot-wire anemometer in 20 °C air. If dynamic pressure is measured using a water-filled micromanometer, determine the smallest manometer deflection for which the pitot–static tube can be considered as accurate without correction for viscous effects.

9.43 For the thermal anemometer in Figures 9.31 and 9.32, determine the decade resistance setting required to set a platinum sensor at 40 °C above ambient if the sensor ambient resistance is $110 \, \Omega$ and $R_3 = 500 \, \Omega$ and $R_4 = 500 \, \Omega$, $\alpha = 0.00395 \, °C^{-1}$.

9.44 Determine the static sensitivity of the output from a constant resistance (hot-wire) anemometer as a function of velocity. Is it more sensitive at high velocities, or at low?

9.45 A laser Doppler anemometer setup in a dual-beam mode uses a 600-mm focal length lens ($\theta = 5.5$ degrees) and an argon-ion laser ($\lambda = 514.4$ nm). Compute the Doppler shift frequency, f_D, expected at 1, 10, and 100 m/s. Repeat for a 300-mm lens ($\theta = 7.3$ degrees).

9.46 A set of 5,000 measurements of velocity at a point in a flow using a dual-beam LDA gives the following results:

$$\overline{U} = 21.37 \text{ m/s} \quad s_U = 0.43 \text{ m/s}$$

For the system, $\theta = 6$ degrees and $\lambda = 623.8$ nm, what is the mean Doppler frequency measured?

9.47 Aircraft airspeed is measured using a pitot (impact) tube and a static port, each mounted somewhere on the external wing or airframe, and connected by tubes across an airspeed indicator. The airspeed indicator is a differential pressure transducer device calibrated to read speed. Explain why rain impacting the opening of the pitot tube during flight is unlikely to flood the airspeed indicator.

Chapter **10**

Flow Measurements

10.1 INTRODUCTION

The rate at which a fluid moves through a conduit is measured in terms of flow rate. This chapter discusses some of the most common and accepted methods for measuring flow rate. Flow rate can be expressed in terms of a flow volume per unit time, known as the *volume flow rate*, or as a mass flow per unit time, known as the *mass flow rate*. Flow rate devices, called flow meters, are used to quantify, totalize, or monitor flowing processes. Type, accuracy, size, pressure drop, pressure losses, capital and operating costs, and compatibility with the fluid are important engineering design considerations for choosing a flow metering device. All methods have both desirable and undesirable features that necessitate compromise in the selection of the best method for the particular application, and many such considerations are discussed in this chapter. Inherent uncertainties in fluid properties, such as density, viscosity, or specific heat, can affect the accuracy of a flow measurement. However, some techniques, such as those incorporated into coriolis mass flow meters, do not require knowledge of exact fluid properties, allowing for highly accurate mass flow measurements in demanding engineering applications. The chapter objective is to present both an overview of basic flow metering techniques for proper meter selection, as well as those design considerations important in the integration of a flow rate device with the process system it will meter.

Upon completion of this chapter, the reader will be able to

- Relate velocity distribution to flow rate within conduits
- Use engineering test standards to select and specify size, specify installation considerations, and use common obstruction meters
- Understand the differences between volume flow rate and mass flow rate and the means required to measure each
- Describe the physical principles employed in various commercially available types of flow meters and the engineering terms common to their use
- Describe calibration methods for flow meters

10.2 HISTORICAL BACKGROUND

The importance to engineered systems give flow measurement methods their rich history. The earliest available accounts of flow metering were recorded by Hero of Alexandria (ca. 150 B.C.) who proposed a scheme to regulate water flow using a siphon pipe attached to a constant head reservoir.

442

The early Romans developed elaborate water systems to supply public baths and private homes. In fact, Sextus Frontinius (A.D. 40–103), commissioner of water works for Rome, authored a treatise on design methods for urban water distribution systems. Evidence suggests that Roman designers understood correlation between volume flow rate and pipe flow area. Weirs were used to regulate bulk flow through aqueducts, and the cross-sectional area of terracotta pipe was used to regulate fresh running water supplies to individual buildings.

After a number of experiments conducted using olive oil and water, Leonardo da Vinci (1452–1519) first formally proposed the modern continuity principle: that duct area and fluid velocity were related to flow rate. However, most of his writings were lost until centuries later, and Benedetto Castelli (ca. 1577–1644), a student of Galileo, has been credited in some texts with developing the same steady, incompressible continuity concepts in his day. Isaac Newton (1642–1727), Daniel Bernoulli (1700–1782), and Leonhard Euler (1707–1783) built the mathematical and physical bases on which modern flow meters would later be developed. By the nineteenth century, the concepts of continuity, energy, and momentum were sufficiently understood for practical exploitation. Relations between flow rate and pressure losses were developed that would permit the tabulation of the hydraulic coefficients necessary for the quantitative engineering design of many modern flow meters.

10.3   FLOW RATE CONCEPTS

The flow rate through a conduit, be it a pipeline, duct, or channel, depends on fluid density, average fluid velocity, and conduit cross-sectional area. Consider fluid flow through a circular pipe of radius r_1 and having a velocity profile at some axial pipe cross section given by $u(r, \theta)$. The mass flow rate depends on the average mass flux, $\rho u(r, \theta)$ flowing through a cross-sectional area. Then the mass flow rate is given by

$$\dot{m} = \iint\limits_{A} \rho u(r, \theta) dA = \overline{\rho U} A \qquad (10.1)$$

The pipe cross-sectional area is $A = \pi r_1^2$. To directly measure mass flow rate, a device must be sensitive in some way to the area-averaged mass flux, $\overline{\rho U}$, or to the fluid mass passing through the device per unit time. Mass flow rate has the dimensions of mass per unit of time (e.g., units of kg/s, lb_m/s, etc.).

The volume flow rate depends only on the area-averaged velocity over a cross section of flow as given by

$$Q = \iint\limits_{A} u \, dA = \overline{U} A \qquad (10.2)$$

So to directly measure volume flow rate requires a device that is in some way sensitive either to the average velocity, $\overline{U}$, or to the fluid volume passing through it per unit time. Volume flow rate has dimensions of volume per unit time (e.g., units of m^3/s, ft^3/s).

The difference between Equations 10.1 and 10.2 is quite significant in that either requires a very different approach to its measurement: one sensitive to the product of density and velocity or to mass rate and the other sensitive only to the average velocity or to volume rate. In the simplest case, where density is a constant, the mass flow rate can be inferred by multiplying the measured volume flow rate by the density. But in the metering of many fluids, this assumption may not be good enough to achieve necessary accuracy. This can be because the density changes or may not be well known, such as in the transport of polymers or petrochemicals, or because small errors in the assumed density accumulate into large errors, such as in the transport of millions of cubic meters of product per day.

The flow character can affect the accuracy of a flow meter. The flow through a pipe or duct can be characterized as being laminar, turbulent, or a transition between the two. Flow character is determined through the engineering parameter known as the Reynolds number, defined by

$$Re_{d_1} = \frac{\overline{U}d_1}{v} = \frac{4Q}{\pi d_1 v} \tag{10.3}$$

where v is the fluid kinematic viscosity and d_1 is the diameter for circular pipes. In pipes, the flow is laminar when $Re_{d_1} < 2000$ and turbulent at higher Reynolds number. The Reynolds number is a necessary parameter in estimating flow rate when using most of the types of flow meters discussed. In estimating the Reynolds number in noncircular conduits, the hydraulic diameter, $d_H = 4r_H$, is used in place of diameter d_1, where r_H is the wetted conduit area divided by its wetted perimeter.

10.4 VOLUME FLOW RATE THROUGH VELOCITY DETERMINATION

Volume flow rate can be determined with direct knowledge of the velocity profile as indicated by Equation 10.2. This requires measuring the velocity at multiple points along a cross section of a conduit to estimate the velocity profile. For highest accuracy, several traverses should be made at differing circumferential locations to account for flow nonsymmetry. Methods for determining the velocity at a point include any of those previously discussed in Chapter 9. Because it is a tedious method, this procedure is most often used for the one-time verification or calibration of system flow rates. For example, the procedure is often used in ventilation system setup and problem diagnosis, where the installation of an inline flow meter is not necessary because operation does not require continuous monitoring.

When using this technique in circular pipes, a number of discrete measuring positions n are chosen along each of m flow cross sections spaced at $360/m$ degrees apart, as shown in Figure 10.1.

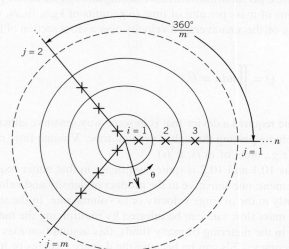

Figure 10.1 Location of n measurements along m radial lines in a pipe.

A velocity probe is traversed along each flow cross section, with readings taken at each of several measurement positions. There are options in selecting the measuring positions for different-shaped ducts, and such details are specified in available engineering test standards (1,2,4,15). The simplest method is to divide the flow area into smaller equal areas, making measurements at the centroid of each small area and assigning the measured velocity to that area. Regardless of the option selected, the average flow rate is estimated along each cross section traversed using Equation 10.2 and the pooled mean of the flow rates for the m cross sections calculated to yield the best estimate of the duct flow rate. Uncertainty is assessed both by repeating the measurements and by analyzing for spatial variation effects. Example 10.1 illustrates this method for estimating volume flow rate.

Example 10.1

A steady flow of air at 20 °C passes through a 25.4-cm inside diameter (i.d.) circular pipe. A velocity-measuring probe is traversed along three cross-sectional lines ($j = 1, 2, 3$) of the pipe, and measurements are made at four radial positions ($i = 1, 2, 3, 4$) along each traverse line, such that $m = 3$ and $n = 4$. The locations for each measurement are selected at the centroids of equally spaced areal increments as indicated below (1,3). Determine the volume flow rate in the pipe.

Radial Location, i	r/r_1	U_{ij} (m/s) Line 1 ($j = 1$)	Line 2 ($j = 2$)	Line 3 ($j = 3$)
1	0.3536	8.71	8.62	8.78
2	0.6124	6.26	6.31	6.20
3	0.7906	3.69	3.74	3.79
4	0.9354	1.24	1.20	1.28

KNOWN $U_{ij}(r/r_1)$ for $i = 1, 2, 3, 4$; $j = 1, 2, 3$

$$d_1 = 25.4 \text{ cm} \left(A = \pi d_1^2/4 = 0.05067 \text{ m}^2\right)$$

ASSUMPTIONS   Constant and steady pipe flow during all measurements; incompressible flow

FIND   Volume flow rate, Q

SOLUTION   The flow rate is found by integrating the velocity profile across the duct along each line and subsequently averaging the three values. For discrete velocity data, Equation 10.2 is written along each line, $j = 1, 2, 3 \ldots m$ as

$$Q_j = 2\pi \int_0^{r_1} Ur\,dr \approx 2\pi \sum_{i=1}^{4} U_{ij} r \Delta r$$

where Δr is the radial distance separating each position of measurement. This can be further simplified since the velocities are located at positions that make up the centroids of equal areas:

$$Q_j = \frac{A}{4} \sum_{i=1}^{4} U_{ij}$$

Then, the mean flow rate along each line of traverse is

$$Q_1 = 0.252 \text{ m}^3/\text{s} \quad Q_2 = 0.252 \text{ m}^3/\text{s} \quad Q_3 = 0.254 \text{ m}^3/\text{s}$$

The average pipe flow rate $\overline{Q}$ is the pooled mean of the individual flow rates (repeated tests)

$$\overline{Q} = \langle Q \rangle = \frac{1}{3} \sum_{j=1}^{3} Q_j = 0.253 \text{ m}^3/\text{s}$$

Example 10.2

Referring to Example 10.1, determine a value of the systematic standard uncertainty in mean flow rate due to the measured spatial variation.

KNOWN   Data over three ($m = 3$) traverse sections.

SOLUTION   The systematic standard uncertainty in mean flow rate due to error introduced by the measured spatial variations is estimated by

$$b_{\overline{Q}} = \frac{s_{\langle Q \rangle}}{\sqrt{m}} = \frac{\sqrt{\sum_{j=1}^{3} (Q_j - \langle Q \rangle)^2 / 2}}{\sqrt{3}} = \frac{0.0012}{\sqrt{3}} = 0.0007 \text{ m}^3/\text{s} \quad \text{with} \quad \nu = 2$$

COMMENT   Here we used statistical information to estimate a systematic uncertainty; it estimates the possible offset in $\overline{Q}$ and is about 0.3% of the mean flow rate.

10.5   PRESSURE DIFFERENTIAL METERS

The operating principle of a pressure differential flow rate meter is based on the relationship between volume flow rate and the pressure difference $\Delta p = p_1 - p_2$, between two defined locations along the flow path,

$$Q \propto (p_1 - p_2)^n \tag{10.4}$$

where $n = 1$ for laminar flow occurring between the pressure measurement locations and $n = 1/2$ for fully turbulent flow. Under steady flow conditions, an intentional reduction in flow area between locations 1 and 2 causes a measurable local pressure drop across this flow path. This reduced flow area leads to a concurrent local increase in velocity due to flow continuity (conservation of mass) principles. The pressure change is in part due to the so-called Bernoulli effect, the inverse relationship between local velocity and pressure, but is also due to flow energy losses. Pressure differential flow rate meters that use area reduction methods are commonly called *obstruction meters*.

Obstruction Meters

Three common obstruction meters are the *orifice plate*, the *venturi*, and the *flow nozzle*. Flow area profiles of each are shown in Figure 10.2 and detailed construction dimensions can be found in test standards (1,3,4). These meters are usually inserted in-line with a pipe, such as between pipe flanges. This class of meters as a whole operates using similar physical reasoning to relate volume flow rate to pressure drop. Referring to Figure 10.3, consider an energy balance between two control surfaces for an incompressible fluid flow through the arbitrary control volume shown. Under the assumptions of incompressible, steady and one-dimensional flow with no external energy transfer, the energy equation is

$$\frac{p_1}{\gamma} + \frac{\overline{U}_1^2}{2g} = \frac{p_2}{\gamma} + \frac{\overline{U}_2^2}{2g} + h_{L_{1-2}} \tag{10.5}$$

where $h_{L_{1-2}}$ denotes the head losses occurring between control surfaces 1 and 2.

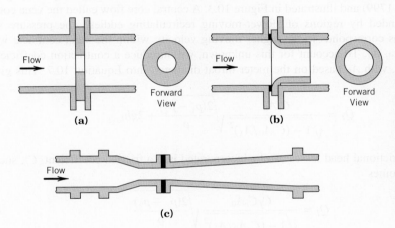

(a)   **(b)**

Figure 10.2 Flow area profiles of common obstruction meters.
(a) Square-edged orifice plate meter. **(b)** American Society of Mechanical Engineers (ASME) long radius nozzle. **(c)** ASME Herschel venturi meter.

(c)

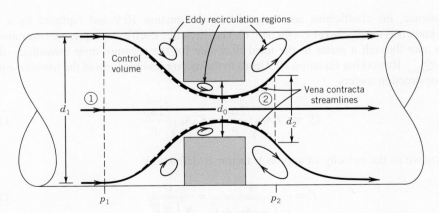

Figure 10.3 Control volume concept as applied between two streamlines for flow through an obstruction meter.

For incompressible flows, conservation of mass between cross-sectional areas 1 and 2 gives

$$\overline{U}_1 = \overline{U}_2 \frac{A_2}{A_1} \tag{10.6}$$

Then, substituting Equation 10.6 into Equation 10.5 and rearranging yields the incompressible volume flow rate

$$Q_I = \overline{U}_2 A_2 = \frac{A_2}{\sqrt{1 - (A_2/A_1)^2}} \sqrt{\frac{2(p_1 - p_2)}{\rho} + 2gh_{L_{1-2}}} \tag{10.7}$$

where the subscript I emphasizes that Equation 10.7 gives an incompressible flow rate. Later we drop the subscript.

When the flow area changes abruptly, the effective flow area immediately downstream of the area reduction is not necessarily the same as the pipe flow area. This was originally investigated by Jean Borda (1733–1799) and illustrated in Figure 10.3. A central core flow called the vena contracta forms that is bounded by regions of slower-moving recirculating eddies. The pressure sensed with pipe wall taps corresponds to the higher moving velocity within the vena contracta with its unknown flow area, A_2. To account for this unknown, we introduce a contraction coefficient C_c, where $C_c = A_2/A_0$, with A_0 based on the meter throat diameter, into Equation 10.7. This gives

$$Q_I = \frac{C_c A_0}{\sqrt{1 - (C_c A_0/A_1)^2}} \sqrt{\frac{2(p_1 - p_2)}{\rho} + 2gh_{L_{1-2}}} \tag{10.8}$$

Furthermore, the frictional head losses can be incorporated into a friction coefficient, C_f, such that Equation 10.8 becomes

$$Q_I = \frac{C_f C_c A_0}{\sqrt{1 - (C_c A_0/A_1)^2}} \sqrt{\frac{2(p_1 - p_2)}{\rho}} \tag{10.9}$$

For convenience, the coefficients are factored out of Equation 10.9 and replaced by a single coefficient known as the *discharge coefficient*, C. The discharge coefficient represents the ratio of the actual flow rate through a meter to the ideal flow rate for the pressure drop measured—that is, $C = Q_{I_{\text{actual}}}/Q_{I_{\text{ideal}}}$. Reworking Equation 10.9 leads to the incompressible form of the working equation used with obstruction meters

$$Q_I = CEA_0 \sqrt{\frac{2\Delta p}{\rho}} = K_0 A_0 \sqrt{\frac{2\Delta p}{\rho}} \tag{10.10}$$

where E, known as the velocity of approach factor, is defined by

$$E = \frac{1}{\sqrt{1 - (A_0/A_1)^2}} = \frac{1}{\sqrt{1 - \beta^4}} \tag{10.11}$$

with the beta ratio defined as $\beta = d_0/d_1$, and where $K_0 = CE$ is called the *flow coefficient*.

The discharge coefficient and the flow coefficient are tabulated quantities found in test standards (1,3,4). Each is a function of the Reynolds number, β ratio, and pressure tap locations for each particular obstruction flow meter design, $C = f\left(\mathrm{Re}_{d_1}, \beta\right)$ and $K_0 = f\left(\mathrm{Re}_{d_1}, \beta\right)$.

Compressibility Effects

In compressible gas flows, compressibility effects in obstruction meters can be accounted for by introducing the compressible adiabatic expansion factor, Y. Here Y is defined as the ratio of the actual compressible volume flow rate, Q, divided by the assumed incompressible flow rate Q_I. Combining with Equation 10.10 yields

$$Q = YQ_I = CEA_0Y\sqrt{2\Delta p/\rho_1} \tag{10.12}$$

where ρ_1 is the upstream fluid density. When $Y = 1$, the flow is incompressible, and Equation 10.12 reduces to Equation 10.10. Equation 10.12 *represents the most general form of the working equation for volume flow rate determination when using an obstruction meter.*

The expansion factor, Y, depends on several values: the β ratio, the gas specific heat ratio, k, and the relative pressure drop across the meter, $(p_1 - p_2)p_1$, for a particular meter type–that is, $Y = f\left[\beta, k, (p_1 - p_2)/p_1\right]$. As a general rule, compressibility effects should be considered whenever $(p_1 - p_2)/p_1 \geq 0.1$.

Standards for Obstruction Meters

The flow behaviors of the most common obstruction meters, namely the orifice plate, venturi, and flow nozzle, have been studied to such an extent that these meters are used extensively without calibration. Values for the discharge coefficients, flow coefficients, and expansion factors are tabulated and available in standard U.S. and international flow handbooks along with standardized construction, installation, and operation techniques (1,3,4,16). Equations 10.10 and 10.12 are very sensitive to pressure tap location. For steam or gas flows, pressure taps should be oriented on the top or side of the pipe; for liquids, pressure taps should be oriented on the side. We discuss the recommended standard tap locations with each meter (1,4). A nonstandard installation or design requires an in situ calibration.

Orifice Meter

An orifice meter consists of a circular plate having a central hole (orifice). The plate is inserted into a pipe so as to effect a flow area change. The orifice hole is smaller than the pipe diameter and arranged to be concentric with the pipe's i.d. The square-edged orifice plate is shown in Figure 10.4. Installation is simplified by housing the orifice plate between two pipe flanges. With this installation technique, any particular orifice plate is interchangeable with others of different β value. The simplicity of the design allows for a range of β values to be maintained on hand at modest expense.

For an orifice meter, plate dimensions and use are specified by engineering standards (1,4). Equation 10.12 is used with values of A_0 and β based on the orifice (hole) diameter, d_0. The plate thickness should be between $0.005 \, d_1$ and $0.02 \, d_1$; otherwise, a taper must be added to the downstream side (1,4). The exact placement of pressure taps is crucial when using standard

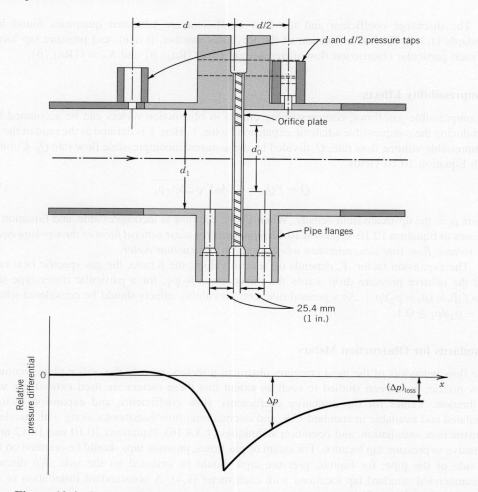

Figure 10.4   Square-edged orifice meter installed in a pipeline with optional 1 *d* and ½ *d*, and flange pressure taps shown. Relative flow pressure drop along pipe axis is shown.

coefficients. Standard pressure tap locations include (1) flange taps where pressure tap centers are located 25.4 mm (1 in.) upstream and 25.4 mm (1 in.) downstream of the nearest orifice face, (2) *d* and *d*/2 taps located one pipe diameter upstream and one-half diameter downstream of the upstream orifice face, and (3) vena contracta taps. Nonstandard tap locations will require *in situ* meter calibration.

Values for the flow coefficient, $K_0 = (Re_{d_1}, \beta)$ and for the expansion factor, $Y = f[\beta, k, (p_1 - p_2)/p_1]$, for a square-edged orifice plate are given in Figures 10.5 and 10.6 based on the use of flange taps. The relative instrument systematic uncertainty in the discharge coefficient (3) is ~0.6% of *C* for $0.2 \leq \beta \leq 0.6$ and $\beta\%$ of *C* for all $\beta > 0.6$. The relative instrument systematic uncertainty for the expansion factor is about $[4(p_1 - p_2)/p_1]\%$ of *Y*. Realistic estimates of

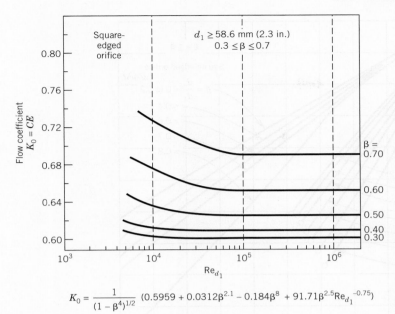

$$K_0 = \frac{1}{(1-\beta^4)^{1/2}} \left(0.5959 + 0.0312\beta^{2.1} - 0.184\beta^8 + 91.71\beta^{2.5}\mathrm{Re}_{d_1}^{-0.75}\right)$$

Figure 10.5 Flow coefficients for a square-edged orifice meter having flange pressure taps. (Compiled from data in reference 1.)

the overall systematic uncertainty in estimating Q using an orifice meter are between 1% (high β) and 3% (low β) at high Reynolds numbers when using standard tables[1]. Although the orifice plate represents a relatively inexpensive flow meter solution with an easily measurable pressure drop, it introduces a large permanent pressure loss, $(\Delta p)_{\mathrm{loss}} = \rho g h_L$, into the flow system. The pressure drop is illustrated in Figure 10.4 with the pressure loss estimated from Figure 10.7.

Rudimentary versions of the orifice plate meter have existed for several centuries. Both Torricelli and Newton used orifice plates to study the relation between pressure head and efflux from reservoirs, although neither ever got the discharge coefficients quite right (5).

Venturi Meter

A venturi meter consists of a smooth converging (21 degrees $\pm$ 1 degree) conical contraction to a narrow throat followed by a shallow diverging conical section, as shown in Figure 10.8. The engineering standard venturi meter design uses either a 15-degree or 7-degree divergent section (1,4). The meter is installed between two flanges intended for this purpose. Pressure taps are located just ahead of the upstream contraction and at the throat. Equation 10.12 is used with values for both A and β based on the throat diameter, d_0.

The quality of a venturi meter ranges from cast to precision-machined units. The discharge coefficient varies little for pipe diameters above 7.6 cm (3 in.). In the operating range $2 \times 10^5 \leq \mathrm{Re}_{d_1} \leq 2 \times 10^6$ and $0.4 \leq \beta \leq 0.75$, use a value of $C = 0.984$ with a systematic uncertainty of 0.7% (95%) for cast units and $C = 0.995$ with a systematic uncertainty of 1% (95%) for machined

---

[1] These uncertainties assume proper placement of the flow meter, i.e. sufficiently far away from out of plane elbows in the flow path.

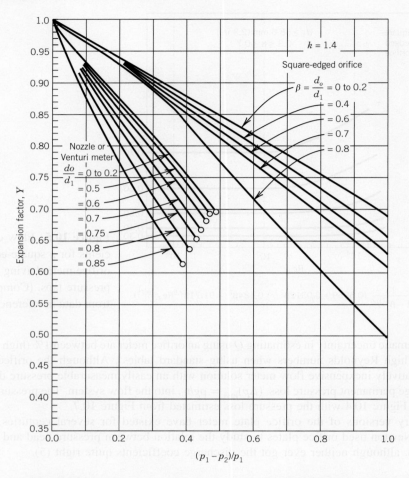

Figure 10.6 Expansion factors for common obstruction meters with $k = c_p/c_v = 1.4$. (Compiled from data in reference 1.)

units (1,3,4). Values for expansion factor are shown in Figure 10.6 and have an instrument systematic uncertainty of $[(4 + 100\beta^2)(p_1 - p_2)/p_1]\%$ of Y (3). Although a venturi meter presents a much higher initial cost over an orifice plate, Figure 10.7 demonstrates that the meter shows a much smaller permanent pressure loss for a given installation. This translates into lower system operating costs for the pump or blower used to move the flow.

The modern venturi meter was first proposed by Clemens Herschel (1842–1930). Herschel's design was based on his understanding of the principles developed by several men, most notably those of Daniel Bernoulli. However, he cited the studies of contraction/expansion angles and their corresponding resistance losses by Giovanni Venturi (1746–1822), and, later, those by James Francis (1815–1892), as being instrumental to his design of a practical flow meter.

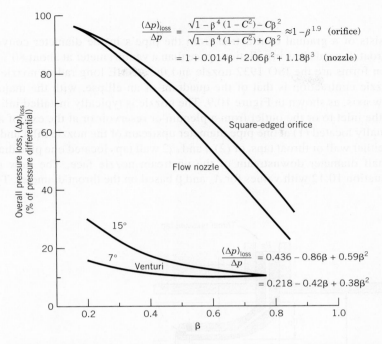

$$\frac{(\Delta p)_{loss}}{\Delta p} = \frac{\sqrt{1 - \beta^4 (1 - C^2)} - C\beta^2}{\sqrt{1 - \beta^4 (1 - C^2)} + C\beta^2} \approx 1 - \beta^{1.9} \quad \text{(orifice)}$$

$$= 1 + 0.014\beta - 2.06\beta^2 + 1.18\beta^3 \quad \text{(nozzle)}$$

$$\frac{(\Delta p)_{loss}}{\Delta p} = 0.436 - 0.86\beta + 0.59\beta^2$$

$$= 0.218 - 0.42\beta + 0.38\beta^2$$

Figure 10.7   The permanent pressure loss associated with flow through common obstruction meters. (Compiled from data in reference 1.)

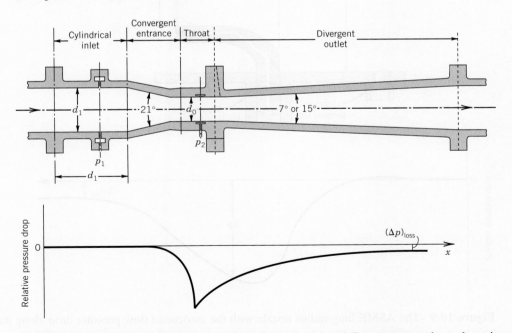

Figure 10.8   The Herschel venturi meter with the associated flow pressure drop along its axis.

Flow Nozzles

A flow nozzle consists of a gradual contraction from the pipe's inside diameter converging down to a narrow throat. It needs less installation space than a venturi meter at about 80% of the initial cost. Common forms are the ISO 1932 nozzle and the ASME long radius nozzle (1,4). The long radius nozzle contraction is that of the quadrant of an ellipse, with the major axis aligned with the flow axis, as shown in Figure 10.9. The nozzle is typically installed inline but can also be used at the inlet to or the outlet from a plenum or reservoir or at the exit of a pipe. Pressure taps are usually located (1) at one pipe diameter upstream of the nozzle inlet and at the nozzle throat using either wall or throat taps, or (2) d and $d/2$ wall taps located one pipe diameter upstream and one-half diameter downstream of the upstream nozzle face. The flow rate is determined from Equation 10.12 with values for A_0 and β based on the throat diameter. Typical

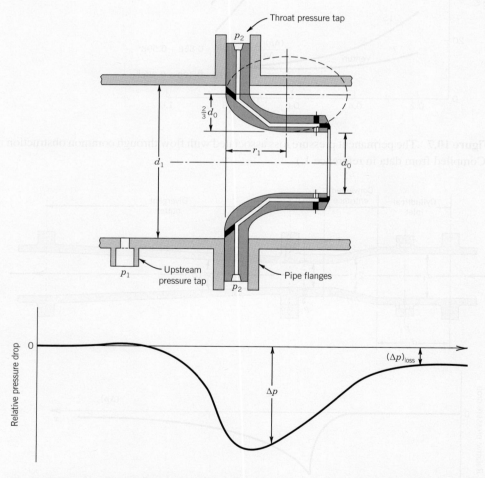

Figure 10.9 The ASME long-radius nozzle with the associated flow pressure drop along its axis.

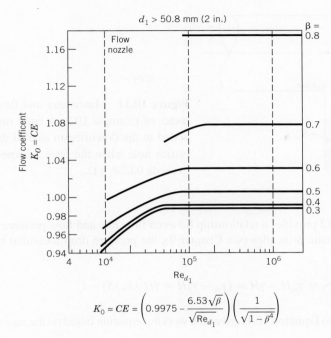

$$K_0 = CE = \left(0.9975 - \frac{6.53\sqrt{\beta}}{\sqrt{Re_{d_1}}}\right)\left(\frac{1}{\sqrt{1 - \beta^4}}\right)$$

Figure 10.10 Flow coefficients for an ASME long-radius nozzle with a throat pressure tap. (Compiled from data in reference 1.)

values for the flow coefficient and expansion factor are given in Figures 10.10 and 10.6. The relative instrument systematic uncertainty at 95% confidence for the discharge coefficient is about 2% of C and for the expansion factor is about $[2(p_1 - p_2)/p_1]\%$ of Y (3). The permanent loss associated with a flow nozzle is larger than for a comparable venturi but significantly smaller than for an orifice (Fig. 10.7) for the same pressure drop.

The idea of using a nozzle as a flow meter was first proposed in 1891 by John Ripley Freeman (1855–1932), an inspector and engineer employed by a fire insurance firm. His work required tedious tests to quantify pressure losses in pipes, hoses, and fittings. He noted a consistent relationship between pressure drop and flow rate through fire nozzles at higher flow rates.

Example 10.3

The U-tube manometer filled with manometer fluid (of specific gravity S_m) of Figure 10.11 is used to measure the pressure drop across an obstruction meter. A fluid having specific gravity S flows through the meter. Determine a relationship between the meter flow rate and the measured manometer deflection, H.

KNOWN   Fluid (S and γ)

Manometer fluid (S_m and γ_m)

FIND $Q = f(H)$

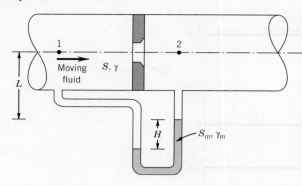

Figure 10.11 Manometer and flow meter of Example 10.3. A taper must be added to the downstream side of the orifice hole when the plate thickness exceeds $0.02d_1$ (1).

SOLUTION   Equation 10.12 provides a relationship between flow rate and flow pressure drop. From Figure 10.11 and hydrostatic principles (see Chapter 9), the pressure drop measured by the manometer is

$$\Delta p = p_1 - p_2 = \gamma_m H - \gamma H = (\gamma_m - \gamma)H = \gamma H\left[(S_m/S) - 1\right]$$

Substituting this relation into Equation 10.12 gives the working equation based on the equivalent pressure head:

$$Q = CEA_o Y\sqrt{2gH\left[(S_m/S) - 1\right]} \tag{10.13}$$

Example 10.4

A 10-cm-diameter, square-edged orifice plate is used to meter the steady flow of 16 °C water through an 20-cm pipe. Flange taps are used and the pressure drop measured is 50 cm Hg. Determine the pipe flow rate. The specific gravity of mercury is 13.5.

KNOWN $d_1 = 20$ cm $H = 50$ cm Hg $d_0 = 10$ cm
Water properties (properties are from Appendix B)
$$\mu = 1.08 \times 10^{-3} \text{ N-s/m}^2 \quad \rho = 999 \text{ kg/m}^3$$

ASSUMPTIONS   Incompressible flow of a liquid ($Y = 1$)

FIND   Volume flow rate, Q

SOLUTION   Equation 10.12 is used with an orifice plate, and it requires knowledge of the product CE. The beta ratio is $\beta = d_0/d_1 = 0.5$, so the velocity of approach factor is calculated from Equation 10.11:

$$E = \frac{1}{\sqrt{1 - \beta^4}} = \frac{1}{\sqrt{1 - 0.5^4}} = 1.0328$$

We know that $C = f(\text{Re}_{d_1}, \beta)$. The flow Reynolds number is estimated using Equation 10.3, which can be modified with the relation, $v = \mu/\rho$,

$$\text{Re}_{d_1} = \frac{4Q}{\pi d_1 v} = \frac{4\rho Q}{\pi d_1 \mu}$$

We see that without information concerning Q, we cannot determine the Reynolds number, and so C cannot be determined explicitly.

Instead, a trial-and-error iteration is undertaken: Guess a value for K_0 (or for C) and iterate. A good start is to guess a value at a high value of Re_{d_1}. This is the flat region of the curve ($\beta = 0.5$) in Figure 10.5, so choose a value of $K_0 = CE = 0.625$.

Based on the manometer deflection and Equation 10.13 (see Ex. 10.2),

$$
\begin{aligned}
Q &= CEA_0 Y \sqrt{2gH\left[(S_m/S) - 1\right]} = K_0 A_0 Y \sqrt{2gH\left[(S_m/S) - 1\right]} \\
&= (0.625)(\pi/4)(0.10\,\text{m})^2(1)\sqrt{2(9.8\,\text{m/s}^2)(0.50\,\text{m})\left[(13.5/1) - 1\right]} \\
&= 0.054\,\text{m}^3/\text{s}
\end{aligned}
$$

Next we have to test the guessed value for K_0 to determine if it was correct. For this value of Q,

$$\text{Re}_{d_1} = \frac{4\rho Q}{\pi d_1 \mu} = \frac{4\left(999\,\text{kg/m}^3\right)\left(0.054\,\text{m}^3/\text{s}\right)}{\pi\left(0.20\,\text{m}\left(1.08 \times 10^{-3}\,\text{N-s/m}^2\right)\right)} = 3.2 \times 10^5$$

From Figure 10.5, at this Reynolds number, $K_0 \approx 0.625$. The solution is converged, so we conclude that $Q = 0.054\,\text{m}^3/\text{s}$.

Example 10.5

Air flows at 20 °C through a 6-cm pipe. A square-edged orifice plate with $\beta = 0.4$ is chosen to meter the flow rate. A pressure drop of 250 cm H_2O is measured at the flange taps with an upstream pressure of 93.7 kPa abs. Find the flow rate.

KNOWN $d_1 = 6\,\text{cm}$ $p_1 = 93.7\,\text{kPa abs}$

$\beta = 0.4$ $T_1 = 20°\text{C} = 293\,\text{K}$

$H = 250\,\text{cm}\,H_2O$

Properties (found in Appendix B):

$$\text{Air}: v = 1.0 \times 10^{-5}\,\text{m}^2/\text{s}$$
$$\text{Water}: \rho_{H_2O} = 999\,\text{kg/m}^3$$

ASSUMPTIONS Treat air behavior as an ideal gas ($p = \rho RT$)

FIND Volume flow rate, Q

SOLUTION The orifice flow rate is found from Equation 10.12, which requires information about E, C, and Y. From the given information, we calculate both the orifice area, $A_{d_0} = 4.52 \times 10^{-4}$ m^2, and, from Equation 10.11, the velocity of approach factor, $E = 1.013$. The air density is found from the ideal gas equation of state:

$$\rho_1 = \frac{p_1}{RT_1} = \frac{93,700 \text{ N/m}^2}{(287 \text{ N-m/kg})(293 \text{ K})} = 1.114 \text{ kg/m}^3$$

or use air-property tables. The pressure drop is $p_1 - p_2 = \rho_{H_2O}gH = 24,500$ N/m^2.

The pressure ratio for this gas flow, $(p_1 - p_2)/p_1 = 0.26$. For pressure ratios greater than 0.1 the compressibility of the air should be considered. From Figure 10.6, $Y = 0.92$ for $k = 1.4$ (air) and a pressure ratio of 0.26.

As in the previous example, the discharge coefficient cannot be found explicitly unless the flow rate is known, since $C = f\left(\text{Re}_{d_1}, \beta\right)$. So a trial-and-error iterative approach is used. From Figure 10.5, we start with a guess of $K_0 = CE \approx 0.61$ (or $C = 0.60$). Then,

$$Q = CEYA_o\sqrt{\frac{2\Delta p}{\rho_1}} = (0.60)(1.013)(0.92)\left(4.52 \times 10^{-4} \text{ m}^2\right)\sqrt{\frac{(2)\left(24,500 \text{ N/m}^2\right)}{1.114 \text{ kg/m}^3}}$$

$$= 0.053 \text{ m}^3/\text{s}$$

CHECK For this flow rate, $\text{Re}_{d_1} = 4\,Q/\pi\,d_1\upsilon = 7.4 \times 10^4$ and from Figure 10.5, $K_0 \approx 0.61$, as assumed. The flow rate through the orifice is taken to be 0.053 m^3/s.

Example 10.6

A pump can often impose pressure oscillations into a pipe flow that are sensed in the pressure measurements. Figure 10.12 plots separate pressure measurements taken upstream, p_1, and downstream, p_2, of an orifice meter, as well as pressure differential measurements, $\Delta p = p_1 - p_2$, taken across the meter. The systematic effect of the oscillations on the internal pipe pressure is seen in comparing p_1 and p_2, which rise and fall together with a correlation coefficient of $r_{p_1 p_2} = 0.998$.

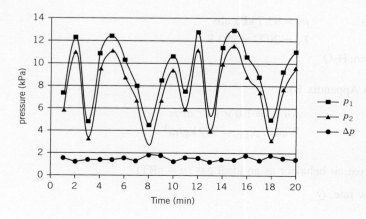

Figure 10.12 Upstream and downstream pressures and differential pressures across an orifice meter for Example 10.6.

Estimate the contribution to random standard uncertainty in estimating pressure differential due to the data scatter. The following information is known:

$$\bar{p}_1 = 9.25 \text{ kPa} \quad s_{p_1} = 2.813 \text{ kPa} \quad N = 20$$
$$\bar{p}_2 = 7.80 \text{ kPa} \quad s_{p_2} = 2.870 \text{ kPa} \quad N = 20$$
$$\overline{\Delta p} = 1.45 \text{ kPa} \quad s_{\Delta p} = 0.188 \text{ kPa} \quad N = 20$$

SOLUTION As seen in Figure 10.12, pump-induced oscillations impose a trend on the individual pressure measurements. This is confirmed by the high correlation coefficient (see Chapter 4) between p_1 and p_2. Notably, the pressure differential signal is unaffected by this periodicity, because although the pressure within the pipe may rise and fall with time, the pressure difference across the orifice plate remains largely unaffected. Hence analyzing the individual pressure data directly without accounting for this systematic effect would overstate the random uncertainty, which should be due only to the data scatter; analyzing the pressure differential data eliminates the systematic effect. We address these issues here and show how to compensate for them.

The measured pressure differential is $\overline{\Delta p} = 1.45$ kPa, or, alternately, when calculated from the individual pressure measurements, $\bar{p}_1 - \bar{p}_2 = 1.45$ kPa. So it is independent of how we measure it.

We can estimate the random standard uncertainty in pressure differential directly from the data variation in Δp as

$$s_{\overline{\Delta p}} = \frac{s_{\Delta p}}{\sqrt{20}} = 0.042 \text{ kPa} \quad \text{with } \nu = 19.$$

Alternatively, we could analyze the individual pressure data and use these to estimate the random standard uncertainty in $\overline{\Delta p}$. With $\Delta p = p_1 - p_2$, the sensitivities are

$$\frac{\partial \Delta p}{\partial p_1} = 1 \qquad \frac{\partial \Delta p}{\partial p_2} = -1$$

The elemental random standard uncertainties are

$$s_{\bar{p}_1} = \frac{s_{p_1}}{\sqrt{20}} = 0.629 \text{ kPa} \qquad s_{\bar{p}_2} = \frac{s_{p_2}}{\sqrt{20}} = 0.642 \text{ kPa}$$

Then, accounting for the correlated error influence on the pressures, we write

$$s_{\overline{\Delta p}} = \left[\left(\frac{\partial \Delta p}{\partial p_1} s_{\bar{p}_1} \right)^2 + \left(\frac{\partial \Delta p}{\partial p_2} s_{\bar{p}_2} \right)^2 + 2 \frac{\partial \Delta p}{\partial p_1} \frac{\partial \Delta p}{\partial p_2} r_{p_1 p_2} s_{\bar{p}_1} s_{\bar{p}_2} \right]^{1/2} = 0.042 \text{ kPa} \qquad (10.14)$$

with $\nu = 19$. Here the last term under the radical corrects for the correlated error between p_1 and p_2.

Thus, either method gives the same result for $\overline{\Delta p}$ and for $s_{\overline{\Delta p}}$. However, the first method is preferred because it directly estimates the desired measured variable, Δp, and its random uncertainty is unaffected by the systematic effect.

If we neglected the correlated effect of the pump oscillation, we would instead be tempted to calculate

$$s_{\overline{\Delta p}} = \left[\left(\frac{\partial \Delta p}{\partial p_1} s_{\overline{p}_1} \right)^2 + \left(\frac{\partial \Delta p}{\partial p_2} s_{\overline{p}_2} \right)^2 \right]^{1/2} = 0.899 \text{ kPa}$$

which overstates the random standard uncertainty significantly and is wrong.

COMMENT Equation 10.14 revisits a method to correct for imposed correlated effects on uncertainty estimates as discussed in Chapter 5. This example shows the importance of reviewing data and data trends in a measurement to uncover test procedure tendencies that affect test interpretation.

Sonic Nozzles

Sonic nozzles are used both to measure and to regulate the flow rate of compressible gases (6). They usually take the form of either a flow nozzle or venturi meter. The sonic condition refers to achieving the speed of sound of the fluid at the throat of the meter. With sonic operating conditions within the flow meter, flow rate will depend only on upstream pressure and temperature. Sonic meters are often used as a transfer standard with uncertainties as low as 0.5%. Both large pressure drops and system pressure losses must be tolerated with the technique (1).

The theoretical basis for such a meter stems from the early work of Bernoulli, Venturi, and Saint-Venant (1797–1886). In 1866, Julius Weisbach (1806–1871) developed a direct relation between pressure drop and a maximum mass flow rate. When the pressure drop across an obstruction meter becomes sufficiently high, the sonic condition will be reached at the meter throat as the fluid accelerates. At that point the throat is considered to be choked and the mass flow rate through the meter is at the maximum for the given inlet conditions. Lowering the downstream pressure will not increase the mass flow rate further.

For a perfect gas undergoing an isentropic process, the pressure ratio corresponding to the onset of the choked flow condition at the meter's minimum area, the meter throat, is given by the *critical pressure ratio*:

$$\left(\frac{p_{th}}{p_t} \right)_{\text{critical}} = \left(\frac{2}{k+1} \right)^{k/(k-1)} \tag{10.15}$$

where p_{th} is the throat pressure and subscript "t" denotes stagnation properties. If $(p_{th}/p_t) \le (p_{th}/p_t)_{\text{critical}}$ then the meter throat is choked. This is the maximum mass flow rate for the prevailing conditions existing upstream of the throat. The mass flow rate can be found from

$$\dot{m} = CA_{th} \frac{p_t}{\sqrt{RT_t}} \sqrt{\frac{2k}{k+1} \left(\frac{2}{k+1} \right)^{2/(k-1)}} = CC^* A_{th} \frac{p_t}{\sqrt{RT_t}} \tag{10.16}$$

with

$$C^* = \sqrt{\frac{2k}{k+1} \left(\frac{2}{k+1} \right)^{2/(k-1)}} \tag{10.17}$$

where k is the specific heat ratio of the gas, C^* is the critical ratio, and C is the discharge coefficient (1,3). For air and other diatomic gases, $k = 1.4$ and so $C^* = 0.6847$. Commercial units are supplied with a discharge coefficient, otherwise $C = 0.98 \pm 2\% \,(95\%)$ is used (1). Corrections (e.g., equation 9.42) can be applied to use measured upstream static properties when $A_{th} \ll A_1$.

Obstruction Meter Selection

Selecting between obstruction meter types depends on a number of factors and engineering compromises. Primary considerations include meter placement, overall pressure loss, overall (capital and operating) costs, accuracy, and operating range (turndown).

Placement

It is important to provide sufficient straight pipe lengths to either side of a flow meter to allow for proper flow development. The flow development dissipates swirl, promotes a symmetric velocity distribution, and allows for proper pressure recovery downstream of the meter so as to reduce systematic error. The available standards provide detailed recommendations regarding installation placement for best practice (1,3,4). Even with recommended lengths, an additional 0.5% systematic uncertainty should be added to the uncertainty in the discharge coefficient to account for swirl effects downstream of turns (3). Unorthodox installations require meter calibration to determine flow coefficients or the use of special correction procedures (3,4,7). The physical size of a meter could become an important consideration in itself, since venturi and nozzle meters can take up a considerable length of pipe.

Pressure Loss

The unrecoverable overall pressure loss, Δp_{loss}, associated with a flow meter depends on the β ratio and flow rate (Fig. 10.7). These losses must be overcome by the system prime mover (e.g., a pump, fan, or compressor driven by a motor) in addition to any other pipe system losses. In established systems, flow meters must be chosen on the basis of the pressure loss that the prime mover can accommodate and still maintain a desired system flow rate. The power, $\dot{W}$, required to overcome any loss in a system for flow rate Q is given by

$$\dot{W} = Q\frac{\Delta p_{\text{loss}}}{\eta} \qquad (10.18)$$

where η is the prime mover efficiency.

Costs

The actual cost of a meter depends on its initial capital cost, the costs of installing it (including system down time), calibration costs, and the added capital and operating costs associated with flow meter pressure losses to the system. Indirect costs may include product losses resulting from the meter flow rate uncertainty.

Accuracy

The ability of an obstruction flow meter to accurately estimate the volume flow rate in a pipeline depends on both the method used to calculate flow rate and factors inherent to the meter. If standard tabulated values for meter coefficients are used, factors that contribute to the overall uncertainty of the measurement enter through the following data-acquisition elemental errors: (1) actual β ratio error and pipe eccentricity, (2) pressure and tap position errors, (3) temperature effects leading to relative thermal expansion of components, and (4) actual upstream flow profile. These errors are in addition to the errors inherent in the coefficients themselves and errors associated with the estimation of the upstream fluid density (1). Direct in situ calibration of any flow meter installation can assess these effects and their contributions to overall uncertainty. For example, the actual β ratio value, pressure tap locations, and other installation effects would be included in a system calibration and accommodated in the computed flow coefficient.

Turndown

The relative range, $(Q_{max} - Q_{min})/Q_{min}$, over which a meter is designed to be used is known as the meter *turndown*. The selection of a meter should consider whether the system into which the meter is to be installed will be used at more than one flow rate, and its effect on system performance over the entire anticipated flow rate range should be included.

Example 10.7

For the orifice meter in Example 10.3 with β = 0.5 and a pressure drop of 50 cm Hg, calculate the permanent pressure loss due to the meter that must be overcome by a pump.

KNOWN $H = 50$ cm Hg
$\beta = 0.5$

ASSUMPTION   Steady flow

FIND Δp_{loss}

SOLUTION   For a properly designed and installed orifice meter, the permanent pressure loss of the meter is about 75% of the pressure drop across the meter (Fig. 10.7). Hence

$$\Delta p_{loss} = 0.75 \times \gamma_{Hg}H = 49.9 \text{ kPa} \quad (\text{or } H_{loss} = 38 \text{ cm Hg})$$

COMMENT   In comparison, for $Q = 0.053$ m³/s a typical venturi (say a 15-degree outlet) with a β = 0.5 would provide a pressure drop equivalent to 19.6 cm Hg with a permanent loss of only 3.1 cm Hg. Since the pump needs to supply enough power to create the flow rate and to overcome losses, a smaller pump could be used with the venturi.

Example 10.8

For the orifice of Example 10.7, estimate the operating costs required to overcome these permanent losses if electricity is available at $0.08/kW-h, the pump is used 6,000 h/year, and the pump motor efficiency is 60%.

KNOWN   Water at 16°C $Q = 0.053 \text{ m}^3/\text{s}$

$\Delta p_{\text{loss}} = 49.9 \text{ kPa}$ $\eta = 0.60$

ASSUMPTIONS   Meter installed according to standards (1)

FIND   Annual cost due to Δp_{loss} alone

SOLUTION   The pump power required to overcome the orifice meter permanent pressure loss is estimated from Equation 10.18,

$$\dot{W} = Q \frac{\Delta p_{\text{loss}}}{\eta}$$

$$= \frac{(0.55 \text{ m}^3/\text{s})(49,900 \text{ Pa})(3600 \text{ s/h})(1 \text{ N/m}^2/\text{Pa})(1 \text{ W/N-m/s})}{0.6} = 4530 \text{ W} = 4.53 \text{ kW}$$

The additional annual pump operating cost required to overcome this is

$$\text{Cost} = (4.53 \text{ kW})(6000 \text{ h/year})(\$0.08/\text{kW-h}) = \$2175/\text{year}$$

COMMENT   In comparison, the venturi discussed in the comment to Example 10.7 would cost about \$180/year to operate. If installation space allows, the venturi meter may be the less expensive long-term solution.

Laminar Flow Elements

Laminar flow meters take advantage of the linear relationship between volume flow rate and the pressure drop over a length for laminar pipe flow ($\text{Re}_d < 2000$). For a pipe of diameter d with pressure drop over a length L, the governing equations of motion lead to

$$Q = \frac{\pi d^4}{128 \, \mu} \frac{p_1 - p_2}{L} \quad \text{where } \text{Re}_d < 2000 \tag{10.19}$$

This relation was first demonstrated by Jean Poiseulle (1799–1869), who conducted meticulous tests documenting the resistance of flow through capillary tubes.

The simplest type of laminar flow meter consists of two pressure taps separated by a length of piping.[2] However, because the Reynolds number must remain low, the flow rate range is limited. This limitation is overcome in commercial units by using bundles of small-diameter tubes or passages placed in parallel and called laminar flow elements (Figure 10.13). The strategy of a laminar flow element is to divide the total flow by passing it through the smaller tubes within the tube bundle The flow rate within each small tube is much less than the total flow in a way that keeps $\text{Re} < 2000$.

---

[2] Alternately, if flow rate can be measured or is known, Equation 10.19 provides the basis for a capillary tube viscometer. The method also achieves a linear resistance.

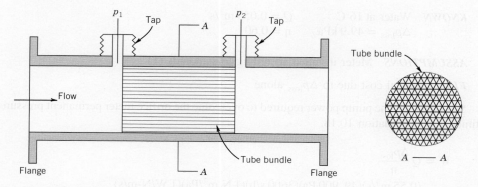

Figure 10.13 Laminar flow element flow meter concept using pipe flanges for installation. For noncircular tubes, Reynolds number is based on $d = 4r_H$.

Pressure drop is measured between the entrance and the exit of the laminar flow element, and a flow coefficient is used to account for inefficiency, such that

$$Q = K_1 \Delta p \qquad (10.20)$$

where coefficient K_1 is the meter static sensitivity and is called its *K-factor*. Commercial units come calibrated with K_1 a constant over the useful meter range.

A laminar flow element has an upper limit on usable flow rate. Various meter sizes and designs are available to accommodate user needs. Turndowns of up to 100:1 are available.

Laminar flow elements offer some distinct advantages over other pressure differential meters. These include (1) a high sensitivity even at low flow rates, (2) an ability to measure flow from either meter direction, (3) a wide usable flow range, and (4) the ability to indicate an average flow rate in pulsating flows. The instrument systematic uncertainty in flow rate determination is as low as 0.25% (95%) of the flow rate. However, these meters are limited to clean fluids for considerations of clogging potential. The measured pressure drop remains a system pressure loss.

10.6   INSERTION VOLUME FLOW METERS

Dozens of volume flow meter types based on a number of different principles have been proposed, developed, and sold commercially. A large group of meters is based on some phenomenon that is actually sensitive to the average velocity across a control surface of known area, that is, $Q = f(\overline{U}, A) = \overline{U}A$. Several of these designs are included in the discussion below. Another common group, called *positive displacement meters*, actually measure parcels of a volume of fluid per unit time-that is, $Q = f(\forall, t) = \forall/t$.

Electromagnetic Flow Meters

The operating principle of an *electromagnetic flow meter* (8) is based on the fundamental principle that an electromotive force (emf) of electric potential E is induced in a conductor of

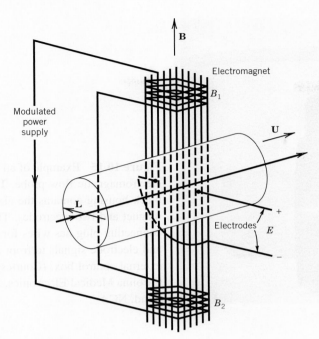

Figure 10.14 Electromagnetic principle as applied to a working flow meter.

length L that moves with velocity U through a magnetic field of magnetic flux, B. Simply, when an electrically conductive liquid moves through a magnetic field, a voltage is induced in the liquid at a right angle to the field. The voltage is detected by metal electrode sensors with the voltage magnitude and polarity directly proportional to the volume flow rate and the flow direction, respectively. This physical behavior was first recorded by Michael Faraday (1791–1867). In principle,

$$E = U \times B \cdot L \tag{10.21}$$

where U, B and L are vector quantities.

A practical use of the principle is shown in Figure 10.14. From Equation 10.21 the magnitude of E is affected by the average velocity, $\overline{U}$, as

$$E = \overline{U}BL \sin \alpha = f(\overline{U})$$

where α is the angle between the mean velocity vector and the magnetic flux vector, usually at 90 degrees. In general, electrodes are located either in or mounted on the pipe wall in a diametrical plane that is normal to the known magnetic field. The electrodes are separated by a scalar length, L. The average magnitude of the velocity, $\overline{U}$, across the pipe is thus inferred through the measured emf. The flow rate is found by

$$Q = \overline{U}\frac{\pi d_1^2}{4} = \frac{E}{BL}\frac{\pi d_1^2}{4} = K_1 E \tag{10.22}$$

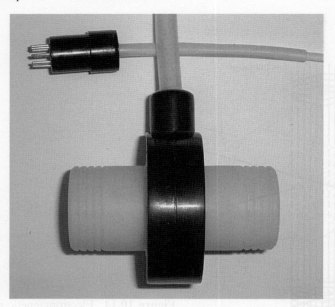

Figure 10.15 Example of an in-line electromagnetic flow probe. The black housing contains the electro-magnet and two electrodes. The connecting plug has wires for power and electrode signals to/from an external control box. (Courtesy of Carolina Medical Electronics, East Bend, NC.)

The value of L is on the order of the pipe diameter, the exact value depending on the meter construction and magnetic flux lines. The static sensitivity (or meter K-factor) K_1 is found by calibration and supplied by a manufacturer such that the relationship between the flow rate and the measured potential is linear.

The electromagnetic flow meter comes commercially as a packaged flow device, which is installed directly inline and connected to an external electronic output unit. Special designs include an independent flow sensor unit that can clamp over a nonmagnetic pipe, a design favored to monitor blood flow rate through major arteries during surgery. The sensor is connected to a control unit by wiring. External sensors are also available as an inline union that can measure flow rate in either direction with instantaneous changes (Figure 10.15).

The electromagnetic flow meter has a very low pressure loss associated with its use due to its open tube, no obstruction design, and is suitable for installations that can tolerate only a small pressure drop. It can be used for either steady or unsteady flow measurements, providing either time-averaged or instantaneous data in the latter. This absence of internal parts is very attractive for metering corrosive and "dirty" fluids. The operating principle is independent of fluid density and viscosity, responding only to average velocity, and there is no difficulty with measurements in either laminar or turbulent flows, provided that the velocity profile is reasonably symmetrical. Uncertainty down to 0.25% (95%) of the measured flow rate can be attained, although values from 1% to 5% (95%) are more common for these meters in industrial settings. The fluid must be conductive, but the minimum conductivity required depends on a particular meter's design. Fluids with values as low as 0.1 microsieman (μsieman)/cm have been metered. Adding salts to a fluid increases its conductivity. Stray electronic noise is perhaps the most significant barrier in applying this type of meter. Grounding close to the electrodes and increasing fluid conductivity reduce noise.

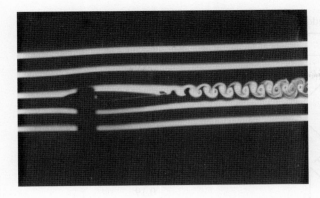

Figure 10.16 Smoke lines in this photograph reveal the vortex shedding behind a streamlined wing-shaped body in a moving flow. (Photograph courtesy of Richard Figliola.)

Vortex Shedding Meters

An oscillating street sign and the "singing" of power lines on a windy day are examples of the effects induced by vortex shedding from bluff-shaped bodies, a natural phenomenon wherein alternating vortices are shed in the wake of the body. The vortices formed on opposite sides of the body are carried downstream in the body's wake, forming a "vortex street," with each vortex having an opposite sign of rotation. This behavior is seen in Figure 10.16, a photograph that captures the vortex shedding downstream of a section of an aircraft wing. The aerodynamicist Theodore von Karman (1881–1963) first deduced the existence of a vortex street, although Leonardo da Vinci appears to have been the first to actually record the phenomenon (9).

A *vortex flow meter* operates on the principle that the frequency of vortex shedding depends on the average velocity of the flow past the body and the body shape. The basic relationship between shedding frequency, f, where $f[Hz] = \omega/2\pi$, and average velocity, $\overline{U}$, for a given shape is given by the Strouhal number,

$$\text{St} = \frac{fd}{\overline{U}} \tag{10.23}$$

where d is a characteristic length for the body.

A typical design is shown in Figure 10.17. The shedder spans the pipe, so its length $\ell \approx d_1$, and $d/\ell \approx 0.3$, so as to provide for strong, stable vortex strength. Although the Strouhal number is a function of the Reynolds number, various geometrical shapes, known as shedders, can produce a

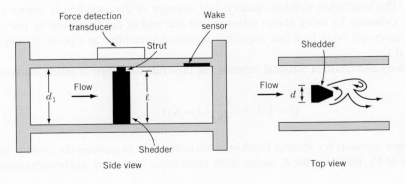

Figure 10.17 Vortex shedding flow meter. Different shedder shapes are available.

Table 10.1 Shedder Shape and Strouhal Number

Cross Section	Strouhal Number[a]
	0.16
	0.19
	0.16
	0.15
	0.12

[a]For Reynolds number $Re_d \geq 10^4$. Strouhal number $St = fd/\overline{U}$.

stable vortex flow that has a constant Strouhal number over a broad range of flow Reynolds numbers (for $Re_d > 10^4$). The oscillation stability, quality, and strength of the shedding is improved over common circular cylinders by using abrupt edges on the rear end of the shape and by providing a slightly concave upstream body face that traps the stagnation streamline at a point. Examples are given in Table 10.1.

For a fixed body and constant Strouhal number, the flow rate for a pipe of inside diameter d_1 is

$$Q = \overline{U}A = c\frac{\pi d_1^2}{4St}fd = K_1 f \tag{10.24}$$

where the constant c accounts for shedder blockage effects that tend to increase the average velocity sensed. The value of K_1, known as the K-factor, is the meter static sensitivity and remains essentially constant for $10^4 < \text{Re}_d < 10^7$. Shedding frequency can be measured in many ways. The shedder strut can be instrumented to detect the force oscillation by using strut-mounted strain gauges or capacitance sensor, for example, or a piezoelectric crystal wall sensor can be used to detect the pressure oscillations in the flow.

The lower flow rate limit on vortex meters is at Reynolds numbers (for Re based on d in Fig. 10.17) near 10,000, below which the Strouhal number can vary nonlinearly with flow rate and shedding becomes unstable. This can be a limitation in metering high-viscosity pipe flows ($\mu > 20$ centipoise). The upper flow bound is limited either by the onset of cavitation in liquids or by the onset of compressibility effects in gases at Mach numbers exceeding 0.2. Property variations affect meter performance only indirectly. Density variations affect the strength of the shed vortex, and this places a lower limit on fluid density, which is based on the sensitivity of the vortex-shedding detection equipment. Viscosity affects the operating Reynolds number. Otherwise, within bounds, the meter is insensitive to property variations.

The meter has no moving parts and relatively low pressure losses compared to obstruction meters. A single meter can operate over a flow range of up to 20:1 above its minimum with a linearity in K_1 of 0.5%. Because d can be replaced by a constant times d_1 in Equation 10.24, we see that the strength of the shedding frequency sensed over the pipe area decreases as the pipe diameter cubed. This indicates that a meter's accuracy decreases as pipe diameter increases, placing an upper limit on the meter size.

Rotameters

The rotameter is a widely used insertion meter for volume flow rate indication. As depicted in Figure 10.18, the meter consists of a float within a vertical tube, tapered to an increasing cross-sectional area at its outlet. Flow entering through the bottom passes over the float, which is free to move. The equilibrium height of the float indicates the flow rate.

The operating principle of a rotameter is based on the balance between the drag force, F_D, and the weight, W, and buoyancy forces, F_B, acting on the float in the moving fluid. It is the drag force that varies with the average velocity over the float.

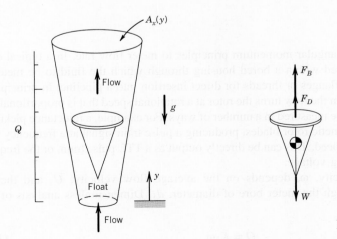

Figure 10.18 Concept of a rotameter.

The force balance in the vertical direction y yields

$$\sum F_y = 0 = +F_D - W + F_B$$

with $F_D = \frac{1}{2} C_D \rho \overline{U}^2 A_x$, $W = \rho_b g \forall_b$, and $F_B = \rho g \forall_b$. The average velocity sensed by the float depends on its height in the tube and is given by

$$\overline{U} = \overline{U}(y) = \sqrt{2(\rho_b - \rho) g \forall_b / C_D \rho A_x} \qquad (10.25)$$

where
ρ_b = density of float (body)
ρ = density of fluid
C_D = drag coefficient of the float
A_x = tube cross-sectional area
$\overline{U}$ = average velocity past the float
$\forall_b$ = volume of float (body)

In operation, the float rises to an equilibrium position. The height of this position increases with flow velocity and, hence, flow rate. This flow rate is found by

$$Q = \overline{U} A_a(y) = K_1 A_a(y) \qquad (10.26)$$

where $A_a(y)$ is the annular area between the float and the tube, and K_1 is a meter constant. Both the average velocity and the annular area depend on the height of the float in the tube. So the float's vertical position is a direct measure of flow rate, which can be read from a graduated scale, electronically sensed with an optical cell, or detected magnetically. Floats with sharp edges are less sensitive to fluid viscosity changes with temperature. A typical meter turndown is 10:1 with an instrument systematic uncertainty of ~2% (95%) of flow rate.

Turbine Meters

Turbine meters make use of angular momentum principles to meter flow rate. In a typical design (Fig. 10.19), a rotor is encased within a bored housing through which the fluid to be metered is passed. Its housing contains flanges or threads for direct insertion into a pipeline. In principle, the exchange of momentum within the flow turns the rotor at a rotational speed that is proportional to the flow rate. Rotor rotation can be measured in a number of ways. For example, a reluctance pickup coil can sense the passage of magnetic rotor blades, producing a pulse train signal at a frequency that is directly related to rotational speed. This can be directly output as a TTL pulse train, or the frequency can be converted to an analog voltage.

The rotor angular velocity, ω, depends on the average flow velocity, $\overline{U}$, and the fluid kinematic viscosity, ν, through the meter bore of diameter, d_1. Dimensionless analysis of these parameters (10) yields

$$Q = K_1 \omega \qquad (10.27)$$

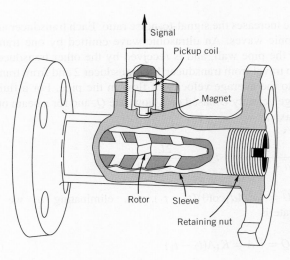

Figure 10.19 Cutaway view of a turbine flow meter.

where K_1 is the meter's K-factor but the relation is a function of Reynolds number. In practice, there is a region in which the rotor angular velocity varies linearly with flow rate, and this region becomes the meter operating range.

Turbine meters offer a low-pressure drop and very good accuracy. A typical instrument's systematic uncertainty in flow rate is ~0.25% (95%) with a turndown of 20:1. The measurements are exceptionally repeatable, making these meters good candidates for local flow rate standards. However, their use must be restricted to clean fluids to avoid fouling their rotating parts. The turbine meter rotational speed is sensitive to temperature changes, which affect fluid viscosity and density and thus K_1. Some compensation for viscosity variations can be made electronically (11). The turbine meter is very susceptible to installation errors caused by pipe flow swirl (7), and a careful selection of installation position must be made.

Transit Time and Doppler (Ultrasonic) Flow Meters

Ultrasonic meters use sound waves to determine flow rate. *Transit time flow meters* use the travel time of ultrasonic waves to estimate average flow velocity. Figure 10.20 shows a pair of transducers, separated by some distance, fixed to the outside of a pipe wall. A reflector applied

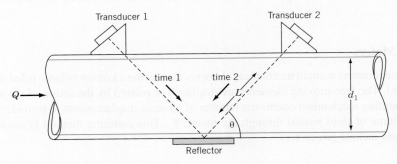

Figure 10.20 Principle of a transit time (ultrasonic) flow meter.

to the opposite outside wall of the pipe increases the signal-to-noise ratio. Each transducer acts as a transmitter and a receiver for ultrasonic waves. An ultrasonic wave emitted by one transducer passes through the fluid, reflects off the pipe wall, and is received by the other transducer. The difference in transit time for a wave to travel from transducer 1 to transducer 2 and from transducer 2 to transducer 1 is directly related to the average velocity of flow in the pipe. For a fluid with speed of sound a, for flow with average velocity $\overline{U}$ based on a flow rate Q, and for a beam oriented at angle θ relative to the pipe flow axis,

$$t_1 = \frac{2L}{a + \overline{U}\cos\theta} \quad t_2 = \frac{2L}{a - \overline{U}\cos\theta}$$

with $L = d_1/\sin\theta$. When $\overline{U} \ll a$, $\overline{U} = (a^2/4d_1\cot\theta)(t_2 - t_1)$. Or, eliminating a, we obtain $\overline{U} = (L/\cos\theta)(t_2 - t_1)/t_2 t_1$. The flow rate is

$$Q = \overline{U}A = K_1 A(t_2 - t_1) \tag{10.28}$$

where K_1 is a constant called the meter K-factor. The method requires accurate measurements of t_1 and t_2.

Transit time meters are noninvasive and thus offer no pressure drop. Portable models are strapped to the outside of the pipe, making them useful for field diagnostics. They can be set up to measure time-dependent velocity, including flow direction with time. Instrument relative systematic uncertainty ranges from $\sim$1% to 5% (95%) of the flow rate.

Doppler flow meters use the Doppler effect to measure the average velocity of fluid particles (contaminants, particulate, or small bubbles) suspended in the pipe flow as introduced in Chapter 9. In one approach for a volume flow meter, a single ultrasonic transducer is mounted to the outside of a pipe wall. A wave of frequency f on the order of 100 kHz to 1 MHz is emitted. The emitted wave is bounced back by particles suspended in the moving fluid, and this scattered wave is detected by the transducer. The scattered waves have a slightly different frequency, the difference being the Doppler frequency f_D. The flow rate is

$$Q = \overline{U}A = \frac{\pi d_1^2 a f_D}{8f\cos\theta} = K_1 f_D \tag{10.29}$$

These devices can measure time-dependent flow rates. The value K_1 is sensitive to pipe wall material and thickness and fluid properties. Instrument systematic uncertainty is $\sim$2% (95%) of the flow rate.

Positive Displacement Meters

Positive displacement meters contain mechanical elements that define a known volume filled with the fluid being metered. The free-moving elements are displaced or rotated by the action of the moving fluid. A geared counting mechanism counts the number of element displacements to provide a direct reading of the volume of fluid passed through the meter, $\forall$. This metering method is common to water, gasoline, and natural gas meters.

To be used as volume flow rate meters, volume measurement can be used in conjunction with a timer so that

$$Q = \forall / t \qquad (10.30)$$

These units serve applications needing ruggedness and accuracy in steady flow metering. Common applications are metering related to domestic water and natural gas consumption.

Diaphragm meters contain two opposing flexible bellows coupled by linkages to exhaust valves that alternately displace a known volume of fluid through the meter. Through the coupled linkage, the flow of fluid through one bellows drives the motion of the opposite bellows. Thus, as entering gas causes one bellows to expand within its chamber sealed by a closed valve, the gas in the opposite bellows exhausts through an open valve to allow its bellows to collapse; this alternating motion cycles the gears of its counter register (output display). Such "dry gas" meters are common to natural gas or propane lines for volume measurement. Thomas Glover invented the first diaphragm meter in 1843, and modern meters retain the major elements of his design.

Wobble meters contain a disk that is seated in a chamber. The flow of liquid through the chamber causes the disk to oscillate so that a known volume of fluid moves through the chamber with each oscillation. The disk is connected to a counter that records each oscillation as volume is displaced. Wobble meters are used for domestic water applications, where they must have an uncertainty of no greater than 1% of actual delivery.

Frequently found on oil trucks and at the gasoline pump, *rotating vane meters* use rotating cups or vanes that move about an annular opening, displacing a known volume with each rotation.

With any of these meters, uncertainty can be as low as 0.2% (95%) of actual delivery. Because of the good accuracy of these meters, they are often used as local working standards to calibrate other types of volume flow meters.

Example 10.9: Focus on the Vortex Flow Meter

A modern vortex flow meter finds applications in liquids, gases and steam measurements. Most meters use a proprietary shedder mounted on an instrumented strut on which a piezoelectric or capacitance sensor responds to the strut vibrations induced by vortex shedding. The sensor signal is passed through a charge amplifier, then into a low-pass filter and amplifier having enough gain to boost the signal into the mV to V range.

An example of a conditioned and strong vortex shedding signal from the sensor is shown in Figure 10.21(a). A high/low trigger level within the circuitry detects when the sensor signal crosses the threshold. As the signal passes through the high and then low trigger levels, this generates a TTL signal of corresponding duration. With regular vortex shedding, the TTL pulses per unit time will equal the shedding frequency. The flow meter applies the preprogrammed K-factor to the counted frequency. The output signal (e.g., LCD display, 4-20 mA, voltage, or pulse) will correspond to the measured flow rate.

If the meter is used outside of its range, the vortex signal will be erratic, noisy, at too high of a frequency, or of insufficient magnitude to even be detected by the sensor. This is illustrated in Figure 10.21(b), in which the TTL signal is erratic or missing due to too low a flow rate. To avoid such mishaps, the manufacturer will specify the flow rate range of the meter for both liquids and gases, the pipe size range that matches the meter size, and the permanent pressure loss for the meter.

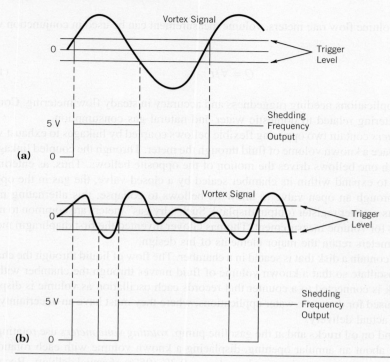

Figure 10.21 Triggering of vortex shedding signal creates a TTL pulse sequence. (**a**) Good signal. (**b**) Improperly sized meter signal.

10.7   MASS FLOW METERS

There are many situations in which the mass flow rate is the quantity of interest. Mass flow rate demands measuring the mass flux, $\overline{\rho U}$, to yield $\dot{m} = \overline{\rho U}A$.

If the density of the metered fluid is known under the exact conditions of measurement, then a direct estimate can be made based on volume flow rate measurements. But not all fluids have a constant, known value of density (e.g., petroleum products, polymers, and cocoa butter), and many processes are subject to significant changes in density. The direct measurement of mass flow rate is desirable because it eliminates the uncertainties associated with estimating or measuring actual density.

The difference between a meter that is sensitive to Q as opposed to $\dot{m}$ is not trivial. Before the 1970s, reliable commercial mass flow meters with sufficient accuracy to circumvent volume flow rate corrections were generally not available, even though the basic principles and implementation schemes for such meters had been understood in theory for several decades. U.S. patents dating back to the 1940s record schemes for using heat transfer, Coriolis forces, and momentum methods to infer mass flow rate directly.

Thermal Flow Meter

The rate at which energy, $\dot{E}$, must be added to a flowing fluid to raise its temperature between two control surfaces is directly related to the mass flow rate by

$$\dot{E} = \dot{m}c_p\Delta T \tag{10.31}$$

where c_p is the fluid specific heat. Methods to utilize this effect to directly measure mass flow rate incorporate an in-line meter having some means to input energy to the fluid over the meter length. The passing of a current through an immersed filament is a common method. Fluid temperatures are measured at the upstream and downstream locations of the meter. This type of meter is quite easy to use and appears to be reliable. It is widely used for gas flow applications. In fact, in the 1980s, the technique was adapted for use in automobile fuel injection systems to provide an exact air–fuel mixture to the engine cylinders despite short-term altitude, barometric, and seasonal environmental temperature changes.

The operating principle of this meter assumes that c_p is known and remains constant over the length of the meter. For common gases, such as air, this assumption is quite good. Flow rate turndown of up to 100:1 is possible with uncertainties of 0.5% (95%) of flow rate with very little pressure drop. But the assumptions become restrictive for liquids and for gases for which c_p may be a strong function of temperature.

A second type of thermal mass flow meter is a velocity-sensing meter and thermal sensor together in one direct insertion unit. The meter uses both hot-film anemometry methods to sense fluid velocity through a conduit of known diameter and an adjacent resistance temperature detector (RTD) sensor for temperature measurement. For sensor and fluid temperatures, T_s and T_f, respectively, mass flow rate is inferred from the correlation

$$\dot{E} = \left[C + B(\rho \overline{U})^{1/n} \right] (T_s - T_f) \tag{10.32}$$

where C, B, and n are constants that depend on fluid properties (12) and are determined through calibration. In a scheme to reduce the fluid property sensitivity of the meter, the RTD may be used as an adjacent resistor leg of the anemometer Wheatstone bridge circuit to provide a temperature-compensated velocity output over a wide range of fluid temperatures with excellent repeatability (0.25%). Gas velocities of up to 12,000 ft/min and flow rate turndown of 50:1 are possible with uncertainties down to 2% of the flow rate and very little pressure drop.

Coriolis Flow Meter

The term "Coriolis flow meter" refers to the family of insertion meters that meter mass flow rate by inducing a Coriolis acceleration on the flowing fluid and measuring the resulting developed force (13). The developed force is directly related to the mass flow rate independent of the fluid properties. The Coriolis effect was proposed by Gaspard de Coriolis (1792–1843) after his studies of accelerations in rotating systems. Coriolis meters pass a fluid through a rotating or vibrating pipe system to develop the Coriolis force. A number of methods using this effect have been proposed since the first U.S. patent for a Coriolis effect meter was issued in 1947. This family of meters has seen steady growth in market share since the mid-1980s.

In the most common scheme for commercially available units, the pipe flow is diverted from the main pipe and divided between two bent, parallel, adjacent tubes of equal diameter, such as shown for the device in Figure 10.22, a device developed in part by the first author of this text.[3] The tubes themselves are mechanically vibrated in a relative out-of-phase sinusoidal oscillation by an electro-magnetic driver. In general, a fluid particle passing through the meter tube, which is rotating (due to

---

[3] In its initial beta test, the unit shown in Figure 10.22 was installed in a chocolate factory to meter the mass flow of cocoa butter.

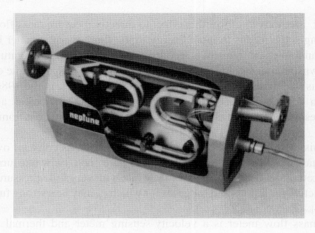

Figure 10.22 Cutaway view of a Coriolis mass flow meter. (Courtesy of Red Seal Measurement, Greenwood, SC.)

the oscillating tube) relative to the fixed pipe, experiences an acceleration at any arbitrary position S. The total acceleration at S, $\ddot{\mathbf{r}}$, is composed of several components (Fig. 10.23),

$$\ddot{\mathbf{r}} = \ddot{\mathbf{R}}o' + \dot{\boldsymbol{\omega}} \times \mathbf{r}_{S/O'} + \boldsymbol{\omega} \times \boldsymbol{\omega} \times \mathbf{r}_{S/O'} + \ddot{\mathbf{r}}_{S/O'} + 2\boldsymbol{\omega} \times \dot{\mathbf{r}}_{S/O'} \qquad (10.33)$$

where boldface refers to vector quantities and

$$\ddot{\mathbf{R}}o' = \text{translation acceleration of rotating origin } O' \text{ relative to fixed origin } O$$
$$\boldsymbol{\omega} \times \boldsymbol{\omega} \times \mathbf{r}_{S/O'} = \text{centripetal acceleration of } S \text{ relative to } O'$$
$$\dot{\boldsymbol{\omega}} \times \mathbf{r}_{S/O'} = \text{tangential acceleration of } S \text{ relative to } O'$$
$$\ddot{\mathbf{r}}_{S/O'} = \text{translational acceleration of } S \text{ relative to } O'$$
$$2\boldsymbol{\omega} \times \dot{\mathbf{r}}_{S/O'} = \text{Coriolis acceleration at } S \text{ relative to } O'$$

with ω the angular velocity of point S relative to O'. In these meters, the tubes are rotated but not translated, so that the translational accelerations are zero. A fluid particle experiences forces because of the remaining accelerations that cause equal and opposite reactions on the meter tube

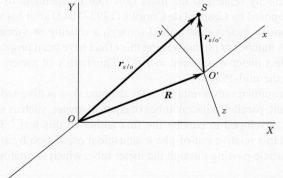

Figure 10.23 Fixed and rotating reference frames.

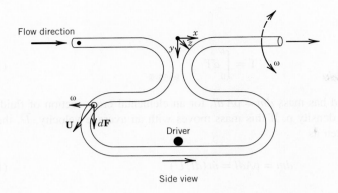

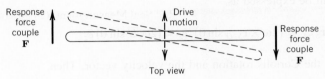

Figure 10.24 Concept of the operating principle of a Coriolis mass flow meter. Use the right-hand rule to relate the vectors of $\boldsymbol{\omega}$, $\mathbf{U}$, and $d\mathbf{F}$.

walls. The Coriolis acceleration distinguishes itself by acting in a plane perpendicular to the tube axes and develops a force gradient that creates a twisting motion or oscillating rotation about the tube plane.

The use of the Coriolis force depends on the shape of the meter. However, the basic principle is illustrated in Figure 10.24. Rather than rotating the tubes a complete 360 degrees about the pipe axis, the meter tubes are vibrated continuously at a drive frequency, ω, with amplitude displacement, z, about the pipe axis. This eliminates rotational seal problems. The driving frequency is selected at the tube resonant frequency that places the driven tube into what is called a limit cycle, a continuous, single-frequency oscillation. This configuration is essentially a self-sustaining tuning fork, because the meter naturally responds to any disturbance at this frequency with a minimum in input energy. By driving the tube up and down, the mass flow causes the tube to wobble. The magnitude of wobble is directly related to the mass flow rate.

As a fluid particle of elemental mass, dm, flows through a section of the flow meter, it experiences the Coriolis acceleration, $2\boldsymbol{\omega} \times \dot{\mathbf{r}}_{S/O'}$, and an inertial Coriolis force

$$d\mathbf{F} = \left(2\boldsymbol{\omega} \times \dot{\mathbf{r}}_{S/O'}\right) dm \qquad (10.34)$$

acting in the z direction. For the flow meter design of Figure 10.22 and as depicted in Figure 10.24, as the particle travels along the meter the direction of the velocity vector changes. This results, using the right-hand rule, in a change in direction of vector force, $d\mathbf{F}$, between the left and right sides. The resultant forces experienced by the tube are of equal magnitude but the opposite sign of those experienced by the particle. Each tube segment senses a corresponding differential torque, $d\mathbf{T}$, a rotation about the y axis at a frequency $\boldsymbol{\omega}_c$,

$$d\mathbf{T} = x' \times d\mathbf{F} = x' \times \left(2\boldsymbol{\omega} \times \dot{\mathbf{r}}_{S/O'}\right) dm \qquad (10.35)$$

where x' refers to the x distance between the elemental mass and the y axis, $\mathbf{x}' = \mathbf{x}'\hat{e}_x$. The total magnitude of torque experienced by each tube is found by integration along the total path length of the tube, L,

$$T = \int_0^L d\mathrm{T} \tag{10.36}$$

A differential element of fluid has mass $dm = \rho A\, dl$, for an elemental cross section of fluid A, of differential length dl, and of density ρ. If this mass moves with an average velocity, $\overline{U}$, then the differential mass can be written as

$$dm = \rho A\, dl = \dot{m}\,(dl/\overline{U}) \tag{10.37}$$

The Coriolis cross-product can be expressed as

$$2\boldsymbol{\omega} \times \dot{\mathbf{r}}_{S/O'} = \left(2\boldsymbol{\omega} \times \dot{\mathbf{r}}_{S/O'} \sin\theta\right)\hat{e}_z = (2\omega_c \overline{U} \sin\theta)\,\hat{e}_z \tag{10.38}$$

where θ is the angle between the Coriolis rotation and the velocity vector. Then,

$$\dot{m} = \frac{T}{2\displaystyle\int_0^L (\rho x' \omega_c \sin\theta)dl}\,\hat{e}_y \tag{10.39}$$

Because the velocity direction changes by 180 degrees, the Coriolis-developed torque acts in opposite directions on each side of the tube.[4] The meter tubes twist about the y axis of the tube (i.e., wobbles) at an angle δ. For small angles of rotation the twist angle is related to torque by

$$\delta = k_s T = \text{constant} \times \dot{m} \tag{10.40}$$

where k_s is related to the stiffness of the tube. The objective becomes to measure the twist angle, which can be accomplished in many ways. For example, by driving the 2-m tubes 180 degrees out of phase, the relative phase at any time between the two tubes is directly related to the mass flow rate. The exact relationship is linear over a wide flow range and determined by calibration with any fluid (14).

The tangential and centripetal accelerations remain of minor consequence due to the tube stiffness in the directions in which they act. However, at high mass flow rates they can excite modes of vibration in addition to the driving mode, ω, and response mode, ω_c. In doing so, they affect the meter linearity and zero error (drift), which affects mass flow uncertainty and meter turndown. Essentially, the magnitude of these undesirable effects is inherent to the particular meter shape and is controlled, if necessary, through tube-stiffening members.

Another interesting problem occurs mostly in meters of small tube diameter where the tube mass may approach the mass of the fluid in the tube. At flow rates that correspond to flow transition from a laminar to turbulent regime, the driving frequency can excite the flow instabilities responsible for the flow transition. The fluid and tube can go out of phase, reducing the response amplitude and its

---

[4] We can envision a design in which the velocity does not reverse direction along the flow path but ω does.

corresponding torque. This affects the meter's calibration linearity, but a good design can contain this effect to within 0.5% of the meter reading.

The meter principle is unaffected by changing fluid properties, but temperature changes affect the overall meter stiffness, an effect that can be compensated for electronically. A very desirable feature is an apparent insensitivity to installation position. Commercially available Coriolis flow meters can measure mass flow rate with an instrument systematic uncertainty in liquids to 0.10% (95%), gases to 0.5% (95%). Turndown is about 20:1. The meter is also used as an effective densitometer that can measure to within 0.0005 g/cm^3.

Example 10.10: The automobile mass air flow sensor

The mass air flow sensor (MAF) in a modern automobile engine measures the mass flow rate (grams/s) of air entering through the engine intake and delivered to the engine cylinders. This sensor sends a signal to the electronic control unit (ECU) that makes corresponding adjustments to maintain a desirable combustion air–fuel ratio for the applied throttle (power requirement) setting.

A common scheme for an MAF unit is depicted in Figure 10.25. The device uses a heated element, which can take on various forms such as a metallic hot-film probe (Chapter 9) or a platinum hot-wire, and an air temperature sensor, such as a thermistor. These form two active arms of a resistance bridge used to monitor the sensor temperature in relation to the incoming air temperature (IAT). With the engine operating, the element is heated to a pre-set temperature above the ambient temperature. As air flow moves through the intake duct, it will cool the heated sensor down. As a conductor's resistance is a nearly linear function of temperature (Chapter 8), the sensor resistance changes with this cooling. The resulting resistance change unbalances the bridge and so the bridge circuitry acts to increase the current to the element to bring its temperature (and resistance) back to the pre-set value. The resulting bridge deflection voltage is directly proportional to the mass flow rate, and this information is sent to the controlling ECU for appropriate action.

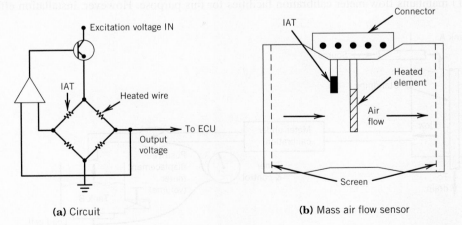

(a) Circuit   **(b)** Mass air flow sensor

Figure 10.25 Schematic of an automobile mass air flow sensor and its bridge circuit.

10.8   FLOW METER CALIBRATION AND STANDARDS

While a fundamental primary standard for flow rate does not exist per se, there are a number of calibration test code procedures in place, and government and private bureaus that perform calibrations. The general procedure for the calibration of inline flow meters requires establishing a steady flow in a calibration flow loop and then determining the volume or mass of flowing fluid passing through the flow meter in an accurately determined time interval. Such flow loop calibration systems are known as *provers*. Often calibration is by comparison to a meter of proven accuracy. Several methods to establish the flow rate are discussed.

In liquids, variations of a "catch-and-weigh" technique are often employed in flow provers. One variation of the technique consists of a calibration loop with a catch tank as depicted in Figure 10.26. Tank A is a large tank from which fluid is pumped back to a constant head reservoir, which supplies the loop with a steady flow. Tank B is the catch-and-weigh tank into which liquid can be diverted for an accurately determined period of time. The liquid volume is measured, either directly using a positive displacement meter, or indirectly through its weight, and the flow rate deduced through time. The ability to determine the volume and the uncertainty in the initial and final time of the event are fundamental limitations to the accuracy of this technique. Neglecting installation effects, the ultimate limits of uncertainty (at 95%) in the flow rate of liquids are on the order of 0.03%, a number based on $u_{\text{meter}} \sim 0.02\%$, $u_{\forall} \sim 0.02\%$, and $u_t \sim 0.01\%$.

Flow meter calibration by measured pipe velocity profile is particularly effective for *in situ* calibration in both liquids and gases, provided that the gas velocity does not exceed about 70% of the sonic velocity. Velocity traverses at any cross-sectional location some 20 to 40 pipe diameters downstream of any pipe fitting in a long section of straight pipe are preferred.

Comparison calibration against a local standard flow meter is another common means of establishing the flow rate. Flow meters are installed in tandem with the standard and directly calibrated against it. Turbine and vortex meters and Coriolis mass flow meters have consistent, highly accurate calibration curves and are often used as local standards. Other provers use an accurate positive displacement meter to determine flow volume over time. Of course, these standards must be periodically recalibrated. In the United States, the National Institute of Standards and Technology (NIST) maintains flow meter calibration facilities for this purpose. However, installation effects in

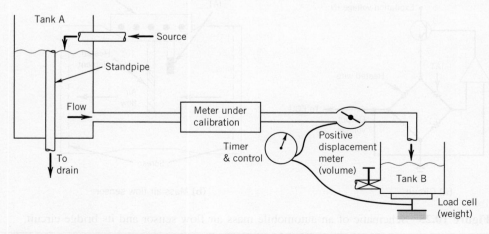

Figure 10.26 Flow diagram of a flow meter prover for liquids.

the end-user facility are not accounted for in an NIST calibration. This last point accounts for much of the uncertainty in a flow meter calibration and use.

In any of these methods, the calibration uncertainty is limited by the standard used, installation effects, end-use variations, and the inherent limitations of the flow meter calibrated.

10.9   ESTIMATING STANDARD FLOW RATE

When the range of pressures and temperatures for flow processes vary in use, the measured "actual" flow rates can be adjusted to a standard temperature and pressure for comparison. This adjusted flow rate is called the *standard flow rate*. Standard flow rates are often reported in units such as standard cubic meters per minute (SCMM) or standard cubic feet per minute (SCFM). They are found by converting from the actual conditions at the measurement point, with the actual flow rate noted with corresponding units, such as ACMM or ACFM, in which A stands for "actual." The mass flow rate remains the same regardless of conditions, so using subscript s for the standard conditions and a for the actual measured conditions, we have $\dot{m}_s = \dot{m}_a$, so that

$$Q_s = Q_a \frac{\rho_a}{\rho_s} \tag{10.41}$$

For example, the actual flow rates of common gases are "standardized" to 760 mm Hg absolute and 20 °C. Assuming ideal gas behavior, $\rho = p/RT$, the standard flow rate is found from Equation 10.41 as

$$Q_s = Q_a \frac{T_s\, p_a}{T_a\, p_s} = Q_a \left(\frac{293}{273 + T_a} \right) \left(\frac{760 + p_a}{760} \right) \tag{10.42}$$

10.10   SUMMARY

Flow quantification has been an important engineering task for well over two millennia. This chapter presents methods to determine the volume rate of flow and the mass rate of flow. The engineering decision involving the selection of a particular meter depends on a number of constraining factors. In general, flow rate can be determined to within about 0.25% of actual flow rate with the best of present technology but practical values for industrial installations are more nearly 3–6% for obstruction meters and 1–3% for insertion meters. However, new methods may push these lower limits even further, provided that calibration standards can be developed to document method uncertainties and the effects of installation.

REFERENCES

1. American Society of Mechanical Engineers, *PTC 19.5—Flow Measurements*, ASME International, New York, 2004 (Rev. 2013).
2. American Society of Heating, Refrigeration and Air Conditioning Engineers (ASHRAE), *ASHRAE Fundamentals*, rev. ed., ASHRAE, Atlanta, 2009.
3. American Society of Mechanical Engineers, *Measurement of Fluid Flow in Pipes Using Orifice, Nozzle and Venturi*, ASME Standard MFC-3M-2004, ASME International, New York, 2004.

4. International Organization for Standardization (ISO), *Measurement of fluid flow by means of pressure differential devices inserted in circular cross-section conduits: Parts 1 through 4*, ISO 5167, ISO, Geneva, 2003.

5. Rouse, H., and S. Ince, *History of Hydraulics*, Dover, New York, 1957.

6. Amberg, B. T., A review of critical flowmeters for gas flow measurements, *Transactions of the ASME*, **84**, 1962, pp. 447–460.

7. Mattingly, G., *Fluid measurements: Standards, Calibrations and Traceabilities*, Proceedings of the ASME/AIChE National Heat Transfer Conference, Philadelphia, PA, 1989.

8. Shercliff, J. A., *Theory of Electromagnetic Flow Measurement*, Cambridge University Press, New York, 1962.

9. da Vinci, L., *Del Moto e Misura Dell'Aqua* (English translation), E. Carusi and A. Favaro, eds., Zanichelli, Bologna, 1923.

10. Hochreiter, H. M., Dimensionless correlation of coefficients of turbine-type flow-meters, *Transactions of the ASME*, **80**, 1958, pp. 1363–1368.

11. Lee, W. F., and H. Karlby, A study of viscosity effects and its compensation on turbine flow meters, *Journal of Basic Engineering*, **82**, 1960, pp. 717–728.

12. Hinze, J. O., *Turbulence*, McGraw-Hill, New York, 1953.

13. American Society of Mechanical Engineers, *Measurement of Fluid Flow by Means of Coriolis Mass Flow Meters*, MFC-11M, ASME International, 2003.

14. Corwon, M., and R. Oliver, Omega-shaped Coriolis-type Mass Flow Meter System, U.S. Patent 4,852,410, 1989.

15. Jorgensen, R. (ed), *Fan Engineering*, 9th ed., Buffalo Forge Co., Buffalo, New York, 1999.

16. Crane Company, *Flow of Fluids through Valves and Fittings*, Technical Paper No. 410, Spiral Edition, Crane Co., Chicago, 2009.

NOMENCLATURE

c_p, c_v	specific heats (l^2/t^2-°)		H	manometer deflection (l)
d	diameter (l)		K_0	flow coefficient ($= CE$)
d_0	flow meter throat or minimum diameter (l)		K_1	flow meter constant; K-factor; static sensitivity
d_1	pipe diameter (l)		L	length (l)
d_2	vena contracta diameter (see Fig. 10.3) (l)		Q	volume flow rate ($l^3 t^{-1}$)
			R	gas constant
f	cyclical frequency ($\omega/2\pi$) (t^{-1})		Re_{d_1}	Reynolds number (based on d_1)
g	gravitational acceleration constant (lt^{-2})		$\dot{m}$	mass flow rate (mt^{-1})
			p	pressure ($ml^{-1}t^{-2}$)
$h_{L_{1-2}}$	energy head loss between points 1 and 2 (l)		$p_1 - p_2, \Delta p$	pressure differential ($ml^{-1}t^{-2}$)
			Δp_{loss}	permanent pressure loss ($ml^{-1}t^{-2}$)
k	ratio of specific heats, c_p/c_v		r	radial coordinate (l)
C	discharge coefficient		r_1	pipe radius (l)
C_D	drag coefficient		r_H	hydraulic radius (l)
E	velocity of approach factor; voltage		A	area (l^2)
Gr	Grashof number		A_0	area based on d_0 (l^2)

A_1 area based on pipe diameter d_1 (l^2)
A_2 vena contracta area (l^2)
$\boldsymbol{B}$ magnetic field flux vector
S specific gravity
T temperature (°); torque (ml^2t^{-2})
U velocity (lt^{-1})
$\forall$ volume (l^3)

Y expansion factor
β diameter ratio
ρ density (ml^{-3})
δ twist angle
ω frequency (t^{-1}); angular velocity (t^{-1})
μ absolute viscosity $(ml^{-1}t^{-1})$
v kinematic viscosity $(= \mu/\rho)$ (l^2t^{-1})

PROBLEMS

10.1 Determine the average mass flow rate of 5 °C air at 1 bar abs through a 5-cm-i.d. pipe whose velocity profile is found to be symmetric and described by $U(r) = 25\left[1 - (r/r_1)^2\right]$ cm/s

10.2 A 20-cm-i.d. pipe through which 10 °C air flows is traversed along three radial lines with measurements taken at five equidistant stations along each radial. Determine the average pipe flow rate.

Radial		$U(r)$ (cm/s)		
Location	r (cm)	Line 1	Line 2	Line 3
1	1.0	25.31	24.75	25.10
2	3.0	22.48	22.20	22.68
3	5.0	21.66	21.53	21.79
4	7.0	15.24	13.20	14.28
5	9.0	5.12	6.72	5.35

10.3 What is the best estimate of the pipe flow rate for Problem 10.2, accounting for spatial variation and if the systematic standard uncertainty of the instrument used is 1% of the reading?

10.4 A mercury-filled ($S = 13.57$) manometer is used in a 5.1-cm water line to measure the pressure drop across a flow meter. If the manometer deflection is 10.16 cm Hg, determine the pressure drop in N/m^2.

10.5 A capacitance pressure transducer is used to measure the pressure differential across an orifice meter located in a 25.4-cm pipe. For air flowing at 32 °C, if the pressure differential is 69 kPa, find the pressure head in cm H$_2$O.

10.6 Estimate the expansion factor in measuring the flow of O$_2$ at 16 °C through a 100-mm-i.d. pipe using a square-edged orifice plate. The upstream pressure is 6.9 bar and the downstream pressure is 5.24 bar. The orifice diameter is 50 mm. $R_{O_2} = 260$ J/kg-K.

10.7 The Reynolds number of a fluid flowing through a 5-cm-diameter, square-edged orifice plate meter within a 15-cm pipe is 250,000. Find the discharge coefficient for flange taps.

10.8 At what flow rate of 20 °C water through a 10-cm-i.d. pipe would the discharge coefficient of a square-edged orifice plate ($\beta = 0.4$) become essentially independent of the Reynolds number?

10.9 Water (25 °C) flows through a square-edged orifice plate ($\beta = 0.5$) housed within a 12-cm-i.d. pipe at 50 L/s. Estimate the power required to overcome the orifice plate's permanent pressure loss. Assume flange taps.

10.10 An orifice plate is installed to meter air flow in a 0.25 m diameter pipe. A flow rate of up to 1 m^3/s is expected. A transducer with maximum range of 300 mm H$_2$O is available to measure the pressure drop. Specify an acceptable diameter for the orifice plate to maximize the signal. Assume the flow remains incompressible ($Y = 1$).

10.11 Determine the flow rate of 38 °C air through a 6-cm pipe if an ASME flow nozzle with a 3-cm throat and throat taps are used, the pressure drop measured is 75 cm H$_2$O, and the upstream pressure is 94.4 kPa abs.

10.12 A square-edged orifice ($\beta = 0.5$) is used to meter N$_2$ at 520 °R through a 4-in. pipe. If the upstream pressure is 20 psia and the downstream pressure is 15 psia, determine the flow rate through the pipe. Flange taps are used. $R_{N_2} = 55.13$ ft-lb/lbm-°R.

10.13 Size a suitable orifice plate to meter the steady flow of water at 20 °C flowing through a 38-cm pipe if the nominal flow rate expected is 200 kg/s and the pressure drop must not exceed 15 cm Hg.

10.14 An in-line flow nozzle is to be used to measure the flow of ethyl alcohol ($\mu = 1.19 \times 10^{-3}$ N-s/m^2; $\rho = 789$ kg/m^3) through a 60 mm diameter pipeline. In selecting the nozzle diameter, the engineer wants about a 4,000 Pa pressure drop using a through pressure tap when the flow rate is 0.180 m^3/min. Size the nozzle diameter.

10.15   A cast venturi meter is to be used to meter the flow of 15 °C water through a 10-cm pipe. For a maximum differential pressure of 76 cm H_2O and a nominal 0.5-m^3/min flow rate, select a suitable throat size.

10.16   For 120 ft^3/m of 60 °F water flowing through a 6-in. pipe, size a suitable orifice, venturi, and nozzle flow meter if the maximum pressure drop cannot exceed 20 in. Hg. Estimate the permanent losses associated with each meter. Compare the annual operating cost associated with each if the power cost is $0.10/kW-h, a 60% efficient pump motor is used, and the meter is operated 6,000 hr/year.

10.17   Estimate the flow rate of water through a 15-cm-i.d. pipe that contains an ASME long radius nozzle ($\beta = 0.6$) if the pressure drop across the nozzle is 25.4 cm Hg. Water temperature is 27 °C.

10.18   A 2-in. (50.8 mm) diameter orifice plate is installed within a 3-in. (76.2 cm) diameter pipe and uses flange taps. The engineer needs to monitor water flow rate through the pipe. She anticipates that water will flow through the pipe at 54 ft^3/min (1.53 m^3/min) and needs to select a pressure transducer. Estimate the pressure drop in bars.

10.19   In order to measure the flow rate in a 2 m × 2 m air conditioning duct, an engineer uses a pitot-static probe to measure dynamic head. The duct is divided into nine equal rectangular areas with the pressure measured at the center of each. Based on the results below, estimate the flow rate for air at 15 °C and 1 atm. Flow rate is average velocity times area.

Position	1	2	3	4	5	6	7	8	9
H (mm H_2O)	5.0	6.0	6.5	6.0	5.0	6.5	7.5	7.0	5.0

10.20   A flow nozzle is to be used at choked conditions to regulate the flow of air at 1.3 kg/s through a 6-cm-i.d. pipe. Upstream stagnation conditions are 690 kPa abs and 20 °C. Determine the maximum β ratio nozzle that can be used. $R_{N_2} = 297$ N-m/kg-K. Use C = 1.

10.21   Compute the flow rate of 20 °C air through a 0.5-m square-edged orifice plate that is situated in a 1.0-m-i.d. pipe. The pressure drop across the plate is 90 mm H_2O, and upstream pressure is 2 atm.

10.22   An ASME long radius nozzle ($\beta = 0.5$) is to be used to meter the flow of 20 °C water through a 8-cm-i.d. pipe. If flow rates range from 0.6 to 1.6 L/s, select the range required of a single pressure transducer if that transducer is used to measure pressure drop over the flow rate range. If the transducer to be used has a typical error of 0.25% of full scale, estimate the uncertainty in flow rate at the design stage.

10.23   A square-edged orifice plate is selected to meter the flow of water through a 2.3-in.-i.d. pipe over the range of 10 to 50 gal/min at 60 °F. It is desired to operate the orifice in the range where C is independent of the Reynolds number for all flow rates. A pressure transducer having accuracy to within 0.5% of reading is to be used. Select a suitable orifice plate and estimate the design-stage uncertainty in the measured flow rate at 10, 25, and 50 gal/min. Assume flange pressure taps and reasonable values for the systematic uncertainty on pertinent parameters.

10.24   Estimate the error contribution to the uncertainty in flow rate due to the effect of the relative humidity of air on air density. Consider the case of using an orifice plate to meter air flow if the relative humidity of the 70 °F air can vary from 10% to 80% but a density equivalent to 45% relative humidity is used in the computations.

10.25   For Problem 10.24, suppose the air flow rate is 17 m^3/hr at 20 °C through a 6-cm-i.d. pipe, a square-edged orifice ($\beta = 0.4$) is used with flange taps, and the pressure drop can be measured to within 0.5% of reading for all pressures above 5 cm H_2O using a manometer. If basic dimensions are maintained to within 0.1 mm, estimate a design-stage uncertainty in the flow rate. Use $p_1 = 96.5$ kPa abs, $R = 287$ N-m/kg-K.

10.26   An application uses water flowing at up to 0.2 m^3/min of 20 °C through a 10-cm-diameter pipe. The engineer wants to select from a set of orifice plates using flange taps to measure and monitor the flow. It is desired that the meter develop a nominal 250 mm Hg pressure head. Specify an appropriate size (d_o) for the orifice meter. $S_{Hg} = 13.6$.

10.27   Dry air at a static pressure and temperature of 2.07 MPa abs and 38 °C moves at 0.50 kg/s through a 76 mm inside diameter pipe. A sonic nozzle with throat diameter of 11.3 mm is used to meter the flow. Estimate the velocity in the upstream pipe and the dynamic pressure. Estimate the difference between using the static and stagnation conditions to estimate the flow rate. Use $C = 1$, $k = 1.4$, $R = 287$ N-m/kg-K.

10.28 Dry air at a stagnation pressure and temperature of 105 kPa abs and 43 °C moves through a 76 mm inside diameter pipe. A sonic nozzle with throat diameter of 11.3 mm is used to meter the flow. Estimate the flow rate. $C = 0.98$, $k = 1.4$, $R = 287$ N-m/kg-K.

10.29 A sonic nozzle can be used to regulate flow rate provided that the critical pressure ratio is maintained. While independent of downstream conditions, the upstream properties adjust themselves to the given set of choked flow conditions. Fortunately, the mass flow rate can be computed from the measured stagnation properties. For a range of 101.3 ± 7 kPa abs and $10° \pm 5°$C, estimate the uncertainty in regulating a mass flow rate at the critical pressure ratio due to these variations alone.

10.30 Select an appropriate range for a differential pressure transducer to be used with an ASME long radius nozzle with $\beta = 0.5$. This system is to be used to measure the flow of 20 °C water through a 20-cm diameter pipe. The operating flow rate expected is between 5,000 cm³/s and 50,000 cm³/s. Estimate the permanent pressure loss associated with this nozzle. As an optional assignment, go to a vendor catalog or website and select an appropriate pressure transducer from its listings that has a 0- to 5-V DC output.

10.31 From a vendor catalog or online site, select a differential pressure transducer to measure between 203 Pa and 20,100 Pa across an obstruction meter. The transducer must be compatible with water. The output signal is to be measured using a data-acquisition system with 0 to 5 V, 12-bit analog-to-digital converter. Estimate the percent relative quantization error at the low and high flow rates. Based on your selection, do you recommend any signal conditioning?

10.32 A vortex flow meter uses a shedder having a Strouhal number of 0.20. Estimate the mean duct velocity if the shedding frequency indicated is 77 Hz and the shedder characteristic length is 1.27 cm.

10.33 A thermal mass flow meter is used to meter 30 °C air flow through a 2-cm-i.d. pipe. If 25 W of power are required to maintain a 1 °C temperature rise across the meter, estimate the mass flow rate through the meter. Clearly state any assumptions. $c_p = 1.006$ kJ/kg-K.

10.34 Research available thermal mass flow meters. Discuss the assumptions necessary for their use. Describe situations where they would be appropriate and not appropriate.

10.35 Fuel oil used in large sea vessels is known as bunker. A typical container ship travels at 24 knots and burns 225 metric tons of bunker per day. In 2014, bunker sold for about $600/ton. Because of the large volumes used, the "tricks of the trade" can often hide fuel volume under measurement error and by manipulating transducers. Fuel is transferred using volumetric metering devices having 2% uncertainty with temperature devices with 1% uncertainty to correct for density. It is proposed to change the metering device to automated coriolis mass flow methods having 0.5% uncertainty, eliminating the temperature devices as well. Estimate the potential savings on a single ocean voyage.

10.36 Estimate an uncertainty in the determined flow rate in Example 10.5, assuming that dimensions are known to within 0.025 mm, the systematic uncertainty in pressure is 0.25 cm H_2O, and the pressure drop shows a standard deviation of 0.5 cm H_2O in 31 readings. Upstream pressure is constant. State your reasonable assumptions.

10.37 A thermal mass flow meter is used to meter air in a 1-cm-i.d. tube. The meter adds 10 W of energy to the air passing through the meter, from which the meter senses a 3 °C temperature gain. What is the mass flow rate? $c_p = 1.006$ kJ/kg-K.

10.38 A vortex meter is to use a shedder having a profile of a forward facing equilateral triangle (Table 10.1) with a characteristic length of 10 mm. Estimate the shedding frequency developed for 20 °C air at 30 m/s in a 10-cm-i.d. pipe. Estimate the meter constant and measured flow rate.

10.39 An engineer has an application of water at 20 °C flowing through DN100, Schedule 40 pipe. Available is a Rosemount 8800D flanged vortex flow meter, which specifies a usable range of 6.86 to 225 m³/hr in liquids. Suppose it uses a proprietary shedder design (St = 0.19) with a characteristic shedder diameter d of 26 mm. What range of shedding frequency should be expected?

10.40   The flow of air is measured to be 30 m³/min at 50 mm Hg and 15 °C. What is the flow rate in SCMM?

10.41   A 6 in. × 4 in. i.d. cast venturi is used to measure the flow rate of water. Estimate the flow rate and its uncertainty at 95%. The following values are known from a large sample:

x	Value	$b_{\bar{x}}$	$s_{\bar{x}}$
C	0.984	0.00375	0
d_o	3.995 in.	0.0005	0
d_1	6.011 in.	0.001	0
ρ	62.369 lb$_m$/ft³	0.002	0.002
H	100 in H$_2$O @ 68 °F	0.15	0.4

10.42   A simple method to measure volume flow rate is to catch a volume of liquid over a period of time, $Q = \forall/t$, Method 1 is to catch a known fixed volume over a measured time; Method 2 is to measure the volume captured in a fixed time. For Method 1, use 10 L. For Method 2, use 10 s. Estimate the volumes captured and times required for the two methods. Use flow rates of 3, 30, 300 L/min.

10.43   In the problem 10.42, suppose volume can be measured to an uncertainty of 0.010 L and time to an uncertainty of 0.10 s. Estimate the uncertainties in the measured flow rates. Discuss the preferred method for use with different flow rates? Assume 95% confidence.

Chapter **11**

Strain Measurement

11.1 INTRODUCTION

The design of safe load-carrying components for machines and structures requires information concerning the distribution of forces within the particular component. Proper design of devices such as shafts, pressure vessels, and support structures must consider load-carrying capacity and allowable deflections. Mechanics of materials provides a basis for predicting these essential characteristics of a mechanical design, and provides the fundamental understanding of the behavior of load-carrying parts. However, theoretical analysis is often not sufficient, and experimental measurements are required to achieve a final design.

Design criteria are based on stress levels within a part. In most cases stress cannot be measured directly. But the length of the rod in Figure 11.1 changes when the load is applied, and such changes in length or shape of a material can be measured. This chapter discusses the measurement of physical displacements in engineering components. The stress is calculated from these measured deflections.

Upon completion of this chapter, the reader will be able to

- Define strain and delineate the difficulty in measuring stress
- State the physical principles underlying mechanical strain gauges
- Analyze strain gauge bridge circuits
- Describe methods for optical strain measurement

11.2 STRESS AND STRAIN

Before we proceed to develop techniques for strain measurements, we briefly review the relationship between deflections and stress. The experimental analysis of stress is accomplished by measuring the deformation of a part under load, and inferring the existing state of stress from the measured deflections. Again, consider the rod in Figure 11.1. If the rod has a cross-sectional area of A_c, and the load is applied only along the axis of the rod, the *normal stress* is defined as

$$\sigma_a = F_N/A_c \qquad (11.1)$$

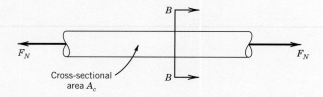

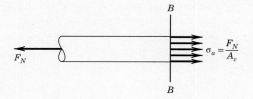

Figure 11.1 Free-body diagram illustrating internal forces for a rod in uniaxial tension.

where A_c is the cross-sectional area and F_N is the tension force applied to the rod normal to the area A_c. The ratio of the change in length of the rod (which results from applying the load) to the original length is the *axial strain*, defined as

$$\varepsilon_a = \delta L / L \tag{11.2}$$

where ε_a is the average strain over the length L, δL is the change in length, and L is the original unloaded length. For most engineering materials, strain is a small quantity; strain is usually reported in units of 10^{-6} in./in. or 10^{-6} m/m. These units are equivalent to a dimensionless unit called a microstrain ($\mu\varepsilon$).

Stress–strain diagrams are very important in understanding the behavior of a material under load. Figure 11.2 is such a diagram for mild steel (a ductile material). For loads less than that required to permanently deform the material, most engineering materials display a linear relationship between

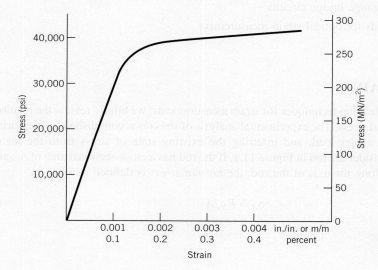

Figure 11.2 A typical stress–strain curve for mild steel.

stress and strain. The range of stress over which this linear relationship holds is called the *elastic region*. The relationship between uniaxial stress and strain for this elastic behavior is expressed as

$$\sigma_a = E_m \varepsilon_a \tag{11.3}$$

where E_m is the modulus of elasticity, or Young's modulus, and the relationship is called Hooke's law. Hooke's law applies only over the range of applied stress where the relationship between stress and strain is linear. Different materials respond in a variety of ways to loads beyond the linear range, largely depending on whether the material is ductile or brittle. For almost all engineering components, stress levels are designed to remain well below the elastic limit of the material; thus, a direct linear relationship may be established between stress and strain.

Lateral Strains

Consider the elongation of the rod shown in Figure 11.1 that occurs as a result of the load F_N. As the rod is stretched in the axial direction, the cross-sectional area will, for most materials, decrease. In the elastic range, there is a constant rate of change in the lateral strain as the axial strain increases. In the same sense that the modulus of elasticity is a property of a given material, the ratio of lateral strain to axial strain is also a material property. This property is called Poisson's ratio, defined as

$$v_P = -\frac{\text{Lateral strain}}{\text{Axial strain}} = -\frac{\varepsilon_L}{\varepsilon_a} \tag{11.4}$$

For the case of axial tension and for most materials the lateral strain will be negative and Poisson's ratio a positive number. Also for the case of uniaxial tension as in Figure 11.1 the volume of the rod will change. The dilatation, e, is defined as

$$e = \frac{\Delta V}{V_o} = \varepsilon_L(1 - 2v_p) \tag{11.5}$$

The total volume is only conserved for the case of $v_p = 0.5$.

Engineering components are seldom subject to one-dimensional axial loading. The relationship between stress and strain must be generalized to a multidimensional case. Consider a two-dimensional geometry, as shown in Figure 11.3, subject to tensile loads in both the x and y directions, resulting in normal stresses σ_x and σ_y. In this case, for a biaxial state of stress, the stresses and strains are

$$\varepsilon_y = \frac{\sigma_y}{E_m} - v_p \frac{\sigma_x}{E_m} \qquad \varepsilon_x = \frac{\sigma_x}{E_m} - v_p \frac{\sigma_y}{E_m}$$

$$\sigma_x = \frac{E_m(\varepsilon_x + v_p\varepsilon_y)}{1 - v_p^2} \qquad \sigma_y = \frac{E_m(\varepsilon_y + v_p\varepsilon_x)}{1 - v_p^2} \tag{11.6}$$

$$\tau_{xy} = G\gamma_{xy}$$

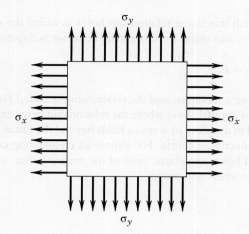

Figure 11.3   Biaxial state of stress.

In this case, all the stress and strain components lie in the same plane. The state of stress in the elastic condition for a material is similarly related to the strains in a complete three-dimensional situation (1,2). Since stress and strain are related, it is possible to determine stress from measured strains under appropriate conditions. However, strain measurements are made at the surface of an engineering component. The measurement yields information about the state of stress on the surface of the part. The analysis of measured strains requires application of the relationship between stress and strain at a surface. Such analysis of strain data is described elsewhere (3), and an example is provided in this chapter.

11.3   RESISTANCE STRAIN GAUGES

The measurement of the small displacements that occur in a material or object under mechanical load can be accomplished by methods as simple as observing the change in the distance between two scribe marks on the surface of a load-carrying member or as advanced as optical holography. In any case, the ideal sensor for the measurement of strain would (1) have good spatial resolution, implying that the sensor would measure strain at a point, (2) be unaffected by changes in ambient conditions, and (3) have a high-frequency response for dynamic (time-resolved) strain measurements. A sensor that closely meets these characteristics is the bonded resistance strain gauge.

In practical application, the bonded resistance strain gauge is secured to the surface of the test object by an adhesive so that it deforms as the test object deforms. The resistance of a strain gauge changes when it is deformed, and this is easily related to the local strain. Both metallic and semiconductor materials experience a change in electrical resistance when they are subjected to a strain. The amount that the resistance changes depends on how the gauge is deformed, the material from which it is made, and the design of the gauge. Gauges can be made quite small for good resolution and with a low mass to provide a high-frequency response. With some ingenuity, ambient effects can be minimized or eliminated.

In an 1856 publication in the *Philosophical Transactions of the Royal Society* in England, Lord Kelvin (William Thomson) (4) laid the foundations for understanding the changes in electrical resistance that metals undergo when subjected to loads, which eventually led to the strain gauge concept. Two individuals began the modern development of strain measurement in the late 1930s—Edward Simmons at the California Institute of Technology and Arthur Ruge at the Massachusetts Institute of Technology. Their development of the bonded metallic wire strain gauge led to commercially available strain gauges. The resistance strain gauge also forms the basis for a variety of other transducers, such as load cells, pressure transducers, and torque meters.

Metallic Gauges

To understand how metallic strain gauges work, consider a conductor having a uniform cross-sectional area A_c and length L made of a material having an electrical resistivity, ρ_e. For this electrical conductor, the resistance, R, is given by

$$R = \rho_e L / A_c \tag{11.7}$$

If the conductor is subjected to a normal stress along the axis of the wire, the cross-sectional area and the length change resulting in a change in the total electrical resistance, R. The total change in R is due to several effects, as illustrated in the total differential:

$$dR = \frac{A_c \left(\rho_e dL + L d\rho_e\right) - \rho_e L dA_c}{A_c^2} \tag{11.8}$$

which may be expressed in terms of Poisson's ratio as

$$\frac{dR}{R} = \frac{dL}{L}\left(1 + 2v_p\right) + \frac{d\rho_e}{\rho_e} \tag{11.9}$$

Hence, the changes in resistance are caused by two basic effects: the change in geometry as the length and cross-sectional area change and the change in the value of the resistivity, ρ_e. The dependence of resistivity on mechanical strain is called piezoresistance and may be expressed in terms of a piezoresistance coefficient, π_1, defined by

$$\pi_1 = \frac{1}{E_m} \frac{d\rho_e/\rho_e}{dL/L} \tag{11.10}$$

With this definition, the change in resistance may be expressed

$$dR/R = dL/L\left(1 + 2v_p + \pi_1 E_m\right) \tag{11.11}$$

Example 11.1

Determine the total resistance of a copper wire having a diameter of 1 mm and a length of 5 cm. The resistivity of copper is 1.7×10^{-8} Ω-m.

KNOWN $D = 1$ mm

$\qquad\qquad L = 5$ cm

$\qquad\qquad \rho_e = 1.7 \times 10^{-8}$ Ω-m

FIND The total electrical resistance

SOLUTION The resistance may be calculated from Equation 11.7 as

$$R = \rho_e L / A_c$$

where

$$A_c = \frac{\pi}{4} D^2 = \frac{\pi}{4}\left(1 \times 10^{-3}\right)^2 = 7.85 \times 10^{-7} \text{ m}^2$$

The resistance is then

$$R = \frac{\left(1.7 \times 10^{-8} \text{ Ω m}\right)\left(5 \times 10^{-2} \text{ m}\right)}{7.85 \times 10^{-7} \text{ m}^2} = 1.08 \times 10^{-3} \text{ Ω}$$

COMMENT If the material were nickel ($\rho_e = 7.8 \times 10^{-8}$ Ω-m) instead of copper, the resistance would be 5×10^{-3} Ω for the same diameter and length of wire.

Example 11.2

A very common material for the construction of strain gauges is the alloy constantan (55% copper with 45% nickel), which has a resistivity of 49×10^{-8} Ω-m. A typical strain gauge might have a resistance of 120 Ω. What length of constantan wire of diameter 0.025 mm would yield a resistance of 120 Ω?

KNOWN The resistivity of constantan is 49×10^{-8} Ω-m.

FIND The length of constantan wire needed to produce a total resistance of 120 Ω

SOLUTION From Equation 11.7, we may solve for the length, which yields in this case

$$L = \frac{R A_c}{\rho_e} = \frac{(120 \text{ Ω})\left(4.91 \times 10^{-10} \text{ m}^2\right)}{49 \times 10^{-8} \text{ Ω-m}} = 0.12 \text{ m}$$

The wire would then be 12 cm long to achieve a resistance of 120 Ω.

COMMENT As shown by this example, a single straight conductor is normally not practical for a local strain measurement with meaningful resolution. Instead, a simple solution is to bend the wire

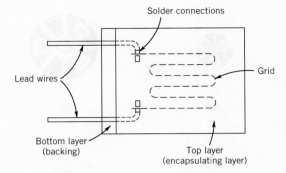

Solder connections

Lead wires

Grid

Bottom layer
(backing)

Top layer
(encapsulating layer)

Figure 11.4   Detail of a basic strain gauge construction. (Courtesy of Micro-Measurements, Raleigh, NC, USA.)

conductor so that several lengths of wire are oriented along the axis of the strain gauge, as shown in Figure 11.4.

Strain Gauge Construction and Bonding

Figure 11.5 illustrates the construction of a typical metallic-foil bonded strain gauge. Such a strain gauge consists of a metallic foil pattern that is formed in a manner similar to the process used to produce printed circuits. This photoetched metal foil pattern is mounted on a plastic backing material. The gauge length, as illustrated in Figure 11.5, is an important specification for a particular application. Because strain is usually measured at the location on a component where the stress is a maximum and the stress gradients are high, the strain gauge averages the measured strain over the gauge length. Because the maximum strain is the quantity of interest and the gauge length is the resolution, errors from averaging can result from improper choice of a gauge length (5).

The variety of conditions encountered in particular applications require special construction and mounting techniques, including various backing materials, grid configurations, bonding techniques,

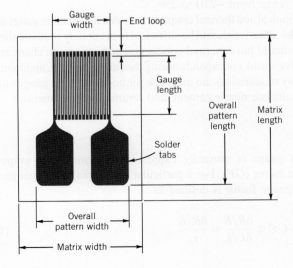

Gauge
width

End loop

Gauge
length

Overall
pattern
length

Matrix
length

Solder
tabs

Overall
pattern width

Matrix width

Figure 11.5   Construction of a typical metallic foil strain gauge. (Courtesy of Micro-Measurements, Raleigh, NC, USA.)

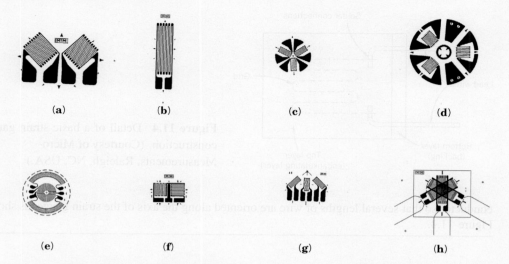

(a) (b) (c) (d)

(e) (f) (g) (h)

Figure 11.6 Strain gauge configurations. (a) Torque Rosette; (b) Linear Pattern; (c) Delta Rosette; (d) Residual Stress Pattern; (e) Diaphragm Pattern; (f) Tee Pattern; (g) Rectangular Rosette; (h) Stacked Rosette. (Courtesy of Micro-Measurements, Raleigh, NC, USA.)

and total gauge electrical resistance. Figure 11.6 shows a variety of strain gauge configurations. The adhesives used in the bonding process and the mounting techniques for a particular gauge and manufacturer vary according to the specific application. However, some fundamental aspects are common to all bonded resistance gauges.

The strain gauge backing serves several important functions. It electrically isolates the metallic gauge from the test specimen and transmits the applied strain to the sensor. A bonded resistance strain gauge must be appropriately mounted to the specimen for which strain is to be measured. The backing provides the surface used for bonding with an appropriate adhesive. Backing materials are available that are useful over temperatures that range from −270 to 290 °C.

The adhesive bond serves as a mechanical and thermal coupling between the metallic gauge and the test specimen. As such, the strength of the adhesive should be sufficient to accurately transmit the strain experienced by the test specimen, and should have thermal conduction and expansion characteristics suitable for the application. If the adhesive shrinks or expands during the curing process, apparent strain can be created in the gauge. A wide array of adhesives are available for bonding strain gauges to a test specimen. Among these are epoxies, cellulose nitrate cement, and ceramic-based cements.

Gauge Factor

The change in resistance of a strain gauge is normally expressed in terms of an empirically determined parameter called the gauge factor (*GF*). For a particular strain gauge, the gauge factor is supplied by the manufacturer. The gauge factor is defined as

$$GF \equiv \frac{\delta R/R}{\delta L/L} = \frac{\delta R/R}{\varepsilon_a} \qquad (11.12)$$

Relating this definition to Equation 11.11, we see that the gauge factor is dependent on Poisson's ratio for the gauge material and its piezoresistivity. For metallic strain gauges, Poisson's ratio is approximately 0.3, and the resulting gauge factor is $\sim$2.

The gauge factor represents the total change in resistance for a strain gauge, under a calibration loading condition. The calibration loading condition generally creates a biaxial strain field, and the lateral sensitivity of the gauge influences the measured result. Strictly speaking then, the sensitivity to normal strain of the material used in the gauge and the gauge factor are not the same. Generally gauge factors are measured in a biaxial strain field that results from the deflection of a beam having a value of Poisson's ratio of 0.285. Thus, for any other strain field there is an error in strain indication due to the transverse sensitivity of the strain gauge. The percentage error due to transverse sensitivity for a strain gauge mounted on any material at any orientation in the strain field is

$$e_L = \frac{K_t(\varepsilon_L/\varepsilon_a + v_{p0})}{1 - v_{p0}K_t} \times 100 \tag{11.13}$$

where

$\varepsilon_a, \varepsilon_L$ = axial and lateral strains, respectively (with respect the axis of the gauge)

v_{p0} = Poisson's ratio of the material on which the manufacturer measured GF (usually 0.285 for steel)

e_L = error as a percentage of axial strain (with respect to the axis of the gauge)

K_t = lateral (transverse) sensitivity of the strain gauge

The uncorrected estimate can be used as the uncertainty.

Typical values of the transverse sensitivity for commercial strain gauges range from -0.19 to 0.05. Figure 11.7 shows a plot of the percentage error for a strain gauge as a function of the ratio of lateral loading to axial loading and the lateral sensitivity. It is possible to correct for the lateral sensitivity effects (6).

Semiconductor Strain Gauges

When subjected to a load, a semiconductor material exhibits a change in resistance, and therefore can be used for the measurement of strain. Silicon crystals are the basic material for semiconductor strain gauges; the crystals are sliced into very thin sections to form strain gauges. Mounting such gauges in a transducer, such as a pressure transducer, or on a test specimen requires backing and adhesive techniques similar to those used for metallic gauges. Because of the large piezoresistance coefficient, the semiconductor gauge exhibits a very large gauge factor, as large as 200 for some gauges. These gauges also exhibit higher resistance, longer fatigue life, and lower hysteresis under some conditions than metallic gauges. However, the output of the semiconductor strain gauge is nonlinear with strain, and the strain sensitivity or gauge factor may be markedly dependent on temperature.

Semiconductor materials for strain gauge applications have resistivities ranging from 10^{-6} to 10^{-2} Ω-m. Semiconductor strain gauges may have a relatively high or low density of charge carriers (3,7). Semiconductor strain gauges made of materials having a relatively high density of charge carriers ($\sim$1,020 carriers/cm^3) exhibit little variation of their gauge factor with strain or

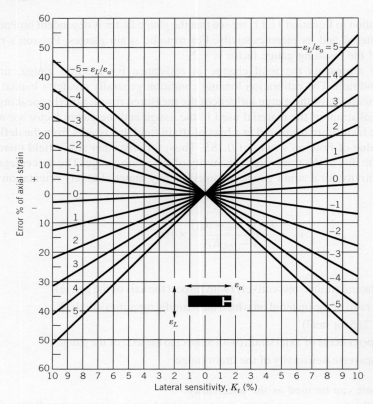

Figure 11.7 Strain measurement error due to strain gauge transverse sensitivity. (Courtesy of Micro-Measurements, Raleigh, NC, USA.)

temperature. On the other hand, for the case where the crystal contains a low number of charge carriers ($<1,017$ carriers/cm^3), the gauge factor may be approximated as

$$GF = \frac{T_0}{T} GF_0 + C_1 \left(\frac{T_0}{T}\right)^2 \varepsilon \tag{11.14}$$

where GF_0 is the gauge factor at the reference temperature T_0, under conditions of zero strain (8), and C_1 is a constant for a particular gauge. The behavior with temperature of a high-resistivity P-type semiconductor is shown in Figure 11.8.

Semiconductor strain gauges find their primary application in the construction of transducers, such as load cells and pressure transducers. Because of the capability for producing small gauge lengths, silicon semiconductor strain gauge technology provides for the construction of very small transducers. For example, flush-mount pressure transducers having diameters of less than 8 mm provide pressure measurements up to 15,000 psi, with excellent frequency response characteristics. However, silicone diaphragm pressure transducers require special construction and procedures for measuring in liquid environments. Semiconductor strain gauges are somewhat limited in the maximum strain that they can measure, approximately 5,000 με for tension, but larger in compression. Because of the possibility of an inherent sensitivity to temperature, careful consideration must

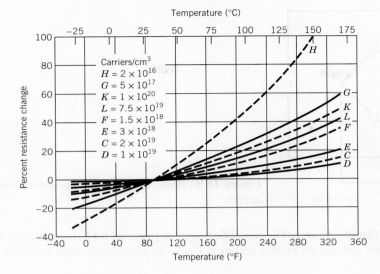

Figure 11.8 Temperature effect on resistance for various impurity concentrations for P-type semiconductors (reference resistance at 81 °F). (Courtesy of Kulite Semiconductor Products, Inc. Used with permission.)

be given to each application to provide appropriate temperature compensation or correction. Temperature effects can result, for a particular measurement, in zero drift for the duration of a measurement.

11.4 STRAIN GAUGE ELECTRICAL CIRCUITS

A bridge circuit is generally used to detect the small changes in resistance that are the output of a strain gauge measurement circuit. A typical strain gauge measuring installation on a steel specimen has a sensitivity of 10^{-6} Ω/(kPa). Thus a high-sensitivity device such as a Wheatstone bridge is desirable for measuring resistance changes for strain gauges. The fundamental relationships for the analysis of such bridge circuits are discussed in Chapter 6. Equipment is commercially available that can measure changes in gauge resistance of less than 0.0005 Ω (0.000001 με).

A simple strain gauge Wheatstone bridge circuit is shown in Figure 11.9. The bridge output under these conditions is given by Equation 6.12:

$$E_0 + \delta E_0 = E_i \frac{(R_1 + \delta R)R_4 - R_3 R_2}{(R_1 + \delta R + R_2)(R_3 + R_4)} \tag{6.12}$$

where E_0 is the bridge output at initial conditions, δE_0 is the bridge deflection associated with the change in the strain gauge resistance δR. Consider the case where all the fixed resistors and the strain gauge resistance are initially equal, and the bridge is balanced such that $E_0 = 0$. If the strain gauge is then subjected to a state of strain, the change in the output voltage, δE_0, from Equation 6.12 reduces to (as shown in Equation 6.13)

$$\frac{\delta E_o}{E_i} = \frac{\delta R/R}{4 + 2(\delta R/R)} \approx \frac{\delta R/R}{4} \tag{11.15}$$

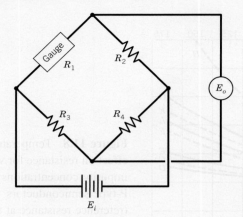

Figure 11.9 Basic strain gauge Wheatstone bridge circuit.

under the assumption that $\delta R/R \ll 1$. This simplified form of Equation 6.12 is suitable for all but those measurements that demand the highest accuracy. Using the relationship from Equation 11.12 that $\delta R/R = GF\varepsilon$,

$$\frac{\delta E_o}{E_i} = \frac{GF\varepsilon}{4 + 2GF\varepsilon} \approx \frac{GF\varepsilon}{4} \tag{11.16}$$

Equations 11.15 and 11.16 yield two practical equations for strain gauge measurements using a single gauge in a Wheatstone bridge.

The Wheatstone bridge has several distinct advantages for use with electrical resistance strain gauges. The bridge may be balanced by changing the resistance of one arm of the bridge. Therefore, once the gauge is mounted in place on the test specimen under a condition of zero loading, the output from the bridge may be zeroed. Two schemes for circuits to accomplish this balancing are shown in Figure 11.10. Shunt balancing provides the best arrangement for strain gauge applications, since the

Circuit arrangement for shunt balance

Differential shunt balance arrangement

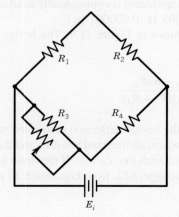

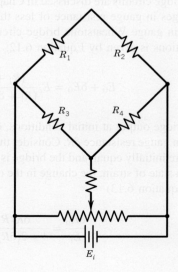

Figure 11.10 Balancing schemes for bridge circuits.

changes in resistance for a strain gauge are small. Also, the strategic placement of multiple gauges in a Wheatstone bridge can both increase the bridge output and cancel out certain ambient effects and unwanted components of strain as discussed in the next two sections.

Example 11.3

A strain gauge, having a gauge factor of 2, is mounted on a rectangular steel bar $\left(E_m = 200 \times 10^6 \text{ kN/m}^2\right)$, as shown in Figure 11.11. The bar is 3 cm wide and 1 cm high and is subjected to a tensile force of 30 kN. Determine the resistance change of the strain gauge if the resistance of the gauge was 120 Ω in the absence of the axial load.

KNOWN   GF = 2 $E_m = 200 \times 10^6 \text{ kN/m}^2$ $F_N = 30 \text{ kN}$

$$R = 120 \,\Omega \quad A_c = 0.03 \text{ m} \times 0.01 \text{ m}$$

FIND   The resistance change of the strain gauge for a tensile force of 30 kN

SOLUTION   The stress in the bar under this loading condition is

$$\sigma_a = \frac{F_N}{A_c} = \frac{30 \text{ kN}}{(0.03 \text{ m})(0.01 \text{ m})} = 1 \times 10^5 \text{ kN/m}^2$$

and the resulting strain is

$$\varepsilon_a = \frac{\sigma_a}{E_m} = \frac{1 \times 10^5 \text{ kN/m}^2}{200 \times 10^6 \text{ kN/m}^2} = 5 \times 10^{-4} \text{ m/m}$$

For strain along the axis of the strain gauge, the change in resistance from Equation 11.12 is

$$\delta R/R = \varepsilon GF$$

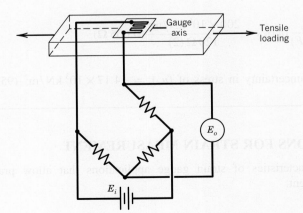

Figure 11.11 Strain gauge circuit subject to uniaxial tension.

or

$$\delta R = R \varepsilon GF = (120\,\Omega)(5 \times 10^{-4})(2) = 0.12\,\Omega$$

Example 11.4

Suppose the strain gauge described in Example 11.3 is to be connected to a measurement device capable of determining a change in resistance with a stated uncertainty of $\pm 0.005\,\Omega$ (95%). This stated uncertainty includes a resolution of $0.001\,\Omega$. What uncertainty in stress would result when using this resistance measurement device?

KNOWN A stress is to be inferred from a strain measurement using a strain gauge having a gauge factor of 2 and a zero load resistance of $120\,\Omega$. The measurement of resistance has a stated uncertainty of $\pm 0.005\,\Omega$ (95%).

FIND The design-stage uncertainty in stress

SOLUTION The design-stage uncertainty in stress, $(u_d)_\sigma$, is given by

$$(u_d)_\sigma = \frac{\partial \sigma}{\partial(\delta R)}(u_d)_{\delta R}$$

with

$$\sigma = \varepsilon E_m = \frac{\delta R/R}{GF} E_m$$

Then with $(u_d)_{\delta R} = 0.005\,\Omega$ and

$$\frac{\partial \sigma}{\partial(\delta R)} = \frac{E_m}{R(GF)}$$

we can express the uncertainty as

$$(u_d)_\sigma = \frac{E_m}{R(GF)}(u_d)_{\delta R} = \frac{200 \times 10^6 \text{ kN/m}^2}{120\,\Omega\,(2)}(0.005\,\Omega)$$

This results in a design-stage uncertainty in stress of $(u_d)_\sigma = \pm 4.17 \times 10^3 \text{ kN/m}^2$ (95%) or $\sim 2.4\%$ of the expected stress.

11.5 PRACTICAL CONSIDERATIONS FOR STRAIN MEASUREMENT

This section describes some characteristics of strain gauge applications that allow practical implementation of strain measurement.

The Multiple Gauge Bridge

The output from a bridge circuit can be increased by the appropriate use of more than one strain gauge. This increase can be quantified by employing a bridge constant as illustrated in the following discussion. In addition, multiple gauges can be used to compensate for unwanted effects, such as temperature or specific strain components. Consider the case when all four resistances in the bridge circuit of Figure 11.9 represent strain gauges. In general, the bridge output is given by

$$E_0 = E_i \left[\frac{R_1}{R_1 + R_2} - \frac{R_3}{R_3 + R_4} \right] \tag{11.17}$$

The strain gauges R_1, R_2, R_3, and R_4 are assumed initially to be in a state of zero strain. If these gauges are now subjected to strains such that the resistances change by dR_i, where $i = 1, 2, 3$, and 4, then the change in the bridge output voltage can be expressed as

$$dE_0 = \sum_{i=1}^{4} \frac{\partial E_0}{\partial R_i} dR_i \tag{11.18}$$

Evaluating the appropriate partial derivatives from Equation 11.17 yields

$$dE_0 = E_i \left[\frac{R_2 dR_1 - R_1 dR_2}{(R_1 + R_2)^2} + \frac{R_3 dR_4 - R_4 dR_3}{(R_3 + R_4)^2} \right] \tag{11.19}$$

Then, from Equations 11.2 and 11.12, $dR_i = R_i \varepsilon_i GF_i$, and the value of dE_0 can be determined. Assuming that $dR_i \ll R_i$, the resulting change in the output voltage, δE_0, may now be expressed as

$$\delta E_0 = E_i \left[\frac{R_1 R_2}{(R_1 + R_2)^2} (\varepsilon_1 GF_1 - \varepsilon_2 GF_2) + \frac{R_3 R_4}{(R_3 + R_4)^2} (\varepsilon_4 GF_4 - \varepsilon_3 GF_3) \right] \tag{11.20}$$

If $R_1 = R_2 = R_3 = R_4$, then

$$\frac{\delta E_0}{E_i} = \frac{1}{4} (\varepsilon_1 GF_1 - \varepsilon_2 GF_2 + \varepsilon_4 GF_4 - \varepsilon_3 GF_3) \tag{11.21}$$

It is possible and desirable to purchase matched sets of strain gauges for a particular application, so that $GF_1 = GF_2 = GF_3 = GF_4$, and

$$\frac{\delta E_0}{E_i} = \frac{GF}{4} (\varepsilon_1 - \varepsilon_2 + \varepsilon_4 - \varepsilon_3) \tag{11.22}$$

Equation 11.22 is important and forms the basic working equation for a strain gauge bridge circuit using multiple gauges (compare this equation with Eq. 11.16).

Equation 11.22 shows that for a bridge containing two or four strain gauges, equal strains on opposite bridge arms sum, whereas equal strains on adjacent arms of the bridge cancel. These characteristics can be used to increase the output of the bridge, to provide temperature compensation, or to cancel unwanted components of strain. Practical means of achieving these desirable characteristics will be explored further after the concept of the bridge constant is developed.

Bridge Constant

Commonly used strain gauge bridge arrangements may be characterized by a bridge constant, κ, defined as the ratio of the actual bridge output to the output that would result from a single gauge sensing the maximum strain, ε_{max}. The output for a single gauge experiencing the maximum strain may be expressed as

$$\frac{\delta R}{R} = \varepsilon_{max} GF \tag{11.23}$$

so that, again for a single gauge,

$$\frac{\delta E_0}{E_i} \cong \frac{\varepsilon_{max} GF}{4} \tag{11.24}$$

The bridge constant, κ, is found from the ratio of the actual bridge output given by Equation 11.22 to the output for a single gauge given by Equation 11.24. When more than one gauge is used in the bridge circuit, Equation 11.16 becomes

$$\frac{\delta E_0}{E_i} = \frac{\kappa \delta R/R}{4 + 2\delta R/R} = \frac{\kappa GF\varepsilon}{4 + 2GF\varepsilon} \approx \frac{\kappa GF\varepsilon}{4} \tag{11.25}$$

Application of the bridge constant is illustrated in Example 11.5.

Example 11.5

Determine the bridge constant for two strain gauges mounted on a structural member, as shown in Figure 11.12. The member is subject to uniaxial tension, which produces an axial strain ε_a and a lateral strain $\varepsilon_L = -v_p\varepsilon_a$. Assume that all the resistances in Figure 11.12 are initially equal so that the bridge is initially balanced. Let $GF_1 = GF_2$.

KNOWN Strain gauge installation shown in Figure 11.12

FIND The bridge constant for this installation

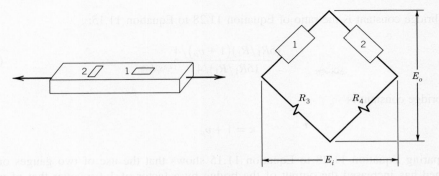

Figure 11.12   Bridge circuit with two arms active; strain gauge installation for increased sensitivity.

SOLUTION   The gauges are mounted so that gauge 1 is aligned with the axial tension and gauge 2 is mounted transversely on the member. The changes in resistance for the gauges may be expressed

$$\frac{\delta R_1}{R_1} = \varepsilon_a (GF)_1 \tag{11.26}$$

and

$$\frac{\delta R_2}{R_2} = -v_p \varepsilon_a (GF)_2 = -v_p \frac{\delta R_1}{R_1} \tag{11.27}$$

With only one gauge active, Equation 11.15 would be applicable and yields

$$\frac{\delta E_0}{E_i} = \frac{\delta R_1 / R_1}{4 + 2(\delta R_1 / R_1)} \approx \frac{\delta R_1 / R_1}{4} \tag{11.15}$$

But with both gauges installed and active, as shown in Figure 11.12, the output of the bridge is determined from an analysis of the bridge response, which results in

$$\frac{\delta E_0}{E_i} = \frac{(\delta R_1 / R_1)(1 + v_p)}{4 + 2(\delta R_1 / R_1)(1 + v_p)}$$

In most practical applications, changes in resistance are small in comparison to the resistance values, that is, $\delta R / R \ll 1$; thus

$$\frac{\delta E_0}{E_i} = \frac{(\delta R_1 / R_1)(1 + v_p)}{4} \tag{11.28}$$

Thus the bridge constant is the ratio of Equation 11.28 to Equation 11.15:

$$\frac{(\delta R_1/R_1)(1+\upsilon_p)/4}{(\delta R_1/R_1)/4} \tag{11.29}$$

And the bridge constant is

$$\kappa = 1 + \upsilon_p$$

Comparing Equation 11.28 to Equation 11.15 shows that the use of two gauges oriented as described has increased the output of the bridge by a factor of $1+\upsilon_p$ over that of using a single gauge.

11.6   APPARENT STRAIN AND TEMPERATURE COMPENSATION

Apparent strain is caused by any change in gauge resistance that is not due to the component of strain being measured. Techniques for accomplishing temperature compensation, eliminating certain components of strain, and increasing the value of the bridge constant can be devised by examining more closely Equation 11.22. The bridge constant is influenced by (1) the location of strain gauges on the test specimen and (2) the gauge connection positions in the bridge circuit.

Let us examine how a component of strain can be removed (compensation) from the measured signal. Consider a beam having a rectangular cross section and subject to the loading condition shown in Figure 11.13, where the beam is subject to an axial load F_N and a bending moment M. The stress distribution in this cross section is given by

$$\sigma_x = \frac{-12My}{bh^3} + \frac{F_N}{bh} \tag{11.30}$$

To remove the effects of bending strain, identical strain gauges are mounted to the top and bottom of the beam as shown in Figure 11.13, and they are connected to bridge locations 1 and 4 (opposite bridge arms). The gauges experience equal but opposite bending strains (Eq. 11.30), and

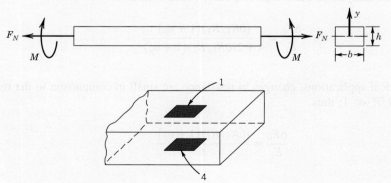

Figure 11.13   Strain gauge installation for bending compensation.

both strain gauges are subject to the same axial strain caused by F_N. The bridge output under these conditions is

$$\frac{\delta E_0}{E_i} = \frac{GF}{4}(\varepsilon_1 + \varepsilon_4) \tag{11.31}$$

where $\varepsilon_1 = \varepsilon_{a1} + \varepsilon_{b1}$ and $\varepsilon_4 = \varepsilon_{a4} - \varepsilon_{b4}$, with subscripts a and b referring to axial and bending strain, respectively. Hence, because $|\varepsilon_{a1}| = |\varepsilon_{a4}|$ and $|\varepsilon_{b1}| = |\varepsilon_{b4}|$, the bending strains cancel, but the axial strains sum, giving

$$\frac{\delta E_0}{E_i} = \frac{GF}{2}\varepsilon_a \tag{11.32}$$

For a single gauge experiencing the maximum strain,

$$\frac{\delta E_0}{E_i} = \frac{GF}{4}\varepsilon_a \tag{11.33}$$

The ratio of the output represented in Equation 11.32 to that represented in Equation 11.33 has a value of 2. This is the bridge constant ($\kappa = 2$) for this strain gauge arrangement. The bending strain cancels from Equation 11.32, indicating that this arrangement compensates for the bending strain.

A guide for some practical bridge-gauge configurations is provided in Table 11.1.

Temperature Compensation

Differential thermal expansion between the gauge and the specimen on which it is mounted creates an apparent strain in the strain gauge. So temperature sensitivity of strain gauges is caused by temperature changes in the gauge itself and the strain experienced by the gauge as a result of differential thermal expansion between the gauge and the material on which it is mounted. Using gauges of identical alloy composition as the specimen minimizes this latter effect. However, even keeping the specimen at a constant temperature may not be enough to eliminate the effect of gauge thermal expansion. Heating of the strain gauge as a result of current flow from the measuring device may be a source of significant error since the gauge is also a temperature-sensitive element. The temperature sensitivity of a strain gauge is an obstacle to accurate mechanical strain measurement that must be overcome. Fortunately, there are effective ways to deal with it.

Figure 11.14 shows two circuit arrangements that provide temperature compensation for a strain measurement. The strain gauge mounted on the test specimen experiences changes in resistance caused by temperature changes and by applied strain, whereas the compensating gauge experiences resistance changes caused only by temperature changes. As long as the compensating gauge, shown in Figure 11.14, experiences an identical thermal environment to that experienced by the measuring gauge, temperature effects will be eliminated from the circuit. To show this, consider the case in which all bridge resistances are initially equal and the bridge is thus balanced. If the temperature of the strain gauges now changes, their resistance changes as a result of thermal expansion, creating an

Table 11.1   Common Gauge Mountings

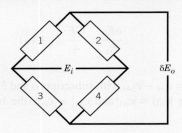

Arrangement	Compensation Provided	Bridge Constant κ
Single gauge in uniaxial stress	None	$\kappa = 1$
Two gauges sensing equal and opposite strain—typical bending arrangement	Temperature	$\kappa = 2$
Two gauges in uniaxial stress	Bending only	$\kappa = 2$
Four gauges with pairs sensing equal and opposite strains	Temperature and bending	$\kappa = 4$
One axial gauge and one Poisson gauge		$\kappa = 1 + \nu_p$
Four gauges with pairs sensing equal and opposite strains—sensitive to torsion only; typical shaft arrangement.	Temperature and axial	$\kappa = 4$

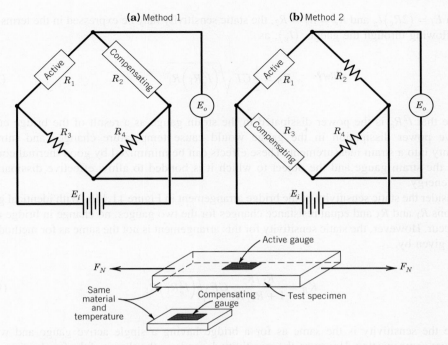

Figure 11.14 Bridge arrangements for temperature compensation.

apparent thermal strain. Under an applied axial load, the output of the bridge is derived from Equation 11.22,

$$\delta E_0 = E_i \frac{GF(\varepsilon_1 - \varepsilon_2)}{4} = E_i \frac{GF\varepsilon_a}{4} \tag{11.34}$$

where $\varepsilon_1 = \varepsilon_a + \varepsilon_T$ and $\varepsilon_2 = \varepsilon_T$ (ε_T) refers to the temperature-induced apparent strain). Under this condition, the output value is not affected by changes in temperature. The result is the same if the compensating gauge is mounted in arm 3 instead, as shown in Figure 11.14b. Furthermore, any two active gauges mounted on adjacent bridge arms compensate for temperature changes.

Looking back, the arrangement in Figure 11.13 does not provide temperature compensation; gauges 1 and 4 are on opposite bridge arms. However, temperature compensation could be provided for that installation by having two additional strain gauges that are at the same temperature as gauges 1 and 4 and that occupy arms 2 and 3 of the bridge.

Bridge Static Sensitivity

The static sensitivity of the bridge arrangement in Figure 11.14a (method 1) is

$$K_B = \frac{E_0}{\varepsilon} = E_i \frac{R_1 R_2}{(R_1 + R_2)^2} GF \tag{11.35}$$

and with $E_i = (2R_g)I_g$ and $R_g = R_1 = R_2$, the static sensitivity may be expressed in the terms of the current flowing through the gauge, (I_g), as

$$K_B = \frac{1}{2} GF \sqrt{\left(I_g^2 R_1\right) R_1}$$ (11.36)

Note that $I_g^2 R_g$ is the power dissipated in the strain gauge as a result of the bridge current. Excessive power dissipation in the gauge would cause temperature changes and introduce uncertainty into a strain measurement. These effects can be minimized by good thermal coupling between the strain gauge and the object to which it is bonded to allow effective dissipation of thermal energy.

Consider the static sensitivity of the bridge arrangement in Figure 11.14b. With identical gauges at positions R_1 and R_3 and equal resistance changes for the two gauges, no change in bridge output would occur. However, the static sensitivity for this arrangement is not the same as for method 1 but rather is given by

$$K_B = \frac{R_1/R_2}{1 + R_1/R_2} GF \sqrt{\left(I_g^2 R_1\right) R_1}$$ (11.37)

Here the sensitivity is the same as for a bridge having a single active gauge and without temperature compensation. However, the sensitivity depends on the choice of the fixed resistor R_2. If $R_1 = R_2$, the resulting sensitivity is the same as for method 1. However, resistor R_2 can be chosen to provide the desired static sensitivity for the circuit, within the limitations of measurement capability and allowable bridge current.

Practical Considerations

An assumption in the definition of the gauge factor is that the change in resistance of the gauge is linear with applied strain for a particular gauge. However, a strain gauge can exhibit some nonlinearity. Also, in cycling between a loaded and unloaded condition, there is some degree of hysteresis and a shift in the resistance for a state of zero strain. A typical cycle of loading and unloading is shown in Figure 11.15. The strain gauge typically indicates lower values of strain during unloading than are measured as the load is increased. The extent of these behaviors is determined not only by the strain gauge characteristics but also by the characteristics of the adhesive and by the previous strains that the gauge has experienced. For properly installed gauges, the deviation from linearity should be on the order of 0.1% (3). On the other hand, first-cycle hysteresis and zero shift are difficult to predict. The effects of first-cycle hysteresis and zero shift can be minimized by cycling the strain gauge between zero strain and a value of strain above the maximum value to be measured before taking measurements.

In dynamic measurements of strain, the dynamic response of the strain gauge itself is generally not the limiting factor for such dynamic measurements. The rise time (90%) of a bonded resistance strain gauge may be approximated (9) as

$$t_{90} \approx 0.8(L/a) + 0.5 \; \mu s$$ (11.38)

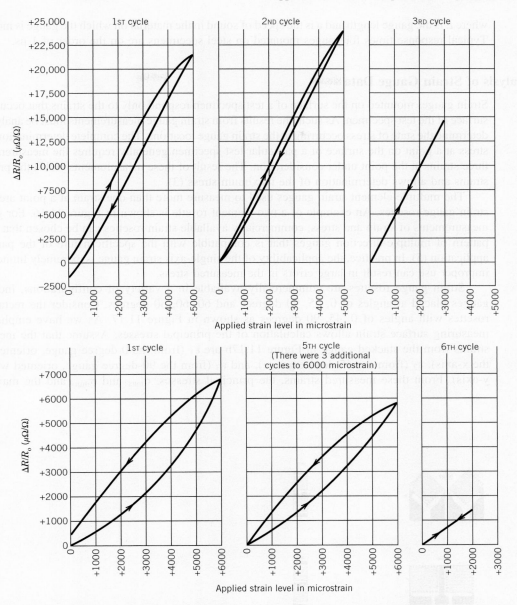

Figure 11.15 Hysteresis in initial loading cycles for two strain gauge materials. (Courtesy of Micro-Measurements, Raleigh, NC, USA.)

where L is the gauge length and a is the speed of sound in the material on which the gauge is mounted. Typical response times for gauges mounted on steel specimens are on the order of 1 μs.

Analysis of Strain Gauge Data

Strain gauges mounted on the surface of a test specimen respond only to the strains that occur at the surface of the test specimen. As such, the results from strain gauge measurements must be analyzed to determine the state of stress occurring at the strain gauge locations. The complete determination of the stress at a point on the surface of a particular test specimen generally requires the measurement of three strains at the point under consideration. The result of these measurements yields the principal strains and allows determination of the maximum stress (3).

The multiple-element strain gauges used to measure more than one strain at a point are called strain gauge rosettes. An example of a two-element rosette is shown in Figure 11.16. For general measurements of strain and stress, commercially available strain rosettes can be chosen that have a pattern of multiple-direction gauges that is compatible with the specific nature of the particular application (5). In practice, the applicability of the single-axis strain gauge is extremely limited, and improper use can result in large errors in the measured stress.

Strain gauge rosettes are commercially available in a variety of configurations, including gauges placed at angles of 0, 45, 90 degrees and 60, 60, 60 degrees. Consider the rectangular rosettes with angles of 0, 45, 90 degrees as shown in Figure 11.17. As we have emphasized, measuring surface strain allows calculation of the principal stresses. Assume that the measured strains from the stacked gauge in Figure 11.17b are ε_1 (from the 0-degree gauge, oriented with the x-axis), ε_2 (from the 45-degree gauge), and ε_3 (from the 90-degree gauge, oriented with the y-axis). From these measured strains, the principal stresses, σ_{max} and σ_{min}, and the maximum

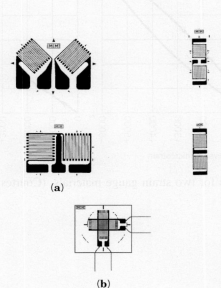

(a)

(b)

Figure 11.16 Biaxial strain gauge rosettes. (a) Single-plane type. (b) Stacked type. (Courtesy of Micro-Measurements, Raleigh, NC, USA.)

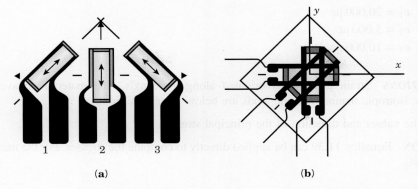

Figure 11.17 Rectangular (0, 45, 90 degree) strain gauge rosettes. (a) Planar configuration. (b) Stacked configuration.

shear stress, τ_{max}, may be calculated as

$$\sigma_{max} = \frac{E_m}{2} \left[\frac{\varepsilon_1 + \varepsilon_3}{1 - v_p} + \frac{1}{1 + v_p} \sqrt{(\varepsilon_1 - \varepsilon_3)^2 + [2\varepsilon_2 - (\varepsilon_1 + \varepsilon_3)]^2} \right]$$

$$\sigma_{min} = \frac{E_m}{2} \left[\frac{\varepsilon_1 + \varepsilon_3}{1 - v_p} - \frac{1}{1 + v_p} \sqrt{(\varepsilon_1 - \varepsilon_3)^2 + [2\varepsilon_2 - (\varepsilon_1 + \varepsilon_3)]^2} \right] \qquad (11.39)$$

$$\tau_{max} = \frac{E_m}{2(1 + v_p)} \sqrt{(\varepsilon_1 - \varepsilon_3)^2 + [2\varepsilon_2 - (\varepsilon_1 + \varepsilon_3)]^2}$$

The angle between the x-axis and the maximum principal stress is given by

$$\phi = \frac{1}{2} \tan^{-1} \frac{2\varepsilon_2 - (\varepsilon_1 + \varepsilon_3)}{\varepsilon_1 - \varepsilon_3} \qquad (11.40)$$

Example 11.6

A rectangular strain gauge rosette is composed of strain gauges oriented at relative angles of 0, 45, and 90 degrees, as shown in Figure 11.17b. The rosette is used to measure strain on an aluminum structural member ($E_m = 69$ MPa, $v_p = 0.334$). The measured values of strain are

$$\varepsilon_1 = 20,000 \ \mu\varepsilon$$
$$\varepsilon_2 = 5,000 \ \mu\varepsilon$$
$$\varepsilon_3 = 10,000 \ \mu\varepsilon$$

Determine the principal stresses and the angle of the maximum principle stress relative to the x-axis.

KNOWN $\varepsilon_1 = 20{,}000\ \mu\varepsilon$

$\varepsilon_2 = 5{,}000\ \mu\varepsilon$

$\varepsilon_3 = 10{,}000\ \mu\varepsilon$

ASSUMPTIONS   Strain gauge 1 is oriented along the *x*-axis. The material behaves in a homogeneous, isotropic manner and the loads are below the elastic limit.

FIND   The values and direction of the principal stresses

SOLUTION   Equation 11.39 can be applied directly to compute the stresses. For the maximum principal stress,

$$\sigma_{max} = \frac{E_m}{2}\left[\frac{\varepsilon_1 + \varepsilon_3}{1 - v_p} + \frac{1}{1 + v_p}\sqrt{(\varepsilon_1 - \varepsilon_3)^2 + [2\varepsilon_2 - (\varepsilon_1 + \varepsilon_3)]^2}\right]$$

$$\sigma_{max} = \frac{69\ \mathrm{MPa}}{2}\left[\frac{(0.02 + 0.01)}{1 - 0.334} + \frac{1}{1 + 0.334}\sqrt{(0.02 - 0.01)^2 + [2 \times 0.005 - (0.02 + 0.01)]^2}\right]$$

$$= 2.13\ \mathrm{MPa}$$

Equation 11.39 and Equation 11.40 yield the remaining values

$$\sigma_{min} = -0.976\ \mathrm{MPa}$$

$$\tau_{max} = 0.578\ \mathrm{MPa}$$

$$\phi = -63.4°$$

COMMENT   Other orientations of strain gauge rosettes result in different equations relating stress and measured strain.

Signal Conditioning

The most common form of signal conditioning in strain gauge bridge circuits is amplification. For an amplifier of gain G_A, Equation 11.25 becomes

$$\frac{\delta E_0}{E_i} = \frac{G_A\kappa(\delta R/R)}{4 + 2(\delta R/R)} = \frac{G_A\kappa GF\varepsilon}{4 + 2(\delta R/R)} \approx \frac{G_A\kappa GF\varepsilon}{4} \tag{11.41}$$

A common means of recording strain gauge bridge circuit signals is the automated data acquisition system. A schematic diagram of such a setup is shown in Figure 11.18.

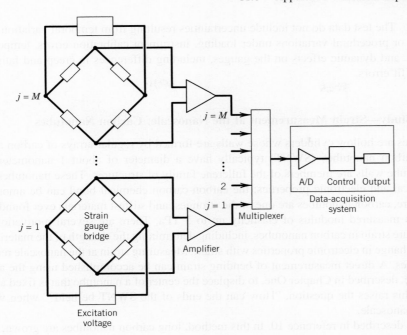

Figure 11.18 Data acquisition and reduction system for Example 11.7.

Example 11.7

The strain developed at M different locations about a large test specimen is to be measured at each location using a setup similar to that shown in Figure 11.18. At each location, similar strain gauges are appropriately mounted and connected to a Wheatstone bridge that is powered by an external supply. The bridge deflection voltage is amplified and measured. A similar setup is used at all M locations. The output from each amplifier is applied as an input through an M-channel multiplexer to an automated data acquisition system. If N (say 30) readings for each of the M setups are taken while the test specimen is maintained in a condition of uniform zero strain, what information is obtained?

   KNOWN   Strain setup of Figure 11.18

$$M(j = 1, 2, \ldots, M) \text{ setups}$$
$$N(i = 1, 2, \ldots, N) \text{ readings per setup}$$

   ASSUMPTIONS   Each setup is operated in a similar manner.

   SOLUTION   Each setup is exposed to the same strain and should indicate the same strain. We can calculate the mean strain at each of the M locations. Any difference between the measured mean strain and the applied strain at each position is an offset error for that channel. For zero applied strain, we can use this information to correct each channel to zero. We can estimate the uncertainty in the correction from the standard deviation of the means for the offset deviations (see Example 5.13). The variation in measured strain at each of the M locations is a measure of the noise.

COMMENT The test data do not include uncertainties resulting from temporal variation of the measured strain or procedural variations under loading, instrument calibration errors, temperature variation effects, and dynamic effects on the gauges, including differences in creep and fatigue or reduction curve fit errors.

Example 11.8: Case Study—Strain Measurement at the Nanoscale: Carbon Nanotubes

Carbon nanotubes are hollow cylinders whose walls are formed by regular arrays of carbon atoms; single-walled carbon nanotubes (SWNT) typically have a diameter of about 1 nanometer. The structure of the tube walls are members of the fullerene family of structures. These nanotubes have unique mechanical and electronic properties; the carbon-carbon chemical bond can be among the strongest in nature, carbon nanotubes are one of the strongest and stiffest materials ever found, with SWNT having a measured modulus of elasticity up to 5 TPa. There are several motivations for wanting to measure strain in carbon nanotubes, including determining the strength of the material and quantifying the change in electronic properties with strain. Measuring strain at the nanoscale requires unique approaches. A direct measurement of bending strain can be accomplished using the atomic force microscope, described in Chapter One, to displace the center of a nanotube that is fixed at each end. However, this raises the question, "How can the ends of the SWNT be held?" when we are working at the nanoscale.

A method is described in reference 10. In this method, long carbon nanotubes are grown, using chemical vapor deposition, over H-shaped slits in a silicone wafer. The width of the slits were approximately 100 μm. The process yields numerous SWNT, and, again at the nanoscale, one clean separate fiber was identified. Then a layer of gold was deposited on the end of the SWNT to "clamp" it to the silicone wafer. This assembly was glued to a steel substrate, with appropriate cutouts and the steel heated to induce thermal expansion and apply a strain.

COMMENT Future applications of nanoscale materials and devices will increase and affect diverse areas, including sports equipment and delivery of medicine in the bloodstream. There is a critical need for measurement technology to keep up with the challenges to measure the properties of the new materials created at this small scale.

11.7 OPTICAL STRAIN MEASURING TECHNIQUES

Optical methods for experimental stress analysis can provide fundamental information concerning directions and magnitudes of the stresses in parts under design loading conditions. Optical techniques have been developed for the measurement of stress and strain fields, either in models made of materials having appropriate optical properties, or through coating techniques for existing specimens. Photoelasticity takes advantage of the changes in optical properties of certain materials that occur when these materials are strained. A second optical method of stress analysis is based on the development of a moiré pattern, which is an optical effect resulting from the transmission or reflection of light from two overlaid grid patterns. The fringes that result from relative displacement of the two grid patterns can be used to measure strain; each fringe corresponds to the locus of points of equal displacement.

Recent developments in strain measurement include the use of lasers and holography to very accurately determine whole field displacements for complex geometries.

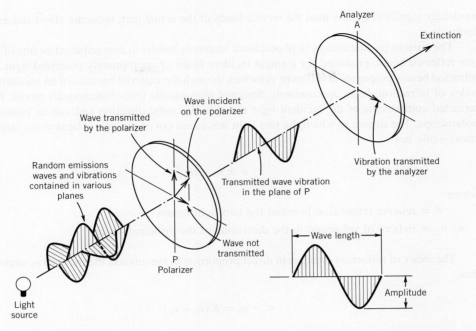

Figure 11.19   Polarization of light. (Courtesy of Micro-Measurements, Raleigh, NC, USA.)

Basic Characteristics of Light

To use optical strain measurement techniques, we must first examine some basic characteristics of light. Electromagnetic radiation, such as light, may be thought of as a transverse wave with sinusoidally oscillating electric and magnetic field vectors that are at right angles to the direction of propagation. In general, a light source emits a series of waves containing vibrations in all perpendicular planes, as illustrated in Figure 11.19. A light wave is said to be plane-polarized if the transverse oscillations of the electric field are parallel to each other at all points along the direction of propagation of the wave.

Figure 11.19 illustrates the effect of a polarizing filter on an incident light wave; the transmitted light is plane polarized, with a known direction of polarization. Complete extinction of the light beam could be achieved by introduction of a second polarizing filter, with the axis of polarization at 90 degrees to the first filter (labeled an Analyzer in Fig. 11.19). These behaviors of light are employed to measure direction and magnitude of strain in photoelastic materials.

Photoelastic Measurement

Photoelastic methods of stress analysis take advantage of the anisotropic optical characteristics of some materials, notably plastics, when subject to an applied load to determine the strain field. Stress analysis may be accomplished either by constructing a model of the part to be analyzed from a material selected for its optical properties, or by coating the actual part or prototype with a photoelastic coating. If a model is constructed from a suitable plastic, the required loads for the model are significantly less than the service loads of the actual part, reducing effort and expense in testing.

The anisotropy that occurs in photoelastic materials results in two refracted beams of light and one reflected beam, produced for a single incident beam of appropriately polarized light. The two refracted beams propagate at different velocities through the material because of an anisotropy in the index of refraction. In an appropriately designed photoelastic (two-dimensional) model, these two refracted components of the incident light travel in the same direction and can be examined in a polariscope. The degree to which the two light waves are out of phase is related to the stress by the stress–optic law:

$$\delta \propto n_x - n_y \tag{11.42}$$

where

δ = relative retardation between the two light beams

n_x, n_y = indices of refraction in the directions of the principal strains

The index of refraction changes in direct proportion to the amount the material is strained, such that

$$n_x - n_y = K(\varepsilon_x - \varepsilon_y) \tag{11.43}$$

The strain optical coefficient, K, is generally assumed to be a material property independent of the wavelength of the incident light. Figure 11.20 shows the use of a plane polariscope to examine the strain in a photoelastic model. Plane polarized light enters the specimen and emerges with two planes of polarization along the principal strain axes. This light beam is then passed through a polarizing filter, called the analyzer, which transmits only the component of each of the light waves that is parallel to the plane of polarization. The transmitted waves interfere, since they are out of phase, and the resulting light intensity is a function of the angle between the analyzer and the principal strain direction and the phase shift between the beams. The variations in strain in the specimen produce a pattern of fringes that can be related to the strain field through the strain optic relation.

When a photoelastic model is observed in a plane polariscope, a series of fringes is observed. The complete extinction of light occurs at locations where the principal strain directions coincide with the axes of the analyzer or where either the strain is zero or $\varepsilon_x - \varepsilon_y = 0$. These fringes, termed isoclinics, are used to determine the principal strain directions at all points in the photoelastic model. Figure 11.21 shows the isoclinics in a ring subject to a compression load (as shown in the figure). A reference direction is selected along the horizontal compression load and labeled 0 degrees. For each measurement angle, one of the principal strains at a point on an isoclinic is parallel to the specified angle and the other is perpendicular. For the 0-degree isoclinics, the principal strains are oriented at 0 and 90 degrees.

Using the facts that the direction of the principal axes is known at a free surface and that the shear stress is zero on a free surface, the magnitude of the stress on the boundary can be determined. The primary applications for photoelasticity, especially in a historical sense, have been in the study of stress concentrations around holes or reentrant corners. In these cases, the maximum stress is at the boundary and corresponds to one of the principal stresses. This maximum stress can be obtained directly by the optical method, because the shear stress is zero on the boundary.

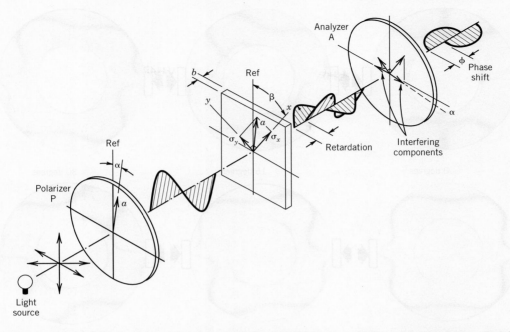

Figure 11.20   Construction of a plane polariscope. (Courtesy of Micro-Measurements, Raleigh, NC, USA.)

Moiré Methods

A moiré pattern results from two overlaid, relatively dense patterns that are displaced relative to each other. This observable optical effect occurs, for example, in color printing, where patterns of dots form an image. If the printing is slightly out of register, a moiré pattern results. Another common example is the striking shimmering effect that occurs with some patterned clothing on television. This effect results when the size of the pattern in the fabric is essentially the same as the resolution of the television image.

In experimental mechanics, moiré patterns are used to measure surface displacements, typically in a model constructed specifically for this purpose. The technique uses two gratings, or patterns of parallel lines spaced equally apart. Figure 11.22 shows two line gratings. There are two important properties of line gratings for moiré techniques. The pitch is defined as the distance between the centers of adjacent lines in the grating, and for typical gratings has a value of from 1 to 40 lines/mm. The second characteristic of gratings is the ratio of the open, transparent area of the grating to the total area, or, for a line grating the ratio of the distance between adjacent lines to the center-to-center distance, as illustrated in Figure 11.22. Clearly, a greater density of lines per unit width allows a greater sensitivity of strain measurement, but as line densities increase, coherent light is required for practical measurement.

To determine strain using the moiré technique, a grating is fixed directly to the surface to be studied. This can be accomplished through photoengraving, cementing film copies of a grating to the

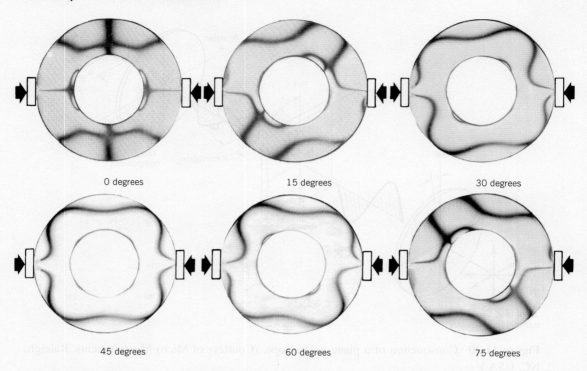

0 degrees   15 degrees   30 degrees

45 degrees   60 degrees   75 degrees

Figure 11.21 Construction of a plane polariscope. (Courtesy of Micro-Measurements, Raleigh, NC, USA.)

surface, or interferometric techniques. The master or reference grating is next placed in contact with the surface, forming a reference for determining the relative displacements under loaded conditions. A series of fringes result when the gratings are displaced relative to each other; the bright fringes are the loci of points where the projected displacements of the surface are integer multiples of the pitch. The technique then is two-dimensional, providing information concerning the projection of the

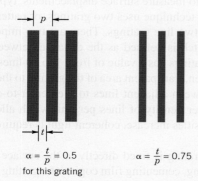

$\alpha = \dfrac{t}{p} = 0.5$
for this grating

$\alpha = \dfrac{t}{p} = 0.75$

Figure 11.22 Moiré gratings.

displacements into the plane of the master grating. Once a fringe pattern is recorded, data reduction techniques are employed to determine the stress and strain field. Graphical techniques exist that allow the strain components in two orthogonal directions to be determined. Further information on moiré techniques, along with additional references, may be found in the review article by Sciammarella (11).

Recently, techniques such as moiré-fringe multiplication have greatly increased the sensitivity of moiré techniques, with possible grating density of 1,200 lines/mm. Moiré interferometry is an extension of moiré-fringe multiplication that uses coherent light and has sensitivities on the order of 0.5 mm/fringe (12). A reflective grating is applied to the specimen, which experiences deformation under load conditions. The technique provides whole field readings of in-plane strain, with a four-beam optical arrangement currently in use (13).

Fiber Bragg Strain Measurement

Associated with recent progress in optical fiber technology is an emerging area of sensor technology based on fiber-optic Bragg gratings. A fiber-optic Bragg grating is illustrated in Figure 11.23.

The Bragg grating itself is an area of the fiber-optic core having a periodic variation of the refractive index of the material. As shown in Figure 11.23, a broadband light input interacts with the Bragg grating in two ways. Some of the incoming light is transmitted, and some of reflected by the Bragg grating. The Bragg wavelength corresponds to the largest intensity of the reflected light. If the Bragg condition is met (14) and the Bragg grating is of infinite length, there will be one Bragg frequency with maximum reflection given by

$$\lambda_B = 2n_{eff}L \tag{11.44}$$

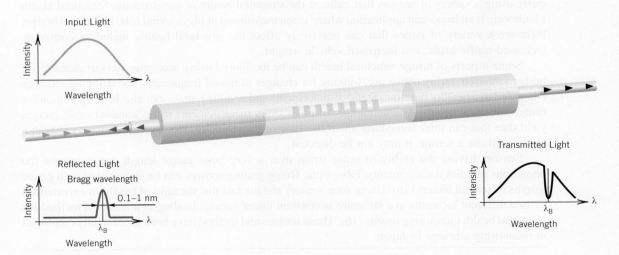

Figure 11.23 Fiber optic Bragg grating

where

λ_B - Bragg wavelength

n_{eff} - effective refractive index of the Bragg grating

L - spacing of the periodic variations in refractive index in the Bragg grating

A representative value of n_{eff} is 1.45 (15). Fiber-optic Bragg grating sensors are sensitive to temperature because thermal expansion changes the spacing, L.

Example 11.9

Determine the required spacing of the periodic variations in a fiber Bragg grating if the wavelength of the peak in reflected light is to be 1,545 nm.

Assuming that the refractive index is 1.45 (15), the resulting spacing is given by

$$L = \frac{\lambda_B}{2n_{eff}} = \frac{1545 \text{ nm}}{2(1.45)} = 533 \text{ nm}$$

COMMENT   During manufacture, fiber-optic Bragg gratings are created by exposing the fiber to an ultraviolet interference pattern that creates the periodic variation in refractive index. Spacings down to several hundred nanometers are easily achievable.

Example 11.10

A steel bridge is a structure that requires continuous maintenance and periodic inspection. However, increasingly bridges and other structures such as high-rise buildings are being monitored continuously using a variety of sensors that indicate the structural health of the structure. Structural health monitoring is an important application where strain measurement plays a vital role. For a steel bridge, there are a variety of issues that can negatively affect the structural health, including corrosion, increased traffic loads, and increased vehicle weight.

Some aspects of bridge structural health can be monitored using accelerometers to characterize bridge vibration during active use, looking for changes in modal frequencies and damping ratios as compared to a healthy baseline. Static displacements and resulting strain can also be used to monitor bridge structural health. A large number of point strain measurements using standard strain gauges yield data that can infer something about the state of the structure. However, if damage exists in a region without a sensor, it may not be detected.

Sensors having the ability to sense strain over a very large gauge length can overcome the limitations of point measurements. Fiber-optic Bragg grating sensors can be constructed with gauge lengths of several meters (16). Using such sensors and the fact that the ratio of the strain experienced by two different locations in a structure is constant under various loading conditions, a method for structural health monitoring results (16). These sensors and method have been successfully employed in monitoring a bridge in Japan.

11.8   SUMMARY

Experimental stress analysis can be accomplished through several practical techniques, including electrical resistance, photoelastic, and moiré strain measurement techniques. Each of these methods yields information concerning the surface strains for a test specimen. The design and selection of an appropriate strain measurement system begins with the choice of a measurement technique.

The bonded electrical resistance strain gauge provides a versatile means of measuring strain at a specific location on a test specimen. Strain gauge selection involves the specification of strain gauge material, the backing or carrier material, and the adhesive used to bond the strain gauge to the test specimen, as well as the total electrical resistance of the gauge. Other considerations include the orientation and pattern for a strain gauge rosette, and the temperature limit and maximum allowable elongation. In addition, for electrical resistance strain gauges, appropriate arrangement of the gauges in a bridge circuit can provide temperature compensation and elimination of specific components of strain.

Optical methods are useful in the initial determination of a stress field for complex geometries, and the determination of whole-field information in model studies. Such whole-field methods provide the basis for design and establish information necessary to make detailed local strain measurements.

The techniques for strain measurement described in this chapter provide the basis for determining surface strains for a test specimen. Although the focus here has been the measurement techniques, the proper placement of strain gauges and the interpretation of the measured results requires further analysis. The inference of load-carrying capability and safety for a particular component is a result of the overall experimental program.

REFERENCES

1. Hibbeler, R. C., *Mechanics of Materials*, 9th ed., Prentice-Hall, Upper Saddle River, NJ, 2013.
2. Timoshenko, S. P., and J. M. Goodier, *Theory of Elasticity*, 3rd ed., Engineering Society Monographs, McGraw-Hill, New York, 1970.
3. Dally, J. W., and W. F. Riley, *Experimental Stress Analysis*, 3rd ed., McGraw-Hill, New York, 1991.
4. Thomson, W. (Lord Kelvin), On the electrodynamic qualities of metals, *Philosophical Transactions of the Royal Society (London)* **146**: 649–751 1856.
5. Micro-Measurements Division, Measurements Group, Inc., Strain Gauge Selection: Criteria, Procedures, Recommendations, Technical Note 505-1, Raleigh, NC, 1989.
6. Micro-Measurements Division, Measurements Group, Inc., Strain Gauge Technical Data, Catalog 500, Part B, and TN-509: Errors Due to Transverse Sensitivity in Strain Gauges, Raleigh, NC, 2011.
7. Kulite Semiconductor Products, Inc, Bulletin KSG-5E: *Semiconductor Strain Gauges*, Leonia, NJ.
8. Weymouth, L. J., J. E. Starr, and J. Dorsey, Bonded resistance strain gauges, *Experimental Mechanics* **6** (4): 19A, 1966.
9. Oi, K., Transient response of bonded strain gauges, *Experimental Mechanics* **6** (9): 463, 1966.

10. Huang, M., Wu, Y., Chandra, B., Yan, H., Shan, Y., Heinz, T., and J. Hone, Direct measurement of strain-induced changes in the band structure of carbon nanotubes, *Physical Review Letters* **100**, 136803, April 2008.

11. Sciammarella, C. A., The moiré method—a review, *Experimental Mechanics* **22** (11): 418, 1982.

12. Post, D., Moiré interferometry at VPI & SU, *Experimental Mechanics* **23** (2): 203, 1983.

13. Post, D., Moiré interferometry for deformation and strain studies, *Optical Engineering* **24** (4): 663, 1985.

14. Yao, J., A tutorial on microwave photonics, *IEEE Photon. Soc. Newsletter*, **26**, 4 2012.

15. Julich, F. and J. Roths, Determination of the effective refractive index of various single mode fibres for fibre Bragg grating sensor applications, Sensor&Test Conference 2009, OPTO2009 Proceedings.

16. Serker, K. and Z. Wu, Structural health monitoring using distributed macro-strain response, *Journal of Applied Sciences* **9** (7): 1276–1284, 2009.

NOMENCLATURE[1]

b	width (l)	
c	speed of sound ($l\,t^{-1}$)	
h	height (l)	
n_i	index of refraction in the direction of the principal strain in the i direction	
$(u_d)_x$	design-stage uncertainty in the variable x	
A_c	cross-sectional area (l^2)	
D	diameter (l)	
E_m	modulus of elasticity ($ml^{-1}t^{-2}$)	
E_i	input voltage (V)	
E_o	output voltage (V)	
e_i	strain gauge lateral sensitivity error as a percentage of axial strain	
F_N	force normal to A_c ($m\,l\,t^{-2}$)	
G	shear modulus ($m\,t^{-2}$)	
G_A	amplifier gain	
GF	gauge factor	
K	strain optical coefficient	
K_B	bridge sensitivity (V)	
K_l	strain gauge lateral sensitivity	
L	length (l)	
δE_o	voltage change (V)	
δL	change in length (l)	

δR	resistance change (Ω)	
M	bending moment ($m\,l^2t^{-2}$)	
R	electrical resistance (Ω)	
T	temperature ($°$)	
T_o	reference temperature ($°$)	
α	moiré grating width to spacing parameter	
γ_{xy}	shear strain in the xy plane ($m\,l^{-1}t^{-2}$)	
δ	relative retardation between two light beams in a photoelastic material	
ε_a	axial strain	
ε_b	bending strain	
ε_i	strain in the i coordinate direction (i.e., x direction)	
ε_l	lateral or transverse strain	
ε_t	apparent strain due to temperature	
κ	bridge constant	
ν_p	Poisson ratio	
ν_{po}	Poisson ratio for gauge factor calibration test	
π_1	piezoresistance coefficient ($t^2l\,m^{-1}$)	
ρ_e	electrical resistivity ($\Omega\,l$)	
σ	stress ($m\,l^{-1}t^{-2}$)	
σ_a	axial stress ($m\,l^{-1}t^{-2}$)	
τ_{xy}	shear stress in the xy plane ($m\,l^{-1}t^{-2}$)	

---

[1] Note that for this development, it is assumed that $\delta R/R \ll 1$.

PROBLEMS

11.1 Calculate the change in length of a steel rod $(E_m = 30 \times 10^6 \text{ psi})$ that has a circular cross section, a length of 10 in., and a diameter of 1/4 in. The rod supports a mass of 50 lb_m in such a way that a state of uniaxial tension is created in the rod.

11.2 Calculate the change in length of a steel rod $(E_m = 20 \times 10^{10} \text{ Pa})$ that has a length of 0.3 m and a diameter of 5 mm. The rod supports a mass of 60 kg in a standard gravitation field in such a way that a state of uniaxial tension is created in the rod. Make the same calculation for a rod made of aluminum $(E_m = 70 \text{ GPa})$.

11.3 For the steel rod in Problem 11.1 calculate the lateral strain and the dilatation assuming that $v_p = 0.3$ for steel.

11.4 For the steel $(v_p = 0.3)$ and aluminum $(v_p = 0.334)$ rods in Problem 11.2, calculate the lateral strain and the dilatation.

11.5 An electrical coil is made by winding copper wire around a core. What is the resistance of 20,000 turns of 16-gauge wire (0.051 in. diameter) at an average radius of 2.0 in.?

11.6 Compare the resistance of a volume of $\pi \times 10^{-5} \text{ m}^3$ of aluminum wire having a diameter of 2 mm, with the same volume of aluminum formed into 1-mm-diameter wire. (The resistivity of aluminum is $2.66 \times 10^{-8} \, \Omega \, \text{m}$.)

11.7 A conductor made of nickel $(\rho_e = 6.8 \times 10^{-8} \, \Omega \, \text{m})$ has a rectangular cross section 5×2 mm and is 5 m long. Determine the total resistance of this conductor. Calculate the diameter of a 5-m-long copper wire that has a circular cross section that yields the same total resistance.

11.8 Consider a Wheatstone bridge circuit that has all resistances equal to 100 Ω. The resistance R_1 is a strain gauge that cannot sustain a power dissipation of more than 0.25 W. What is the maximum applied voltage that can be used for this bridge circuit? At this level of bridge excitation, what is the bridge sensitivity?

11.9 Consider a Wheatstone bridge circuit that has all resistances equal to 100 Ω. What applied voltage would yield a bridge sensitivity of 0.05 V/Ω? What current would flow through each resistor? What power would each resistor need to dissipate?

11.10 A resistance strain gauge with $R = 120 \, \Omega$ and a gauge factor of 2 is placed in an equal-arm Wheatstone bridge in which all the resistances are equal to 120 Ω. If the maximum gauge current is to be 0.05 A, what is the maximum allowable bridge excitation voltage?

11.11 A strain gauge that has a nominal resistance of 350 Ω and a gauge factor of 1.8 is mounted in an equal-arm bridge, which is balanced at a zero applied strain condition. The gauge is mounted on a 1-cm^2 aluminum rod, with $E_m = 70 \text{ GPa}$. The gauge senses axial strain. The bridge output is 1 mV for a bridge input of 5 V. What is the applied load, assuming the rod is in uniaxial tension?

11.12 Suppose the bridge in Problem 11.11 was operated in balanced mode. What change in resistance would be required to balance the bridge? Assume the loading conditions are the same.

11.13 A single strain gauge is mounted as shown in Figure 11.11. If the gauge indicates a strain of 4×10^{-4} m/m, what is the corresponding axial stress for the following materials?

a. Steel

b. Aluminum

c. Platinum

11.14 A single strain gauge is mounted on a structural member as shown in Figure 11.11. The material is titanium, and the structural member is subject to an axial stress of 3.2×10^6 kPa. What is the axial strain?

11.15 Equation 11.15 provides an expression for $\delta E_o / E_i$ for a single strain gauge mounted in a Wheatstone bridge.

a. List all the assumptions made in developing this equation from Equation 6.12

b. Plot the error resulting from assuming $\delta R / R \ll 1$ that allows Equation 6.12 to be reduced to Equation 11.15

11.16 Consider a structural member subject to loads that produce both axial and bending stresses, as shown in Figure 11.13. Two strain gauges are to be mounted on the member and connected in a Wheatstone bridge in such a way that the bridge output indicates the axial component of strain only (the installation is bending compensated). Show that the installation of the gauges shown in Figure 11.13 will not be sensitive to bending.

11.17 A steel beam member ($v_p = 0.3$) is subjected to simple axial tensile loading. One strain gauge aligned with the axial load is mounted on the top and center of the beam. A second gauge is similarly mounted on the bottom of the beam. If the gauges are connected as arms 1 and 4 in a Wheatstone bridge, determine the bridge constant for this installation. Is the measurement system temperature compensated? (Explain why, or why not.) If $\delta E_0 = 10\,\mu V$ and $E_i = 10\,V$, determine the axial and transverse strains. The gauge factor for each gauge is 2, and all resistances are initially equal to 120 Ω.

11.18 Consider the analysis of the bridge constant for an axial and transverse strain gauge described in Example 11.5. For the structural member described in Example 11.3, plot the error in bridge constant that results from assuming $\delta R/R \ll 1$ for loads ranging from 15 kN to 150 kN.

11.19 An axial strain gauge and a transverse strain gauge are mounted to the top surface of a steel beam that experiences a uniaxial stress of 2,222 psi. The gauges are connected to arms 1 and 2 (Fig. 11.12) of a Wheatstone bridge. With a purely axial load applied, determine the bridge constant for the measurement system. If $\delta E_0 = 250\,\mu V$ and $E_i = 10\,V$, estimate the average gauge factor of the strain gauges. For this material, Poisson's ratio is 0.3 and the modulus of elasticity is 29.4×10^6 psi.

11.20 A strain gauge is mounted on a steel cantilever beam of rectangular cross section. The gauge is connected in a Wheatstone bridge; initially $R_{gauge} = R_2 = R_3 = R_4 = 120\,\Omega$. A gauge resistance change of 0.1 Ω is measured for the loading condition and gauge orientation shown in Figure 11.24. If the gauge factor is 2.05 ± 1% (95%) estimate the strain. Suppose the uncertainty in each resistor value is 1% (95%). Estimate an uncertainty in the measured strain due to the uncertainties in the bridge resistances and gauge factor. Assume that the bridge operates in a null mode, which is detected by a galvanometer. Also assume reasonable values for other necessary uncertainties and parameters, such as input voltage or galvanometer sensitivity.

11.21 Two strain gauges are mounted so that they sense axial strain on a steel member in uniaxial tension. The 120 Ω gauges form two legs of a Wheatstone bridge, and are mounted on opposite arms. For a bridge

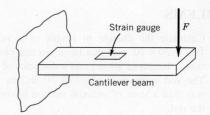

Figure 11.24 Loading for Problem 11.20.

excitation voltage of 4 V and a bridge output voltage of 120 μV under load, estimate the strain in the member. What is the resistance change experienced by each gauge? The gauge factor for each of the strain gauges is 2 and E_m for this steel is 29×10^6 psi.

11.22 A rectangular bar is instrumented with strain gauges and subjected to a state of uniaxial tension. The bar has a cross-sectional area of 2 in.^{2}, and the bar is 12 in. long. The two strain gauges are mounted such that one senses the axial strain, while the other senses the lateral strain. For an axial load of 1,500 lb, the axial strain is measured as 1,500 με (μin./in.), and the lateral gauge indicates a strain of −465 με. Determine the modulus of elasticity and Poisson's ratio for this material.

11.23 A round member having a cross-sectional area of 3 cm^2 experiences an axial load of 10 kN. Two strain gauges are mounted on the member, one measuring an axial strain of 600 με (m/m), and the other measuring a lateral strain of −163 με. Determine the modulus of elasticity and Poisson's ratio for this material.

11.24 A square structural member having a cross sectional area of 4 in.^{2} is loaded in uniaxial tension with a force of 2,300 lb$_f$. The axial strain is measured as 100 με and the lateral strain as −21 με. Determine the modulus of elasticity and Poisson's ratio for this material. Is this Poisson's ratio possible? For what materials?

11.25 Show that the use of a dummy gauge together with a single active gauge compensates for temperature but not for bending. Consider the case in which the active gauge is subjected to axial loading with minimal bending and both gauges experience the same temperature.

11.26 Show that four strain gauges mounted to a shaft such that the gauge pairs measure equal and opposite strain can be used to measure torsional twist (as suggested in Table 11.1). Show that this method compensates for axial and bending strains and for temperature.

11.27 For each bridge configuration described below, determine the bridge constant. Assume that all of the active gauges are identical, and that all of the fixed resistances are equal. The locations of the gauges in the bridge are shown in Figure 11.25.

Bridge Configuration	Description
(a) 1: active gauge; 2–4: fixed resistors	Single gauge in uniaxial tension
(b) 1: active gauge; 2: Poisson gauge; 3, 4: fixed resistors	Two active gauges in uniaxial stress field
	Gauge 1 aligned with maximum axial stress; gauge 2 lateral
(c) 1: active gauge; 3: active gauge; 2, 4: fixed resistors	Equal and opposite strains applied to the active gauges (bending compensation)
(d) Four active gauges	Gauges 1 and 4 aligned with uniaxial stress; gauges 2 and 3 transverse
(e) Four active gauges	Gauges 1 and 2 subject to equal and opposite strains; gauges 3 and 4 subject to the same equal but opposite strains

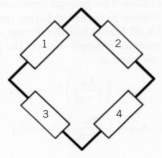

Figure 11.25 Bridge arrangement for Problems 11.27 to 11.29.

11.28 A bathroom scale uses four internal strain gauges to measure the displacement of its diaphragm as a means for determining load. Four active gauges are used in a bridge circuit. The gauge factor is 2, and gauge resistance is 120 Ω for each gauge. If an applied load to the diaphragm causes a compression strain on R_1 and R_4 of 20 με while gauges R_2 and R_3 experience a tensile strain of 20 με, estimate the bridge deflection voltage. The supply voltage is 9 V. Refer to Figure 11.25 for gauge position.

11.29 Suppose in Problem 11.28 that the lead wires of two gauges (R_2 and R_4) are accidentally interchanged on the assembly line such that a compression strain is now sensed on R_1 and R_2 of 20 με while gauges R_4 and R_3 experience a tensile strain of 20 με. Should this really matter? Explain. Estimate the bridge deflection voltage. Refer to Figure 11.25 for gauge position.

Problems 11.30 through 11.32 address measuring strains associated with a thin-walled pressure vessel. The pressure vessel is constructed of steel and has a circular cross-section. The tangential and longitudinal stresses in the wall of the pressure vessel are

$$\sigma_t = pD/2t \qquad \sigma_l = pD/4t$$

where

σ_t = tangential stress
σ_l = longitudinal stress
p = pressure
t = wall thickness

11.30 A single-strain gauge is mounted on the surface of a thin-walled pressure vessel that has a diameter of 1 m. For a strain gauge oriented along the tangential stress direction, determine the percentage error in tangential strain measurement caused by lateral sensitivity as a function of pressure and vessel wall thickness. The lateral sensitivity is 0.03.

11.31 Consider a Wheatstone bridge circuit that has all fixed resistances equal to 100 Ω and with a strain gauge located at the R_1 position, which has a value of 100 Ω under conditions of zero strain. The strain gauge is mounted so as to sense longitudinal strain for a thin-walled pressure vessel made of steel that has a wall thickness of 2 cm and a diameter of 2 m. The strain gauge has a gauge factor of 2 and cannot sustain a power dissipation of more than 0.25 W. What is the maximum static sensitivity that can be achieved with this proposed measurement system? Is this static sensitivity constant with input pressure? If not, under what conditions would it be reasonable to assume a constant static sensitivity? The static sensitivity should be expressed in units of V/kPa.

11.32 Design a Wheatstone bridge measurement system to measure the tangential strain in the wall of the pressure vessel described above, and develop a reasonable estimate of the resulting uncertainty. You may assume that the strain does not vary with time and that the bridge is operated in a balanced mode. Your selection of specified values for the input voltage, the fixed resistances in the bridge and the galvanometer sensitivity, and their associated uncertainties will allow completion of this design.

11.33 A steel cantilever beam is fixed at one end and free to move at the other. A load F of 980 N is applied to the free end. Four axially aligned strain gauges ($GF = 2$) are mounted to the beam a distance L from the applied load, two on the upper surface, R_1 and R_4, and two on the lower surface, R_2 and R_3. The bridge deflection output is passed through an amplifier (gain, $G_A = 1000$) and measured. For a cantilever, the relationship between applied load and strain is

$$F = \frac{2E_m I \varepsilon}{Lt}$$

where I is the beam moment of inertia ($= bt^3/12$), t is the beam thickness, and b is the beam width. Estimate the measured output for the applied load if $L = 0.1$ m, $b = 0.03$ m, $t = 0.01$ m, and the bridge excitation voltage is 5 V. $E_m = 200$ GPa.

11.34 A cantilever beam is to be used as a scale. The beam, made of 2024-T4 aluminum, is 21 cm long, 0.4 cm thick, and 2 cm wide. The scale load of between 0 and 200 g is to be concentrated at a point along the beam centerline 20 cm from its fixed end. Strain gauges are to be used to measure beam deflection and mounted to a Wheatstone bridge to provide an electrical signal that is proportional to load. Either 1/4-, 1/2-, or full-bridge gauge arrangements can be used. Design the sensor arrangement and its location on the beam. Specify appropriate signal conditioning to excite the bridge and to measure its output on a data acquisition system using a ±5 V, 12-bit A/D converter. Can the system achieve a 4% uncertainty at the design-

stage? The following specifications and system choices are stated at the 95% confidence level:

Sensors: one, two, or four axial gauges at 120 Ω ± 0.2% (selectable)

Gauge factor of 2.0 ± 1%

Bridge excitation: 1, 3, 5 V ± 0.5% (selectable)

Bridge null: within 5 mV

Signal conditioning:
  Amplifier gain: 1X, 10X, 100X, 1000X ± 1% (selectable)
  Low-pass filter: f_c at 0.5, 5, 50, 500 Hz @ −12dB/octave (selectable)

Data acquisition system:
  Conversion errors: < 0.1% reading
  Sample rate: 1, 10, 100, 1,000 Hz (selectable)

11.35 A rectangular strain gauge rosette is composed of strain gauges oriented at relative angles of 0, 45, and 90 degrees, as shown in Figure 11.17b. The rosette is used to measure strain on a steel structural member having $E_m = 200$ GPa and $\nu_p = 0.3$. The measured values of strain are

$$\varepsilon_1 = 10000 \; \mu\varepsilon$$
$$\varepsilon_2 = 5000 \; \mu\varepsilon$$
$$\varepsilon_3 = 50000 \; \mu\varepsilon$$

Determine the principal stresses and the angle that the maximum principal stress makes with the 0-degree gauge.

11.36 Repeat previous problem for the same conditions but change the material to titanium. Research applications for titanium, and comment on the results.

11.37 Research the equivalent of Equation 11.39 for a 60–60–60 degree strain rosette.

11.38 If the directions of the principal stresses are known a priori a strain gauge rosette having two gauges oriented at 90 degrees to each other are sufficient to determine the principal stresses. The principal stresses are given as

$$\sigma_{\max} = \left(\frac{E_m}{1 - v_p^2}\right)(\varepsilon_1 + v_p \varepsilon_2)$$

$$\sigma_{\min} = \left(\frac{E_m}{1 - v_p^2}\right)(\varepsilon_2 + v_p \varepsilon_1)$$

where $\varepsilon_1 > \varepsilon_2$. If $E_m = 180$ GPa and $v_p = 0.3$ what are the principal stresses if $\varepsilon_1 = 8000 \; \mu\varepsilon$ and $\varepsilon_2 = -6000 \; \mu\varepsilon$.

Chapter 12

Mechatronics[1]: Sensors, Actuators, and Controls

12.1 INTRODUCTION

Rapid advances in microprocessors have led to a dramatic increase in electronically controlled devices and systems. All these systems require sensors and actuators to be interfaced with the electronics. Understanding the operating principles and limitations of sensors is essential to selecting and interfacing sensors for linear motion, rotary motion, and engineering variables such as force, torque, and power. *Actuators* are required to produce motion, such as to move an electric car seat, for a fly-by-wire throttle for automotive applications, or for positioning a precision laser welder. This chapter discusses sensors and actuators and provides a brief introduction to linear control theory.

Upon completion of this chapter, the reader will be able to

- Describe and analyze methods for displacement measurement
- State the physical principles underlying velocity and acceleration measurements
- Describe various load cells and their appropriate applications
- Describe various methods for measuring torque and power
- Describe various actuators and their role in mechatronic systems
- Analyze proportional-integral-derivative (PID) control schemes as part of a mechatronic system

12.2 SENSORS

The previous chapters described a wide variety of measurement sensors and the fundamentals of their operation. Thermocouples, strain gauges, flow meters, and pressure sensors represent the means to measure the very important engineering process variables of temperature, strain, flow rate, and pressure. In this chapter we add to this base by introducing methods and sensors for the measurement of linear and rotary displacement, acceleration and vibration, velocity measurement, force or load, torque, and mechanical power.

---

[1] The term "mechatronics" is derived from the terms "mechanical" and "electronic" and refers to the integration of mechanical and electronic devices.

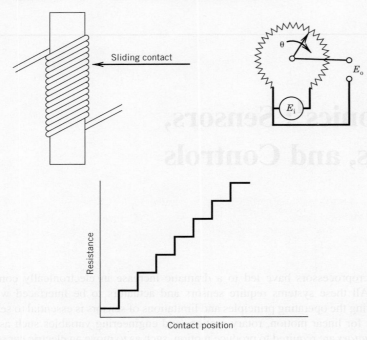

Figure 12.1 Potentiometer construction.

Displacement Sensors

Methods to measure position or displacement are a key element in many mechatronic systems. Often potentiometric or linear variable differential transformer (LVDT) transducers are employed in these applications.

Potentiometers

A potentiometer is a device employed to measure linear or rotary displacement. The principle of operation relies on a change in electrical resistance with displacement.

A wire-wound potentiometer[2] or variable electrical resistance transducer is depicted in Figure 12.1. This transducer is composed of a sliding contact and a winding. The winding is made of many turns of wire wrapped around a non-conducting substrate. Output signals from such a device can be realized by imposing a known voltage across the total resistance of the winding and measuring the output voltage, which is proportional to the fraction of the distance the contact point has moved along the winding. Potentiometers can also be configured in a rotary form, with numerous total revolutions of the contact possible in a helical arrangement. The output from the sliding contact as it moves along the winding is actually discrete, as illustrated in Figure 12.1; the resolution is limited by the number of turns per unit distance. The loading errors associated with voltage-dividing circuits, discussed in Chapter 6, should be considered in choosing a measuring device for the output voltage from the potentiometer.

---

[2] The potentiometer–transducer should not be confused with the potentiometer–instrument. Although both are based on voltage-divider principles, the latter measures low-level voltages (10^{-6} to 10^{-3} V) as described in Chapter 6.

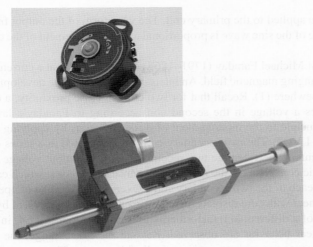

Figure 12.2 Conductive plastic potenti-ometer. (Courtesy of Novotechnic U.S., Inc. Used with permission).

Wire-wound potentiometers display a stepwise output as the wiper contacts successive turns of the wire winding. Conductive plastic potentiometers were developed to eliminate this stepwise output and are now widely employed in mechatronic systems. Two such potentiometers are shown in Figure 12.2. The key components are the structural support and mechanical interface, the conductive plastic resistor, and the wiper where electrical contact occurs. A linear output with displacement is most often the design objective for these sensors. Typical linearity errors introduce instrument uncertainty from 0.2% to 0.02% of the reading.

Linear Variable Differential Transformers

The LVDT, as shown in Figure 12.3, produces an AC output with an amplitude that is proportional to the displacement of a movable core. The movement of the core causes a mutual inductance in the

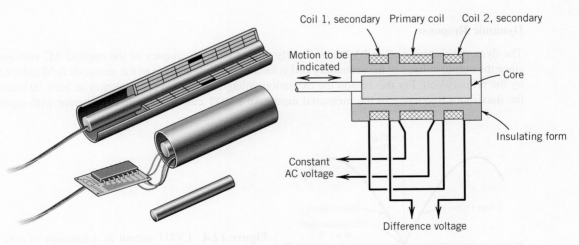

Figure 12.3 Construction of a linear variable differential transformer (LVDT).

secondary coils for an AC voltage applied to the primary coil. The waveform of the output from the LVDT is sinusoidal; the amplitude of the sine wave is proportional to the displacement of the core for a limited range of core motion.

In 1831, the English physicist Michael Faraday (1791–1867) demonstrated that a current could be induced in a conductor by a changing magnetic field. An interesting account of the development of the transformer may be found elsewhere (1). Recall that for two coils in close proximity, a change in the current in one coil induces a voltage in the second coil according to Faraday's law. The application of this inductance principle to the measurement of distance begins by applying an AC voltage to the primary coil of the LVDT. The two secondary coils are connected in a series circuit, such that when the iron core is centered between the two secondary coils the output voltage is zero (Fig. 12.3). Motion of the magnetic core changes the mutual inductance of the coils, which causes a different voltage to be induced in each of the two secondary coils. Over a limited range of operation, the output voltage is essentially linear with core displacement, as first noted in a U.S. patent by G. B. Hoadley in 1940 (2). The output of a differential transformer is illustrated in Figure 12.4, in which both the linear range and nonlinear behavior are observed. Beyond the linear range, the output amplitude rises in a nonlinear manner to a maximum, and eventually falls to zero. The output voltages on either side of the zero displacement position are 180 degrees out of phase. Thus with appropriate phase measurement it is possible to determine positive or negative displacement of the core. However, note that owing to harmonic distortion in the supply voltage and to the two secondary coils' not being identical, the output voltage with the coil centered is not zero but instead reaches a minimum. Resolution of an LVDT strongly depends on the resolution of the measurement system used to determine its output. Resolutions down to the nanometer range are possible.

The differential voltage output of an LVDT, as shown in Figure 12.4, may be analyzed by assuming that the magnetic field strengths are uniform along the axis of the coils, neglecting end effects, and limiting the analysis to the case where the core does not move beyond the ends of the coils (2). Under these conditions, the differential voltage may be expressed in terms of the core displacement. The sensitivity of the LVDT in the linear range is a function of the number of turns in the primary and secondary coils, the root-mean-square (rms) current in the primary coil, and the physical size of the LVDT.

Dynamic Response

The dynamic response of an LVDT is directly related to the frequency of the applied AC voltage, since the output voltage of the secondary coil is induced by the variation of the magnetic field induced by the primary coil. For this reason the excitation voltage should have a frequency at least 10 times the maximum frequency in the measured input. An LVDT can be designed to operate with input

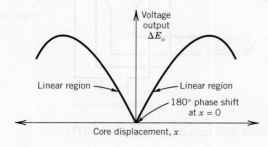

Figure 12.4 LVDT output as a function of core position.

frequencies ranging from 60 Hz up to 25kHz (for specialized applications, frequencies in the megahertz range can be used).

The maximum allowable applied voltage for an LVDT is determined by the current-carrying capacity of the primary coil, typically in the 1- to 10-V range. A constant current source is preferable for an LVDT to limit temperature effects. For other than a sine wave input voltage form, harmonics in the input signal increase the voltage output at the null position of the core. The appropriate means of measuring and recording the output signal from an LVDT and the AC frequency applied to the primary coil should be chosen based on the highest frequencies present in the input signal to the LVDT. For example, for static measurements and signals having frequency content much lower than the excitation frequency of the primary coil, an AC voltmeter may be an appropriate choice for measuring the output signal. In this case, it is likely that the frequency response of the measuring system would be limited by the averaging effects of the AC voltmeter. For higher-frequency signals, it is possible to create a DC voltage output that follows the input motion to the LVDT through demodulation and amplification of the resulting signal using a dedicated electronic circuit. Alternatively, the output signal can be sampled at a sufficiently high frequency using a computer data acquisition system to allow signal processing for a variety of purposes.

The measurement of distance using an LVDT is accomplished using an assembly known as an LVDT gauge head. Such devices are widely used in machine tools and various types of gauging equipment. Control applications have similar transducer designs. The basic construction is shown in Figure 12.5. The gauge head can reduce instrument errors to as low as 0.05% with repeatability of 0.0001 mm.

Angular displacement can also be measured using inductance techniques employing a rotary variable differential transformer (RVDT). The output curve of an RVDT and a typical construction are shown in Figure 12.6, in which the linear output range is approximately ±40 degrees.

Measurement of Acceleration and Vibration

The measurement of acceleration is required for a variety of purposes, ranging from machine design to guidance systems to structural health. Because of the range of applications for acceleration and vibration measurements, there exists a wide variety of transducers and measurement techniques, each associated with a particular application. This section addresses some fundamental aspects of these measurements along with some common applications.

Displacement, velocity, and acceleration measurements are also referred to as shock or vibration measurements depending on the waveform of the forcing function that causes the acceleration. A forcing function that is periodic in nature generally results in accelerations that are analyzed as vibrations. On the other hand, a force input having a short duration and large amplitude would be classified a shock load.

The fundamental aspects of acceleration, velocity, and displacement measurements can be discerned through examination of the most basic device for measuring acceleration and velocity, a seismic transducer.

Seismic Transducer

A seismic transducer consists of three basic elements, as shown in Figure 12.7: a spring–mass–damper system, a protective housing, and an appropriate output transducer. Through the appropriate design of the characteristics of this spring–mass–damper system, the output is a direct indication of

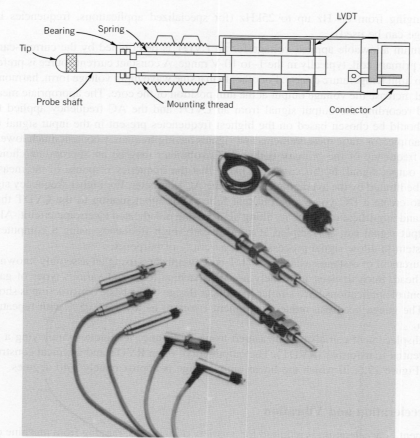

Figure 12.5 LVDT gauge head. Top: Cross section of a typical LVDT gauge head. (Courtesy of Measurement Specialties, Inc.; from reference 2. Used with permission.)

either displacement or acceleration. To accomplish a specific measurement, this basic seismic transducer is rigidly attached to the object experiencing the motion that is to be measured.

Consider the case where the output transducer senses the position of the seismic mass; a variety of transducers could serve this function. Under some conditions, the displacement of the seismic mass serves as a direct measure of the acceleration of the housing and the object to which it is attached. To illustrate the relation between the relative displacement of the seismic mass and acceleration, consider the case in which the input to the seismic instrument is a constant acceleration. The response of the instrument is illustrated in Figure 12.8. At steady-state conditions, under this constant acceleration, the mass is at rest with respect to the housing. The spring deflects an amount proportional to the force required to accelerate the seismic mass, and since the mass is known, Newton's second law yields the corresponding acceleration. The relationship between a constant acceleration and the displacement of the seismic mass is linear for a linear spring (where $F = ky$).

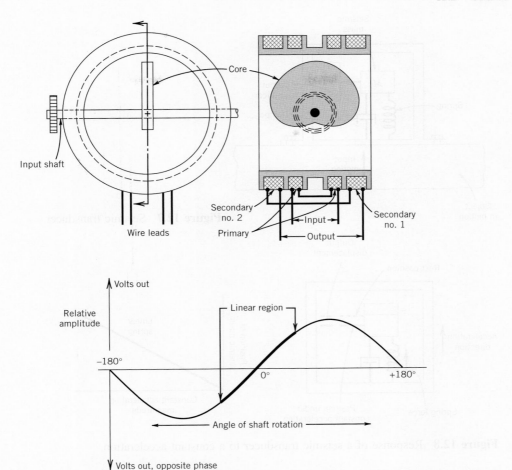

Figure 12.6 Rotary variable differential transformer. (Courtesy of Measurement Specialties, Inc.; from reference 6. Used with permission.)

We might want to measure not only constant accelerations, but also complex acceleration waveforms. Consider an input to our seismic instrument such that the displacement of the housing is a sine wave, $y_{hs} = A \sin \omega t$, and the absolute value of the resulting acceleration of the housing is $A\omega^2 \sin\omega t$. If we consider a free-body diagram of the seismic mass, the spring force and the damping force must balance the inertial force for the mass. (Notice that gravitational effects could play an important role in the analysis of this instrument—for instance, if the instrument were installed in an aircraft.) The spring force and damping force are proportional to the relative displacement and velocity between the housing and the mass, whereas the inertial force is dependent only on the absolute acceleration of the seismic mass.

We define a relative displacement, y_r, as the difference between the housing displacement, y_{hs}, and the displacement of the seismic mass, y_m, given by

$$y_r = y_m - y_{hs} \tag{12.1}$$

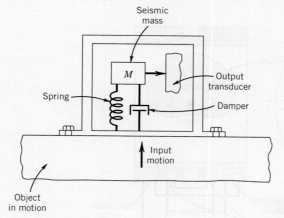

Figure 12.7   Seismic transducer.

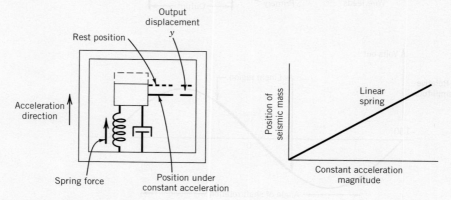

Figure 12.8   Response of a seismic transducer to a constant acceleration.

Newton's second law may then be expressed

$$m\frac{d^2 y_m}{dt^2} + c\frac{dy_r}{dt} + ky_r = 0 \tag{12.2}$$

Substituting Equation 12.1 into Equation 12.2 yields

$$m\left(\frac{d^2 y_{hs}}{dt^2} + \frac{d^2 y_r}{dt^2}\right) + c\frac{dy_r}{dt} + ky_r = 0 \tag{12.3}$$

But we know that $y_{hs} = A\sin\omega t$ and

$$\frac{d^2 y_{hs}}{dt^2} = -A\omega^2 \sin \omega t \tag{12.4}$$

Thus,

$$m\frac{d^2 y_r}{dt^2} + c\frac{dy_r}{dt} + ky_r = mA\omega^2 \sin \omega t \tag{12.5}$$

This equation is identical in form to Equation 3.12. As in the development for a second-order system response, we examine the steady-state solution to this governing equation. The transducer senses the relative motion between the seismic mass and the instrument housing. Thus, for the instrument to be effective, the value of y_r must provide indication of the desired output.

For the input function $y_{hs} = A\sin\omega t$, the steady-state solution for y_r is

$$(y_r)_{steady} = \frac{(\omega/\omega_n)^2 A \cos(\omega t - \phi)}{\left\{ \left[1 - (\omega/\omega_n)^2\right]^2 + \left[2\zeta(\omega/\omega_n)\right]^2 \right\}^{1/2}} \tag{12.6}$$

where

$$\omega_n = \sqrt{\frac{k}{m}} \qquad \zeta = \frac{c}{2\sqrt{km}} \qquad \phi = \tan^{-1} - \frac{2\zeta(\omega/\omega_n)}{1 - (\omega/\omega_n)^2} \tag{12.7}$$

The characteristics of this seismic instrument can be discerned by examining Equations 12.6 and 12.7. The natural frequency and damping are fixed for a particular design. We wish to examine the motion of the seismic mass and the resulting output for a range of input frequencies.

For vibration measurements, it is desirable to measure the amplitude of the displacements associated with the vibrations; thus, the desired behavior of the seismic instrument would be to have an output that gives a direct indication of y_{hs}. For this to occur, the seismic mass should remain stationary in an absolute frame of reference, and the housing and output transducer should move with the vibrating object. To determine the conditions under which this behavior would occur, the amplitude of y_r at steady state can be examined. The ratio of the maximum amplitude of the output divided by the equivalent static output is $(y_r)_{max}/A$. For vibration measurements, this ratio should have a value of 1. Figure 12.9 shows $(y_r)_{max}/A$ as a function

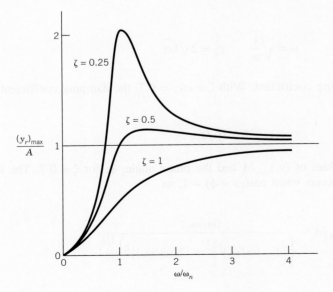

Figure 12.9 Displacement amplitude at steady state as a function of input frequency for a seismic transducer.

of the ratio of the input frequency to the natural frequency as determined from Equation 12.6. Clearly, as the input frequency increases, the output amplitude, y_r, approaches the input amplitude, A, as desired. Thus a seismic instrument that is used to measure vibration displacements should have a natural frequency smaller than the expected input frequency. Damping ratios near 0.7 are common for such an instrument. The seismic instrument designed for this application is called a *vibrometer*.

Example 12.1

A seismic instrument like the one shown in Figure 12.7 is to be used to measure a periodic vibration having an amplitude of 0.05 in. and a frequency of 15 Hz.

1. Specify an appropriate combination of natural frequency and damping ratio such that the dynamic error in the output is less than 5%.
2. What spring constant and damping coefficient would yield these values of natural frequency and damping ratio?
3. Determine the phase lag for the output signal. Would the phase lag change if the input frequency were changed?

KNOWN Input function $y_{hs} = 0.05 \sin 30\pi t$

FIND Values of ω_n, ζ, k, m, and c to yield a measurement with less than 5% dynamic error. Examine the phase response of the system.

SOLUTION Numerous combinations of the mass, spring constant, and damping coefficient would yield a workable design. Let's choose $m = 0.05 \ \text{lb}_m$ (22.7 g) and $\zeta = 0.7$. We know that for a spring–mass–damper system

$$\omega = \sqrt{\frac{k}{m}} \qquad c_c = 2\sqrt{km}$$

where c_c is the critical damping coefficient. With $\zeta = c/c_c = 0.7$, the damping coefficient, c, is found as

$$c = 2(0.7)\sqrt{km}$$

We can now examine the values of $(y_r)_{max}/A$ and the phase angle, ϕ, for $\zeta = 0.7$. The largest magnitude in Equation 12.6 occurs when $\cos(\omega t - \phi) = 1$, so

$$(y_r)_{max}/A = \frac{(\omega/\omega_n)^2}{\left\{\left[1 - (\omega/\omega_n)^2\right]^2 + \left[2\zeta(\omega/\omega_n)\right]^2\right\}^{1/2}}$$

and ϕ is found from Equation 12.7. Here is sample of the results:

$\dfrac{\omega}{\omega_n}$	$\dfrac{(y_r)_{max}}{A}$	ϕ(degrees)
10	1.000	172
8	1.000	169.9
6	1.000	166.5
4	0.999	159.5
3	0.996	152.3
2	0.975	137.0
1.7	0.951	128.5

Using the dynamic error as $\left| \dfrac{(y_r)_{max}}{A} - 1 \right|$, acceptable behavior is achieved for values of $\omega/\omega_n \geq 1.7$ for $\zeta = 0.7$, and an acceptable maximum value of the natural frequency is

$$\omega_n = \omega/1.7 = 30\pi/1.7 = 55.4 \text{ rad/s}$$

with[3]

$$\omega_n = \sqrt{k/m}$$

for $m = 0.05 \text{ lb}_m$ (22.7 g), then the value of k is 4.8 lb/ft (70 N/m). So the value of the damping coefficient is

$$c = 2(0.7)\sqrt{km} = 0.12 \text{ lb-s/ft} = 1.76 \text{ N-s/m}$$

COMMENTS   We should consider other constraints, including size, cost, and operating environment. The phase shift–frequency behavior for this combination of design parameters could allow distortion of complex waveform signals so if that is important, another set of parameters may be preferable. This and other issues can be explored using the program file *seismic_transducer.vi*.

Acceleration Measurement with a Seismic Instrument

If it is desired to measure acceleration, the behavior of the seismic mass must be quite different from that in Example 12.1. The amplitude of the acceleration input signal is $A\omega^2$. To have the output value y_r represent the acceleration, it is clear from Equation 12.6 that the value of $y_r/A(\omega/\omega_n)^2$ must be a constant over the design range of input frequencies. If this is true, the output will be proportional to the acceleration. The amplitude of $y_r/A(\omega/\omega_n)^2$ is

$$\frac{(y_r)_{steady}}{A(\omega/\omega_n)^2} = \frac{\cos(\omega t - \phi)}{\left\{ \left[1 - (\omega/\omega_n)^2 \right]^2 + \left[2\zeta(\omega/\omega_n) \right]^2 \right\}^{1/2}} \tag{12.8}$$

---

[3] Note: 1 lb = 32.2 lb_m-ft/s^2, whereas 1 N = 1 kg-m/s^2.

and

$$M(\omega) = \frac{1}{\left\{ \left[1 - (\omega/\omega_n)^2 \right]^2 + \left[2\zeta(\omega/\omega_n) \right]^2 \right\}^{1/2}} \tag{12.9}$$

where $M(\omega)$ is the magnitude ratio as given by Equation 3.22 in Chapter 3. The magnitude ratio is plotted as a function of input frequency and damping ratio in Figure 3.16.

From Figure 3.16, we find that the desired flat magnitude response is achieved over a range of input frequency ratios from $0 \leq \omega/\omega_n \leq 0.4$. Typically, in an accelerometer the damping ratio is designed to be near 0.7 so that the phase shift is linear with frequency such that distortion is minimized.

In summary, the seismic instrument can be designed so that the output can be interpreted in terms of either the input displacement or the input acceleration. Acceleration measurements may be integrated to yield velocity information; the differentiation of displacement data to determine velocity or acceleration introduces significantly more difficulties than the integration process.

Transducers for Shock and Vibration Measurement

In general, the destructive forces generated by vibration and shock are best quantified through the measurement of acceleration. Although a variety of accelerometers are available, strain gauge and piezoelectric transducers are widely employed for the measurement of shock and vibration (3).

A piezoelectric accelerometer employs the principles of a seismic transducer through the use of a piezoelectric element to provide a portion of the spring force. Figure 12.10 illustrates one basic construction of a piezoelectric accelerometer. A preload is applied to the piezoelectric element simply by tightening the nut that holds the mass and piezoelectric element in place. Upward or downward motion of the housing changes the compressive forces in the piezoelectric element, resulting in an appropriate output signal. Instruments are available with a range of frequency response from 0.03 to 10,000 Hz. Depending on the piezoelectric material used in the transducer construction, the static sensitivity can range from 1 to 100 mV/g. Steady accelerations cannot be effectively measured using such a piezoelectric transducer.

Strain gauge accelerometers are generally constructed using a mass supported by a flexure member, with the strain gauge sensing the deflection that results from an acceleration of the mass. The frequency response and range of acceleration of these instruments are related such that instruments designed for higher accelerations have a wider bandwidth but significantly lower static sensitivity. Table 12.1 provides typical performance characteristics for two strain gauge accelerometers that employ semiconductor strain gauges.

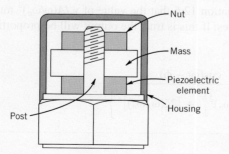

Figure 12.10 Basic piezoelectric accelerometer.

Table 12.1   Representative Performance Characteristics for
Piezoresistive Accelerometers[a]

Characteristic	25-g Range	2,500-g Range
Sensitivity [mV/g]	50	0.1
Resonance frequency [Hz]	2,700	30,000
Damping ratio	0.4–0.7	0.03
Resistance [Ω]	1,500	500

[a]Adapted from reference 7.

Many other accelerometer designs are available, including potentiometric or reluctive or accelerometers that use closed-loop servo systems to provide a high output level. Piezoelectric transducers have the highest frequency response and range of acceleration but have relatively lower sensitivity; amplification can overcome this drawback to some degree. Semiconductor strain gauge transducers have a lower frequency response limit than do piezoelectric transducers, but they can be used to measure steady accelerations.

Velocity Measurements

Linear and angular velocity measurements use a variety of approaches ranging from radar and laser systems for speed measurement to mechanical counters to provide an indication of a shaft rotational speed. For many applications, the sensors employed provide a scalar output of speed. However, sensors and methods exist that can provide indication of both speed and direction when properly employed. Here we consider techniques for the measurement of linear and angular speed.

Displacement, velocity, and acceleration measurements are made with respect to some frame of reference. Consider the case of a game of billiards in a moving railway car. Observers on the platform and on the train would assign different velocity vectors to the balls during play. The velocity vectors would differ by the relative velocity of the two observers. Simply differentiating the vector velocity equation, however, shows that the accelerations of the balls are the same in all reference frames moving relative to one another with constant relative velocity.

Linear Velocity Measurements

As previously discussed, measurements of velocity require a frame of reference. For example, the velocity of a conveyor belt might be measured relative to the floor of the building where it is housed. On the other hand, advantages may be realized in a control system if the velocity of a robotic arm, which "picks" parts from this same conveyor belt, is measured relative to the moving conveyor. In the present discussion, it is assumed that velocity is measured relative to a ground state, which is generally defined by the mounting point of a transducer.

Consider the measurement of velocity relative to a fixed frame of reference. Typically, if the measurement of linear velocity is to be made on a continuous basis, an equivalent angular rotational speed is measured, and the data is analyzed in such a way so as to produce a measured linear velocity. For example, a speedometer on an automobile provides a continuous record of the speed of the car, but the output is derived from measuring the rotational speed of the drive shaft or transmission.

Velocity from Displacement or Acceleration

Velocity, in general, can be directly measured by mechanical means only over very short times or small displacements, due to limitations in transducers. However, if the displacement of a rigid body is measured at identifiable time intervals, the velocity can be determined through differentiation of the time-dependent displacement. Alternatively, if acceleration is measured, the velocity may be determined from integration of the acceleration signal. The following example demonstrates the effect of integration and differentiation on the uncertainty of velocities computed from acceleration or displacement.

Example 12.2

Suppose our goal is to assess the merits of measuring velocity through the integration of an acceleration signal as compared to differentiating a displacement signal. The following conditions are assumed to apply.

For both $y(t)$ and $y''(t)$, the data acquisition system and transducers have the characteristics described in Table 12.2. Note that the uncertainties are derived directly from the analog-to-digital (A/D) resolution error and accuracy statements. Assume that the signals for both acceleration and displacement are sampled at 10 Hz and numerical techniques are used to differentiate or integrate the resulting signals.

SOLUTION   Consider first determining the velocity through differentiation of the displacement signal. Displacement is measured digitally by the data acquisition system, with a digitized value of displacement recorded at time intervals δt. The velocity at any time $n\delta t$ can be approximated as

$$v(t) = y'(t) = \frac{y_{n+1} - y_n}{\delta t} \tag{12.10}$$

Table 12.2   Specification and Uncertainty Analyses for Displacement and Acceleration

Measured Variable	Functional Form	Full-Scale Output Range
Displacement (cm)	$y(t) = 20 \sin 2t$	0 to 10 V
Acceleration (cm/s^2)	$y''(t) = -80 \sin 2t$	−5 to 5 V

Uncertainty Values for Displacement and Acceleration	
Measured Variable	Uncertainty
Displacement	Accuracy: 1% full scale = 0.2 cm
	A/D 8 bit (0.04 V) = 0.08 cm
	Total uncertainty = ±0.22 cm
Acceleration	Accuracy: 1% full scale = 0.05 cm/s^2
	A/D 8 bit (0.04 V) = 0.04 cm/s^2
	Total uncertainty = ±0.064 cm/s^2

where

y_{n+1} = the $(n + 1)$ measurement of displacement at time $(n + 1)\delta t$

y_n = the nth measurement of displacement at time $n\delta t$

$v(t)$ = velocity at time t

If the signal is sampled at 10 Hz, δt is 0.1 s. For the present, we assume that the uncertainty in time is negligible, so that the uncertainty in v, u_v can be expressed as

$$u_v = \left\{ \left[\frac{\partial v}{\partial y_{n+1}} u_{y_{n+1}} \right]^2 + \left[\frac{\partial v}{\partial y_n} u_{y_n} \right]^2 \right\}^{1/2} \tag{12.11}$$

The uncertainties in the measured displacements, y_n and y_{n+1}, will be equal and are listed in Table 12.2. Substituting these values for uncertainty in Equation 12.11 yields an uncertainty of ± 3 cm/s in the velocity measurement. Notice that this uncertainty magnitude is not a function of time or the measured velocity. This corresponds to a minimum uncertainty of 30% in the velocity measurement.

To determine velocity from acceleration, the measured values of acceleration must be integrated. Because we have a digital signal, the integration can be accomplished numerically as

$$v(t) = y'(t) = \sum_i y_i'' \delta t \tag{12.12}$$

Assuming that the uncertainty in time is negligible, the uncertainty in velocity at any time t is simply

$$u_v = u_y'' t \tag{12.13}$$

Clearly, the integration process tends to accumulate error as the calculation of velocity proceeds in time, as illustrated in Figure 12.11, but can produce acceptable results for very short time periods.

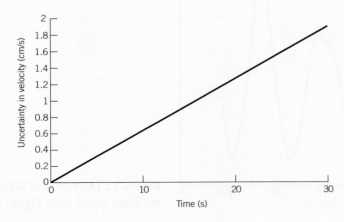

Figure 12.11 Uncertainty in velocity as a function of time.

COMMENT This example can be used to illustrate several useful principles for data analysis. Consider the effects of adding noise, or a degree of random error, to measured data; such a signal is shown in Figure 12.12. In general, noise tends to be minimized through a process of integration and amplified through differentiation. Figure 12.13 shows the result of integrating the noisy data from Figure 12.12, with the result that the effects of noise are reduced. However, because of the accumulation of error, the technique must balance the integration period against error accumulation. Differentiation tends to be extremely sensitive to low amplitude, high-frequency noise, which can create very large errors in derivatives, especially at high sampling frequencies. This is apparent

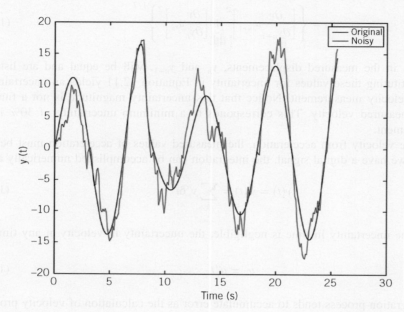

Figure 12.12 Deterministic signal with added noise.

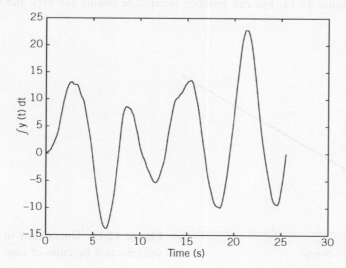

Figure 12.13 Result of integrating the noisy signal from Figure 12.12.

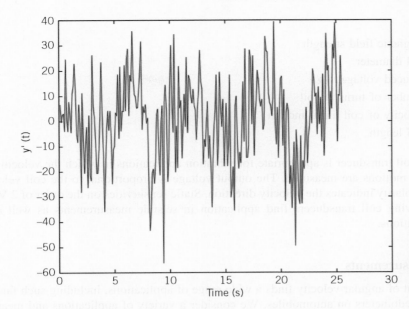

Figure 12.14 Result of numerical differentiation of the noisy signal from Figure 12.12.

in Figure 12.14, in which the signal in Figure 12.12 has been differentiated numerically. Appropriate filtering or smoothing techniques are generally effective at reducing errors associated with noise in derivatives in cases in which the noise has low amplitude.

Moving Coil Transducers

Moving coil transducers take advantage of the voltage generated when a conductor experiences a displacement in a magnetic field. An illustration of a moving coil velocity pickup is provided in Figure 12.15. Recall that a current carrying conductor experiences a force in a magnetic field and that a force is required to move a conductor in a magnetic field. For the latter case, a voltage is induced in the conductor. Consider the case in which the magnetic field strength is at right angles to the conductor. The induced voltage is given by

$$E = \pi B D_c l N \frac{dy}{dt} \tag{12.14}$$

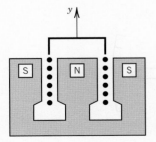

Figure 12.15 Moving coil transducer.

where

B = magnetic field strength
D_c = coil diameter
E = induced voltage
N = number of turns in coil
dy/dt = velocity of coil linear motion
l = coil length

A moving coil transducer is appropriate for vibration applications in which the velocities of small amplitude motions are measured. The output voltage is proportional to the coil velocity, and the output polarity indicates the velocity direction. Static sensitivities on the order of 2 V-s/m are typical. Moving coil transducers find application in seismic measurements as well as in vibration applications.

Angular Velocity Measurements

The measurement of angular velocity finds a wide range of applications, including such familiar examples as speedometers on automobiles. We consider a variety of applications and measurement techniques.

Mechanical Measurement Techniques

Mechanical means of measuring angular velocity or rotational speed were developed primarily to provide feedback for control of engines and steam turbines. Mechanical governors and centrifugal tachometers (4) operate on the principle illustrated in Figure 12.16. Here the centripetal acceleration of the flyball masses result in a steady-state displacement of the spring, which provides a control

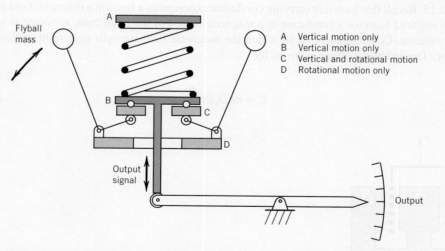

Flyball mass

A   Vertical motion only
B   Vertical motion only
C   Vertical and rotational motion
D   Rotational motion only

Output signal

Output

Figure 12.16 Mechanical angular velocity sensor.

signal or is a direct indication of rotational speed. For this arrangement, the spring force is proportional to the square of the angular velocity.

Stroboscopic Angular Velocity Measurements

A stroboscopic light source provides high-intensity flashes of light, which can be caused to occur at a precise frequency. A stroboscope is illustrated in Figure 12.17. Stroboscopes permit the intermittent observation of a periodic motion in a manner that appears to stop or slow the motion. Figure 12.18 illustrates the use of a strobe to measure rotational speed. A timing mark on the rotating object is illuminated with the strobe, and the strobe frequency is adjusted such that the mark appears to remain motionless, as shown in Figure 12.18a. Thus the highest synchronous speed is the actual rotational speed. At this speed, the output value is available from a calibration of the stroboscopic lamp flash frequency, with uncertainties to less than 0.1%. Clearly, a timing mark would appear motionless for integer multiples of the actual rotational speed and for integral submultiples, as illustrated in Figure 12.18.

The synchronization of images at flashing rates other than the actual speed requires some practical approaches to ensuring the accurate determination of speed. Spurious images can easily result for symmetrical objects, and some asymmetric marking is necessary to prevent misinterpretation of stroboscopic data. To distinguish the actual speed from a submultiple, the flashing rate can be decreased until another single synchronous image appears. If this flashing rate corresponds to half the original rate, then the original rate is the actual speed. If it does not occur at half the original value, then the original value is a submultiple.

Figure 12.17 Stroboscope. (Courtesy of Nidec-Shimpo America Corp. Used with permission.)

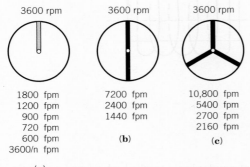

Figure 12.18 Images resulting from harmonic and subharmonic flashing rates for stroboscopic angular speed measurement; rotation rate being measured is 3,600 rpm. Note: fpm indicates *flashes per minute*.

The upper limit of the flash rate of the strobe does not limit the ability of the stroboscope to measure rotational speed. For high speeds, synchronization can be achieved N times, with ω_1 representing the maximum achievable synchronization speed, and $\omega_2, \omega_3, \ldots$ representing successively lower synchronization speeds. The measured rotational speed is then calculated as

$$\omega = \frac{\omega_1 \omega_N (N - 1)}{\omega_1 - \omega_N} \tag{12.15}$$

The program file *stroboscope.vi* on the companion software disk illustrates the basic operation and interesting stroboscopic effects.

Electromagnetic Techniques

Several measurement techniques for rotational velocity use transducers that generate electrical signals, which are indicative of angular velocity. One of the most basic is illustrated in Figure 12.19. This transducer consists of a toothed wheel and a magnetic pickup; the pickup consists of a magnet and a coil. As the toothed wheel rotates, a voltage is induced in the coil as a result of changes in the magnetic field. As each ferromagnetic tooth passes the pickup, the reluctance of the magnetic circuit changes in time, yielding a voltage in the coil given by

$$E = C_B N_t \omega \sin N_t \omega t \tag{12.16}$$

where

$\quad E$ = output voltage
$\quad C_B$ = proportionality constant
$\quad N_t$ = number of teeth
$\quad \omega$ = angular velocity of the wheel

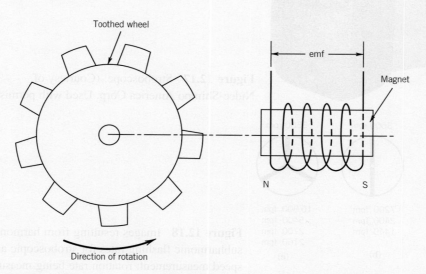

Figure 12.19 Angular velocity measurement employing a toothed wheel and magnetic pickup.

The angular velocity can be found either from the amplitude or the frequency of the output signal. The voltage amplitude signal is susceptible to noise and loading errors. Thus less error is introduced if the frequency is used to determine the angular velocity; typically, some means of counting the pulses electronically is employed. This frequency information can be transmitted digitally for recording, which eliminates the noise and loading error problems associated with voltage signals.

Force Measurement

The measurement of force is most familiar as the process of weighing, ranging from weighing micrograms of a medicine to weighing trucks on the highway. Force is a quantity derived from the fundamental dimensions of mass, length, and time. Standards and units of measure for these quantities are defined in Chapter 1. The most common techniques for force measurement are described in this section.

Load Cells

"Load cell" is a term used to describe a transducer that generates a voltage signal as a result of an applied force, usually along a particular direction. Such force transducers often consist of an elastic member and a deflection sensor. These deflection sensors may employ changes in capacitance, resistance, or the piezoelectric effect to sense deflection. A technology overview for such devices is provided elsewhere (5). Consider first load cells that are designed using a linearly elastic member instrumented with strain gauges.

Strain Gauge Load Cells   Strain gauge load cells are most often constructed of a metal, and have a shape such that the range of forces to be measured results in a measurable output voltage over the desired operating range. The shape of the linearly elastic member is designed to meet the following goals: (1) provide an appropriate range of force-measuring capability with necessary accuracy, (2) provide sensitivity to forces in a particular direction, and (3) have low sensitivity to force components in other directions.

A variety of designs of linearly elastic load cells are shown in Figure 12.20. In general, load cells may be characterized as beam-type load cells, proving rings, or columnar-type designs. Beam-type load cells may be characterized as bending beam load cells or shear beam load cells.

A bending beam load cell as shown in Figure 12.21 is configured such that the sensing element of the load cell functions as a cantilever beam. Strain gauges are mounted on the top and bottom of the beam to measure normal or bending stresses. Figure 12.21 provides qualitative indication of the shear and normal stress distributions in a cantilever beam. In the linear elastic range of the load cell, the bending stresses are linearly related to the applied load.

In a shear beam load cell, the beam cross section is that of an I-beam. The resulting shear stress in the web is nearly constant, allowing placement of a strain gauge essentially anywhere in the web with reasonable accuracy. Such a load cell is illustrated schematically in Figure 12.22, along with the shear stress distribution in the beam. In general, bending beam load cells are less costly because of their construction, but the shear beam load cells have several advantages, including lower creep and faster response times.

Piezoelectric Load Cells   Piezoelectric materials are characterized by their ability to develop a charge when subject to a mechanical strain. The most common piezoelectric material is

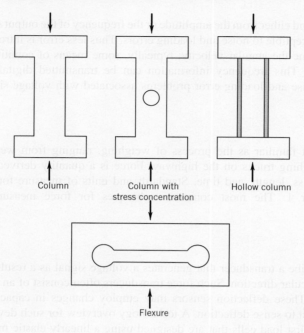

Column

Column with
stress concentration

Hollow column

Flexure

Figure 12.20 Elastic load cell designs.

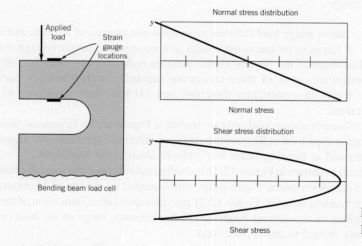

Applied
load

Strain
gauge
locations

Bending beam load cell

Normal stress distribution

Normal stress

Shear stress distribution

Shear stress

Figure 12.21 Bending beam
load cell and stress distributions.

single-crystal quartz. The basic principle of transduction that occurs in a piezoelectric element may best be thought of as a charge generator and a capacitor. The frequency response of piezoelectric transducers is very high, because the frequency response is determined primarily by the size and material properties of the quartz crystal. The modulus of elasticity of quartz is approximately 85 GPa, yielding load cells with typical static sensitivities ranging from 0.05 to 10 mV/N, and a frequency response up to 15,000 Hz. A typical piezoelectric load cell construction is shown in Figure 12.23.

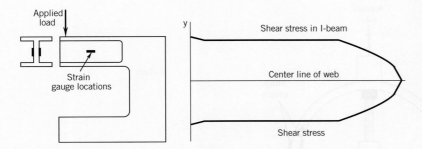

Figure 12.22 Shear beam load cell and shear stress distribution.

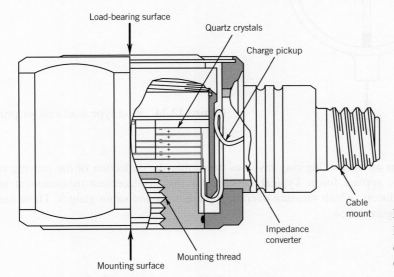

Figure 12.23 Piezoelectric load cell design. (Courtesy of the Kistler Instrument Co. Used with permission.)

Proving Ring A ring-type load cell can be employed as a local force standard. Such a ring-type load cell, as shown in Figure 12.24, is often employed in the calibration of materials testing machines because of the high degree of precision and accuracy possible with this arrangement of transducer and sensor. If the sensor is approximated as a circular right cylinder, the relationship between applied force and deflection is given by

$$\delta y = \left(\frac{\pi}{2} - \frac{4}{\pi}\right) \frac{F_n D^3}{16EI} \qquad (12.17)$$

where

δ_y = deflection along the applied force
F_n = applied force
D = diameter
E = modulus of elasticity
I = moment of inertia

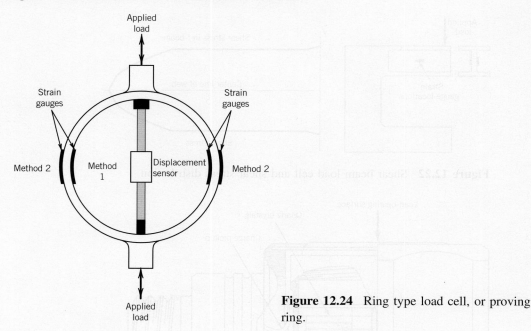

Figure 12.24   Ring type load cell, or proving ring.

The application of the proving ring involves measuring the deflection of the proving ring in the direction of the applied force. Typical methods for this displacement measurement include displacement transducers, which measure overall displacement, and strain gauges. These methods are illustrated in Figure 12.24.

Torque Measurements

Torque and mechanical power measurements are often associated with the energy conversion processes that serve to provide mechanical and electrical power to our industrial world. Such energy conversion processes are largely characterized by the mechanical transmission of power produced by prime movers such as internal combustion engines. From automobiles to turbine-generator sets, mechanical power transmission occurs through torque acting through a rotating shaft.

The measurement of torque is important in a variety of applications, including sizing of load-carrying shafts. This measurement is also a crucial aspect of the measurement of shaft power, such as in an engine dynamometer. Strain gauge–based torque cells are constructed in a manner similar to load cells, in which a torsional strain in an elastic element is sensed by strain gauges appropriately placed on the elastic element. Figure 12.25 shows a circular shaft instrumented with strain gauges for the purpose of measuring torque.

Consider the stresses created in a shaft of radius R_0 subject to a torque T. The maximum shearing stress in a circular shaft occurs on the surface and may be calculated from the torsion formula[4]

$$\tau_{\max} = TR_0/J \tag{12.18}$$

---

[4] Coulomb developed the torsion formula in 1775 in connection with electrical instruments.

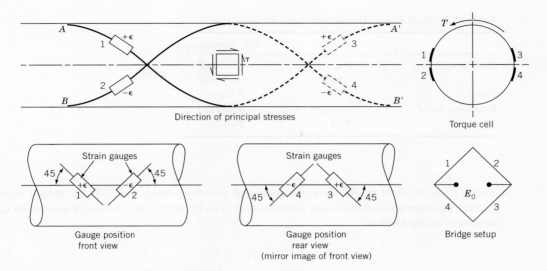

Figure 12.25 Shaft instrumented for torque measurement.

where

τ_{max} = maximum shearing stress

T = applied torque

J = polar moment of inertia ($\pi R_0^2/2$ for a solid circular shaft)

For a shaft in pure torsion, there are no normal stresses, σ_x, σ_y, or σ_z. The principal stresses lie along a line that makes a 45-degree angle with the axis of the shaft, as illustrated in Figure 12.25, and have a value equal to τ_{max}. Strains that occur along the curve labeled A–A' are opposite in sign from those that occur along B–B'. These locations allow placement of four active strain gauges in a Wheatstone bridge arrangement and the direct measurement of torque in terms of bridge output voltage.

Mechanical Power Measurements

Almost universally, prime movers such as internal combustion (IC) engines and gas turbines convert chemical energy in a fuel to thermodynamic work transmitted by a shaft to the end use. The resulting power is transmitted through a mechanical coupling. This section discusses the measurement of such mechanical power transmission.

Rotational Speed, Torque, and Shaft Power

Shaft power is related to rotational speed and torque as

$$\vec{P_s} = \vec{\omega} \times \vec{T} \qquad (12.19)$$

Table 12.3 Shaft Power, Torque, and Speed Relationships

	SI	U.S. Customary
Shaft power, P	$P = \omega T$	$P = \dfrac{2\pi n T}{550}$
Power	P (W)	P (hp)
Rotational speed	ω (rad/s)	n (rev/s)
Torque	T (N m)	T (ft lb)

where $\overrightarrow{P_s}$ is the shaft power, $\overrightarrow{\omega}$ is the rotational velocity vector, and $\overrightarrow{T}$ is the torque vector. In general, the orientation of the torque and rotational velocity vectors are such that the equation may be written in scalar form as

$$P_s = \omega T \tag{12.20}$$

Table 12.3 provides a summary of useful equations related to shaft power, torque, and speed as employed in mechanical measurements. Historically, a device called a Prony brake was used to measure shaft power. A typical Prony brake arrangement is shown in Figure 12.26. Consider using the Prony brake to measure power output for an IC engine. The Prony brake serves to provide a well-defined load for the engine, with the power output of the engine dissipated as thermal energy in the braking material. By adjusting the load, the power output over a range of speeds and throttle settings can be realized. The power is measured by recording the torque acting on the torque arm and the rotational speed of the engine. Clearly, this device is limited in speed and power, but it does demonstrate the operating principles of power measurement, and is historically significant as the first technique for measuring power.

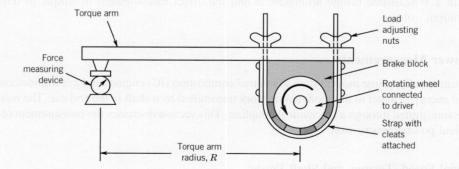

Figure 12.26 Prony brake. (Courtesy of the American Society of Mechanical Engineers, New York, NY. Reprinted from PTC 19.7-1980 6.)

Cradled Dynamometers

A Prony brake is an example of an absorbing dynamometer. The term "dynamometer" refers to a device that absorbs and measures the power output of a prime mover. Prime movers are large mechanical power-producing devices such as gasoline or diesel engines or gas turbines. Several methods of energy dissipation are used in various ranges of power, but the measurement techniques are governed by the same underlying principles. Thus we consider first the measurement of power and then discuss means for dissipating the sometimes large amounts of power generated by prime movers.

The cradled dynamometer measures mechanical power by measuring the rotational speed of the shaft, which transmits the power, and the reaction torque required to prevent movement of the stationary part of the prime mover. This reaction torque is impressively illustrated by the so-called wheelies performed by motorcycle riders. A cradled dynamometer is supported in bearings, called trunnion bearings, such that the reaction torque is transmitted to a torque or force-measuring device.

In principle, the operation of the dynamometer involves the steady-state measurement of the load F_r created by the reaction torque and the measurement of shaft speed. Using Equation 12.19, the transmitted shaft power can be calculated directly.

The American Society of Mechanical Engineers (ASME) Performance Test Code (PTC) 19.7 (6) provides guidelines for the measurement of shaft power. According to PTC 19.7, overall uncertainty in the measurement of shaft power by a cradled dynamometer results from (1) trunnion-bearing friction error uncertainty, (2) force measurement uncertainty (F_r), (3) moment arm length uncertainty (L_r), (4) static unbalance of dynamometer error uncertainty, and (5) uncertainty in rotational speed measurement.

A means of supplying a controllable load to the prime mover and dissipating the energy absorbed in the dynamometer are integral parts of the design of any dynamometer. Several techniques are described for providing an appropriate load.

Eddy Current Dynamometers: A direct current field coil and a rotor allow shaft power to be dissipated by eddy currents in the stator winding. The resulting conversion to thermal energy by Joulian heating of the eddy currents necessitates that some cooling be supplied, typically from cooling water.

Alternating Current and DC Generators: Cradled AC and DC machines are employed as power-absorbing elements in dynamometers. The AC applications require variable frequency capabilities to allow a wide range of power and speed measurements. The power produced in such dynamometers may be dissipated as thermal energy using resistive loads.

Waterbrake Dynamometers: A waterbrake dynamometer employs fluid friction and momentum transport to create a means of energy dissipation. Two representative designs are provided in Figure 12.27. The viscous shear type brake is useful for high rotational speeds, and the agitator type unit is used over a range of speeds and loads. Waterbrakes may be employed for applications up to 10,000 HP (7,450 kW). The load absorbed by waterbrakes can be adjusted using water level and flow rates in the brake. Figure 12.28 shows a state-of-the-art waterbrake engine dynamometer. Figure 12.28a shows the dynamometer without an engine installed, and Figure 12.28b illustrates the installed engine and test setup.

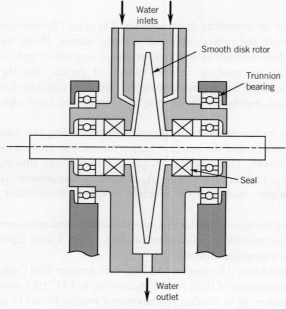

(a) Viscous shear type

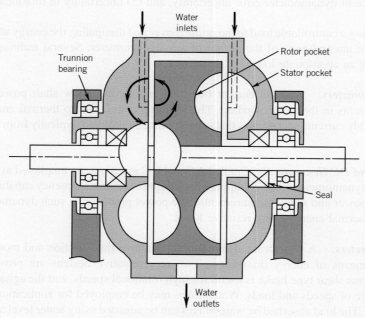

(b) Momentum exchange type

Figure 12.27 Waterbrake dynamometers. (Courtesy of the American Society of Mechanical Engineers, New York, NY. Reprinted from PTC 19.7-1980 6.)

(a)

(b)

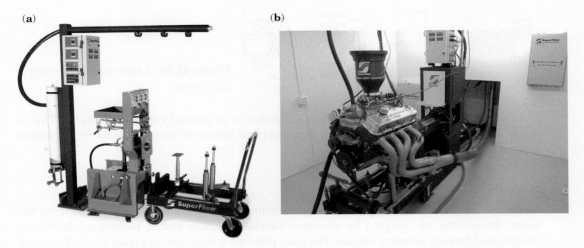

Figure 12.28 Waterbrake engine dynamometer. **(a)** Dynamometer without test engine installed. **(b)** Dynamometer with test engine in place. (Photo credit SuperFlow Dynamometers & Flowbenches.)

12.3 ACTUATORS

Linear Actuators

The task of a linear actuator is to provide motion in a straight line. We discuss three ways to achieve linear motion:

1. Conversion of rotary motion into linear motion. This can be accomplished using a linkage, as in the slider-crank mechanism, or using screw threads coupled to a rotary motion source.

2. Use of a fluid pressure to move a piston in a cylinder. When air or another gas is used as the working fluid, the system is called a pneumatic system. When a fluid such as oil is used as the working fluid, the system is termed hydraulic.

3. Electromagnetic

Slider–Crank Mechanism

A common means of generating a reciprocating linear motion, or converting linear motion to rotary motion, is the slider–crank mechanism, as illustrated in Figure 12.29. Such a mechanism is the basis

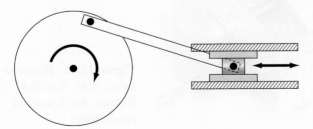

Figure 12.29 Slider–crank mechanism.

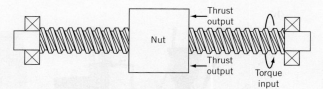

Figure 12.30 Linear actuation using a lead screw.

of transforming the reciprocating motion of the piston in an internal combustion engine. This or similar linkages could also be applied in pick-and-place operations or in a variety of automation applications.

Screw-Drive Linear Motion

A common means for translating rotary motion into linear motion is a lead screw. A lead screw has helical threads that are designed for minimum backlash to allow precise positioning. Numerous designs exist for such actuating threads. The basic principle is illustrated in Figure 12.30. The rotary motion of the lead screw is translated into linear motion of the nut, with the torque required to drive the lead screw directly related to the thrust the particular application requires.

Common applications that employ a lead screw include the worktable for a mill, as well as a variety of other precision positioning translation tables, such as the one shown in Figure 12.31.

Pneumatic and Hydraulic Actuators

The term "pneumatic" implies a component or system that uses compressed air as the energy source. On the other hand, a hydraulic system or component uses incompressible oil as the working fluid. An example of a hydraulic system is the power steering on an automobile; such a system is illustrated in Figure 12.32. Hydraulic fluid is supplied at an elevated pressure from the power steering pump.

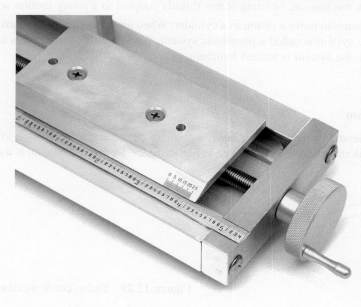

Figure 12.31 Precision translation table. (UniSlide® from Velmex, Inc. Used with permission.)

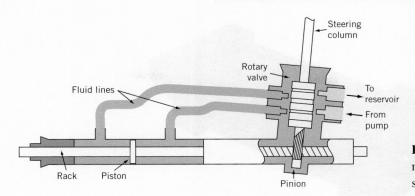

Figure 12.32 Schematic diagram of a power steering system.

When a steering input is made from the driver, the rotary valve allows high-pressure fluid to enter the appropriate side of the piston, and aid in turning the wheels. By maintaining a direct connection between the steering column and the rack and pinion, the car can be steered even if the hydraulic system fails.

Pneumatic Actuators

When compressed air is the energy source of choice, a pneumatic cylinder can create linear motion. In general, the purpose of a pneumatic cylinder is to provide linear motion between two fixed locations. Figure 12.33 shows a pneumatic cylinder and a cutaway of such a cylinder. By applying high-pressure compressed air to either side of the piston, linear actuation between two defined positions can easily be accomplished.

Solenoids

"Solenoid" is a term used to describe an electromagnetic device that is employed to create linear motion of a plunger, as shown in Figure 12.34. The initial force available from a solenoid can be determined from

$$F = \frac{1}{2}(NI)^2 \frac{\mu A}{\delta^2} \tag{12.21}$$

where

F = force on plunger

N = number of turns of wire in the electromagnet

I = current

μ = magnetic permeability of air $(4\pi \times 10^{-7} \text{H/m})$

δ = size of the air gap

A = plunger cross-sectional area

When the electromagnet is actuated, the resulting magnetic force pulls the plunger into the C-frame. Because the air gap is largest when the electromagnet is actuated, the minimum force occurs at actuation and the force increases as the air gap decreases.

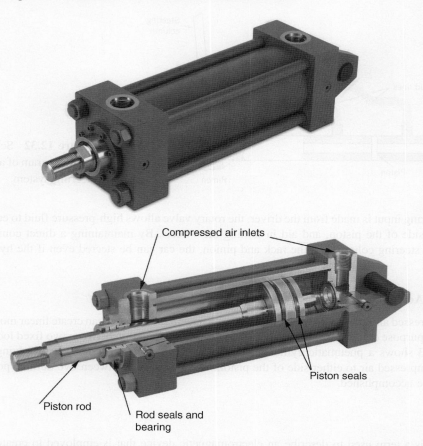

Figure 12.33 Construction of a pneumatic cylinder. (Courtesy of Parker Hannifin Corp. Used with permission.)

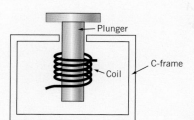

Figure 12.34 Construction of a solenoid linear actuator.

Rotary Actuators

Stepper Motors

There is a class of electric motors that has the primary purpose of providing power to a process. An example would be the electric motor that drives an elevator, an escalator, or a centrifugal blower.

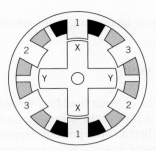

Figure 12.35 Variable reluctance stepper motor design.

In these applications the electric motor serves as a prime mover, with clear and specific requirements for rotational speed, torque, and power. However, some applications have stringent requirements for positioning.

Rotary positioning presents a significant engineering challenge, but one that is so ubiquitous that it has been addressed through a variety of design strategies. One design strategy is to employ a free-rotating DC motor to supply the motive power and impose precise control on the resulting motion through gearing and some control scheme. DC motors that are subject to feedback control are generally described as servo-motors. Although this may be appropriate and necessary for some applications, the stepper motor has found wide-ranging applications in precision rotary motion control and is a better choice for many applications.

Stepper or stepping motors are capable, as their name implies, of moving a fraction of a rotation with a great degree of precision. This is accomplished by the design of a rotor that aligns with the magnetic field generated by energized coils. The step size can range from 90 degrees to as little as 0.5 degrees or less. Two common types of stepper motors are variable reluctance and unipolar designs. The design of a variable reluctance stepping motor is illustrated in Figure 12.35. Let's consider the operation of this motor. There are three sets of windings, labeled 1, 2, and 3 in the figure, and there are two sets of teeth on the rotor, labeled X and Y. When the windings labeled 1 are energized, the rotor snaps to a position where one set of the teeth are aligned with the windings. This motion is a result of the magnetic field generated by the windings. Suppose that winding 1 is de-energized and winding 2 is energized. The rotor will turn until the teeth marked Y are aligned with winding 2. This produces a 30-degree step.

A useful characteristic of stepper motors is holding torque. As long as one of the windings is energized, the rotor resists motion, until the torque produced by the winding to rotor interaction is overcome.

The motor shown in Figure 12.36 is a variable reluctance design. Unipolar motors incorporate permanent magnets as the rotor. Figure 12.36 shows a rotor having six magnetic

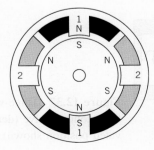

Figure 12.36 Variable reluctance stepper motor design having six poles and two windings.

poles and two sets of windings. The motor moves in 30-degree increments as the windings are alternately energized.

Flow-Control Valves

Valves are mechanical devices intended to allow, restrict, throttle, or meter fluid flow through pipes or conduits. Flow-control valves are used to regulate either the flow or the pressure of a fluid by their actuation. They generally function by allowing flow while in their open position, stopping flow when closed, and metering flow or fluid pressure to a desired value with a position that is somewhere between these settings, which is called proportional control. These valves contain a valve positioning element that is driven by an actuator, such as a solenoid. Any valve type can be controlled. The common control valve design offers either a single chamber body containing a poppet with valve seat or a multichamber body containing a sliding spool with multiple poppets. Flow-control valves are used to transfer gases, liquids, and hydraulic fluids. The application ratings are as follows: general service, for working with common liquids and gases; cryogenic service, for fluids such as liquid oxygen; vacuum service, for low-pressure applications; and oxygen service, for contamination-free flow of oxygen.

The control valve can respond to signals from any type of process variable transducer. The signal determines the position of the actuating solenoid. A specific characteristic of any control valve refers to whether its nonenergized operating state is open or closed. This is referred to as its "fail position." The fail position of a control valve is determined by the nonenergized solenoid plunger position. This position is an important consideration for process safety.

These valves come in various configurations reflecting their number of ports. A two-way valve has two ports. Two-way position control takes on one of two values: open or closed. A two-way valve has two connections: supply port (P) and service port (A). Most common household valves fall into this category. A three-way valve has three port connections: supply (P), exhaust (T), and service (A). The service port may be switched between the supply and the exhaust. A four-way has four connections: supply (P), exhaust (T), and two service ports (A and B). The valve connects either P to A and B to T, or P to B and A to T. In general, an N-way valve has N ports, with N number of flow directions available. An example of a three-port sliding spool control valve is shown in Figure 12.37. The solenoid drives the spool, which contains two valve seats. In the fully activated position, port P is open to service port A. When the solenoid is deactivated, port T is open to service port A (shown). For example, in one application this valve can be used to pressurize a system (open the system to port P) for a period of time and then adjust the system pressure to another value (open the system to port A) for a period of time.

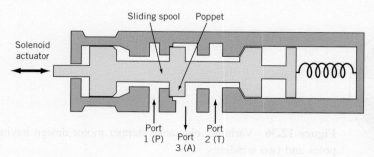

Figure 12.37 Three-way flow control valve (deactivated position shown).

All valve ports offer some level of flow resistance, and this is specified through a flow coefficient, C_v. Flow resistance can be adjusted in design by varying the internal dimensions of the valve chamber and can be set operationally by varying the element position within the chamber. The flow coefficient is found based on the formulation detailed in Chapter 10, or simply as

$$Q = C_v\sqrt{\Delta p} \tag{12.22}$$

where Q is the steady flow rate through the valve and Δp is the corresponding pressure drop. This loss is also expressed in terms of a K-factor based on the average velocity through the ports,

$$\Delta p = K\rho\overline{U}^2/2 \tag{12.23}$$

Flow-control valves are classified in a number of ways: the type of control, the number of ports in the valve housing, the specific function of the valve, and the type of valve element used in the construction of the valve. Directional control valves allow or prevent the flow of fluid through designated ports. Flow can move in either direction. Check valves are a special class of directional valve that allow flow in only one direction. Proportional valves can be positioned to control the amount, pressure, and direction of fluid flow. In a proportional valve, the valve is opened by an amount proportional to the applied current. These valves provide a way to control pressure or flow rate with a high response rate.

In the simplest application, a solenoid is used to turn a valve either on or off. In a more demanding application, the solenoid is expected to cycle rapidly to open and close the valve. The time between each signal cycle coupled with the internal flow loss character of the valve determines the average flow. Valve response time can be defined in several ways but all are consistent with the methods used in Chapter 3. The 90% response time, t_{90}, is the time required to either fill or exhaust a target device chamber through a valve port, in effect a step function response. There is a separate response time for filling or exhausting. Either way,

$$t_{90} = m + F\forall \tag{12.24}$$

where m is the valve lag time between when the signal is applied and steady flow is established at the designated port, F is the reciprocal of the average flow rate through the port, and $\forall$ is the volume of the target device chamber. For example, a valve having an F of 0.54 ms/cc and a lag time of 20 ms requires $t_{90} = 155$ ms to fill a 250-cc chamber. Alternatively, the valve frequency response can be found by cycling the valve with a sine wave electrical signal and measuring the flow rate through the valve. The valve frequency bandwidth is thus established by its -3 dB point.

12.4   CONTROLS

Control of a process or system can be exerted in a wide variety of ways. Suppose our goal is to create a healthy lawn by appropriate watering. Each day we could monitor the weather forecast, take into account the probability of precipitation, and choose whether to water and for how long. We could choose to water all of the lawn or just those parts most subject to stress from heat and lack of moisture. If we choose to water, we could place the sprinklers and turn on the faucet (remembering to shut off the flow at an appropriate later time).

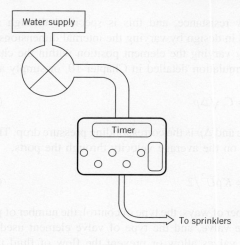

Figure 12.38 Open-loop control of a sprinkler system.

All the functions described above for lawn care are completely reasonable for a person to accomplish, and they represent the functioning of an intelligent controller. Suppose we wish to introduce some automation into the process.

At the simplest level, a timer-based control system could be implemented, as shown in Figure 12.38. The functioning of this system would be to open and close the faucet at predetermined times of the day. At the simplest level, this could be a mechanical timer that watered the lawn once each 24-hour period for a predetermined length of time. This type of control is called an *open loop control*. For this control system, there are no sensors to monitor the amount of water applied to the lawn; in fact, all that the control system is accomplishing is to open and, later, close the faucet.

More advanced automatic control systems implement a *closed-loop control*. For the present example, it might be desired to apply 100 gallons (379 liters) of water to the lawn. A flow meter that sensed the total water flow that had occurred could be used to provide feedback to the control system to allow the faucet to be closed when the flow totaled 100 gallons. The term *closed-loop* or *feedback control* simply means that the variable that is to be controlled is being measured, and that the control system in some way uses this measurement to exert the control.

Figure 12.39 illustrates a control system designed to apply 100 gallons of water to the lawn. There are two inputs to the controller: the time of day and the output of the totalizing flow meter. At the appropriate time of day, the controller opens the valve. The totalizing flow meter output is used by the controller to close the valve after the total flow reaches 100 gallons. This type of binary control scheme is termed *on–off control*. The valve controlling the water flow is either fully open or fully closed.

Probably the most familiar form of a binary on–off control system is the thermostat for a home furnace or air conditioner. Figure 12.40 shows the status of a home furnace and a time trace of the inside temperature during a winter day. A schematic representation of this control system is shown in Figure 12.41. A key element here is that there is the possibility of a *disturbance* that would influence the rate of change of the inside temperature. Suppose a delivery arrives and the door remains open for a period of time. The thermostat must then respond to this disturbance and attempt to maintain the inside temperature at the set point.

Essentially all practical implementations of on-off control systems require a dead band that creates a hysteresis loop in the control action. This is illustrated in Figure 12.42. The dead band is

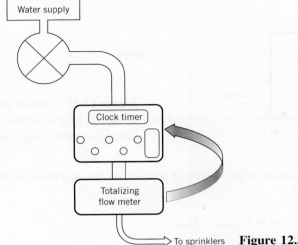

To sprinklers   **Figure 12.39** Feedback control of a sprinkler system.

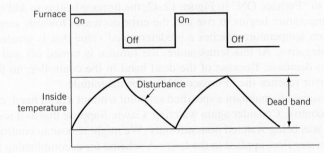

Figure 12.40 Operation of on–off controller with a dead band.

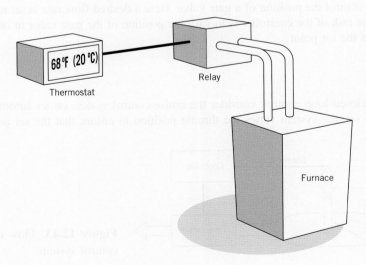

Figure 12.41 Components of a thermostatic control for a home furnace.

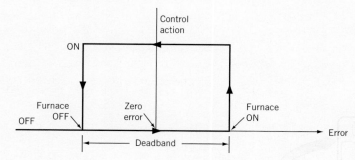

Figure 12.42 Control action and hysteresis loop of a binary on-off controller with a dead band.

centered around zero error, and the action of the controller depends on the magnitude of the error. Here the error is defined as

$$e = T_{\text{setpoint}} - T_{\text{room}}$$

Recall that we are considering a furnace thermostat under winter conditions. As the room temperature falls relative to the set point, the error becomes a larger positive number. When the error reaches the value corresponding to "Furnace ON" in Figure 12.42, the furnace begins to add heat to the conditioned space. Room temperature begins to rise and the error decreases towards zero. The furnace remains on until the room temperature reaches a predetermined value that is greater than the set point. Here the error is negative. At this temperature, the furnace is turned off and room temperature begins once again to decrease. Because of the dead band in the controller, no further control action occurs until the error reaches the "Furnace ON" error magnitude.

Many control systems are designed to maintain a specified set point without a dead band; clearly this is not possible using on–off control. Consider again watering a lawn. Suppose that as a result of varying water pressure, the lawn was being watered nonuniformly. We might choose to control both the water flow rate and the total water flow applied to the lawn. A scheme for accomplishing this is shown in Figure 12.43. The key components of the control system are a flow meter, the controller, and an actuator that can control the position of a gate valve. Here a desired flow rate is set to, say, 2 gallons per minute. The task of the controller is to vary the position of the gate valve in order to maintain the flow rate at the set point.

Dynamic Response

As another example of closed-loop control, consider the cruise-control system on an automobile. After a desired speed is set, the system varies the throttle position to ensure that the set point is

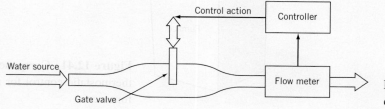

Figure 12.43 Flow rate control system.

maintained. Another issue in the analysis and design of control systems lies in the dynamic response of the physical process, the controller, and the actuators. An automobile does not respond instantaneously to changes in the throttle position, and time is required for a stepper motor to change the position of a valve. The complexities of the combined responses of the physical and control systems, especially in the presence of disturbances, are the subject of the remainder of this section.

Laplace Transforms

In Chapter 3, we used Laplace transforms to model the response of simple systems. Now let us apply Laplace transforms to understand how control systems function. The fundamental basis of the application of Laplace transforms to control systems is the solution of a mathematically well-posed initial value problem.

Consider an initial-value problem (here time is the independent variable) that is described by an ordinary differential equation. If we apply the Laplace transform to a differential equation, we convert the differential equation to an algebraic equation. For partial differential equations in time and one spatial variable, the Laplace transform converts the partial differential equation into an ordinary differential equation in the space variable. Appendix C reviews the application of Laplace transforms and provides a table of Laplace transform pairs.

We illustrate the application of the Laplace transform through solution of first-order and second-order differential equations, which are important for control systems.

Example 12.3

Consider the response of a first-order system to a unit step input forcing function, which we have previously modeled by Equation 3.4. The governing equation is

$$\tau \dot{y} + y = U_s(t) \tag{12.25}$$

Applying the Laplace transform (Table C.1 can be used as a reference) yields

$$\tau s Y(s) - y(0) + Y(s) = \frac{1}{s} \tag{12.26}$$

where we have employed the property of the Laplace transform that

$$\mathcal{L}\left[\tau \frac{dy(t)}{dt}\right] = \tau \mathcal{L}\left[\frac{dy(t)}{dt}\right]$$

The process of solving the differential equation involves the following steps:

1. Apply the Laplace transform to the governing differential equation, in this case Equation 12.25.
2. Solve the resulting equation for $Y(s)$, Equation 12.26.
3. Employ a table of Laplace transform pairs to determine $y(t)$.

We assume that $y(0) = 0$. Solving Equation 12.26 for $Y(s)$ yields

$$Y(s) = \frac{1}{s(\tau s + 1)} \tag{12.27}$$

An important tool for the inversion of Laplace transforms is partial fractions. To accomplish the inversion of Equation 12.27, we start from the assumption

$$\frac{1}{s(\tau s + 1)} = \frac{A}{s} + \frac{B}{(\tau s + 1)} \tag{12.28}$$

Next we find a common denominator for the right-hand side of Equation 12.28:

$$\frac{1}{s(\tau s + 1)} = \frac{A(\tau s + 1)}{s(\tau s + 1)} + \frac{Bs}{s(\tau s + 1)}$$

In this form, we see that

$$1 = A(\tau s + 1) + Bs$$

Equating powers of s yields

$$1 = A$$
$$0 = A\tau + B \Rightarrow B = -\tau$$

We can now express $Y(s)$ as

$$Y(s) = \frac{1}{s} + \frac{-\tau}{(\tau s + 1)} = \frac{1}{s} + \frac{-1}{(s + 1/\tau)} \tag{12.29}$$

We find that (e.g., from transform pairs 1 and 5 in Table C.1)

$$y(t) = 1 - e^{-t/\tau} \tag{12.30}$$

With the appropriate initial condition and steady-state output value, this can be written as Equation 3.6:

$$\Gamma(t) = \frac{y(t) - y_\infty}{y_0 - y_\infty} = e^{-t/\tau}$$

where τ is the time constant for a first-order system and $\Gamma(t)$ is the error fraction

Example 12.4

Employing Laplace transforms, determine the solution of the second-order, ordinary linear differential equation,

$$y'' + 4y' - 5y = 0 \tag{12.31}$$

with the initial conditions

$$y(0) = -1$$
$$y'(0) = 4$$

SOLUTION   We recall that

$$\mathcal{L}\left[y^{(n)}\right] = s^n \mathcal{L}[y] - s^{n-1}y(0) - s^{n-2}y'(0) - \ldots - y^{(n-1)}(0)$$

If we apply this relationship to Equation 12.31, we find that the Laplace transform is

$$s^2 Y(s) - sy(0) - y'(0) + 4[sY(s) - y(0)] - 5Y(s) = 0$$

Substituting the initial conditions yields

$$s^2 Y(s) - s - 4 + 4[sY(s) + 1] - 5Y(s) = 0$$

Solving for $Y(s)$ yields

$$Y(s) = \frac{-s}{s^2 + 4s - 5} \tag{12.32}$$

Now we are faced with the task of determining the inverse Laplace transform of this expression. Factoring the denominator yields

$$Y(s) = \frac{-s}{(s+5)(s-1)} \tag{12.33}$$

Applying partial fractions allows Equation 12.33 to be expressed

$$\frac{-s}{(s+5)(s-1)} = \frac{-5/6}{s+5} + \frac{-1/6}{s-1} \tag{12.34}$$

Finding the inverse Laplace transform (e.g., from Table C.1) provides the solution for Equation 12.31 as

$$y(t) = -\frac{5}{6}e^{-5t} - \frac{1}{6}e^t \tag{12.35}$$

Block Diagrams

A very useful representation of feedback control systems is accomplished using block diagrams. We first describe the basic elements of a block diagram.

Operational Blocks

An operational block performs a defined operation on a signal. Consider a public-address system consisting of a microphone, an amplifier, and speakers. Figure 12.44 provides a single-input, single-output block representing the amplifier. An ideal amplifier would follow exactly the waveform of the input signal from the microphone, and simply multiply the voltage signal by a constant value, κ. This constant value represents a *gain*, so it is called a gain block.

In Figure 12.45 a gain block is shown that supplies a reference voltage signal based on a temperature set point. In most practical applications, signals are transmitted in control systems as voltage or current. The gain represented in Figure 12.45 would have units of volts/°C. We note that a pure linear gain is equivalent to the static sensitivity of a zero-order system.

An important type of block used to represent a control system is a comparator. Figure 12.46 illustrates the operation of a comparator. Two voltage signals are either added or subtracted by the comparator. Figure 12.46 uses a furnace thermostat as an example. The signals represent the desired temperature set point and the measured temperature, in terms of voltages. The difference in these two values represents the error in the temperature value, the difference between the measured and set-point values.

We can now construct a block diagram of the thermostat for the home furnace described in Figures 12.41 and 12.42. Figure 12.47 shows the block diagram of the controller, the furnace, and the house. Together the furnace and the house are usually referred to as the plant or the process.

The detailed design of a control system requires that we consider the time-dependent behavior of both the controller and the process. We propose a process and derive the governing equation for the process. Then a controller is implemented and the dynamic response of the system to a step-change in the set point is derived. A key point in this model is the use of a stationary operating set point.

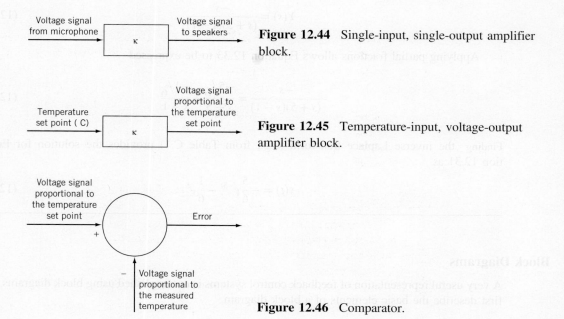

Figure 12.44 Single-input, single-output amplifier block.

Figure 12.45 Temperature-input, voltage-output amplifier block.

Figure 12.46 Comparator.

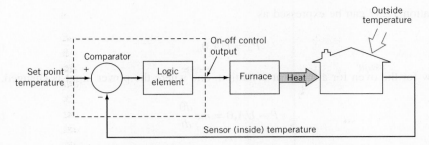

Figure 12.47   Block diagram representation of a furnace thermostat.

Model for Oven Control

Plant Model

The system we wish to control is illustrated in Figure 12.48. An oven is maintained at a temperature above the ambient temperature by power input P from an electric heater. The controller is implemented to maintain a desired oven temperature.

A first law analysis of the oven at steady-state conditions yields

$$\dot{P} = \dot{Q}_{\text{loss}}$$

The energy loss from the oven $\dot{Q}_{\text{loss}}$ may be expressed in terms of the oven temperature, T, the ambient temperature, T_∞, the surface area, A_s, and an overall loss coefficient, U, as

$$\dot{Q}_{\text{loss}} = UA_s(T - T_\infty) \tag{12.36}$$

Consider a steady-state operating point for the oven, designated by the subscript o. At steady state, the power supplied by the heater is exactly balanced by the energy lost to the ambient,

$$\dot{P}_o = UA_s(T_o - T_\infty) \tag{12.37}$$

To aid in the mathematical analysis, we define a new temperature variable as

$$\theta = T - T_\infty \tag{12.38}$$

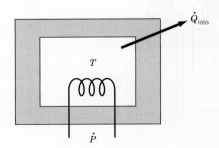

Figure 12.48   Energy flows for an oven.

so that Equation 12.37 can be expressed as

$$\dot{P}_o = UA_s\theta_o \tag{12.39}$$

The first law for the oven for a transient condition, one where the power input is changed, is

$$\dot{P} - UA_s\theta = mc\frac{d\theta}{dt} \tag{12.40}$$

where m is the mass of the oven and c is the average specific heat. This product represents the total heat capacity of the oven. Equation 12.40 now provides the governing equation for the plant in this example.

Controller Model

We now wish to implement a controller. The first controller we implement is termed a proportional controller and is illustrated in Figure 12.49. The error between the set point temperature and the actual temperature of the oven is multiplied by a constant value to determine the power input to the heater. A block diagram of the controller and plant combined system is shown in Figure 12.50; two proportional gains are shown in this figure to emphasize that the error signal will most likely be a voltage.

The governing equation for the temperature of the oven may be expressed

$$\frac{mc}{[UA_s + \kappa_p\kappa_m]}\frac{d\theta}{dt} + \theta = \frac{\kappa_p\kappa_m}{[UA_s + \kappa_p\kappa_m]}\theta_{set} \tag{12.41}$$

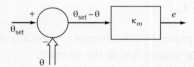

Figure 12.49 Proportional control.

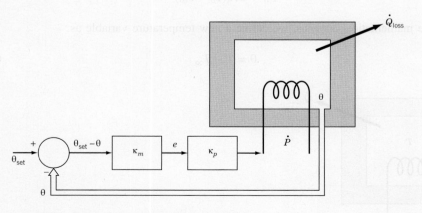

Figure 12.50 Proportional control of an oven.

By comparison with Equation 3.4,

$$\tau \dot{y} + y = KF(t)$$

it is clear that this represents the response of a first-order system having a time constant of

$$\tau = \frac{mc}{[UA_s + \kappa_p \kappa_m]} \tag{12.42}$$

The analysis for this step change is best accomplished by using Laplace transforms.

Laplace Transform Analysis

Taking the Laplace transform of Equation 12.41 yields

$$\frac{mc}{[UA_s + \kappa_p \kappa_m]} s\Theta(s) + \Theta(s) = \frac{\kappa_p \kappa_m \Theta_{set}(s)}{[UA_s + \kappa_p \kappa_m]} \tag{12.43}$$

which is of the form

$$(\tau s + 1)\Theta(s) = KF(s)$$

Solving Equation 12.43 for the ratio $\Theta(s)/\Theta_{set}(s)$ yields

$$\frac{\Theta(s)}{\Theta_{set}(s)} = \frac{\dfrac{\kappa_p \kappa_m}{[UA_s + \kappa_p \kappa_m]}}{\left[\dfrac{mc}{UA_s + \kappa_p \kappa_m} s + 1\right]} = KG(s) \tag{12.44}$$

For convenience, define

$$C_1 = \frac{mc}{UA_s + \kappa_p \kappa_m}$$

$$C_2 = \frac{\kappa_p \kappa_m}{UA_s + \kappa_p \kappa_m} \tag{12.45}$$

With these definitions, Equation 12.44 becomes

$$\frac{\Theta(s)}{\Theta_{set}(s)} = \frac{C_2}{[C_1 s + 1]} = G(s) \tag{12.46}$$

Equation 12.46 represents the transfer function, $G(s)$, for the system consisting of the oven and the proportional controller. For the linear system consisting of the oven and controller, the transfer function represents the ratio of the Laplace transforms of the output to the input assuming a zero initial condition.

Step Response

Suppose we consider the startup of the furnace, with an initial temperature equal to the ambient, or $\theta = 0$. From this condition we impose a value of θ_{set} that is larger than the ambient temperature, say θ_1; this represents a step-change input that is expressed in Laplace transform space as

$$\Theta_{set}(s) = \frac{\theta_1}{s} \tag{12.47}$$

The transform of the governing differential equation becomes

$$\Theta(s) = \frac{\theta_1 C_2}{s[C_1 s + 1]} \tag{12.48}$$

Employing partial fractions yields

$$\Theta(s) = \theta_1 C_2 \left[\frac{1}{s} - \frac{1}{s + 1/C_1} \right] \tag{12.49}$$

From Table C.1, we can determine the time-domain solution corresponding to this Laplace transform as

$$\theta(t) = \theta_1 C_2 \left[1 - e^{-t/C_1} \right] \tag{12.50}$$

Recall that we have assumed that $\theta(0) = 0$. In the limit as $t \to \infty$

$$\theta(t) = \theta_1 C_2 = \frac{\kappa_p \kappa_m \theta_1}{UA_s + \kappa_p \kappa_m}$$

For a step change input, we have found that $\theta(t) \neq \theta_1$ unless $\frac{\kappa_p \kappa_m}{UA_s + \kappa_p \kappa_m} = 1$. This will not in general be the case. Thus we find that a proportional controller, in general, is characterized by a nonzero steady-state error.

Example 12.5

Consider the proportional control of an oven that has a total mass of 20 kg, with an average specific heat of 800 J/kg-K. The oven is initially at room temperature. A step input is supplied to the controller, changing the set-point temperature to 100 °C above the ambient. Plot the temperature of the oven and the power supplied to the oven as a function of time, and determine the steady-state error if the values of the gains in the system are $\kappa_m = 20$ and $\kappa_p = 12$.

SOLUTION Equation 12.50 provides the solution for oven temperature as a function of time, for the case in which there is a step change in the set-point temperature. By examining Figure 12.50, we see that our controller provides a power to the oven given by

$$\kappa_p \kappa_m (\theta_1 - \theta)$$

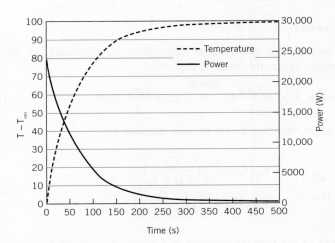

Figure 12.51 Response of oven-controller system.

The temperature response and the supplied power are provided in Figure 12.51.

COMMENT The values of the gains in this problem are directly tied to the power required for the oven. The physical arrangement of heaters and the required electrical service are a strong consideration in a practical implementation of this conceptual design. The effect of the controller gains and the system parameters can be further explored using the program file *First Order PID.vi*.

Proportional-Integral (PI) Control

The steady-state error that exists for a proportional controller is not acceptable in many applications. Adding a control signal that is proportional to the time integral of the error improves the performance of the proportional controller. Let's first examine the behavior of a pure integral controller.

Integral Control

Figure 12.52 provides a block diagram in the Laplace transform domain for an integral controller applied to control the temperature of our oven. In the time domain, this results in

$$\dot{P}(t) = \kappa_I \int_0^t e(t)dt + \dot{P}(0) \tag{12.51}$$

We immediately see that as long as the error remains finite and positive, the power will continue to increase. Clearly a pure integral controller would have very limited application.

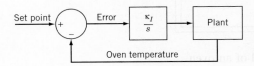

Figure 12.52 Block diagram for an integral controller.

Proportional–Integral (PI) Control

Suppose the actions of the proportional and integral controllers are combined, as shown in Figure 12.53. In Laplace transform space, this may be expressed

$$P(s) = \kappa_p \kappa_m E(s) + \frac{\kappa_I \kappa_m}{s} E(s) = \left[\kappa_p \kappa_m + \frac{\kappa_I \kappa_m}{s} \right] E(s) \tag{12.52}$$

Expressed in terms of the set point, the closed loop transfer function is

$$\frac{\Theta(s)}{\Theta_{set}(s)} = \frac{\kappa_m \left(\kappa_p + \dfrac{\kappa_I}{s} \right) \left(\dfrac{C_2}{C_1 s + 1} \right)}{1 + \kappa_m \left(\kappa_p + \dfrac{\kappa_I}{s} \right) \left(\dfrac{C_2}{C_1 s + 1} \right)} \tag{12.53}$$

Clearing fractions yields a form of the transfer function that allows the inverse transform to be determined:

$$\frac{\Theta(s)}{\Theta_{set}(s)} = \frac{\kappa_m \left(\kappa_p s + \kappa_I \right) C_2}{C_1 s^2 + s \left(1 + C_2 \kappa_m \kappa_p \right) + C_2 \kappa_m \kappa_I} \tag{12.54}$$

Once again we impose a step change in the set-point temperature so that $\Theta_{set}(s) = \frac{\theta_1}{s}$, and

$$\Theta(s) = \frac{\theta_1 \left[\kappa_m \left(\kappa_p s + \kappa_I \right) C_2 \right]}{s \left[C_1 s^2 + s \left(1 + C_2 \kappa_m \kappa_p \right) + C_2 \kappa_m \kappa_I \right]} \tag{12.55}$$

Employing the final value theorem, we multiply by s and take the limit as $s \to 0$:

$$\lim_{s \to 0} \Theta(s) = \lim_{s \to 0} \frac{s \left[\kappa_m \left(\kappa_p s + \kappa_I \right) C_2 \right] \theta_1}{s \left[C_1 s^2 + s \left(1 + C_2 \kappa_m \kappa_p \right) + C_2 \kappa_m \kappa_I \right]} = \frac{C_2 \kappa_m \kappa_I}{C_2 \kappa_m \kappa_I} \theta_1 = \theta_1 \tag{12.56}$$

Clearly, the steady-state error for this PI controller is zero. This is a general result for PI control.

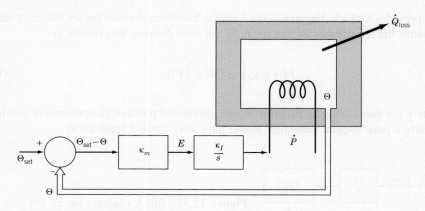

Figure 12.53 Block diagram for PI control of an oven.

Time Response

The time response of this system to a step-change input could be determined by finding the inverse Laplace transform of Equation 12.57:

$$\frac{\Theta(s)}{\Theta_{set}(s)} = \frac{\kappa_m(\kappa_p s + \kappa_I)C_2}{C_1 s^2 + s(1 + C_2\kappa_m\kappa_p) + C_2\kappa_m\kappa_I} \qquad (12.57)$$

However, by comparison with the Laplace transform of the governing differential equation for a second-order system, as provided in Equation 3.13, Palm (7) shows that

$$\varsigma = \frac{1 + C_2\kappa_m\kappa_p}{2\sqrt{C_1 C_2\kappa_m\kappa_I}} \qquad (12.58)$$

and that for $\varsigma < 1$ the equivalent time constant is

$$\tau = \frac{2C_1}{1 + C_2\kappa_m\kappa_p} \qquad (12.59)$$

The behavior of the oven-controller system can be further explored using the program file *First Order PID.vi.*

Proportional Integral–Derivative Control of a Second-Order System

Consider a spring–mass–damper system such as the one described in Figure 3.3. The governing equation describing the position of the mass as a function of time for a given forcing function $f(t)$ is

$$\frac{1}{\omega_n^2}\frac{d^2y}{dt^2} + \frac{2\varsigma}{\omega_n}\frac{dy}{dt} + y = f(t) \qquad (12.60)$$

This system and the governing equation serve as a model to demonstrate the properties of proportional integral–derivative (PID) control when applied to a second-order system.

The goal is to apply a step change in input to the system and have the mass move to the new equilibrium position; in other words, the ideal response of the system would be a step-change. Our first objective is to review the response of the second-order system without a controller in place. The behavior of the system is examined in two examples.

Example 12.6

A spring–mass–damper system has mass of 2 kg, a spring constant of 7,900 N/m, and a damping coefficient of 176 kg/s. Plot the open loop time response of this system to a step-change input in force.

ASSUMPTIONS   The initial conditions for velocity, $\frac{dy}{dt}$, and position, y, are zero.

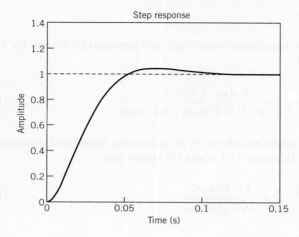

Figure 12.54 Second-order system response to a step-change input with a damping ratio of 0.7. Other parameters are described in Example 12.6.

SOLUTION Although the solution to this example could be determined from the results in Chapter 3, we will determine the response using Laplace transforms. Taking the Laplace transform of Equation 12.60 yields

$$\frac{1}{\omega_n^2} s^2 Y(s) + \frac{2\zeta}{\omega_n} sY(s) + Y(s) = KF(s) \tag{12.61}$$

For the mass, spring constant, and damping ratio given in the problem statement, we compute the natural frequency and damping ratio as

$$\omega_n = \sqrt{\frac{k}{m}} = \sqrt{\frac{7900}{2}} = 62.85 \text{ rad/s} \quad \text{and} \quad \zeta = \frac{c}{2\sqrt{km}} = \frac{176}{2\sqrt{7900 \times 2}} = 0.7$$

The open-loop transfer function defines the dynamic response of the system and is defined as

$$\frac{Y(s)}{KF(s)} = \left[\cfrac{1}{\cfrac{1}{\omega_n^2} s^2 + \cfrac{2\zeta}{\omega_n} s + 1} \right] = \left[\cfrac{1}{\cfrac{1}{3950} s^2 + 0.02235s + 1} \right] \tag{12.62}$$

where K is one for the present case. Options exist for calculating and plotting the open-loop response of this system. The program file *Second order PID.vi* can be used. Figure 12.54 shows the response of this system to a step input; the output has been normalized so that the displacement, as characterized by the amplitude, is from 0 to 1.

Example 12.7

Consider the second-order system described in Example 12.6. Plot the open loop response of the system to a step-change in input for damping ratios of 0.3 and 1.3.

SOLUTION Figures 12.56 and 12.57 show the response of the system to a step-change input for values of the damping coefficient of 0.3 and 1.3, respectively. By comparing Figures 12.55

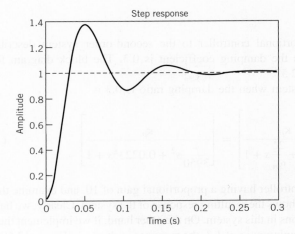

Figure 12.55 Second-order system response to a step-change input with a damping ratio of 0.3. Other parameters are described in Example 12.7.

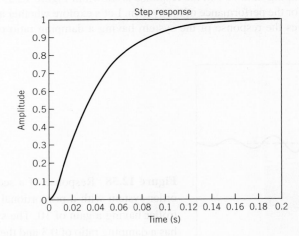

Figure 12.56 Second-order system response to a step-change input with a damping ratio of 1.3. Other parameters are described in Example 12.7.

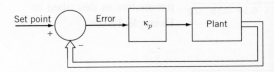

Figure 12.57 Block diagram for proportional control of a second-order plant.

through 12.57, we can learn some important characteristics of second-order systems. Suppose our goal is to have the system respond quickly to the input, but without oscillation and without exceeding the equilibrium value excessively. Let's characterize the response of the system having a damping ratio of 1.3 as being comparatively slow. This system requires 0.2 s to approach the equilibrium value of 1. The system having a damping coefficient of 0.3 first reaches 1 in a time less than 0.05 s, but there is both overshoot and oscillations. The system having a damping ratio of 0.7 has no oscillations, and minimal overshoot. The system with $\zeta = 0.3$ has a shorter rise time but longer settling time than that with $\zeta = 0.7$.

In the previous example, the response for a damping ratio of 0.7 appears to be the most likely candidate to meet our needs. But can we improve the situation by implementing a controller?

Proportional Control

Consider the application of a proportional controller to the second-order system described in Example 12.7 for the case in which the damping coefficient is 0.3. The block diagram for the control system is shown in Figure 12.57.

The transfer function for this system when the damping ratio is 0.3 is

$$\frac{Y(s)}{KF(s)} = \left[\frac{\kappa_p}{\frac{1}{\omega_n^2}s^2 + \frac{2\zeta}{\omega_n}s + 1} \right] = \left[\frac{\kappa_p}{\frac{1}{3950}s^2 + 0.02235s + 1} \right] \quad (12.63)$$

We implement a proportional controller having a proportional gain of 10, and examine the step response of the system. Figure 12.58 shows the resulting response of the system. Clearly, we have not eliminated the overshoot and oscillations in this system. On the other hand, if we implement the same controller for the system having a damping ratio of 1.3, the response is shown in Figure 12.59. Here the proportional controller is helpful for the performance of the system. Let's explore whether a more sophisticated control scheme improves the response of the system having a damping ratio of 0.3.

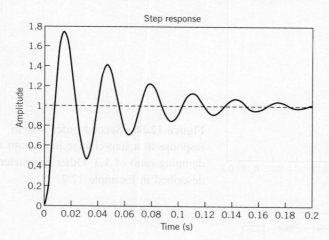

Figure 12.58 Response of a second-order system with a proportional controller having a gain of 10. The system has a damping ratio of 0.3 and the other parameters as described in Example 12.6.

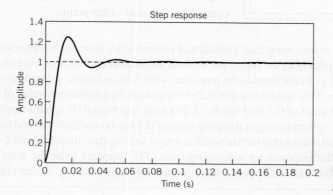

Figure 12.59 Proportional control of second-order system with a controller gain of 10, with damping 1.3.

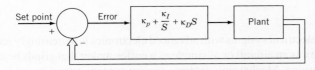

Figure 12.60 Block diagram for proportional integral–derivative (PID) control of a second-order system.

Proportional Integral-Derivative Control

A block diagram representing a PID control for our spring–mass–damper system is shown in Figure 12.60. The transfer function for this system is

$$\frac{Y(s)}{KF(s)} = \left[\frac{\kappa_D s^2 + \kappa_p s + \kappa_I}{\frac{1}{\omega_n^2}s^3 + \left(\frac{2\zeta}{\omega_n} + \kappa_D\right)s^2 + \kappa_p s + \kappa_I} \right] \tag{12.64}$$

Recall that taking a derivative in time space is multiplying by s in Laplace transform space, and integrating in time space is dividing by s in Laplace transform space. The choice of gains for the control actions, $\kappa_p, \kappa_I,$ and κ_D, is a challenging task and one for which much theory has been established (7). However, for our purposes we only wish to demonstrate the possible improvement in the system performance.

We select gains for the controllers of unity and just test the performance. For our second-order system having a damping ratio of 0.3, the result is shown in Figure 12.61. We compare this figure with Figures 12.55 and 12.57 and clearly see that this control scheme has improved the performance of the system. The system behavior with a variety of parameters and control gains can be explored using the program file *Second order PID.vi*. Although there are interactions among the proportional, integral, and derivative actions in a PID controller, the basic trends with increasing gain are provided in Table 12.4.

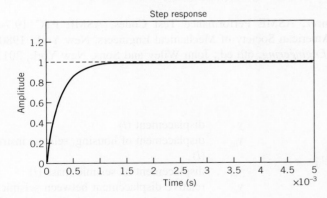

Figure 12.61 Response of a second-order system having a damping ratio of 0.3 when subject to PID control.

Table 12.4 Effect of Increasing Gain on Rise Time, Overshoot, and Steady-State Error for a PID Controller

Control Action	Rise Time	Overshoot	Steady-State Error
Proportional	⇓	⇑	⇓
Integral	⇓	⇑	Eliminate
Derivative		⇑	

12.5 SUMMARY

The integration of mechanical and electrical systems with advanced electronics is increasingly required for the design of systems as complex as an aircraft or as simple as a coffee maker that grinds beans and makes coffee at a set time each morning. Mechatronic systems require sensors, actuators, and control schemes to function as an integrated system. This chapter presented the operating principles and designs of those sensors used most often for linear and rotary displacement in mechatronic systems. "Actuators" is a term that represents a broad range of mechanical, electrical, pneumatic, and hydraulic devices designed to create motion. Selection of an actuator should be done through a system integration approach, so that the sensors and control scheme are appropriate for the actuator chosen.

We have provided an introduction to block-diagram representation of dynamic systems based on Laplace transforms, and have provided a review of the application of Laplace transforms to the solution of ordinary, linear differential equations. A discussion of open- and closed-loop control methods provided examples of proportional, proportional-integral, and proportional-integral-derivative control.

REFERENCES

1. Coltman, J. W., The transformer, *Scientific American*, 86, January 1988.
2. Herceg, E. E., *Schaevitz Handbook of Measurement and Control*, Schaevitz Engineering, Pennsauken, NJ, 1976.
3. Bredin, H., Measuring shock and vibration, *Mechanical Engineering*, 30, February 1983.
4. Measurement of Rotary Speed, ASME Performance Test Codes, ANSI/ASME PTC 19.13–1961, American Society of Mechanical Engineers, New York, 1961.
5. Gindy, S. S., Force and torque measurement, a technology overview, *Experimental Techniques*, 9: 28, 1985.
6. *Measurement of Shaft Power*, ASME Performance Test Codes, ASME PTC 19.7–1980, (Reaffirmed Date: 1988), American Society of Mechanical Engineers, New York, 1980.
7. Nise, N., *Control Systems Engineering*, 6th ed., John Wiley and Sons, New York, 2011.

NOMENCLATURE

a	acceleration ($l\,t^{-2}$)	y	displacement (l)
c	damping coefficient ($m\,t^{-1}$)	y_{hs}	displacement of housing, seismic instrument (l)
c_c	critical damping coefficient ($m\,t^{-1}$)		
g	acceleration of gravity ($l\,t^{-2}$)	y_m	displacement of seismic mass (l)
k	spring constant ($m\,t^{-2}$)	y_r	relative displacement between seismic mass and housing in a seismic instrument (l)
m	mass (m)		
p	pressure ($m\,l^{-1}t^{-2}$)	δy	deflection (l)
$\mathbf{r}$	radius, vector (l)	A	amplitude
t	time (t)	B	magnetic field strength ($m\,C^{-1}t^{-1}$)
δt	time interval for data sampling (t)	C_v	discharge coefficient
u	uncertainty	D_c	coil diameter for magnetic pickup (l)
v	linear velocity ($l\,t^{-1}$)	E_i	input voltage (V)
		E_o	output voltage (V)

F	force, vector $(m\,l\,t^{-2})$		T	torque $(m\,l^2t^{-2})$
J	polar moment of inertia (l^4)		$\forall$	volume (l^3)
L	length (l)		ζ	damping ratio
M	magnitude ratio		τ	time constant (t)
M	moment, vector $(m\,l^2t^{-2})$		τ_{max}	maximum shearing stress $(m\,t^2l^{-1})$
P	power $(m\,l^2t)$		ϕ	phase angle
Q	volumetric flow rate (l^3t^{-1})		κ	controller gain
R_o	outer radius (l)		ω	rotational speed (t^{-1})
T	temperature $(°)$		ω_n	natural frequency (t^{-1})

PROBLEMS

12.1 Consider a linear potentiometer as shown in Figure 12.1. The potentiometer consists of 0.1-mm copper wire $(\rho_e = 1.7 \times 10^{-8}\ \Omega\text{-m})$ wrapped around a core to form a total resistance of 1 kΩ. The sliding contact surface area is very small.

   a. Estimate the range of displacement that could be measured with this potentiometer for a 1.5-cm core.

   b. The circuit shown in Figure 12.62 is used to record position. On a single plot, show the loading error in an indicated displacement as a function of displacement over the range found in (a) for values of the meter resistance, R_m, of 1, 10, and 100 kΩ. For practical meters, would the loading error be significant?

12.2 Mechatronic applications for linear displacement sensors are numerous. Select a particular application and develop specifications that would be required for a linear displacement sensor. Possible applications include automotive seat position, pick-and-place operations, and throttle position sensors.

12.3 Tiltmeters are instruments used near volcanoes and faults in the earth's crust to measure rotation or tilting of the ground. Research and describe the use of LVDTs in tiltmeters. Include a discussion of accuracy requirements.

12.4 Compare and contrast wire-wound and conductive-plastic potentiometers. Are there advantages and disadvantages for either linear or rotary applications?

12.5 Compare and contrast the use of an LVDT and a resistance-based potentiometer for position measurement.

12.6 A seismic instrument, as shown in Figure 12.7, is used to measure a vibration given by

$$y(t) = 0.2\cos 10t + 0.3\cos 20t$$

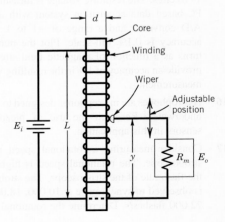

Figure 12.62 Circuit for Problem 12.1.

where

y = displacement in in.

t = time in s

The seismic instrument is to have a damping ratio of 0.7 and a spring constant of 1.2 lb/ft.

a. Select a combination of seismic mass and damping coefficient to yield less than a 10% amplitude error in measuring the input signal. Under what conditions would the output signal experience significant distortion?

b. Plot and describe the phase response of the system, either in a plot or tabular form.

12.7 A seismic instrument is to measure motion given by

$$y(t) = 0.5 \cos 15t + 0.8 \cos 30t$$

where

y = displacement in cm

t = time in s

The seismic instrument is to have a damping ratio of 0.7.

a. Select a combination of seismic mass, spring constant, and damping coefficient to yield less than a 6% amplitude error in measuring the input signal. Under what conditions would the output signal experience significant distortion?

b. Describe the phase response of the system, either in a plot or tabular form.

12.8 A seismic instrument has a natural frequency of 20 Hz and a damping ratio of 0.65. Determine the maximum input frequency for a vibration such that the amplitude error in the indicated displacement is less than 5%.

12.9 A seismic instrument consists of a housing and a seismic mass. To measure vibration, the seismic mass should remain stationary in an absolute frame of reference. To measure acceleration, the magnitude ratio should be unity. Explain in detail the requirements for mass, spring constant, and damping ratio to satisfy these requirements.

12.10 A seismic instrument consists of a housing and a seismic mass. To measure vibration, the seismic mass should remain stationary in an absolute frame of reference. To measure acceleration, the magnitude ratio should be unity. Using Labview, develop a model of the seismic instrument for both vibration and acceleration measurement.

12.11 Seismic instruments are employed to register ground motion during earthquakes. Research current state-of-the-art instruments and data acquisition for earthquake monitoring, such as the force balance accelerometer.

12.12 Determine the bandwidth for a seismic instrument employed as an accelerometer having a seismic mass of 0.2 g and a spring constant of 20,000 N/m, with very low damping. Discuss the advantages of a high natural frequency and a low damping ratio. Piezoelectric sensors are well suited for the construction of accelerometers because they possess these characteristics.

12.13 For the two piezoresistive accelerometers described in Table 12.1, and with reasonable estimates of acceptable errors, determine the useful input frequency range for these two instruments based on the resonance frequency and damping ratio.

12.14 In Example 12.2, integration is identified as a method for reducing the effects of noise in a signal. Discuss how a moving average can be used to reduce the effects of noise in a velocity measurement through the integration of an acceleration signal. Discuss the effects of averaging time on the elimination of noise. Assume that the noise has significant amplitude but has higher-frequency content than the velocity being measured.

12.15 Consider a moving coil transducer having a coil diameter D_c of 0.8 cm and a coil length of 2 cm. The nominal range of velocities to be measured is 1–10 cm/s. The resulting voltage is measured by a PC-based data acquisition system with an 8-bit A/D converter and a range of −1 to 1 V. The accuracy is 0.1% full-scale. Plot the number of turns as a function of magnetic field strength to provide an accuracy of 1% in the resulting velocity measurement.

12.16 Hydrophones are microphones designed to be used underwater. Research the use of piezoelectric sensors in this application.

12.17 Consider measuring a rotational speed using a stroboscope. The rotational speed is higher than the flash rate of the stroboscope. The stroboscope is observed to synchronize at 10,000, 18,000, and 22,000 flashes/s. Determine the rotational speed.

12.18 Write an instructional guide for using a strobe to measure rotational speed. Include both the case when the rotational speed is below the upper limit of the flash rate of the strobe and that when it is above.

12.19 Research the state-of-the art specifications for load cells designed to measure the smallest possible forces. What applications exist for such precise measurements of small forces?

12.20 Research the use of load cells in robotics, both in industrial robots and robotics research.

12.21 Design a proving ring load cell appropriate to serve as a laboratory calibration standard in the range of 250 to 1,000 N. The proving ring material is steel.

12.22 Power transmitted through the drive shaft of a car results in a rotational speed of 1,800 rpm with a power transmission of 40 horsepower. Determine the torque that the driveshaft must support.

12.23 Discuss the importance of a dynamometer for automotive emissions testing.

12.24 Compare and contrast engine dynamometers and chassis dynamometers for automotive applications, and for motorsports.

12.25 Research applications for linear actuators. For each application that you identify, suggest the most appropriate linear actuator. Why was that particular actuator chosen for that particular application? For many applications, a choice is made to convert rotary motion to linear motion.

12.26 Pneumatic actuators span a very large range of size and force. Research the range of commercially available pneumatic cylinders.

12.27 Estimate the flow coefficient for a flow control valve if the rating corresponds to 32 SCFM (0.906 SCMM) with a $\Delta p = 10$ psi (0.69 bar). The test is conducted using air with a line pressure of 100 psi (6.7 bar), $T = 68\,°\text{F}$ (20 °C), and a relative humidity of 36% at the supply port. What is the pressure at the exhaust port? Base your answer on standard conditions.

12.28 A low profile three-port control valve with 6-mm port diameters has an F value of 0.82 ms/cc and a lag time of 8 ms to pressurize through ports P to A. The exhaust path through ports T to A has an F value of 0.70 ms/cc with an 8 ms lag time. The valve is connected to a 500 mL vessel. Supply pressure is 1 bar; exhaust pressure is 0 bar. Estimate the time needed to pressurize this vessel to 0.9 bar. Estimate the time to exhaust this vessel to 0.1 bar.

12.29 Research the design of a totalizing flow meter, with particular emphasis on the meters used to generate water bills.

12.30 The companion software includes a program *stroboscope.vi* that models the behavior of a stroboscope.

    a. Set the bar rotation frequency to 40 rps, and then set the strobe frequency to 20 and 40 Hz. Explain your results.

    b. Set the bar rotation frequency to 100 rps. Using $f_1 = 50$ Hz and $f_N = 10$ Hz, apply Equation 12.15 to show that we can measure rotation frequencies larger than the maximum frequency of the stroboscope. Be careful to find all N of the synchronous frequencies.

12.31 Describe the operating principle of a thermostat for residential applications, and the design of a bimetallic sensor for measuring the temperature. How is a dead band created in such an instrument?

12.32 Show that a proportional controller has a steady-state error. How would you quantify the steady-state error for a controller and a first-order system?

12.33 Using *First order PID.vi*, develop a model of the oven described in Example 12.5 coupled with a PI controller. Vary the proportional and integral gains and discuss the resulting behavior of this first-order system.

12.34 Using *Second order PID.vi*, develop a model for PID control of the oven described in Example 12.5. Vary the gains and discuss the resulting behavior of the system. Quantify the power input required to achieve a given response.

12.35 Consider the block diagram for an automotive speed control system shown in Figure 12.63.

    a. Find the transfer function from $V_r(\text{s})$ to $V(\text{s})$, where V_r is the desired speed and V is the actual speed.

    b. Assume that the desired speed is a constant and in Laplace space is v_o/s. Determine values of the gains such that

$$\lim_{t \to \infty}(t) = v_o$$

12.36 Research automatic controls systems for commercial buildings. Address the functions associated

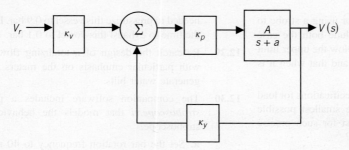

Figure 12.63 Block diagram for
an automotive speed control.

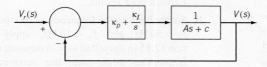

Figure 12.64 Block diagram for Problem 12.37.

with ventilation, heating, cooling and fans. Identify those variables that must be controlled for human comfort.

12.37 Explore the behavior of the system shown in Figure 12.64 using the appropriate Labview vi for values of A (5 < A < 15), and C (2 < C < 6).

12.38 For the fly ball governor shown in Figure 12.65, show that

$$k(x - x_o) = \frac{b}{a} 2MR\omega^2$$

where

k = spring constant
x = spring displacement from equilibrium
x_o = spring equilibrium position
M = flyball mass

And other variables as identified in Figure 12.65.

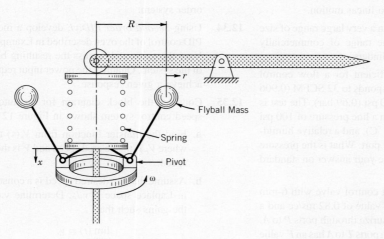

Figure 12.65 Fly ball
governor.

Appendix A

Property Data and Conversion Factors

Table A.1 Properties of Pure Metals and Selected Alloys

	Density (kg/m^3)	Modulus of Elasticity (GPa)	Coefficient of Thermal Expansion (10^{-6} m/m-K)	Thermal Conductivity (W/m-K)	Electrical Resistivity (10^{-6} Ω-cm)
Pure Metals					
Aluminum	2 698.9	62	23.6	247	2.655
Beryllium	1 848	275	11.6	190	4.0
Chromium	7 190	248	6.2	67	13.0
Copper	8 930	125	16.5	398	1.673
Gold	19 302	78	14.2	317.9	2.01
Iron	7 870	208.2	15.0	80	9.7
Lead	11 350	12.4	26.5	33.6	20.6
Magnesium	1 738	40	25.2	418	4.45
Molybdenum	10 220	312	5.0	142	8.0
Nickel	8 902	207	13.3	82.9	6.84
Palladium	12 020	—	11.76	70	10.8
Platinum	21 450	130.2	9.1	71.1	10.6
Rhodium	12 410	293	8.3	150.0	4.51
Silicon	2 330	112.7	5.0	83.68	1×10^5
Silver	10 490	71	19.0	428	1.47
Tin	5 765	41.6	20.0	60	11.0
Titanium	4 507	99.2	8.41	11.4	42.0
Zinc	7 133	74.4	15.0	113	5.9
Alloys					
Aluminum (2024, T6)	2 770	72.4	22.9	151	4.5
Brass (C36000)	8 500	97	20.5	115	6.6
Brass (C86500)	8 300	105	21.6	87	8.3
Bronze (C90700)	8 770	105	18	71	1.5
Constantan annealed (55% Cu 45% Ni)	8 920	—	—	19	44.1
Steel (AISI 1010)	7 832	200	12.6	60.2	20
Stainless Steel (Type 316)	8 238	190	—	14.7	—

Source: Complied from *Metals Handbook*, 9th ed., American Society for Metals, Metals Park, OH, 1978, and from other sources.

Table A.2 Thermophysical Properties of Selected Metallic Solids

Composition	Melting point (K)	Properties at 300 K				Properties at various temperatures (K)							
		ρ (kg/m³)	c_p (J/kg·K)	k (W/m·K)	$\alpha \times 10^4$ (m²/s)	k(W/m·K)				c_p (J/kg·K)			
						100	200	400	600	100	200	400	600
Aluminum													
Pure	933	2702	903	237	97.1	302	237	240	231	482	796	949	1033
Alloy 2024-T6 (4.5% Cu, 1.5% Mg, 0.6% Mn)	775	2770	875	177	73.0	65	163	186	186	473	787	925	1042
Alloy 195, cast (4.5% Cu)	—	2790	883	168	68.2	—	—	174	185	—	—	—	542
Chromium	2118	7160	449	93.7	29.1	159	111	90.9	80.7	192	384	484	542
Copper													
Pure	1358	8933	385	401	117	482	413	393	379	252	356	397	417
Commercial bronze (90% Cu, 10% Al)	1293	8800	420	52	14	—	42	52	59	—	785	460	545
Phosphor gear bronze (89% Cu, 11% Sn)	1104	8780	355	54	17	—	41	65	74	—	—	—	—
Cartridge brass (70% Cu, 30% Zn)	1188	8530	380	110	33.9	75	95	137	149	—	360	395	425
Constantan (55% Cu, 45% Ni)	1493	8920	384	23	6.71	17	19	—	—	237	362	—	—
Iron													
Pure	1810	7870	447	80.2	23.1	134	94.0	69.5	54.7	216	384	490	574
Armco (99.75%)	—	7870	447	72.7	20.7	95.6	80.6	65.7	53.1	215	384	490	574
Carbon steels													
Plain carbon (Mn ≤ 1%, Si ≤ 0.1%)	—	7854	434	60.5	17.7	—	—	56.7	48.0	—	—	487	559
AISI 1010	—	7832	434	63.9	18.8	—	—	58.7	48.8	—	—	487	559
Carbon-silicon (Mn ≤ 1%, 0.1% < Si ≤ 0.6%)	—	7817	446	51.9	14.9	—	—	49.8	44.0	—	—	501	582
Carbon-manganese-silicon (1% < Mn ≤ 1.65%, 0.1% < Si ≤ 0.6%)	—	8131	434	41.0	11.6	—	—	42.2	39.7	—	—	487	559

Table A.1 (continued) — Thermophysical Properties of Selected Metallic Solids

Composition	Melting Point, T (K)	ρ (kg/m³)	c_p (J/kg·K)	k (W/m·K)	$\alpha \cdot 10^7$ (m²/s)	100 K (k / c_p)	200 K (k / c_p)	400 K (k / c_p)	600 K (k / c_p)
Chromium (low) steels									
Cr–1/4 Mo–Si (0.18% C, 0.65% Cr, 0.23% Mo, 0.6% Si)	—	7822	444	37.7	10.9	— / —	— / —	38.2 / 492	36.7 / 575
1 Cr–1/2 Mo (0.16% C, 1% Cr, 0.54% Mo, 0.39% Si)	—	7858	442	42.3	12.2	— / —	— / —	42.0 / 492	39.1 / 575
1 Cr–V (0.2% C, 1.02% Cr, 0.15% V)	—	7836	443	48.9	14.1	— / —	— / —	46.8 / 492	42.1 / 575
Stainless steels									
AISI 302	1670	8055	480	15.1	3.91	— / —	— / —	17.3 / 512	20.0 / 559
AISI 304		7900	477	14.9	3.95	9.2 / 272	12.6 / 402	16.6 / 515	19.8 / 559
AISI 316		8238	468	13.4	3.48	— / —	— / —	15.2 / 504	18.3 / 550
AISI 347		7979	480	14.2	3.71	— / —	— / —	15.8 / 513	18.9 / 559
Lead	601	11 340	129	35.3	24.1	39.7 / 118	36.7 / 125	34.0 / 132	31.4 / 142
Magnesium	923	1740	1024	156	87.6	169 / 649	159 / 934	153 / 1074	149 / 1170
Molybdenum	2894	10 240	251	138	53.7	179 / 141	143 / 224	134 / 261	126 / 275
Nickel									
Pure	1728	8900	444	90.7	23.0	164 / 232	107 / 383	80.2 / 485	65.6 / 592
Nichrome (80% Ni, 20% Cr)	1672	8400	420	12	3.4	— / —	— / —	14 / 480	16 / 525
Inconel X–750 (73% Ni, 15% Cr, 6.7% Fe)	1665	8510	439	11.7	3.1	8.7 / —	10.3 / 372	13.5 / 473	17.0 / 510

Table A.3   Thermophysical Properties of Saturated Water (Liquid)

T (K)	ρ (kg/m^3)	c_p (kJ/kg $\cdot$ K)	$\mu \times 10^6$ (N $\cdot$ s/m^2)	k (W/m $\cdot$ K)	Pr	$\beta \times 10^6$ (K^{-1})
273.15	1000	4.217	1750	0.569	12.97	−68.05
275.0	1000	4.211	1652	0.574	12.12	−32.74
280	1000	4.198	1422	0.582	10.26	46.04
285	1000	4.189	1225	0.590	8.70	114.1
290	999	4.184	1080	0.598	7.56	174.0
295	998	4.181	959	0.606	6.62	227.5
300	997	4.179	855	0.613	5.83	276.1
305	995	4.178	769	0.620	5.18	320.6
310	993	4.178	695	0.628	4.62	361.9
315	991	4.179	631	0.634	4.16	400.4
320	989	4.180	577	0.640	3.77	436.7
325	987	4.182	528	0.645	3.42	471.2
330	984	4.184	489	0.650	3.15	504.0
335	982	4.186	453	0.656	2.89	535.5
340	979	4.188	420	0.660	2.66	566.0
345	977	4.191	389	0.664	2.46	595.4
350	974	4.195	365	0.668	2.29	624.2
355	971	4.199	343	0.671	2.15	652.3
360	967	4.203	324	0.674	2.02	679.9
365	963	4.209	306	0.677	1.90	707.1
370	961	4.214	289	0.679	1.79	728.7
373.15	958	4.217	279	0.680	1.73	750.1
400	937	4.256	217	0.688	1.34	896
450	890	4.40	152	0.678	0.99	
500	831	4.66	118	0.642	0.86	
550	756	5.24	97	0.580	0.88	
600	649	7.00	81	0.497	1.14	
647.3	315	0	45	0.238	∞	

Formulas for interpolation (T = absolute temperature)
$$f(T) = A + BT + CT^2 + DT^3$$

$f(T)$	A	B	C	D	Standard Deviation, σ
			$273.15 < T < 373.15$ K		
ρ	766.17	1.80396	-3.4589×10^{-3}		0.5868
c_p	5.6158	-9.0277×10^{-3}	14.177×10^{-6}		4.142×10^{-3}
k	−0.4806	5.84704×10^{-3}	-0.733188×10^{-5}		0.481×10^{-3}
			$273.15 < T < 320$ K		
$\mu \times 10^6$	0.239179×10^6	-2.23748×10^3	7.03318	-7.40993×10^{-3}	4.0534×10^{-6}
$\beta \times 10^6$	-57.2544×10^3	530.421	−1.64882	1.73329×10^{-3}	1.1498×10^{-6}
			$320 < T < 373.15$ K		
$\mu \times 10^6$	35.6602×10^3	−272.757	0.707777	-0.618833×10^{-3}	1.0194×10^{-6}
$\beta \times 10^6$	-11.1377×10^3	84.0903	−0.208544	0.183714×10^{-3}	1.2651×10^{-6}

Source: From Incropera, F. P., and D. P. DeWitt, *Fundamentals of Heat and Mass Transfer*, New York, Copyright © 1985 John Wiley & Sons, Inc. Used with permission.

Table A.4   Thermophysical Properties of Air

T [K]	ρ [kg/m^3]	c_p [kJ/kg · K]	$\mu \times 10^7$ [N · s/m^2]	$v \times 10^6$ [m^2/s]	$k \times 10^3$ [W/m · K]	$\alpha \times 10^5$ [m^2/s]	Pr
200	1.7458	1.007	132.5	7.590	18.1	10.3	0.737
250	1.3947	1.006	159.6	11.44	22.3	15.9	0.720
300	1.1614	1.007	184.6	15.89	26.3	22.5	0.707
350	0.9950	1.009	208.2	20.92	30.0	29.9	0.700
400	0.8711	1.014	230.1	26.41	33.8	38.3	0.690
450	0.7740	1.021	250.7	32.39	37.3	47.2	0.686
500	0.6964	1.030	270.1	38.79	40.7	56.7	0.684
550	0.6329	1.040	288.4	45.57	43.9	66.7	0.683
600	0.5804	1.051	305.8	52.69	46.9	76.9	0.685
650	0.5356	1.063	322.5	60.21	49.7	87.3	0.690
700	0.4975	1.075	338.8	68.10	52.4	98.0	0.695
750	0.4643	1.087	354.6	76.37	54.9	109.	0.702
800	0.4354	1.099	369.8	84.93	57.3	120.	0.709
850	0.4097	1.110	384.3	93.80	59.6	131.	0.716
900	0.3868	1.121	398.1	102.9	62.0	143.	0.720
950	0.3666	1.131	411.3	112.2	64.3	155.	0.723
1000	0.3482	1.141	424.4	121.9	66.7	168.	0.726

Formulas for Interpolation (T = absolute temperature)

$$\rho = \frac{348.59}{T} \qquad (\sigma = 9 \times 10^4)$$

$$f(T) = A + BT + CT^2 + DT^3$$

$f(T)$	A	B	C	D	Standard Deviation, σ
c_p	1.0507	-3.645×10^{-4}	8.388×10^{-7}	-3.848×10^{-10}	4×10^{-4}
$\mu \times 10^7$	13.554	0.6738	-3.808×10^{-4}	1.183×10^{-7}	0.4192×10^{-7}
$k \times 10^3$	-2.450	0.1130	-6.287×10^{-5}	1.891×10^{-8}	0.1198×10^{-3}
$\alpha \times 10^8$	-11.064	7.04×10^{-2}	1.528×10^{-4}	-4.476×10^{-8}	0.4417×10^{-8}
Pr	0.8650	-8.488×10^{-4}	1.234×10^{-6}	-5.232×10^{-10}	1.623×10^{-3}

Source: From F. P. Incropera and D. P. DeWitt, *Fundamentals of Heat and Mass Transfer*. New York, Copyright © 1985 John Wiley & Sons Inc. Used with permission.

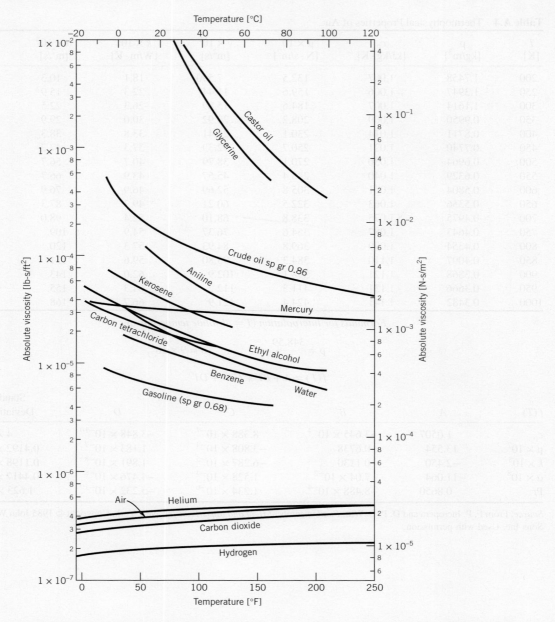

Figure A.1 Absolute viscosities of certain gases and liquids. (Compiled from multiple sources including Pritchard, P.J., *Fox and McDonald's Introduction to Fluid Mechanics – 8th ed*, Hoboken, NJ, Copyright © 2011 John Wiley & Sons, Inc.)

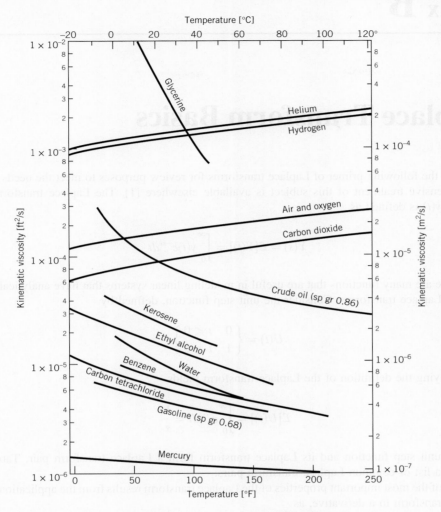

Figure A.2 Kinematic viscosities of certain gases and liquids. The gases are at standard pressure. (Compiled from multiple sources including Pritchard, P.J., *Fox and McDonald's Introduction to Fluid Mechanics – 8th ed*, Hoboken, NJ, Copyright © 2011 John Wiley & Sons, Inc.)

Appendix B

Laplace Transform Basics

We offer the following primer of Laplace transforms for review purposes to meet the needs of this text. Extensive treatment of this subject is available elsewhere [1]. The Laplace transform of a function $y(t)$ is defined as

$$Y(s) = \mathcal{L}[Y(t)] = \int_0^\infty y(t)e^{-st}dt \tag{B.1}$$

There are many functions that are useful in modeling linear systems that have analytical forms for their Laplace transforms. Consider the unit step function, defined by

$$U(t) = \left\{ \begin{matrix} 0 & t < 0 \\ 1 & t > 0 \end{matrix} \right\} \tag{B.2}$$

Applying the definition of the Laplace transform yields

$$\mathcal{L}[U(t)] = \int_0^\infty e^{-st}dt = \frac{1}{s} \tag{B.3}$$

The unit step function and its Laplace transform form a Laplace transform pair. Table B.1 provides a list of common Laplace transform pairs.

One of the most important properties of the Laplace transform results from the application of the Laplace transform to a derivative, as

$$\mathcal{L}\left[\frac{dy(t)}{dt}\right] = sY(s) - y(0) \tag{B.4}$$

Differentiation in the time domain is the same as multiplication in the Laplace transform domain. This relationship is derived through application of integration by parts of the definition of the Laplace transform, and can be generalized as

$$\mathcal{L}\left[\frac{d^n y(t)}{dt^n}\right] = s^n Y(s) - s^{n-1}y(0) - s^{n-2}\frac{dy(0)}{dt} - \cdots - \frac{d^{n-1}y(0)}{dt^{n-1}} \tag{B.5}$$

Table B.1   Laplace Transform Pairs

	$f(t)$	$\mathcal{L}(f)$
1	1	$\dfrac{1}{s}$
2	t	$\dfrac{1}{s^2}$
3	t^2	$\dfrac{2!}{s^3}$
4	t^n	$\dfrac{n!}{s^{n+1}}$

where n is integer

5	e^{-at}	$\dfrac{1}{s+a}$
6	te^{-at}	$\dfrac{1}{(s+a)^2}$
6	$\dfrac{e^{-at}-e^{-bt}}{b-a}$	$\dfrac{1}{(s+a)(s+b)}$
7	$\sin at$	$\dfrac{a}{s^2+a^2}$
8	$\cos at$	$\dfrac{s}{s^2+a^2}$

B.1   FINAL VALUE THEOREM

A useful property of the Laplace transform allows us to determine the value of the function $y(t)$ knowing only the Laplace transform of the function. The final value theorem states

$$\lim_{t \to \infty} y(t) = \lim_{s \to \infty} [sY(s)] \qquad (B.6)$$

Let's apply this to the unit step function

$$\lim_{s \to 0} sY(s) = \lim_{s \to 0} s\frac{1}{s} = 1 \qquad (B.7)$$

The unit step function has a limit of unity.

B.2   LAPLACE TRANSFORM PAIRS

Table B.1 provides Laplace transform pairs for a variety of important engineering functions. More extensive lists and methods for more complex transformations may be found in most advanced engineering mathematics texts.

REFERENCE

1. Schiff, J. L., *The Laplace Transform: Theory and Applications*, Springer, New York, 1999.

Glossary

absolute temperature scale   A temperature scale referenced to the absolute zero of temperature; equivalent to thermodynamic temperature.

accelerometer   An instrument for measuring acceleration.

accuracy   The closeness of agreement between the measured value and the true value.

active filter   A filter circuit that includes one or more active electrical elements, such as an operational amplifier.

actual flow rate   Flow rate specified under the actual temperature and pressure conditions of the measurement. *See* standard flow rate.

actuator   Device that creates a desired linear or rotary motion; classes generally include electrical, mechanical, hydraulic or pneumatic.

A/D converter   A device that converts analog voltages into digital numbers.

advanced-stage uncertainty   The uncertainty estimate beyond the design stage that includes the uncertainty due to measurement system, measurement procedure, and measured variable variation.

aleatory uncertainty   Uncertainty due to the natural variability of the measurand.

alias frequency   A false frequency appearing in a discrete data set that is not in the original signal.

amplitude   Characteristic of a signal or component frequency of a signal that expresses a range of time-dependent magnitude variations.

amplitude spectrum   A plot associating the individual components of frequency comprising a signal with their corresponding amplitudes.

apparent strain   A false output from a strain gauge as a result of input from extraneous variables such as temperature.

astable multivibrator   Switching circuit that toggles on or off continuously in response to an applied voltage command.

band-pass filter   A device that allows a range of signal frequencies to pass unaltered while frequencies outside this range, both higher and lower, are attenuated.

Bessel filter   Filter design having a maximally linear phase response over its passband.

bias   An offset value.

bias error   *See* systematic error. Archaic term no longer used in measurements.

bimetallic thermometer   A temperature measuring device that uses differential thermal expansion of two materials.

bistable multivibrator   *See* flip-flop.

bit   The smallest unit in binary representation.

bit number   The number of bits used to represent a word in a digital device.

bonded resistance strain gauge   A strain sensor that exhibits an electrical resistance change indicative of strain. The sensor is bonded to the test specimen in such a way that the gauge accurately measures the strain at the surface of the test specimen.

Bragg grating   Fiber optic device that has a periodic variation of the refraction of index.

bridge constant   The ratio of the actual bridge voltage to the output of a single strain gauge sensing the maximum strain.

buffer   Memory set aside to be accessed by a specific I/O device, such as to temporarily store data.

bus   A major computer line used to transmit information.

Butterworth filter   Filter design having a maximally flat magnitude ratio over its passband.

byte   A collection of 8 bits.

calibration   Act of applying a known input value to a system so as to observe the system output value.

calibration curve   Plot of the input values versus the output values resulting from a calibration.

cascading filters   Multiple filters arranged in series.

central tendency   Tendency of data scatter to group about a central (mean or most expected) value.

charge amplifier   Amplifier used to convert a high impedance charge into a voltage.

circular frequency   Frequency specified in radians per second; angular frequency.

closed-loop control Type of control action that is based on the output value from a sensor relative to a set-point.

combined standard uncertainty Uncertainty estimate that includes both systematic and random standard uncertainties.

common-mode voltage Voltage difference between two different ground points. *See* ground loop.

concomitant method An alternate method for assessing the value of a variable.

conduction error Error in temperature measurement caused by the diffusion of heat through a temperature sensor either toward or away from the measuring environment so as to change the sensor temperature.

confidence interval Interval that defines the limits about a mean value within which the true value is expected to fall.

confidence level Probability that the true value falls within the stated confidence interval.

control A means of setting and maintaining a value during a test.

control parameter Parameter whose value is set during a test and that has a direct effect on the value of the measured variable.

conversion error The errors associated with converting an analog value into a digital number.

conversion time The duration associated with converting an analog value into a digital number.

coupled system The combined effect of interconnected instruments used to form a measurement system.

critical pressure ratio The pressure ratio necessary to accelerate a fluid to a velocity equal to its speed of sound.

cut-off frequency The 3 dB design point frequency of a filter.

D/A converter A device that converts a digital number into an analog voltage.

damping ratio A measure of system damping—a measure of a system's ability to absorb or dissipate energy.

d'Arsonval movement Device that senses the flow of current through torque on a current carrying loop; the basis for most analog electrical meters.

data acquisition module A device that contains a minimum of components required for data acquisition; often connected to a computer by a link, wired or wireless.

data acquisition system A system that quantifies and stores data.

deflection mode Measuring method that uses signal energy to cause a deflection in the output indicator. *See* loading error.

dependent variable A variable whose value depends on the value of one or more other variables.

design-stage uncertainty The uncertainty due only to instrument error and instrument resolution; an uncertainty value based on the minimum amount of information that is known at the early stage of test design.

deterministic signal A signal that is predictable in time or in space, such as a sine wave or a ramp function.

deviation plot Graphical representation of the difference between a reference and expected value.

diaphragm meter A type of positive displacement flow meter that measures gas volume flow rate by alternately passing isolated but known volumes of the gas between two chambers.

differential-ended connection Dual-wire connection scheme that measures the difference between two signals with no regard to ground.

digital signal Representation of an analog signal that is quantized in both magnitude and time.

digital signal processor Microprocessor used to manage the transfer of digital data information and perform prescribed mathematical manipulation of the data.

dilatation The ratio of the change in volume to the original volume.

direct memory access Transfer protocol that allows a direct transmission path between a device and computer memory.

discharge coefficient Coefficient equating the ratio of actual flow rate across a flow meter to the ideal flow rate if no losses or flow separation occurred.

discrete random variable Random variable whose value is not continuous.

discrete time signal Signal in which information about the magnitude of the represented variable is available only at discrete points in time.

dispersion The scattering of values of a variable around a mean value.

distortion Any undesirable change in the shape of a waveform that occurs during signal conditioning or measurement.

disturbance   An effect that changes the operating condition of a control system away from its setpoint.

Doppler anemometry   Refers to various techniques that use the Doppler shift to measure velocity.

Doppler flowmeter   Flowmeter that uses the Doppler shift in the frequency of ultrasonic waves passing through the fluid flow to indicate volume flow rate.

dynamic calibration   Process of calibration such that the input variable is time-varying.

dynamic pressure   Difference between the total and static pressures at any point.

dynamic response   Response of a system to a well-defined sine wave input.

dynamometer   Device for measuring shaft power.

earth ground   A ground path whose voltage level is zero.

elastic region   An operating range of strain versus load wherein a material returns to its initial length when the load is removed.

electromagnetic flow meter   Device that bases fluid flow rate on the magnitude of the current induced in a magnetic field caused by the velocity of the fluid passing through that magnetic field.

end standard   Length standard described by the distance between two flat square ends of a bar.

epistemic uncertainty   Uncertainty arising from unknowns about the measurand, process, or model.

error   Difference between a measured value and its true (actual) value. In controls, it is the difference between a sensor value and a set-point.

error fraction   A measure of the time-dependent error in the time response of a first-order system.

expanded uncertainty   Uncertainty interval with the combined effects of random and systematic uncertainties at a stated confidence level (typically 95%) other than one standard deviation.

expansion factor   Coefficient used to account for compressibility effects in obstruction flow meters.

extraneous variable   A variable that is not or that cannot be controlled during a measurement, but that affects the measured value.

feedback-control stage   Stage of a measurement system that interprets the measured signal and exerts control over the process.

filter   Circuit or algorithm used to reduce or remove the amplitude (or the effect) of unwanted frequencies from a signal.

firewire   A serial method for data transmission between a computer and peripherals using a thin cable and based on the IEEE 1394 standard.

first-order system   A system characterized as having time-dependent storage or dissipative ability but having no inertia.

fixed point temperature   Temperatures specified through a reproducible physical behavior, such as a melting or boiling point.

flip-flop   A switching circuit that toggles either on or off on command. Also called a bistable multivibrator.

flow coefficient   Product of the discharge coefficient and the velocity of approach factor; used in a flow meter calculation.

flow nozzle   Flow meter consisting of a gradual concentric contraction from a larger diameter to a smaller throat diameter that is inserted into a pipe to create a measurable pressure drop proportional to flow rate.

Fourier analysis   A method for expressing a signal as a series of sines and cosine functions.

Fourier transform   Mathematical function that allows a complex signal to be decomposed into its frequency components. Method to represent a signal in terms of amplitude as a function of frequency.

frequency   Value indicating the rate of change of a signal with time.

frequency bandwidth   The range of frequencies in which the magnitude ratio remains within 3 dB of unity (0 dB).

frequency response   The magnitude ratio and phase shift of a system as a function of applied input frequency.

frequency spectrum   A representation, such as a plot, associating the individual frequency components comprising a signal with amplitude and phase angle.

full-scale output (FSO)   The arithmetic difference between the end points (range) of a calibration expressed in units of the output variable.

fundamental frequency   Smallest frequency component of a signal.

gain   Result of amplification or some multiplier of a signal magnitude; also used to refer to a constant multiplier for a transfer function block.

galvanometer Instrument that is sensitive to current flow through torque on a current loop; typically used to detect a state of zero current flow.

gauge blocks Working length standard for machining and calibration.

gauge factor Property of a strain gauge that relates changes in electrical resistance to strain.

gauge length Effective length over which a strain gauge averages strain during a measurement.

ground Signal return path to earth.

ground loop A signal circuit formed by grounding a signal at more than one point with the ground points having different voltage potentials.

Hall effect Voltage induced across a current-carrying conductor situated within a magnetic field. The magnetic field exerts a transverse force pushing the charged electrons to one side of the conductor.

handshake An interface procedure to control data flow between different devices that includes start and stop (hello and goodbye) signals.

harmonics Frequencies present in a signal that are usually integer multiples of a fundamental frequency

heat flux sensor A device that uses physical phenomena to indicate local heat flux.

high-pass filter A device that allows signal frequencies above the cut-off frequency to pass unaltered while lower frequencies are attenuated.

Hooke's law The fundamental linear relation between stress and strain. The proportionality constant is the modulus of elasticity.

hypothesis test Statistical test based on inferential statistics to accept or reject a statistical hypothesis that two measured parameters are equal, or one is greater or less than the other.

hysteresis A difference in the indicated value for any particular input value when approached going upscale as opposed to going downscale.

immersion errors Temperature measurement errors due to the measuring environment leading to differences between sensor temperature and the temperature being measured.

impedance matching Process of setting either the input impedance or the output impedance in a way to keep signal loading errors to an acceptable level.

independent variable A variable whose value is changed directly to cause a change in the value of a dependent variable.

indicated value The value displayed or indicated by the measurement system.

input value Process information sensed by the measurement system.

insertion error Any error brought on by the interaction between a sensor and the immediate environment in which it is installed.

in situ In measurements, installed in the system as intended for use.

interference Extraneous effect that imposes a deterministic trend on the measured signal.

interferometer Device for measuring length that uses interference patterns of light sources; provides very high resolution.

interrupt A signal to initiate a different procedure; analogous to raising one's hand.

junction For thermocouples, an electrical connection between two dissimilar materials.

laminar flow element Device that measures flow rate by the linear pressure drop versus flow rate relationship occurring when fluid travels through a tube or cluster of tubes under laminar flow conditions.

least squares regression Analysis tool used to generate a curve fit equation that minimizes the sum of the squares of the deviations between the predicted values of the curve fit and the data set on which the curve fit is based.

linearity The closeness of a calibration curve to a straight line.

linear variable differential transformer (LVDT) A sensor/transducer that provides a voltage output as a function of core displacement.

line standard Standard for length made by scribing two marks on a dimensionally stable material.

load cell Sensor for measuring force or load.

loading error Error due to the very act of measurement so as to change the value measured.

low-pass filter A device that allows signal frequencies below a cut-off frequency to pass unaltered while higher frequencies are attenuated.

Mach number Ratio of the local speed to the speed of sound in the medium.

magnitude Indication of the size or level of a quantity.

magnitude ratio The ratio of the values of output amplitude to the input amplitude of a dynamic signal.

mass flow rate   Mass flow per unit time.

measurand   Measured variable.

measured value   The number assigned to the variable measured.

measured variable   A variable whose value is measured; also known as the measurand.

measurement   The act of assigning a specific value to the variable being measured.

mechatronics   Subject area focused on the interaction between electrical and mechanical components of devices and the control of such devices.

metrology   The science of weights and measures.

moiré patterns   An optical effect resulting from overlaying two dense grids that are displaced slightly in space from one another.

monostable   One-shot device that toggles on and then off on a single voltage command.

moving coil transducer   A sensor that uses a displacement of a conductor in a magnetic field to sense velocity.

multiplexer   An input/output (I/O) switch having multiple input signal lines but only one output signal line.

natural frequency   The frequency of the free oscillations of an undamped system.

noise   An extraneous effect that imposes random variations on the measured signal.

notch filter   A filter that only removes or reduces signal amplitude within an narrow frequency band termed as the notch.

null mode   Measuring method that balances a known amount against an unknown amount to determine the measured value.

Nyquist frequency   Half the sample rate or sampling frequency.

obstruction flow meter   Device that measures flow rate by the magnitude of the pressure drop arising by purposefully creating an abrupt area change in the flow.

ohmmeter   An instrument for measuring resistance.

open-loop control   Type of control action that does not use a sensor to measure the variable being controlled, such as by controlling room lights using a light switch timer.

operating conditions   Conditions of the test, including all equipment settings, sensor locations, and environmental conditions.

optical pyrometer   Nonintrusive temperature measurement device. Measurement is made through an optical comparison.

orifice meter   Flow meter consisting of a flat plate with a hole in its center that fits between a pipe flange to create a pressure drop proportional to flow rate; also known as an orifice plate.

oscilloscope   An instrument for measuring a time-varying voltage signal and displaying a trace; very high-frequency signals can be examined.

outlier   A measured data point whose value falls outside the range of reasonable probability.

output stage   Stage of a measurement system that indicates or records the measured value.

output value   The value indicated by the measurement system.

overall error   Instrument uncertainty estimate based on the square root of the sum of the squares (RSS) of the uncertainties of all known instrument errors.

parallel communication   Technique that transmits information in groups of bits simultaneously.

parameter   Fixed value that sets the behavior of a process.

particle image velocimetry   Optical method that measures flow velocity by observing the distance traveled by fluid particles per unit time suspended within the fluid.

passband   Range of frequencies over which the signal amplitude is attenuated by no more than 3 dB. *See* stopband.

passive filter   A filter circuit that only uses passive electrical elements such as resistors, inductors, and capacitors.

Peltier effect   Describes the reversible conversion of energy from electrical to thermal when a current flows through a junction of dissimilar materials.

phase shift   A shift or offset between the input signal and the output signal representing the measured signal; may be expressed in terms of an angle.

phase spectrum   A plot associating the individual components of frequency comprising a signal with their corresponding phase shift.

photoelastic   Describes a change in optical properties with strain.

piezoresistance coefficient   Constant that describes the change in electrical resistance with applied strain.

pitot-static pressure probe   Device used to measure dynamic pressure.

Poisson's ratio   Ratio of lateral strain to axial strain.

pooled value   Statistical value determined from separate data sets, such as replications, that are subsequently grouped together.

population   The set representing all possible values for a variable.

potentiometer   Refers either to a variable resistor or to an instrument for measuring small voltages with high accuracy.

potentiometer transducer   Variable electrical resistance transducer for measurement of length or rotation angle.

power spectrum   A plot associating the individual components of frequency comprising a signal with their corresponding power magnitudes.

precision   The amount of dispersion expected with repeated measurements; archaic term used to describe the random uncertainty.

precision   (instrument specification) Ultimate performance claim of instrument repeatability based on multiple tests conducted at multiple labs and on multiple units.

precision error   Archaic term no longer used in measurements. *See* random error.

precision indicator   Statistical measure that quantifies the dispersion in a reported value.

precision interval   Interval evaluated from data scatter that defines the probable range of a statistical value and ignoring the effects of systematic errors.

primary standard   The defining value of a unit.

Prony brake   A historically significant dynamometer design that uses mechanical friction to dissipate and measure power.

prover   Flow meter calibration device using some method placed in series with the candidate flow meter that accurately measures the volume of fluid passing per unit time.

proving ring   An elastic load cell, which may be used as a local calibration standard.

pyranometer   Optical instrument used to measure irradiation on a plane.

quantization   Process of converting an analog value into a digital number.

quantization error   An error brought on by the resolution of an A/D converter.

radiation error   Temperature measurement insertion error caused by a radiative heat exchange between a temperature sensor and portions of its immediate environment not at the same temperature as the variable being measured.

radiation shield   A device that reduces radiative heat transfer to a temperature sensor to reduce insertion errors associated with radiation.

radiometer   Device for measuring radiative source temperature using a thermopile detector.

random error   The portion of error that varies randomly on repeated measurements of the same variable.

randomization methods   Test methods used to break up the interference effects from extraneous variables by strategies that randomize the effects of those variables on the data set.

random standard uncertainty   The random uncertainty estimated at the one standard deviation level or at the 68% probability level for a normal distribution.

random test   Measurement strategy that sets a random order to the changes in the independent variable.

random uncertainty   Uncertainty assigned to a random error; contribution to combined uncertainty due to random errors.

range   The lower to upper limits of an instrument or test.

recovery errors   Temperature measurement error in a high speed flowing gas stream related to the conversion of kinetic energy into a temperature rise at the sensor.

reference junction   In reference to a thermocouple, the junction at which the temperature is known.

repeatability   Instrument specification estimating the expected random uncertainty based on multiple tests within a given lab on a single unit.

repetition   Repeating measurements during the same test.

replication   Duplicating a test under similar operating conditions using similar instruments.

reproducibility   Random uncertainty estimate based on multiple tests performed in different labs on a single unit.

resistivity   Material property describing the basic electrical resistance of a material.

resolution   The smallest detectable change in a measured value as indicated by the measuring system.

resonance frequency The frequency at which the magnitude ratio reaches a maximum value greater than unity.

result The final value determined from a set or sets of measurements.

resultant A variable calculated from the values of other variables.

ringing frequency The frequency of the free oscillations of a damped system; a function of the natural frequency and damping ratio.

rise time The time required for a first-order system to respond to 90% of a step change in input signal.

rms value Root-mean-square value.

roll-off slope Decay rate of a filter stopband.

root-sum-square value (RSS) The value obtained by taking the square root of the sum of a series of squared values.

rotating cusp meter A type of positive displacement flow meter that measures volume flow rate by the rate at which fluid is displaced within alternating rotating cusps.

RTD Resistance temperature detector; senses temperature through changes in electrical resistance of a conductor.

sample A data point or single value of a population; a dataset.

sample rate The rate or frequency at which data points are acquired; reciprocal of sample time increment.

sample statistics Statistics of a variable determined from a set of data points whose set size is less than the population of that variable.

sample time increment The time interval between two successive data measurements. Reciprocal of sample rate.

saturation error Error due to an input value exceeding the device maximum.

second-order system A dynamic system model whose behavior includes inertia.

Seebeck effect Source of open circuit electromotive force in thermocouple circuits.

seismic transducer Instrument for measuring displacement in time, velocity, or acceleration.

sensitivity The rate of change of a variable (y) relative to a change in some other variable (x), for example, dy/dx.

sensitivity index The sensitivity between two variables evaluated at some operating point and used to weight individual uncertainties in uncertainty propagation.

sensor The portion of the measurement system that senses or responds directly to the process variable being measured.

serial communication Technique that transmits information 1 bit at a time.

settling time The time required for a second-order system to settle to within ±10% of the final value of a step change in input value.

shield A metal foil or wire braid wrapped around signal-carrying wires that intercepts unwanted stray external electrical fields to reduce noise and interference effects.

shielded twisted pair Describes electrical wiring that reduces mutual induction between conductors through twisting the conductors around each other and that reduces noise and interference through shielding.

signal Information passing through a measurement system; often exists in electrical, optical or mechanical form.

signal conditioning stage Stage of a measurement system that modifies the signal to a desired magnitude and form.

signal-to-noise ratio Ratio of signal power to noise power.

single-ended connection Two-wire connection scheme that measures the signal relative to ground.

single-sample uncertainty analysis Method of uncertainty analysis used to evaluate uncertainty when very few repeated measurements are made.

sound pressure level meter (SPL) Stand-alone device with integral microphone and digital signal processor used to measure acoustical sound pressure levels.

span The difference between the maximum and minimum values of operating range of an instrument.

stage Refers to a general component in a measurement signal flow that has a specific purpose.

stagnation pressure Pressure at the point where a flow is brought to rest. *See* total pressure.

standard The known or reference value used as the basis of a calibration; also, a test method set by international agreement for quantifying some outcome or device performance.

static calibration   Calibration method whereby the input value is held constant while the output value is measured.

static pressure   Pressure sensed by a fluid particle as it moves with the flow.

static sensitivity   The rate of change of the output signal relative to a change in a static input signal. The slope of the static calibration curve at a point. Also called the static gain. *See* sensitivity.

steady response   A portion of the time-dependent system response that either remains constant with time or repeats its waveform with time.

step response   System time response to a step change in input value.

stopband   Range of frequencies over which the signal amplitude is attenuated by 3 dB or more. *See* passband

strain   Elongation per unit length of a member subject to an applied force.

strain gauge   Sensor used to measure the strain on the surface of a part when under load.

strain gauge rosette   Two or more strain gauges arranged at precise angles and locations relative to each other so as to measure differing components of strain.

stress   Internal force per unit area.

stroboscope   High-intensity source of light that can be made to flash at a precise rate; used to measure the frequency of rotation.

Student's *t*-distribution   Sampling distribution that arises when estimating the mean of a population when the sample size is small.

systematic error   The portion of error that remains constant in repeated measurements of the same variable; a constant offset between the indicated value and the actual value measured.

systematic standard uncertainty   The systematic uncertainty estimated at the one standard deviation level or at the 68% probability level for a normal distribution. This value is often taken as being half the estimated systematic uncertainty.

systematic uncertainty   Uncertainty assigned to a systematic error.

temperature scale   Establishes a universal means of assigning a quantitative value to temperatures.

thermistor   Temperature-sensitive semiconductor resistor.

thermocouple   Wire sensor consisting of at least two junctions of two dissimilar conductors; used to measure temperature.

thermoelectric   Refers to a type of device that indicates a thermally induced voltage.

thermopile   A multiple junction thermocouple circuit designed to measure a particular temperature, an average temperature, or a temperature difference.

Thomson effect   Describes the presence of a voltage arising through a temperature difference in a homogeneous conductor.

time constant   System property defining the time required for a first-order system to respond to 63.2% of a step input.

time delay   The lag time between an applied input signal and the measured output signal.

time response   The time-dependent system response to an input signal.

total pressure   Pressure sensed at a point where the flow is brought to rest without losses; the sum of the static and dynamic pressures; the isentropic value of stagnation pressure.

total sample period   Duration of the measured signal as represented by the data set.

transducer   The portion of measurement system that transforms the sensed information into a different form; also, a device that houses the sensor, transducer, and, often, signal conditioning stages of a measurement system.

transient response   Portion of the time-dependent system response that decays to zero with time.

transit time flow meter   Device that determines flow rate by measuring the time it takes for an ultrasonic wave to travel through the moving flow along a well-defined path.

trigonometric series   Infinite series of trigonometric functions; applications include representing complex waveforms in terms of frequency content.

true rms   Data-reduction technique that can correctly provide the rms value for a nonsinusoidal signal; a signal conditioning method that integrates the signal.

true value   The actual or exact value of the measured variable.

TTL (true-transistor-logic)   Switched signal that toggles between a high (e.g., 5 V) and a low (e.g., 0 V) state.

turndown Ratio of the highest flow rate to the lowest flow rate that a particular flow meter can measure within specification.

type A uncertainty An uncertainty value estimated from the statistics of a variable.

type B uncertainty An uncertainty value estimated by other than statistical methods.

uncertainty An estimate of the probable error in a reported value.

uncertainty analysis A process of identifying the errors in a measurement and quantifying their effects on the value of a stated result.

uncertainty interval An interval about a variable or result that is expected to contain the true value.

USB (universal serial bus) A serial method of data transmission between a computer and peripherals using a four-wire cable.

validation Process or test to assess whether test model data behave in the same way as the physical process the test simulates; can be used to assign uncertainty in the model.

velocity of approach factor A coefficient used in obstruction flow meter calculations that depends on the ratio of flow meter throat diameter to pipe diameter.

venturi meter In-line flow meter whose internal diameter narrows from the pipe diameter to a smaller diameter throat and then expands back to the pipe diameter so as to create a pressure drop proportional to flow rate.

verification Test designed to ensure that a measurement system as a whole or some portion of the system is working correctly; can be used to assign uncertainty values to aspects of a measurement system.

vernier calipers Tool for measuring both inside and outside dimensions.

vernier scale Technique that allows increased resolution in reading a length scale.

voltage divider circuit An electrical circuit that proportions the input to output voltages using a variable resistance.

volume flow rate Flow volume per unit time.

VOM Acronym for volt-ohm meter.

vortex flow meter Device that measures flow rate by measuring the frequency of vortex shedding from a body placed within the flow.

waveform The shape and form of a signal as plotted with time or space.

Wheatstone bridge Electrical circuit for measuring resistance with a high measure of accuracy and sensitivity.

wobble meter A type of positive displacement flow meter that measures fluid volume flow rate by the rate of fluid displaced within alternating known volumes created by oscillating disks.

word A collection of bits used to represent a number.

zero drift A shift away from a zero output value under a zero input value condition.

zero-order system A system whose behavior is independent of time-dependent characteristics, such as storage or inertia.

zero-order uncertainty Random uncertainty estimate based only on a measurement system's resolution or finest increment of measure.

Index

Note: Page numbers in *italics* refer to illustrations

CONVERSION FACTORS[*]

MASS

$1\ lb_m = 0.4536\ kg$
$1\ slug = 14.5939\ kg$

LENGTH

$1\ inch = 0.0254\ m$
$1\ ft\ \ = 0.3048\ m$
$1\ km\ = 1000\ m = 1 \times 10^3\ m$
$1\ cm\ = 0.01\ m = 0.3937\ in. = 0.0328\ ft$
$1\ mm\ = 0.001\ m = 1 \times 10^{-3}\ m$
$1\ \mu m\ = 0.000001\ m = 1 \times 10^{-6}\ m$
$1\ nm\ = 0.000000001\ m = 1 \times 10^{-9}\ m$
$1\ mi\ \ = 5280\ ft = 1.6093\ km$
$1\ l.y. = 9.4605 \times 10^{15}\ m$

AREA

$1\ ft^2\ \ = 0.0929\ m^2$
$1\ cm^2 = 1 \times 10^{-4}\ m^2 = 0.155\ in.^2$

VOLUME

$1\ L\ \ \ = 1000\ cm^3 = 0.2642\ gal = 1 \times 10^{-3}\ m^3$
$1\ gal\ = 231.0\ in.^3 = 0.1337\ ft^3$
$1\ gal\ = 3.7854\ L = 0.003785\ m^3$
$1\ ft^3\ = 28.3168\ L = 0.0283\ m^3$

TIME

$1\ min = 60\ s$
$1\ h\ \ \ = 60\ min = 3600\ s$
$1\ day = 8.6400 \times 10^4\ s$

FORCE

$1\ N\ \ \ = 1\ kg\text{-}m/s^2 = 1 \times 10^5\ dynes$
$1\ lb\ \ = 4.4482\ N$
$1\ kg_f\ = 9.80665\ N$

PRESSURE OR STRESS

$1\ Pa\ \ \ \ \ \ = 1\ N/m^2$
$1\ lb/in^2\ \ = 6894.757\ Pa$
$1\ atm\ \ \ \ = 14.6959\ lb/in.^2 = 760\ Torr$
$\quad\quad\quad\ = 101,325\ Pa$
$1\ bar\ \ \ \ = 14.5038\ lb/in.^2 = 1 \times 10^5\ Pa$
$\quad\quad\quad\ = 750\ Torr$
$1\ dyn/cm^2\ = 0.10\ Pa$
$1\ inch\ Hg\ = 3386.38\ N/m^2$
$1\ inch\ H_2O = 2.54\ cm\ H_2O = 249.089\ Pa$
$\quad\quad\quad\quad = 0.0361\ lb/in.^2$
$1\ Torr\ \ \ \ = 1\ mm\ Hg = 133.322\ Pa$
$1\ \mu strain\ \ = 10^{-6}\ m/m = 10^{-6}\ in/in$

TEMPERATURE

$K\ \ \ = {}^\circ C + 273.15$
$1{}^\circ C = 1.8{}^\circ F = 1\ K$
${}^\circ F\ \ \ = 1.8{}^\circ C + 32$
${}^\circ R\ \ \ = {}^\circ F + 459.67$

VOLUME FLOW RATE

$1\ gal/min = 0.00223\ ft^3/s = 0.06309\ L/s$
$1\ m^3/min = 35.315\ ft^3/min = 1 \times 10^6\ cm^3/min$

ANGLE

$1^\circ\ \ \ \ \ \ \ \ \ = 0.01745\ rad$
$1'\ \ \ \ \ \ \ \ \ = 2.909 \times 10^{-4}\ rad$
$1\ revolution = 2\pi\ rad$

ROTATION

$1\ rev/s = 2\pi\ rad/s = 6.2832\ rad/s$
$1\ rpm\ \ = 1\ rev/min = 0.1047\ rad/s$

FREQUENCY

$1\ Hz\ \ = 2\pi\ rad/s$

[*]Many of these conversion factors have been rounded off.

MOMENT OR TORQUE

1 lb-ft = 1.3558 N-m

POWER

1 W   = 1.0 J/s = 860.42 cal/hr
1 hp   = 550.0 ft-lb/s = 745.6999 W
1 kW   = 1 × 10^3 W
1 Btu/hr = 778.1692 ft-lb/hr = 0.2931 W

ENERGY

1 J   = 1.0 N-m = 1 × 10^7 ergs
1 erg = 1 dyne-cm
1 cal = 4.1868 J
1 ft-lb = 1.3558 J
1 Btu = 1055.0558 J

VISCOSITY

1 lb-s/ft^2 = 47.880 N-s/m^2
1 centipoise = 0.01 dyne-s/cm^2 = 0.001 Pa-s

SPECIFIC HEAT

1 Btu/lb$_m$-°F = 4.1868 kJ/kg-°C

GAS CONSTANT

1 ft-lb/lb$_m$-°R = 5.380 J/kg-K

THERMAL CONDUCTIVITY

1 Btu/hr-ft-°F = 1.7307 W/m-°C

HEAT TRANSFER COEFFICIENT

1 Btu/hr-ft^2-°F = 5.6786 W/m^2-°C

BULK MODULUS

1 × 10^6 psi = 6.895 × 10^9 Pa

PHYSICAL CONSTANTS

Standard Acceleration of Gravity
  g = 9.80665 m/s^2 = 32.1742 ft/s^2

Speed of Light
  c = 2.998 × 10^8 m/s

Planck's Constant
  h$_p$ = 6.6261 × 10^{-34} J-s

Stefan-Boltzmann Constant
  σ = 5.6704 × 10^{-8} W/m^2-K^4
    = 0.1712 × 10^{-8} Btu/h-ft^2-°R^4

Universal Gas Constant
  R = 8.3143 J/gmole-K
    = 1.9859 Btu/lbmole-°R